Food Frying

Food Frying

Chemistry, Biochemistry, and Safety

Alam Zeb

University of Malakand,
Khyber Pakhtunkhwa, Pakistan

WILEY

This edition first published 2019

Registered Office
John Wiley & Sons, Inc., 111 River Street, Hoboken, NJ 07030, USA
John Wiley & Sons Ltd, The Atrium, Southern Gate, Chichester, West Sussex, PO19 8SQ, UK

Editorial Office
9600 Garsington Road, Oxford, OX4 2DQ, UK

For details of our global editorial offices, customer services, and more information about Wiley products visit us at www.wiley.com.

Wiley also publishes its books in a variety of electronic formats and by print-on-demand. Some content that appears in standard print versions of this book may not be available in other formats.

Library of Congress Cataloging-in-Publication Data

Names: Zeb, Alam, 1979– author.
Title: Food frying : chemistry, biochemistry and safety / Alam Zeb, University of Malakand, Khyber Pakhtunkhwa, Pakistan.
Description: Hoboken, NJ, USA : Wiley, [2019] | Includes bibliographical references and index. |
Identifiers: LCCN 2018042187 (print) | LCCN 2018042489 (ebook) | ISBN 9781119468486 (Adobe PDF) | ISBN 9781119468394 (ePub) | ISBN 9781119468516 (hardcover)
Subjects: LCSH: Fried food. | Food–Analysis. | Biochemistry–Industrial applications. | Frying–Safety measures.
Classification: LCC TX689 (ebook) | LCC TX689 .Z43 2019 (print) | DDC 641.7/7–dc23
LC record available at https://lccn.loc.gov/2018042187

Cover Design: Wiley
Cover Image: © LauriPatterson / Getty Images

Set in 10/12pt Warnock by SPi Global, Pondicherry, India
Printed in Singapore by C.O.S. Printers Pte Ltd

10 9 8 7 6 5 4 3 2 1

To
*My Mother (*BIBI NASEEBA, *died on Friday, 13 October 2017):*
A teacher and most precious gift of my life.
Your true love, purity, and affections for me will always be missed.

Contents

Foreword

The process of deep fat frying has been in use for nearly 4000 years, and today it is one of the most popular ways to prepare food, with an increasing market worldwide. Reasons for this success are that the process is relatively cheap, fast, and easy to use, and imparts good taste and smell characteristics to the product. Although the process itself is very simple, involving immersing the product into hot oil or fat and waiting for some time, deep fat frying is an art, with many pitfalls that must be taken into consideration in order to produce high-quality tasty products that are attractive to the consumer. In addition, the health aspect is coming more and more into focus.

The process occurring in the fryer is very complex and dynamic, with numerous different reactions taking place at the surface of the food and in the oil. Fatty acids are degraded to volatile aroma-active compounds; triacylglycerols are cross-linked, resulting in an increase of viscosity; the polarity of the oil changes due to degradation of triacylglycerols, inducing foaming; and the properties of the product change. The factors that make deep fat frying a complex process include not only temperature and time but also the type of fryer, oil, or food being fried, the use of additives, the amount of frying medium being used, the temperature profile, cleaning of the fryer, and the type of heat transfer employed. Even today, the interplay of these factors and the consequences for the process are not fully understood.

People who are involved in the frying process, such as processors, restaurant operators, and suppliers to these industries (including oil producers, food and ingredient suppliers, equipment manufacturers, and the service trade) should have an advanced and comprehensive knowledge of the dynamics of frying. *Food Frying: Chemistry, Biochemistry, and Safety* by Professor Alam Zeb gives a very good overview of the different aspects of this complex process. The frying process is comprehensively presented in 14 chapters, divided into 4 parts covering the concept of frying, the chemistry of frying, the biochemistry of frying, and the safety of frying.

When a new book on deep fat frying is published, the question arises whether it is needed or not. There are already some standard reference works on the subject, but the content of this book shows that the author has approached the topic in great detail and included many new aspects. That makes it a comprehensive source for information about the frying process. It discusses not only the chemistry of frying but also the analysis and evaluation of used frying oils, the biochemical effects and health-related issues important for the evaluation of the frying process, and issues of food safety, which is one of the most important topics today. The book gives the reader advice on how to protect the consumer from potential health drawbacks of eating fried food. It provides information for persons working in the food production industry, restaurant operations, the basic sciences, official and commercial laboratories, marketing, service provision, and the supply chain who are involved in the production of fried food, research on the frying process, or the assessment of used frying oils, or who are interested in deep fat frying in general.

I believe that this new book on deep fat frying is a wonderful enrichment of the list of existing books on frying and I wish it the success it deserves.

Bertrand Matthäus
Max Rubner-Institut
Federal Research Institute of Nutrition and Food
Department of Safety and Quality of Cereals
Working Group of Lipid Research
Detmold
Germany

Preface

Food frying is a traditional and widely used method of preparation of a variety of foods. Food is fried in edible oils or fats for the purpose of obtaining a characteristic flavour, texture, and acceptability. A wide range of vegetables, meat, fish, snacks, and other products are prepared using frying. Fried foods contribute a huge market worth billions of dollars to the world economy and thus demand continuous research and development. The use of new technologies for the extraction and purification of edible oils and the production of fried foods of the highest quality is evolving daily, and there is a general interest in understanding the process of food frying. Thus, this is a comprehensive book that provides information on all the latest scientific research on the chemistry, biochemistry, and safety of food frying.

The frying of foods involves complex chemical reactions that produce several biochemical changes upon ingestion of the fried product. This book aims to cover them in four main parts. Part I explains the basics of food frying. It consists of three chapters, on food frying, frying techniques, and frying in different cultures. Part II covers the important chemistry of food frying in terms of composition, reactions, and consequent changes in the quality of fried foods. It consists of four chapters giving details of the chemistry of the frying medium and of fried foods and presenting a clear explanation of the fundamental chemical reactions and interactions that take place during frying. This part will be especially helpful for readers in the food industries or academia. Part III covers the biochemical effects and health-related issues of food frying. It consists of three main chapters dealing with the digestion, absorption, metabolism, and health aspects of food frying. It also discusses up-to-date knowledge of the biochemical effects associated with food frying. Part IV covers the topic of food safety, which is one of the key topics of modern food research. It consists of four chapters, which cover the safety assessment of food frying, the toxicity of frying, how to improve the quality of fried foods, and the future of this area.

The book is organized in such a manner as to provide a smooth flow of scientific knowledge from chemistry to biochemistry to safety, while allowing each chapter to be read alone. It should serve as the latest reference resource for food scientists, technologists, food chemists, biochemists, nutritionists, and health professionals working in academia, scientific labs, and industry. It provides fundamental and applied information to benefit those with different backgrounds in science. It can also be used as a textbook for undergraduate, graduate, and postgraduate students in the relevant disciplines.

About the Author

Dr Alam Zeb is a Professor of Biochemistry at the Department of Biotechnology, University of Malakand, Pakistan. He has served for the last 16 years at this university, while teaching chemistry and biochemistry courses to undergraduate, graduate, and postgraduate students of biotechnology. Dr Zeb received his PhD *with distinction* from the Institute of Biochemistry, Technical University of Graz, Austria in 2010, funded by the Higher Education Commission (HEC) of Pakistan. During his studies at Graz, he also taught practical food chemistry courses to postgraduate students. He is one of Pakistan's highest-funded research project winners in the field of food science and technology. Dr Zeb has published more than 100 research articles regarding the subject of food science in various international peer-reviewed journals. He has supervised several PhD and MPhil research students during his service at the University of Malakand. He is a member of the editorial board of several international journals, including *Frontiers in Chemistry*, and *Frontiers in Nutrition*. Dr Zeb has been enlisted as Productive Scientist of Pakistan since 2009, and has received research productivity awards from the Pakistan Council for Science and Technology for the last seven consecutive years, representing his potential and enthusiasm for research in the field of food biochemistry.

Acknowledgement

This work would not have been possible without the help of some important people. Dr Francisco J. Morales and Dr Marquez-Ruiz of the Institute of Food Science, Technology and Nutrition (ICTAN), Spanish National Research Council (CSIC), Spain are highly acknowledged for providing the original copyrighted data used to produce high-quality figures in Chapter 5. Similarly, Prof Dr Michael Murkovic of the Institute of Biochemistry, Technical University of Graz, Austria is thanked for providing an original copyrighted figure for Chapter 5, and support at the university. I am grateful to the Ernst Mach-Nachbetreuungsstipendium (EZA) Fellowship, OeAD Austria for providing an opportunity to complete this work in Graz, Austria. I am also thankful to the University of Malakand for supporting me in finalizing this book, while doing research at the Technical University of Graz.

Finally, I acknowledge my wife, son (*Umar Alam Khan*), and a daughter (*Haya Bibi*) for missing me at home, while I was completing this work in Graz.

Part I

Concept of Food Frying

1

Food Frying

The Concept

1.1 Introduction

Frying is a food preparation technique that involves foods and hot oil or fats. It is one of the fastest and simplest techniques for cooking food with pleasant, attractive properties. The food frying process consists of four main components: first, specific conditions such as temperature and pressure; second, a utensil or fryer; third, foods; and fourth, frying oil (referred to as 'frying medium' from this point onward). The fryer may be a simple pan or a complex industrial technology. The frying medium is usually an oil or an animal-derived fat. The temperatures used for frying are in the 150–200 °C range. High temperatures promote reactions between food components like proteins and carbohydrates, surface dehydration of the crust, and oil uptake (Gertz 2014). In contrast to boiling in hot water, the heating of foods at elevated temperatures provides a desirable appearance (colour), texture (crispness), flavour, and taste (Perkins 2007). Frying is a more efficient process than other cooking methods, and has gained great popularity in both restaurants and industry because of its speed and operational simplicity.

Even though deep frying is an old and very popular process, it is still poorly understood. Proper frying practice and the most appropriate frying oil are generally determined by experience. Good understanding of the frying process helps in optimizing the manufacturing processes with regard to the quality of the food, the use life of the fat, and energy consumption. To guarantee a good quality of the fried end product, it is necessary to install a management system which includes all critical points of the frying process (Gertz 2014).

1.2 History of Frying

Food frying is one of oldest known food preparation techniques. However, the exact date of its invention and who first used it is hard to trace. Some authors propose that it was invented by the ancient Chinese (Rossell 2001). The third book of the Old Testament, Leviticus, chapter 2, verses 4–7, written c. 600 BCE, distinguishes between bread baked in an oven and that cooked on the griddle or in the pan. Pliny (c. 1st century CE) provides a prescription for spleen disease that involves frying eggs in vinegar (Morton 1998). Knowledge of frying was common in the fourteenth century, but scientific writing on the subject was still rare. Table 1.1 provides a brief

Food Frying: Chemistry, Biochemistry, and Safety, First Edition. Alam Zeb.

Table 1.1 Timeline of the frying process.

Year	Event
3000 BCE	Chinese frying of meat
1300 BCE	Hebrews fry flat breads
1537 CE	Potatoes introduced in Europe
1600–1700	French fried potatoes emerged
1853	Potato chip invented in Saratoga Springs, NY by George Crum
1890s	Potato chip industry begins in the United States
1897	Hydrogenation of edible oils invented
1906	Commercial oil roasting of shelled peanuts by Planters
1908	J. P. Dushesues founds Leominster Potato Chip Company
1926	Laura Scudder develops first potato chip bag of waxed paper
1927–1930	Cellophane begins to be used for potato chip bags
1929	Clarence Birdseye develops new commercial freezing technologies
1932	First tortilla chips (Doolin and Filler) produced in San Antonio, TX
1933	Dixie wax paper introduces pre-printed glassine bag. Cracker barrel marketing of chips comes to an end
1930–1935	National Potato Chip Institute tells consumers that chips are not fattening if eaten in small amounts
1938	H. W. Lay Co. founds Lay's Potato Chip Company in Atlanta, GA
1946	First automatic packaging machine for chips developed
1945–1950	Extruded snacks introduced. MacBeth introduces continuous immersion cookers
1950	Under-pan fired cookers introduced
1950–1952	Fryers with external heat exchangers with oil circulation introduced; Pork rinds introduced
1950–1955	Laminate bags with polypropylene/cellophane and polypropylene glassine introduced
1953	Simplot scientists develop a technique for par-frying potato slices
1957	Heat & Control introduces 'Big Goose', a 1600 lb. capacity continuous system
1958	Urshel develops new slicers for potato chip manufacture
1961	Frito-Lay merger
1969	Potato chip controversy develops with the introduction of Pringles and Chippos
1970–1975	7000 lb. capacity fryers introduced
1973	1st DGF Symposium. Germany proposes regulations based on oxidized fatty acids for restaurant frying oil
1979	2nd DGF Symposium. Polar materials provides index of restaurant frying oil quality
1987	Blumenthal publishes surfactant theory of frying
2000	3rd DGF Symposium. Principle quality index should be sensory parameters of food being fried
2004	4th DGF Symposium
2004	4th International Symposium on Deep-Frying
2011	6th International Symposium on Deep-Frying, Germany
2013	7th International Symposium on Deep-Frying, United States
2015	8th International Symposium on Deep-Frying
2016	1st International Symposium on Lipid Oxidation and Antioxidants, Portugal
2017	9th International Symposium on Deep Frying, China

Source: Modified and reproduced with kind permission of John Wiley & Sons (Stier 2004).

history and timeline of the frying process (Stier 2004). The German Fat Society, 'Deutsche Gesellschaft fuer Fettwissenschaft' (DGF) may be considered a pioneer of fat science and frying. The DGF held symposia on frying fats in 1973 and 1979. The first conference on the subject, 'The Frying of Foods', was held in Madrid in 1986. These conferences and symposia were considered the prime motivator for attracting scientists' attention to frying science and technology. The European Society for Lipid Science and Technology has contributed much to the understanding of frying. In recent times, significant scientific breakthroughs have occurred, resulting in a huge sum of knowledge on frying that it is not possible to accommodate in a single book.

1.3 Mechanism of Frying

Frying involves essential components such as foods and frying medium. As a general rule, frying medium is initially preheated to about 160–200 °C and then desired foods are either kept in it or immersed in it. Several physical and chemical reactions occur that result in the formation of fried foods. Figure 1.1 shows the experimental results of oil distribution during the frying and cooling of a potato slice. During frying, a small amount of oil is absorbed by the potato, while during cooling, the internal oil content increases at a fast rate for the first minute, while surface oil decreases until an equilibrium is reached after 4 minutes. In order to understand the fundamentals of frying, the following concepts are important.

1.3.1 Heat and Mass Transfer

Foods can be prepared at an elevated temperature in different ways, such as by pan frying, shallow frying, or deep frying, while temperatures of 100–103 °C can be achieved when cooking in water. With the use of fat or hot air as a heating medium, a higher

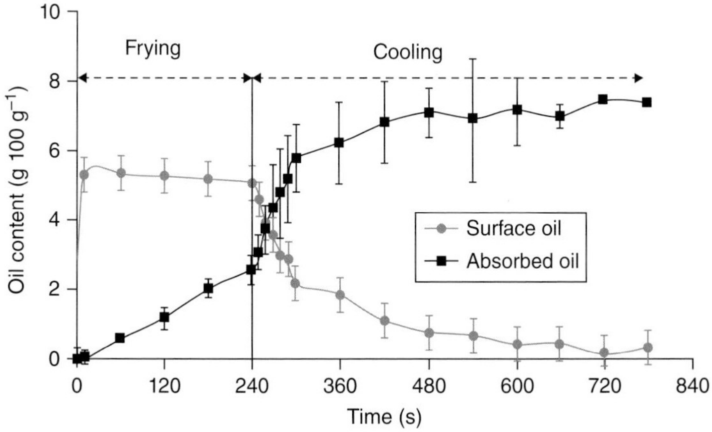

Figure 1.1 Oil content absorbed or remaining on the surface of French fries during frying (170 °C) and cooling (20 °C). *Source:* Reproduced with kind permission of John Wiley & Sons (Gertz 2014).

temperature can be generated without exceeding the boiling temperature of the water inside the food (Vitrac et al. 2000). As the temperature difference between the heating medium and the food increases, the heat transfer within the cooking process runs much faster. Simultaneous with the heat transfer, mass is transferred from the food to the frying medium and vice versa (Gertz 2014).

A food is a solid body with holes and pores filled with water and air. Immediately after its immersion in hot oil, traces of free water at the surface evaporate very rapidly, causing a violent bubbling and drying of the surface. When the vaporization of water is faster than the ability of the surrounding oil to remove the steam by convection, the heat transfer rate from oil to food surface is zero, due to the heat resistance of the steam evaporating from the surface. The introduction of the food into the hot oil and the sudden evaporation of moisture from the food cause a violent bubbling. Bubbling enlarges the contact area between air and oil. Thus, the heat transfer rate between oil and air increases, accelerating the oxidative degradation of the oil (Costa et al. 1999). By lowering the oil temperature, reducing the quantity of food to be fried, or pre-drying the food, rapid evaporation of the water from the surface of the food and intense bubbling can be avoided (Sobukola et al. 2010). With increasing frying time, the bubbling becomes less intense and the evaporating water steam has a more protective effect (Dana et al. 2003), creating a steam blanket above the oil surface and reducing the headspace air flow, providing protection against oxidation by avoiding contact with air.

The heat is transferred by convection from the oil to the surface of the product and by conduction to the centre of the product. The water inside the product is heated to boiling point, resulting in an increased pressure. As a consequence, water at the surface leaves the product and that in the interior of the food migrates from the central position radially outward to the walls (Vitrac et al. 2000). This water transport is responsible for providing cooling in the external region of the product after the first period of frying, ensuring that the food is not burnt or charred. The moisture in the inner part of the food is heated to boiling, inducing gelatinization of starch and denaturation of proteins (Gertz 2014).

It has been shown by Manglik (2006) that by adding small quantities of surface-active soluble agents, the interfacial tension and surface tension (oil–solid interfacial, oil–vapour tension) can be altered and the heat transfer improved. A simple practical test at 170 °C with 10 × 10 × 500 mm potato pieces demonstrates the heat transfer capacities of different oils (Table 1.2). The measured times till the central temperature reaches 100 °C are all different due to the different physical and chemical properties of the oils. In frying tests, it has been observed that potatoes fried in beef tallow and palmolein contain higher amounts of acrylamide than those fried in sunflower or rapeseed oil for the same period, due to the accelerated heat transfer (Gertz et al. 2003).

Palmolein and beef tallow contain more polar components, such as mono- and diacylglycerols and medium-chain triacylglycerols (TAGs), than do oils such as rapeseed oil, sunflower oil, and groundnut oil. To compare the relative polarities, fresh oils were measured with FOS (Food Oil Sensor). The reading units of the FOS measurements are related to their dielectric constants and polarity (Wegmuller 1994). It is possible that these more polar compounds reduce the surface tension between oil and food surface or oil and water steam; however, Gil and Handel (1995) have not observe any effect of diacylglycerols or fatty acids on the surface tension in frying oils. Blumenthal (1991) published a monograph proposing a surfactant theory of frying. As oil degrades, more

Table 1.2 Formation of acrylamide during deep frying in various oils.

	Time needed to reach 100 °C in the score	FOS-unit	Acrylamide level (μg kg^{-1}) of French fries (40 g) prepared in oil (fryer capacity: 2 l); heating time: 3, 5 min, 170 °C
Palmolein	65	3.5	594
Beef tallow	74	2.5	301
Sunflower oil	80	0.8	205
Rapeseed oil	80	0.5	203
Groundnut oil	82	0.4	190
Groundnut oil hardened	84	0.0	192

Source: Reproduced with kind permission of John Wiley & Sons (Gertz 2014).

surfactant materials are formed, causing increased contact between oil and food. Those materials cause a better heat transfer at the oil–food interface and reduce the initially high surface tension between these two immiscible zones. The so-called 'Frying Oil Quality Curve' (Figure 1.2) demonstrate the relationship between the degradation of frying oil and the chemical changes in the oil. This curve shows five stages of oil degradation, and relates them to food quality. The goal must be to extend the optimum frying window.

After the external zone is dehydrated, the crust–crumb interface starts moving towards the centre from the food. This porous dried region (crust) is created and continues to increase in size as long as water is migrating from the central position of the food radially outward to the surface. The temperature profile in the crust is a function of oil temperature, whereas the evolution of the core temperature is independent of the oil temperature so long as water is transferred from the inner part of the food to the surface (Figure 1.3); it cannot be changed to accelerate the frying process.

The thickness of the crust increases with frying time up to about 0.3–2.0 mm (Ngadi et al. 1997; Gertz 2014). When frying potato chips (crisps), the crust region enlarges quickly and the core zone disappears. The low thickness, lack of liquid water to be evaporated, falling pressure, and high heat transfer raise the temperature of the material above 100 °C very quickly.

Models developed to calculate the transient temperature, moisture content, and oil content during the frying process use many simplifying assumptions. Dincer (1996) propose a single-phase model using a discretization of heat and mass transfer. To simplify the calculations, it is assumed that there is no effect of mass transfer on heat transfer and vice versa. Farid and Chen (1998) found a good agreement between the predicted and experimental temperature distribution, except at the end of the frying period, where the central temperature of the potato chips exceeded the boiling temperature of water.

A two-phase or moving-boundary model has been proposed to describe the mechanism of heat and mass transfer during frying by Farkas et al. (1996), among others. These authors proposed the existence of two regions, separated by an interface: the core (unfried) and the crust (fried) regions. During immersion frying, heat is transferred

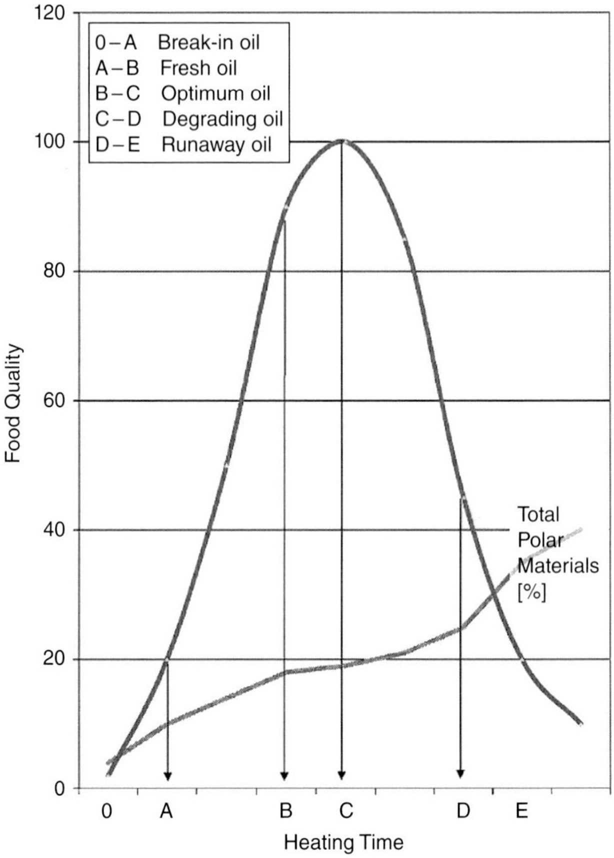

Figure 1.2 Frying oil quality curve according to Blumenthal (1991). *Source:* Reproduced with kind permission of John Wiley & Sons (Gertz 2014).

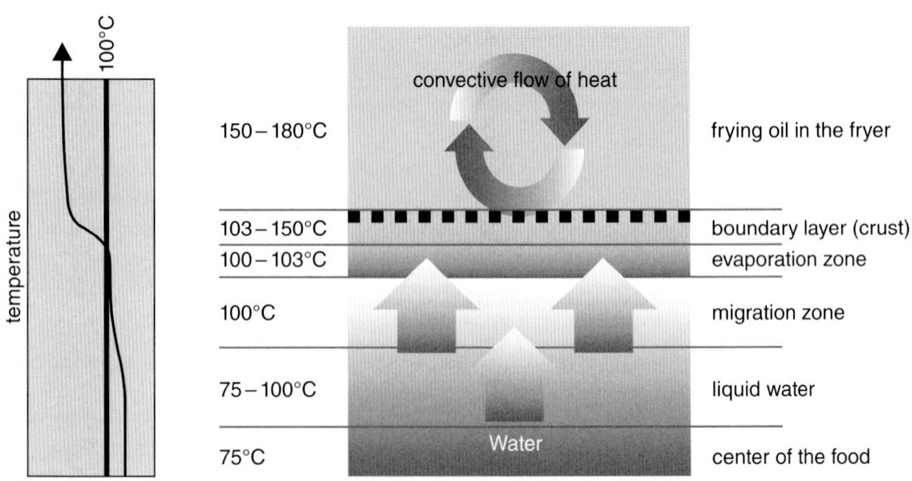

Figure 1.3 Scheme of heat and mass transfer during deep frying. *Source:* Reproduced with kind permission of John Wiley & Sons (Gertz 2014).

from the frying oil to the core via the crust region. Water is evaporated at the moving boundary at 100 °C. The heat conduction equation was used to describe the heat transfer in this region. The temperature difference is the driving force of heat transfer. The changing physical properties due to increasing fat degradation made the system too complex for a realistic model to be developed. Also, the effects of crust formation on physical properties were neglected. This model is comprehensive and helps to explain the frying process. The simplified scheme of heat and mass transfer during the actual frying of food is depicted in Figure 1.3.

1.3.2 Oil Uptake

During the frying process, food takes up oil contents. Oil uptake seems to be independent of frying temperature, but is significantly affected by frying time, moisture loss, and the structure of the products to be fried (Gamble et al. 1987; Alvarez et al. 2000). The initial moisture content of food can be reduced by pre-treatments such as pre-drying at temperature 70–75 °C, air or vacuum drying (Mariscal and Bouchon 2008), osmotic dehydration (Bunger et al. 2003) and blanching (Sobukola et al. 2010), or else in combination with a post-treatment (Mariscal and Bouchon 2008). In the food industry, the use of hydrocolloids such as carboxymethylcellulose, pectin, sodium alginate, powdered cellulose, and modified starch is very common to retard the moisture loss (Holikar et al. 2005; Saha and Bhattacharya 2010; Gertz 2014).

The initial superficial vaporization and subsequent in-depth vaporization create a porous, dried, and overheated region which is generically called 'crust'. The pores can be small voids, molecular interstices, or large caverns which are filled with water and air. They may be interconnected or nonconnected. The water should be able to travel throughout the entire porous structure, as if it were a network of pipes. For this reason, it is important not to overheat the product when food is immersed into hot oil. A good structured crust helps to retard the loss of moisture; otherwise, the pores will be too large or will be destroyed due to the high vapour pressure. When the porosity of the material is low, the increase in pressure can significantly reduce the drying rate. For materials with weak structures due to high water content and/or the absence of cell structure, water transport can be so intense that liquid water escapes the surface without vaporization (Gertz 2014).

Fried foods like yeast-raised doughnuts have more oil accumulated at the surface due to their thin crust layer. Unlike in deep frying, these foods float on the surface during frying (i.e. shallow frying). The heated water in the core starts boiling and increases the volume of the food. Only a little water finds its way through the crust. For this reason, the crust is very thin. Another effect is that volatiles remain in the hot oil, producing off-flavours like rancidity, because not enough steam is leaving the crust and stripping them off (Gertz 2014).

1.3.3 Mechanism of Oil Absorption

Oil uptake is a complex mechanism that is still not clearly understood. The initial product structure, the various interchanges between the product and the heating medium, and the variations of product and oil properties are the factors which complicate this phenomenon (Ziaiifar et al. 2008).

1.3.3.1 Water Escape and Oil Uptake

Most authors agree that during frying, heat and mass transfer are controlled by heat transfer at the surface of the product. The rate of vaporization is proportional to the temperature difference between the oil and the boiling point of water (Vitrac et al. 2002). Numerous works propose a simple description based on a convective mass transfer approach that is too simple. Farkas et al. (1996) were the first scientists to propose a physical description of frying. They stated that this process should be described as a complex Stephan problem, because of the coupled heat and mass transfer resulting in the displacement of a moving vaporization front that separates two dynamic regions: a dehydrated crust and a humid core. As the crust presents low thermal conductivity, it affects heat, and mass transfer and is partly responsible for the decrease in the dehydration rate.

In general, we can say that the more the water is removed from the surface, the more the oil is absorbed. Figure 1.4 plots the relationship between oil content and moisture loss. When mass transfers in deep fat frying are studied, the escape of water is usually linked to oil absorption. Indeed, Gamble et al. (1987) found that moisture loss and oil uptake were inter-related and were both linear functions of the square root of frying time. They hypothesized that the oil entering the slice would lie in the voids left by the escaping water. Hence, in addition to quantitative aspects, water loss can become an explanatory variable for transformation and especially oil uptake, because water escape is at the origin of very diverse material phenomena such as the creation of cavities (Vitrac et al. 2000). Indeed, as dehydration occurs at a temperature above 100 °C, water steam finds selective weaknesses in the cellular adhesion that lead to the formation of capillary pathways, increasing surface porosity. Furthermore, some of this vapour may be trapped within the pores as a result of restrictive intercellular diffusion and expand, becoming superheated, distorting the pore walls, and contributing to product porosity. Accordingly, some studies have examined the increase of porosity during frying and correlated it to the amount of oil uptake (Pinthus and Saguy 1994; Moreira et al. 1997). Characterization of the product microstructure thus appears to be a determining factor in the description of transfers at the macroscopic scale, such as oil uptake.

1.3.3.2 Capillary Pressure and Oil Uptake

Moreira et al. (1999) introduced a physical relation between oil absorption and porosity, stating that the mechanism of oil uptake may be caused by capillary forces. Indeed, when a fluid displacement such as oil absorption occurs in microcanals like crust pores, surface phenomena such as viscosity or capillary forces become very important. Capillarity is the ability of a narrow pore to draw a liquid upwards. It occurs when the adhesive intermolecular forces between a liquid and a solid are stronger than the cohesive intermolecular forces in the liquid. This causes a concave meniscus to form where the liquid is in contact with the vertical surface. This phenomenon creates a difference of pressure between the two sides of the curbed interface, as expressed by the Laplace law (Figure 1.4):

$$P1 - P2 = \frac{2\gamma \cos\theta}{r} \tag{1.1}$$

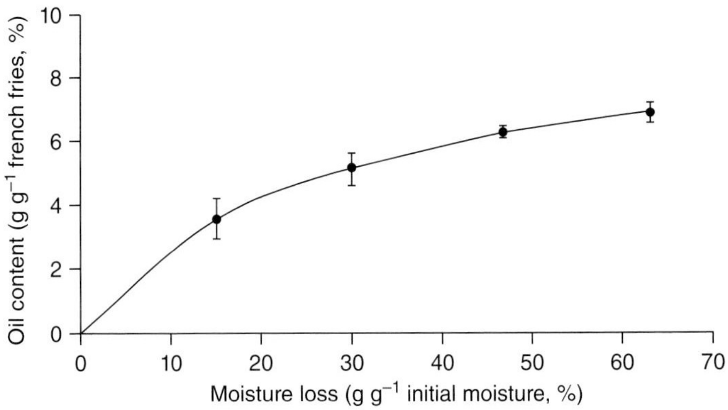

Figure 1.4 Oil content versus moisture loss in French fries during frying at 155 °C. *Source:* Reproduced with kind permission of John Wiley & Sons (Ziaiifar et al. 2008).

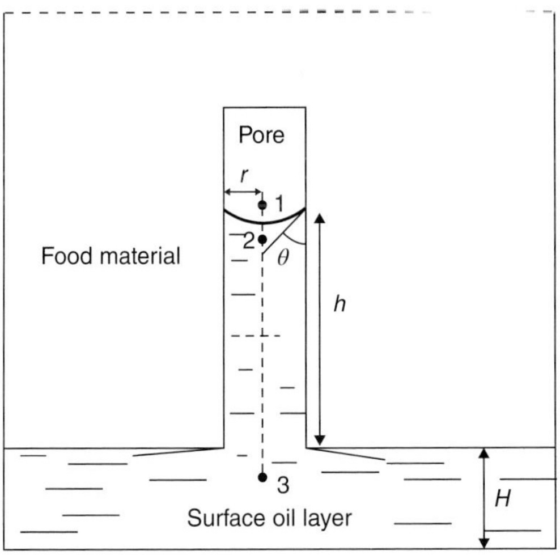

Figure 1.5 Diagram of oil flowing into a pore. *Source:* Reproduced with kind permission of John Wiley & Sons (Ziaiifar et al. 2008).

where P_i is the pressure at the point i (Pa), γ is the surface tension of the oil (N m^{-1}), ϑ is the wetting angle between the oil and the solid (rad), and r is the pore radius (m).

In addition, $P_2 - P_3 = -\rho gh$, according to the hydrostatic pressure difference (Figure 1.5), where ρ is the oil density (kg m^{-3}), g is the acceleration gravity (m s^{-2}), and h is the height of the capillary motion (m).

Therefore, the pressure difference ΔP at the two points 1 and 3 of the pore is:

$$\Delta P = P1 - P3 = \frac{2\gamma \cos\theta}{r} - pgh \tag{1.2}$$

Oil absorption is therefore dependent on pore radius. Small pores cause higher capillary pressures and thus higher oil content (Moreira et al. 1997). In addition, the lower the contact angle between the oil and the product surface, the higher the adhesion forces and the oil uptake. Finally, the higher the surface tension of the liquid, the higher the oil uptake. Moreira and Barrufet (1998) stated that γ decreases with increasing temperature, resulting in a capillary pressure reduction. This fact contributes to limiting oil uptake during frying.

The main difficulty with capillarity motion determination at the end of frying is the determination of pore radii that are nonhomogenous in shape. Furthermore, pores can be filled with liquid, water vapour, or air depending on the conditions at the end of frying. Thus, the wetting property of oil towards the solid matrix is nonhomogenous and difficult to determine in such complex and multiple fluid phase systems. Moreover, capillary motion equations are mostly used in their static form, resulting in the expression of the equilibrium positions of fluids. This simplification comes from the fact that other forces, such as vacuum effect and the weight of absorbed oil, are involved in oil absorption.

1.3.3.3 Vapour Condensation and Vacuum Effect

During frying, intense drying occurs at a temperature greater than the temperature of water ebullition. The solid matrix of the food is an obstacle to water bubble growth, leading to a pressure gradient in the food. Overpressure was evaluated in an experimental work by Vitrac et al. (2000), who measured an inner overpressure of 45 kPa during the frying of an alginate gel containing 10% starch. Overpressure depends on the initial structure of the material. Indeed, the more resistant the structure is towards fluid dilatation, the higher the pressure inside the material. However, some structures are not sufficiently resistant to pressure and can break, allowing liquid water to escape. This phenomenon of water loss in both steam and liquid forms was observed during the deep fat frying of apple slices (Vitrac et al. 2003).

Consequently, during frying, when the product still presents high free water content susceptible to evaporate, the escape of water and the associated overpressure in the material is an obstacle to oil absorption. In opposition, when the product is removed from the fryer, the core temperature decreases, steam condenses, and the pressure in the product abruptly decreases. As a consequence, the important difference between the inner and outer pressures creates a 'vacuum effect', resulting in the penetration of the surface oil into the product (Gamble and Rice 1987). Moreover, Vitrac et al. (2000) found the depression in a food model gel to be 35 kPa a few seconds after the product had been removed from the oil bath; they therefore stated that this vacuum is the most important force acting on oil uptake in the porous media.

1.3.3.4 Adherence and Drainage of Oil

Oil absorption involves a balance between adhesion forces (capillary and water condensation) and drainage of oil during the cooling period (Ufheil and Escher 1996). In the case of frying, adherence, which is an important factor in oil uptake, is the ability of the oil to stick to the outer surface of the product. Drainage is the removal of surface oil as a result of gravity forces.

Theoretically, when a solid is removed from a bath of wetting liquid, it drags out a liquid film the thickness (*H*) of which (Figure 1.4) is given by the Landau–Levich–Derjaguin relation as (Krozel et al. 2000):

$$H = 0.94 \frac{(\mu U)^{2/3}}{\gamma^{\frac{1}{6}} (pg)^{1/2}} \tag{1.3}$$

where μ is the oil viscosity (Pa s), γ is its surface tension ($N\,m^{-1}$), U is the speed of removal ($m\,s^{-1}$), ρ is the oil density ($kg\,m^{-3}$), and g is the gravity due to acceleration ($m\,s^{-2}$).

Experimental data on fibre coating show that the film formed in the case of slow withdrawals from pure viscous liquids such as oils fits the Landau law. At high withdrawal velocities, the thickness decreases with the velocity, because the solid can only drag the viscous boundary layer. Furthermore, the presence of surface-active compounds may cause a thickening of the film (Quéré and de Ryck 1998).

1.3.4 Product Properties Affecting Oil Uptake

1.3.4.1 Size, Shape, and Surface of the Product

As oil uptake is a surface phenomenon, the specific dimensions of a food will determine the amount of oil that can be taken up. Results show that oil absorption increases significantly when product thickness is reduced and product surface is increased (Ziaiifar et al. 2008). For instance, French fries absorb less oil than chips because they have a smaller surface/volume ratio, as shown in Table 1.3 (Paul and Mittal 1997). A linear relationship has been set between surface area and oil content (Gamble and Rice 1987). As most of the fat penetrates the food through the pores in the crust, the structural properties of the outer layer of the food are important. Indeed, cells broken during cutting are a privileged location for oil absorption (Dana and Saguy 2006). Using good-quality blades for cutting can, therefore, reduce the surface roughness of the product and thus the surface area, resulting in lower oil uptake.

1.3.4.2 Composition and Density of the Product

As shown in Table 1.3, from a humid raw product, deep fat frying provides more or less high-fat content products depending on their initial composition and thickness. Indeed, initial solid content in the product is a factor that influences oil uptake during frying because of the relationship between water loss and oil uptake (Pinthus and Saguy 1994; Moreira et al. 1997). For a final fried product that exhibits an intermediary water content (i.e. thick product) such as plantain cylinders, French fries, and eggplant cubes, the higher the initial water content, the higher the final oil uptake (Table 1.3). Similarly, a higher potato density ($1103\,kg\,m^{-3}$ compared with $1093\,kg\,m^{-3}$) can reduce oil content by about 10% (Ufheil and Escher 1996), because of the relationship between a product's density and its initial water content (Paul and Mittal 1997). Table 1.3 shows that the important moisture loss that occurs during frying of thin products (i.e. chips) leads to considerable fat uptake, because of the extensive void volumes created by the water escape (Gamble et al. 1987). Finally, the ability of a raw material to present or develop high porosity during frying, mainly because of a high level of initial water content or extensive water loss, will control oil uptake. However, the typical behaviour of eggplant

Table 1.3 Water and fat content in various raw and deep-fried products.

Food	Water content ($g\,100\,g^{-1}$ wet basis)	Oil content ($g\,100\,g^{-1}$ wet basis)	Water content ($g\,g^{-1}$ nonfat dry matter)	Oil content ($g\,g^{-1}$ nonfat dry matter)	Reference
Raw products					
Plantain	60	0.1	1.53	0.005	(Rojas-Gonzalez et al. 2006)
Potato	80	0.1	4	0.005	(Talburt 1987)
Eggplant	95	0.1	19.38	0.005	(Kalogeropoulos et al. 2006)
Cassava	60	0.1	1.50	0.005	(Vitrac et al. 2000)
Tortilla dough	47	1.5	0.88	0.03	(Moreira et al. 1999)
Fried products					
Plantain cylinders	32	7	0.80	0.15	(Rojas-Gonzalez et al. 2006)
French fries	44	13	1.05	0.30	(Talburt 1987)
Potato chips	2	40	0.03	0.69	(Talburt 1987)
Eggplant cubes	50	40	3.85	2.71	(Kalogeropoulos et al. 2006)
Cassava chips	2	25	0.05	0.35	(Vitrac et al. 2000)
Tortilla chips	2	25	0.025	0.35	(Moreira et al. 1999)

Source: Modified and reproduced with kind permission of John Wiley & Sons (Ziaiifar et al. 2008).

during frying has to be noticed. Indeed, even if the residual water content is still quite high after frying (only 50% of initial water is removed), oil uptake is as high as in totally dehydrated products like chips. This phenomenon can be explained by the fact that eggplant is an aqueous nonstarchy product whose structure is very weak and spongy (Kalogeropoulos et al. 2006).

1.3.5 Frying Oil Properties and Oil Uptake

1.3.5.1 Oil Type

The effect of oil type differs greatly depending on its composition. Kita and Lisińska (2005) wrote that fat absorption is higher when the amount of unsaturated fatty acid increases in oil. On the other hand, Vitrac et al. (2000) showed that oil uptake is weaker with an unsaturated oil such as cotton oil than with palm oil, because of the former's weak viscosity during cooling and its ability to drain easily. These contradictions could be explained by the fact that oil viscosity is very influential in the oil absorption mechanism but is involved in both adhesion and draining dynamics (Eq. 1.3). Moreover, the frying oil may contain a portion of fat that solidifies upon cooling, making it harder to drain or shake from the food, as well as less likely to penetrate deeply into crust pores. Fat content can be considered to be a sum of both fat penetrations into the crust and fat crystallization on the surface.

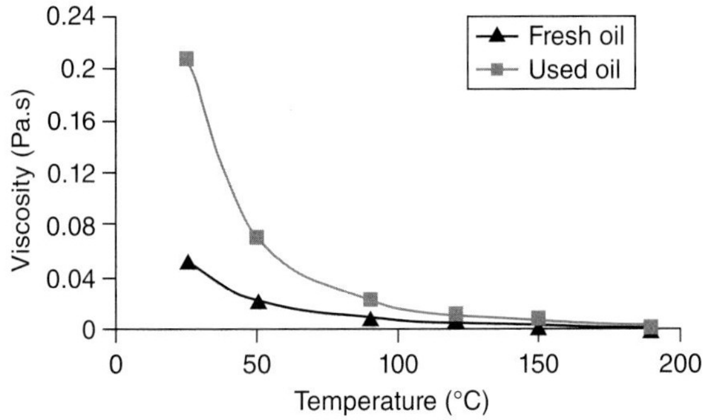

Figure 1.6 Viscosity changes in fresh and used soybean oils. *Source:* Reproduced with kind permission of John Wiley & Sons (Ziaiifar et al. 2008).

The higher the oil viscosity, the slower the oil migration. Initial oil viscosity depends not only on the oil type but also on the temperature and oil quality. As shown in Figure 1.6, oil viscosity decreases with a decreasing temperature, following the Arrhenius equation. The oil's initial superficial tension is also an important factor to consider in the capillary action leading to oil uptake (Eq. 1.1). An increase in interfacial tension leads to an increase in oil uptake. Therefore, the addition of surfactants (surface-active agents or wetting agents) such as Tween80 (polysorbate) and Span80 (sorbitan mono-oleate) in the frying oil could change surface properties and modify oil content in the final product (Pinthus and Saguy 1994). However, these products are not widely approved for food use.

1.3.5.2 Oil Ageing

The result of oil ageing is an increase in viscosity due to polymer formation (Figure 1.6) and a decrease of contact angle due to the formation of polar compounds. The increase in viscosity could contribute to an increase in oil quantity on the food surface, according to Eq. (1.3), while the decrease in contact angle could increase the wetting properties of the oil, both of which would result in higher oil content. For this reason, Tseng et al. (1996) argued that the amount of surface oil on tortilla chips increased with oil degradation. Still, some studies prove that the total final oil content of tortilla chips is not significantly affected by the change in oil quality between the first and the thirtieth frying operation (Mehta and Swinburn 2001). Nevertheless, this chemical evolution, which can have adverse effects on oil quantity and quality, can be slowed by the addition of natural or synthetic antioxidants (Man and Jaswir 2000; Houhoula et al. 2004). In conclusion, oil ageing plays a role in oil uptake, but its effect is less than expected.

1.3.6 Process Factors

A wide spectrum of process factors, including the conditions of pre-processing, frying, and post-frying, has been reported to affect oil absorption in fried foods, as shown in Figure 1.7. Some of these process steps and conditions have been patented as means of decreasing oil uptake.

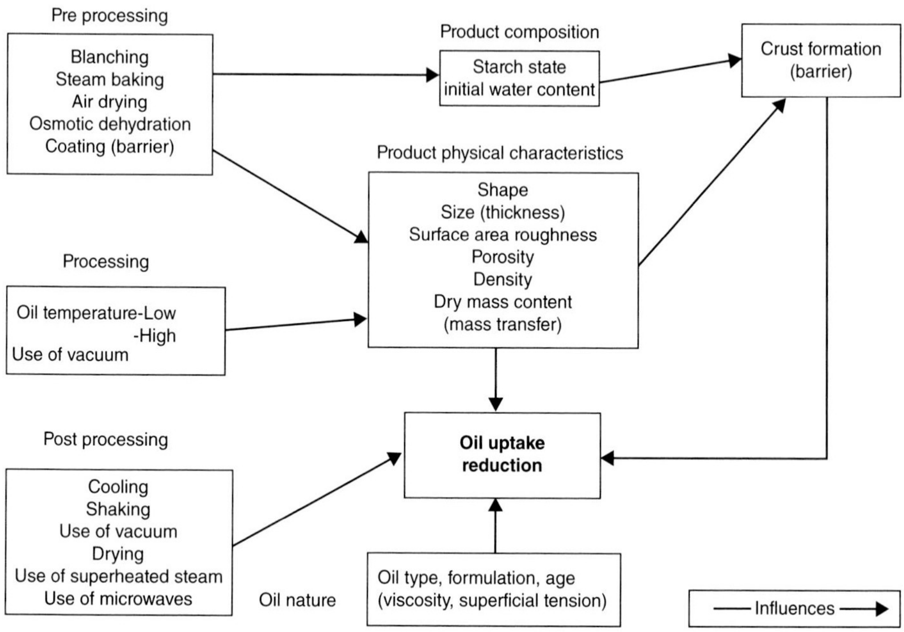

Figure 1.7 Important factors involved in the frying operation affecting oil uptake. *Source:* Reproduced with kind permission of John Wiley & Sons (Ziaiifar et al. 2008).

1.3.6.1 Pre-processing Factors Affecting Oil Uptake

The most popular processes applied before frying of products at the industrial scale are blanching, air drying, osmotic dehydration, steam baking, and surface treatment (coatings). Blanching is a process of food preparation wherein the food substance is plunged into boiling water or steam in order to deactivate its enzyme and micro-organism contents. The effect of this pre-frying process on final oil uptake is quite ambiguous, because of the different conditions applied. Some authors (Rimac-Brnčić et al. 2004) state that blanching before frying decreases the dry mass content of the product because of the migration of water-soluble components from the product to the blanching water. As a result, this phenomenon increases water content and thus final oil content (Alvarez et al. 2000; Pedreschi and Moyano 2005). On the other hand, other studies suggest that the surface starch gelatinization that occurs during blanching could form a firm thin layer that protects the food from oil absorption. In addition, blanching can activate pectin esterase enzymes which make surface cell walls collapse, causing a decrease in the porosity and oil content of the product (Aguilera-Carbó et al. 1999).

1.3.6.2 Post-frying Conditions

When the product is removed from the frying medium, its temperature immediately starts to decrease. Below 100 °C, water vapour condenses and the internal pressure drops, resulting in the creation of a positive pressure vacuum favouring oil uptake. The most important factors affecting oil absorption during post-frying are the cooling conditions. Indeed, temperature influences oil viscosity and the interfacial tension involved in oil uptake phenomena. Matz (2012) stated that if the product is removed from the fryer while its temperature is still increasing, oil uptake will decrease. The hydrodynamics of cooling is also important: vigorously shaking the basket of fried products immediately after removal from the fryer can drain the surface oil if it is still liquid and has not yet been sucked into the pores. The oil that can penetrate the pores is thus limited (Thanatuksorn et al. 2005).

1.3.7 Chemical and Physical Changes of the Frying Medium

During frying, a series of complex changes and reactions take place. The nature and rate of decomposition of the products depend, among other things, on the composition of the oil (fatty acids pattern, unsaponifiable matter content), the mode of frying (intermittent or continuous; shallow, pan, or deep frying), the frying temperature, the length of the frying process, and the type of food being fried. To cover these physical and chemical changes, many analytical methods are proposed in the literature (Croon et al. 1986; Gertz et al. 2000). Increasing acidity, total polar materials (TPM), and polymerized TAGs (PTG), darkening colour, and decreasing iodine value and polyunsaturated fatty acid content are typical indices of oil degradation at elevated temperatures. In 2000, the DGF recommended determining both the sum of all polar compounds and polymerized TAGs:

Analysis of suspect frying fats and oils should use two tests to confirm abuse. Recommended analyses should be total polar materials (TPM) (<24%), polymeric triglycerides (PTG) (<12%). Determination of TPM and PTG are now considered to be the most reliable and objective tests for assessing deterioration (DGF 2000).

A number of nations have set limits for both criteria for evaluating used frying oils. The group of polar compounds includes oxidized fat degradation products, di- and oligomerized TAGs, and naturally occurring polar compounds such as mono- and diacylglycerols (Gertz 2014). The determination of TPM and PTG using the standard methods or by NIR requires laboratory equipment. Thus, these indices are more often used in research and governmental laboratories than in regular in-house quality control. Many quick tests are now available on the market to identify polar materials by measuring the dielectricity constant. Besides free fatty acid (FFA), these quick tests are increasingly being applied in the frying industry for quality control.

1.4 Why We Fry Foods

The basic question is, why are we frying foods? Historically, some foods were not digestible or easily consumable. Normal cooking with water and heat consumes a lot of time, and frying speeds this process. Some of the reasons why we fry our foods include:

- Frying increases the speed of food preparation.
- Frying converts indigestible foods to digestible one.
- Frying is a more efficient process than other cooking methods.
- Frying can form foods with specific tastes and textures.
- Frying is responsible for significant odour formation in foods.
- Frying increases the acceptability of foods.
- Frying is easy and economical compared to other food preparation methods.

1.5 Key Concepts

- Frying is one of the fastest and simplest food preparation techniques.
- Frying consists of four main components: specific conditions, frying equipment, foods, and frying medium.
- Frying is one of oldest food preparation techniques, although it is hard to trace exactly when it was invented.
- The mechanism of frying consists of heat and mass transfer, oil uptake, and physical and chemical reactions.
- Food bodies are made of holes and pores filled with water. The water evaporates according to the temperature and timing of frying. Subsequently, the holes and pores are occupied by oil.
- The dehydration of the food surface in frying produces a crust/crumb, which increases with frying time.
- Oil uptake can be reduced by several pre-treatment procedures to produce foods of acceptable quality.
- The size, shape, composition, density, and surface of the food product can all affect oil uptake.
- Pre- and post-frying conditions significantly affect the frying of foods.
- Physical and chemical changes occur in foods and the frying medium, affecting flavour, taste, and sensory characteristics.

References

Aguilera-Carbó, A., Montañez, J.C., Anzaldúa-Morales, A. et al. (1999). Improvement of color and limpness of fried potatoes by in situ pectinesterase activation. *European Food Research and Technology* 210: 49–52.

Alvarez, M.D., Morillo, M.J., and Canet, W. (2000). Characterization of the frying process of fresh and blanched potato strips using response surface methodology. *European Food Research and Technology* 211: 326–335.

Blumenthal, M. (1991). A new look at the chemistry and physics of deep-fat frying. *Food Technology (USA)* 45: 68–71.

Bunger, A., Moyano, P., and Rioseco, V. (2003). NaCl soaking treatment for improving the quality of french-fried potatoes. *Food Research International* 36: 161–166.

Costa, R.M., Oliveira, F.A.R., Delaney, O., and Gekas, V. (1999). Analysis of the heat transfer coefficient during potato frying. *Journal of Food Engineering* 39: 293–299.

Croon, L.B., Rogstad, A., Leth, T., and Kiutamo, T. (1986). A comparative study of analytical methods for quality evaluation of frying fat. *Fette, Seifen, Anstrichmittel* 88: 87–91.

Dana, D., Blumenthal, M.M., and Saguy, I.S. (2003). The protective role of water injection on oil quality in deep fat frying conditions. *European Food Research and Technology* 217: 104–109.

Dana, D. and Saguy, I.S. (2006). Review: mechanism of oil uptake during deep-fat frying and the surfactant effect-theory and myth. *Advances in Colloid and Interface Science* 128–130: 267–272.

DGF (2000). Recommendations for the 3rd International Symposium on Deep-Fat Frying – Optimal Operation. *European Journal of Lipid Science and Technology* 102: 594.

Dincer, I. (1996). Modelling for heat and mass transfer parameters in deep-frying of products. *Heat and Mass Transfer* 32: 109–113.

Farid, M.M. and Chen, X.D. (1998). The analysis of heat and mass transfer during frying of food using a moving boundary solution procedure. *Heat and Mass Transfer* 34: 69–77.

Farkas, B.E., Singh, R.P., and Rumsey, T.R. (1996). Modeling heat and mass transfer in immersion frying. I, model development. *Journal of Food Engineering* 29: 211–226.

Gamble, M.H. and Rice, P. (1987). Effect of pre-fry drying of oil uptake and distribution in potato crisp manufacture. *International Journal of Food Science and Technology* 22: 535–548.

Gamble, M., Rice, P., and Selman, J. (1987). Relationship between oil uptake and moisture loss during frying of potato slices from cv record UK tubers. *International Journal of Food Science and Technology* 22: 233–241.

Gertz, C. (2014). Fundamentals of the frying process. *European Journal of Lipid Science and Technology* 116: 669–674.

Gertz, C., Klostermann, S., and Kochhar, S.P. (2000). Testing and comparing oxidative stability of vegetable oils and fats at frying temperature. *European Journal of Lipid Science and Technology* 102: 543–551.

Gertz, C., Klostermann, S., and Parkash Kochhar, S. (2003). Deep frying: the role of water from food being fried and acrylamide formation. *OCL* 10: 297–303.

Gil, B. and Handel, A.P. (1995). The effect of surfactants on the interfacial tension of frying fat. *Journal of the American Oil Chemists' Society* 72: 951–955.

Holikar, M.V., Annapure, U.S., Singhal, R.S., and Kulkarni, P.R. (2005). Pectin and calcium chloride treatment for low-fat fried green gram splits. *Journal of the Science of Food and Agriculture* 85: 1677–1680.

Houhoula, D.P., Oreopoulou, V., and Tzia, C. (2004). Antioxidant efficiency of oregano in frying and storage of fried products. *European Journal of Lipid Science and Technology* 106: 746–751.

Kalogeropoulos, N., Grigorakis, D., Mylona, A. et al. (2006). Dietary evaluation of vegetables pan-fried in virgin olive oil following the Greek traditional culinary practice. *Ecology of Food and Nutrition* 45: 105–123.

Kita, A. and Lisińska, G. (2005). The influence of oil type and frying temperatures on the texture and oil content of French fries. *Journal of the Science of Food and Agriculture* 85: 2600–2604.

Krozel, J.W., Palazoglu, A.N., and Powell, R.L. (2000). Experimental observation of dip-coating phenomena and the prospect of using motion control to minimize fluid retention. *Chemical Engineering Science* 55: 3639–3650.

Man, Y.B.C. and Jaswir, I. (2000). Effect of rosemary and sage extracts on frying performance of refined, bleached and deodorized (RBD) palm olein during deep-fat frying. *Food Chemistry* 69: 301–307.

Manglik, R.M. (2006). On the advancements in boiling, two-phase flow heat transfer, and interfacial phenomena. *Journal of Heat Transfer* 128: 1237–1242.

Mariscal, M. and Bouchon, P. (2008). Comparison between atmospheric and vacuum frying of apple slices. *Food Chemistry* 107: 1561–1569.

Matz, S.A. (2012). *Snack Food Technology*. Berlin: Springer Science+Business Media.

Mehta, U. and Swinburn, B. (2001). A review of factors affecting fat absorption in hot chips. *Critical Reviews in Food Science and Nutrition* 41: 133–154.

Moreira, R.G. and Barrufet, M.A. (1998). A new approach to describe oil absorption in fried foods: a simulation study. *Journal of Food Engineering* 35: 1–22.

Moreira, R.G., Castell-Perez, M.E., and Barrufet, M.A. (1999). *Deep Fat Frying: Fundamentals and Applications*. Gaithersburg, MD: Aspen.

Moreira, R.G., Sun, X., and Chen, Y. (1997). Factors affecting oil uptake in tortilla chips in deep-fat frying. *Journal of Food Engineering* 31: 485–498.

Morton, I. (1998). Geography and history of the frying process. *Grasas y Aceites* 49: 247–249.

Ngadi, M.O., Watts, K.C., and Correia, L.R. (1997). Finite element method modelling of moisture transfer in chicken drum during deep-fat frying. *Journal of Food Engineering* 32: 11–20.

Paul, S. and Mittal, G.S. (1997). Regulating the use of degraded oil/fat in deep-fat/oil food frying. *Critical Reviews in Food Science and Nutrition* 37: 635–662.

Pedreschi, F. and Moyano, P. (2005). Effect of pre-drying on texture and oil uptake of potato chips. *LWT – Food Science and Technology* 38: 599–604.

Perkins, E.G. (2007). Volatile odor and flavor components formed in deep frying. In: *Deep Frying: Chemistry, Nutrition and Practical Application* (ed. M.D. Erickson). Urbana, IL: AOCS Press.

Pinthus, E.J. and Saguy, I. (1994). Initial interfacial tension and oil uptake by deep-fat fried foods. *Journal of Food Science* 59: 804–807.

Quéré, D. and de Ryck, A. (1998). Le mouillage dynamique des fibres. *Annals of Physics* 23: 1–149.

Rimac-Brnčić, S., Lelas, V., Rade, D., and Šimundić, B. (2004). Decreasing of oil absorption in potato strips during deep fat frying. *Journal of Food Engineering* 64: 237–241.

Rojas-Gonzalez, J.A., Avallone, S., Brat, P. et al. (2006). Effect of deep-fat frying on ascorbic acid, carotenoids and potassium contents of plantain cylinders. *International Journal of Food Sciences and Nutrition* 57: 123–136.

Rossell, J. (2001). *Frying: Improving Quality*. Sawston: Woodhead Publishing.

Saha, D. and Bhattacharya, S. (2010). Hydrocolloids as thickening and gelling agents in food: a critical review. *Journal of Food Science and Technology* 47: 587–597.

Sobukola, O.P., Awonorin, S.O., Oladimeji, S.L., and Olukayode, B.F. (2010). Optimization of pre-fry drying of yam slices using response surface methodology. *Journal of Food Process Engineering* 33: 626–648.

Stier, R.F. (2004). Frying as a science – an introduction. *European Journal of Lipid Science and Technology* 106: 715–721.

Talburt, W.F. (1987). *Potato Processing*. New York, NY: Van Nostrand Reinhold.

Thanatuksorn, P., Pradistsuwana, C., Jantawat, P., and Suzuki, T. (2005). Effect of surface roughness on post-frying oil absorption in wheat flour and water food model. *Journal of the Science of Food and Agriculture* 85: 2574–2580.

Tseng, Y.C., Moreira, R., and Sun, X. (1996). Total frying-use time effects on soybean-oil deterioration and on tortilla chip quality. *International Journal of Food Science and Technology* 31: 287–294.

Ufheil, G. and Escher, F. (1996). Dynamics of oil uptake during deep-fat frying of potato slices. *LWT – Food Science and Technology* 29: 640–644.

Vitrac, O., Dufour, D., Trystram, G., and Raoult-Wack, A.-L. (2002). Characterization of heat and mass transfer during deep-fat frying and its effect on cassava chip quality. *Journal of Food Engineering* 53: 161–176.

Vitrac, O., Raoult-Wack, A.L., and Trystram, G. (2003). Influence of liquid water transport on heat and mass transfer during deep-fat frying. In: *Transport Phenomena in Food Processing* (ed. J. Welti Chanes, J. Véles-Riuz and G.V. Barbosa-Cànovasn). Boca Raton, FL: CRC Press.

Vitrac, O., Trystram, G., and Raoult-Wack, A.L. (2000). Deep-fat frying of food: heat and mass transfer, transformations and reactions inside the frying material. *European Journal of Lipid Science and Technology* 102: 529–538.

Wegmuller, F. (1994). Polar components of frying fats derived from data of dielectric measurements. *Zeitschrift für Lebensmittel-Untersuchung und -Forschung* 199: 51–54.

Ziaiifar, A.M., Achir, N., Courtois, F. et al. (2008). Review of mechanisms, conditions, and factors involved in the oil uptake phenomenon during the deep-fat frying process. *International Journal of Food Science and Technology* 43: 1410–1423.

2

Frying Techniques

2.1 Introduction

Unlike the art, the science of frying is rather complex, involving multifaceted high-energy interactions among several components, primarily the frying medium, water, air, metals, and the frying food. The relatively uncontrolled chemical reactions arising from these complex interactions often lead to a progressive deterioration of the frying oil, compromising both the sensorial and the safety qualities of the food (Aladedunye 2015). The economic loss associated with the disposal of spent frying oils, wastage of foods due to shortened shelf lives, consumption of fried foods prepared under poor management, and health consequences of toxic compounds generated from frying facilities cannot be overemphasized (Ghidurus et al. 2010; Gueraud et al. 2010; Sebastian et al. 2014).

Research in the past 60 years has been geared towards understanding the science of frying with a view to enhancing the fry life of the oil, improving the safety and quality of the fried food by limiting oil absorption/adsorption and the formation of toxic components, prolonging shelf life, and protecting the health of frying operators by limiting the formation of toxic fumes during the frying process. These are the core principles of the various symposia on deep fat frying held by the Deutsche Gesellschaft für Fettwissenschaft e.V. (DGF), the German Society for Fat Research, and similar organizations (DGF 2000, 2013). Indeed, the versatility of the frying process and the pressing need for improvement have attracted many reviews in the past decades (Przybylski and Eskin 1995; Stier 2013; Adekunle 2014; Gertz 2014; Erickson 2015; Aladedunye et al. 2017). However, the frying process still needs to be studied in more detail.

2.2 Concept of Deep Frying

Deep frying allows food to be fried until specific colour, taste, and properties are attained. During frying, continuous and simultaneous heat and mass transfer occurs. Heat transfer may occur through convection between the oil and the food surface and by conduction within the food structure. When the food is immersed in the hot oil, changes begin with the evaporation of water vapour. The water is evaporated due to the high temperature of the oil, which is above the boiling point of water. Pores and crust are formed following this. The higher the water content, the higher the evaporation,

Food Frying: Chemistry, Biochemistry, and Safety, First Edition. Alam Zeb.

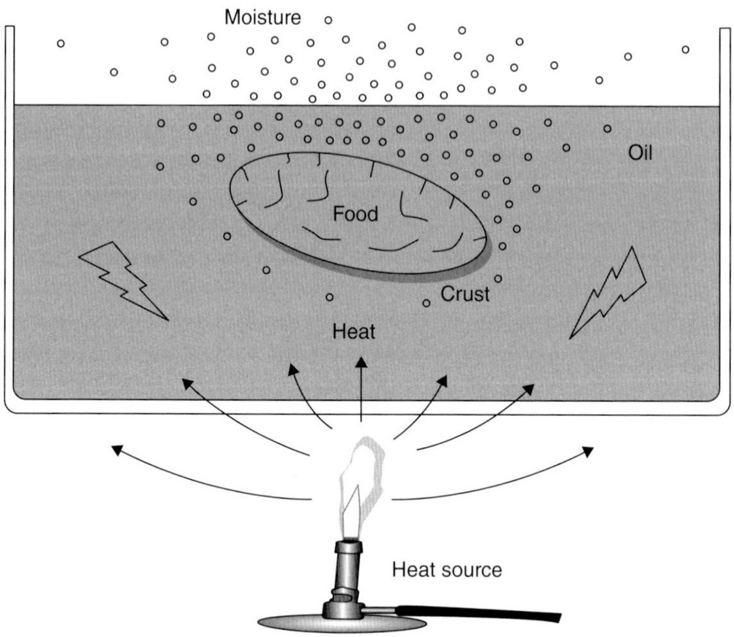

Figure 2.1 Schematic representation of the deep frying of foods.

and thus the more pores and the higher the crust. The oil contents move to the pores and occupy them, in a process called oil uptake (Sahin and Sumnu 2009). Oil uptake is higher in the cooling stages of the frying process.

Several other physical changes occur during frying, such as colour development, changes in the weight and size of the food, flavour development, and density changes. These are mostly the result of chemical reactions in the food and the frying medium. The convective heat transfer coefficient varies with the temperature and degradation of the frying medium. Bubbling on the surface of the oil and both inside and on the surface of the food also significantly affects heat transfer (Figure 2.1). Heat transfer usually occurs alongside mass transfer. The ejection of steam bubbles in high volumes is responsible for a significant stirring close to the exchange surface of the food. Thus, the convective heat transfer coefficient is measured using an active probe. The temperature of the oil is controlled by electrical sensors, which are used to obtain the desired water content.

2.3 Tools Used in Frying

Several tools may be used, depending upon the type of frying being done. Figure 2.2 shows a compact countertop electrical deep fryer with temperature control and timer. Such controlled fryers provide a better quality of fried food. In home kitchens, the following tools are usually required for food frying:

- Countertop compact electrical deep fryer, dual-basket fryer, or frying pan
- Stainless steel oil storage can strainer

Figure 2.2 Countertop compact electric deep fryer. *Source:* Courtesy of Waring Commercial, USA.

- Deep-frying stainless steel thermometer
- Knife
- Colander
- Cutting board
- Egg piercer, poacher, and separator
- Spatula
- Food mill
- Grater/shredder
- Meat grinder
- Peeler
- Poultry shears
- Rolling pin
- Weighing scale
- Sieves
- Spoon thermometer
- Can opener
- Whisk
- Wooden spoon

Restaurants and the food industry also use most of the same tools, but the fryers are usually larger and are mostly automatic or semi-automatic (Figure 2.3). The details of such fryers are reviewed by Gupta et al. (2004). In restaurant-grade fryers, heating is provided by submersible elements controlled by special probes and monitored by computer. The computer control system also provides Hazard Analysis and Critical Control Points (HACCP) control. The fryer and the heating elements are made of high-quality stainless steel. The Italian company FIREX Srl manufactures a number of different automatic fryers with single or double frying vessels, each with the capacity to fry 50 kg of chips per hour.

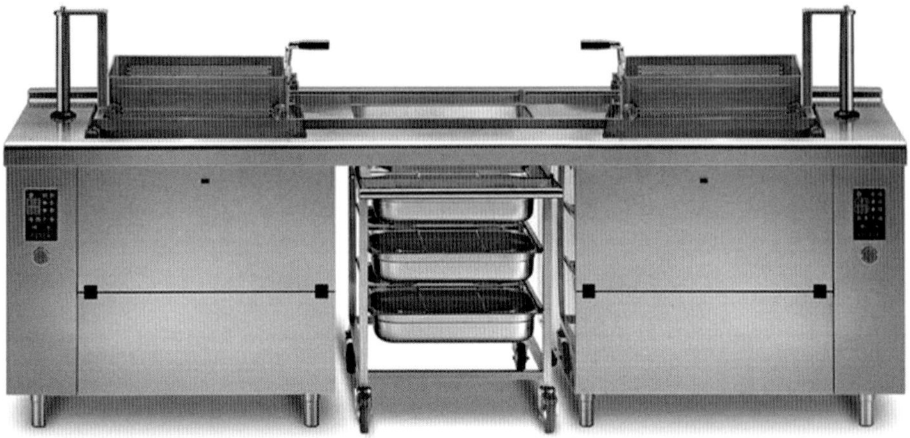

Figure 2.3 Automatic industrial multi-fryer. Firex model No. FAGT 2-070 with two frying vessels and computer controls. *Source:* Courtesy of Firex Srl, Italy.

2.4 Optimized Conditions

The frying stability of oil is a measure of its resistance to several of the degradative reactions that occur during frying. The various factors influencing the stability and performance of a frying oil can be categorized as external and internal, depending on whether they are oil-dependent or operation-dependent.

2.4.1 External Factors

External factors are relatively independent of the inherent quality of the frying oil. They include: accessibility of oxygen; temperature; duration of frying; size, dimensions, and composition of the food; fryer design (shape, size relative to production requirements, surface-to-volume ratio, responsiveness of heating element, and accuracy of temperature controllers); type of frying operation (continuous or discontinuous/intermittent); and frying management (use of active and passive filters and oil replenishment).

2.4.1.1 Accessibility of Oxygen

Lipid oxidation is arguably the single most important factor affecting the life of edible oil. The oxidation of oil increases with the amount of dissolved oxygen in it (Min and Wen 1983). Przybylski and Eskin (1988) reported that the amount of oxygen dissolved in oil is sufficient to provide a peroxide value of 10 meq kg^{-1}. The total amount of volatile aldehydes emitted during thermal treatment of high oleic safflower oil was found to increase with increasing oxygen concentration (Fujisaki et al. 2002). Besides affecting the extent of oxidative degradation, the concentration of oxygen also affects the types of degradation product formed during frying. For instance, in an atmosphere containing 2% oxygen, acetaldehyde was the dominant volatile carbonyl compound formed during oxidative degradation of high oleic safflower oil, whereas hexanal and nonanal were the most abundant when 20% oxygen was present. Because individual volatile carbonyl compounds possess characteristic flavour and threshold values, the concentration of oxygen during deep frying will also affect the flavour of the fried food.

Despite the poor solubility of oxygen at frying temperatures and under steam blanketing, several factors are known to increase the availability of oxygen in the frying medium. The introduction of fresh food and oil into a frying medium increases the level of oxygen in the oil (Warner 2004). Thermal agitation accompanying boiling during the frying of food breaks the oil surface and enhances the accessibility of atmospheric oxygen by enlarging the surface area (Mezouari and Eichner 2006). Furthermore, surface-active compounds naturally occurring in oil or formed during frying can also enhance the access of atmospheric oxygen during frying. Mistry and Min (1987) reported that free fatty acids (FFAs) decrease the surface tension of oil, thereby increasing the diffusion rate of oxygen into the oil and so accelerating oil oxidation. A frying system with a large surface-to-volume ratio also increases the availability of oxygen during frying. According to a study by Totani et al. (2013), oxygen lost at frying temperature due to insolubility or consumption is slowly restored during cooling and standing at room temperature (between 25 and 120 °C), significantly affecting overall oil stability and performance during intermittent frying. In other words, the reduced solubility of oxygen at frying temperature does not significantly hinder its influence on the performance of the frying oil, especially during intermittent frying. Indeed, one of the recommendations of the 7th International Symposium on Deep-Fat Frying is to avoid holding oils at temperatures between 60 and 130 °C during downtimes when an oil is not in use for frying (DGF 2013).

Approaches aimed at reducing the availability of oxygen during frying have been reported as ways of protecting oil from thermo-oxidative degradation. Przybylski and Eskin (1988) evaluated the efficacy of nitrogen and carbon dioxide flushing in preventing canola oil oxidation under frying conditions. The oil samples were heated at 195 °C in a variety of containers of differing dimensions, through which the flow of N_2 and CO_2 was regulated. The peroxide value, the thiobarbituric acid value, and the formation of volatile compounds were used to assess the level of thermo-oxidation in the oils. It was observed that oils heated without prior N_2 or CO_2 flushing underwent more rapid oxidation compared to oils with prior N_2 or CO_2 flushing. Based on their findings, the authors suggested that to prevent oxidation of oils and fats during thermal treatment, the following options should be considered: (i) flush the oil with CO_2 rather than N_2; (ii) flush the oil with N_2 for 15 minutes or CO_2 for 5 minutes prior to heating to eliminating any dissolved oxygen; (iii) maintain the linear flow of gas in the container at 50 cm min^{-1}; and (iv) refrain from filling the vessel to more than 70% of its height. The superior protection offered by CO_2 over N_2 was attributed to its higher density and greater solubility in oil. Although this study indicated that frying under inert gases could inhibit the thermo-oxidative degradation of oils, the effectiveness of this method in actual deep frying was not verified until several years later (Aladedunye and Przybylski 2009b).

Shyu et al. (1998) reduced the oxygen content by conducting frying under vacuum conditions. For 6 consecutive days, carrot slices were fried at 105 °C in palm oil, lard, and soybean oil for 20 minutes each hour in an 8-hour shift. Peroxide value, acid value, viscosity, fatty acid (FA) composition, and total polar components were used to evaluate the extent of thermo-oxidative degradation. The authors concluded that vacuum frying imparted a lower thermo-oxidative degradation on the oils than typical atmospheric frying. Although the better stability of the oils during vacuum frying was attributed to the lower content of oxygen, the lower frying temperature in the study (105 °C) makes a comparison with typical frying (temperature ~180 °C) unrealistic. Several studies (including detailed reviews) on the effects of vacuum frying have been reported, notably by Moreira's group, but the focus has always been more on the sensory attributes of the products than on the frying oils (Mistry and Min 1987; Totani et al. 2013). Thus, investigations into the effect of vacuum on the performance of frying oils under a similar temperature regime to that used for standard frying are required (Aladedunye and Przybylski 2009b). According to Aladedunye and Przybylski (2009b), frying under a CO_2 blanket reduced the total polar compounds by 54%, while a 76% reduction was observed during vacuum frying compared to standard frying conditions. Similarly, lower oxidative degradation was observed when measured by anisidine value. At the end of the frying period, the reduction in unsaturated FA content was 3.8, 1.9, and 12.7% when frying under CO_2 blanketing, vacuum, and standard frying conditions, respectively. The rate of tocopherol degradation was 3 and 12 times slower in vacuum frying when compared to CO_2 blanketing and standard frying conditions, respectively. Evidently, oxygen concentration has a huge impact on the performance of oils even at frying temperature, despite the poor solubility.

Whilst the vacuum system has huge potential in extending the fry life of oils, its application in institutional frying could be costly and labour-intensive, and it would definitely be challenging to apply given the current state of technology. Inert gas blanketing, on the other hand (primarily CO_2, which is several-fold cheaper than N_2 and evidently more effective), could offer a viable alternative – but a technology capable of trapping, purifying, and recycling spent CO_2 gas may be required, mainly due to environmental concerns. Further, considering the superior effectiveness of vacuum at removing dissolved oxygen from frying oil (Aladedunye and Przybylski 2009b), an optimal vacuum pressure would need to be chosen to ensure that sufficient oxygen level was present in the frying medium to produce flavour-impacting compounds.

It must be stressed that the method use to deliver the CO_2 gas to the oil is critical to its effectiveness. It was observed that whereas gently releasing CO_2 longitudinally across the surface of oil through perforated tubing attached 1 cm above it dramatically inhibited thermo-oxidative deterioration, bubbling the CO_2 directly into the oil significantly increased it. In fact, the rate of thermo-oxidative deterioration as measured by TPC, and polymer formation increased with increasing CO_2 flow rate due to the increase in bubbling (Figure 2.4). Thus, while gently blowing CO_2 across the oil surface prevents oxygen from getting into the oil, especially during the crucial moment when food is being fried and the oil is opening up from the boiling process, bubbling with CO_2 has the opposite effect, enhancing oxygen access to the frying oil. Evidently, the competition between the increased oxygen access to the frying oil consequent to CO_2bubbling and the oxygen-purging effect of CO_2 is decided in favour of the former. This is in contrast to the report by Sionek et al. (2013), where no significant differences were found in either blanketing or continuous flushing of CO_2 or N_2 gas during the oxidative storage of refined rapeseed oil at up to 120 °C. Indeed, this observation further supports the general knowledge that the behaviour of oils at frying temperature is dramatically different than at 120 °C and below.

The addition to frying oils of silicone, particularly dimethylpolysiloxane (DMPS), as an antifoaming agent is well known, and its effects on stability, including its dependence on frying parameters, have been adequately reviewed (Freeman et al. 1973; Márquez-Ruiz et al. 2004). Although a number of mechanisms have been proposed for DMPS's

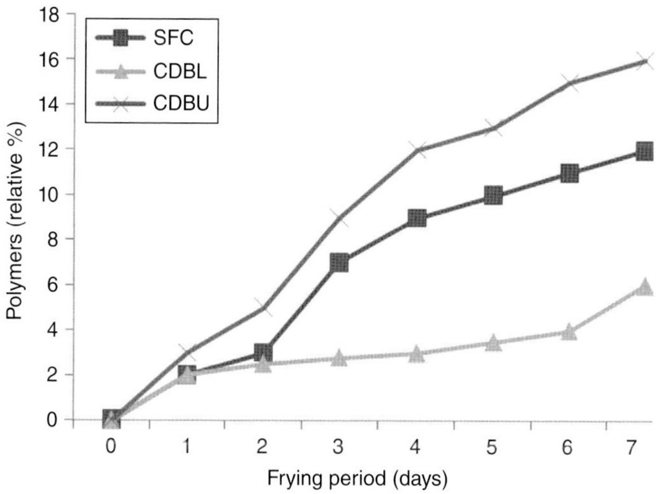

Figure 2.4 Effect of carbon dioxide blanketing (CDBL) and carbon dioxide bubbling (CDBU) on the frying stability of frying oils, compared to standard frying conditions (SFC). *Source:* Reproduced with kind permission of John Wiley & Sons (Aladedunye 2015).

activity in delaying the thermo-oxidative degradation of frying oils (Freeman et al. 1973), the most prominent is related to its low fat solubility (<1 ppm), and thus its ability to accumulate at the oil surface, where it forms a physical barrier against oxygen (Freeman et al. 1973; Ohta et al. 1987). Consequently, the minimum concentration of DPMS that exerts a protective action corresponds to that forming a monolayer on the oil surface. This concentration evidently depends on the surface-to-oil volume ratio and has been reported to be between 0.05 and 0.06 μg cm^{-2} oil-to-air surface (Freeman et al. 1973). Because DPMS's primary mechanism is surface action, a major disturbance to the oil's surface, such as during the continuous frying of food, would be expected to limit its effectiveness; thus, the addition of DPMS would be of benefit only in discontinuous or intermittent frying operations, such as in catering services, fast food outlets, and restaurants, where fryers usually remain without food for significant periods of time (Márquez-Ruiz et al. 2004).

2.4.1.2 Temperature

The various reactions threatening the oxidative stability of oil require some level of energy to proceed. For instance, 50 kcal mol^{-1} of energy is required to break the carbon–hydrogen bond on the carbon 11 of linoleic acid and to initiate free radical formation (Zeb 2015). The oxygen–oxygen bond of alkyl hydroperoxide requires 44 kcal mol^{-1} to break it. This energy requirement is more than fulfilled at the temperature employed during frying. Nawar (1984) reported a peroxide value of 1777 meq kg^{-1} when ethyl linolenate was heated at 70 °C for 6 hours. However, only 13.3% of this value remained when the same oil was heated at 180 °C for 5 hours. At 250 °C, tremendous degradation of hydroperoxide occurred, and only 2.5% of the peroxide value was left after 3 hours of thermal treatment. Therefore, apart from accelerating the initiation step of oxidative degradation, elevated temperature enhances thermal degradation of alkyl hydroperoxides, the primary oxidation product. Consequently, oxidative degradation proceeds more rapidly during deep frying than at room temperature (Blumenthal 1991; Tyagi and Vasishtha 1996).

An increase in frying temperature increases thermal oxidation and oligomerization reactions, not only of the FA or triacylglycerol (TAG) molecules, but also of the unsaponifiable minor components. Thus, antioxidant minor components in oil are either thermally inactivated during frying or have their levels severely reduced (Allam and Mohamed 2002; Réblová 2006; Marmesat et al. 2010). While tocopherol in rice bran oil heated to 100 °C in the absence of air showed a reduction of 29% at the end of 432 hours of heating, this reached 100% when the oil was heated at 180 °C for 240 hours (Bruscatto et al. 2009). Oxidation and polymerization of phytosterols, consequent to an increase in temperature, have also been studied, and the formation of oligomers arising from thermo-oxidation of phytosterols has been reported (Rudzińska et al. 2009, 2010; Sosińska et al. 2014).

The effect of temperature on isomerization reactions has also been reported. Wolff (1993) studied the formation of linolenic acid geometrical isomers in flaxseed oils heated under vacuum at different temperatures. After 16 hours of heating at 190, 220, and 245 °C, the total *trans* linolenic acid increased from an initial 0.2% in fresh oil to 1.3, 9.5, and 28.8%, respectively. Tyagi and Vasishtha (1996) reported an increase in *trans* fat from 1.7 to 2.6% in soybean oil when the frying temperature was increased from 170 to 190 °C. Moreno et al. (1999) utilized Fourier transform infrared (FTIR) spectroscopy to monitor the formation of *trans* isomers during the heating of olive oil, sunflower oil, corn oil, and lard over a wide temperature range of 80–300 °C. Their results showed that the amount of *trans* isomers consistently increased as a function of temperature irrespective of the type of oil. Tsuzuki et al. (2010) compared the amounts of *trans* isomers formed when canola oil was heated at different temperatures (160, 180, and 200 °C) and found a statistically significant increase in the isomerization rate when the heating temperature was increased from 180 to 200 °C. In a similar experiment, 10 batches of potatoes were fried in canola oil every 10 minutes, and the amounts of *trans* fatty acids (TFAs) formed at the end of the frying period (100 minutes) were analyzed by gas chromatography. It was observed that at all tested temperatures, the amounts of *trans* isomers were higher in oils used for deep frying than in heated samples. Although the levels of *trans* isomers observed in the fried potatoes may not have any practical implications related to nutritional claims about 'zero *trans*' content in a serving portion, the duration of frying (1 hour and 40 minutes) was short and may not be representative of a typical institutional frying operation.

The results of this study clearly emphasize the importance of controlling frying temperature. When frying is carried out at a temperature above the optimum (regulated at 165 °C, e.g. in Germany), either due to a faulty temperature controller or deliberately to increase turnover at the peak hours, the rate of oil degradation will increase accordingly, shortening the fry life of the oil and compromising the quality and safety of the fried products (Mehta and Swinburn 2001; Aladedunye and Przybylski 2009a). For instance, a 1999 national survey of fast food outlets in New Zealand reported a wide range of frying temperatures used in institutional frying operations, ranging from 112 to 233 °C, with 26% of the surveyed outlets operating at above 190 °C (Mehta and Swinburn 2001). Aladedunye (2015) reported that the actual oil temperature could be up to 30 °C higher than the temperature controller settings, and unfortunately, only a small percentage of institutional frying operators are consciously making efforts to calibrate and ascertain the correctness of their fryers' temperature controllers. For instance, a survey in Australia showed that only 56% of fast food outlets had correctly calibrated temperature controllers (Mehta and Swinburn 2001).

2.4.1.3 Duration of Frying

Although deep frying is a fast method of food preparation and the contact time between food and oil is relatively short, the length of time spent using the frying oil is usually long, especially during continuous and repeated frying such as in industrial and institutional operations. All things being equal, thermo-oxidative degradation increases with increasing duration of oil use (Xu et al. 2000; Houhoula et al. 2003; Farhoosh and Tavassoli-Kafrani 2010; Rani et al. 2010). Bansal et al. (2010) used three different frying oils – palm olein; sunflower oil; and a blend of palm olein, sesame, and peanut oils – to prepare French fries. The oils were heated for 6 hours daily for 4 days, and 10 batches of French fries were fried each day without oil replenishment. The extent of isomerization of the FAs was quantified by gas chromatography and infrared spectroscopy. Irrespective of the type of oil and the quantification method, the authors reported a direct relationship between the extent of isomerization and the number of frying cycles.

The dwell time, defined as the contact time between the food and the frying oil, is also critical, and must be balanced with the frying temperature. It is generally agreed that the presence of food reduces the deterioration of frying oils (compared to heating without food), due to the temperature-lowering effect of water in the fresh food (at the instance of introduction) and the blanketing effect of the ensuing steam against oxygen during frying. However, over-extended dwell time can nullify this 'gain' due to: (i) the continuous migration and expression of pro-oxidant metals and decomposition products from the food to the frying oil; and (ii) the continuous presence of food (after the loss of the initial moisture/steam), which increases oxygen penetration into the oil by opening the oil surface (Aladedunye 2015).

Periodically replenishing frying oil is a very crucial oil management program; however, this exercise is futile and unhealthy once the oil reaches the 'degrading' and 'runaway' stages (Stier 2004).

2.4.1.4 Size, Dimensions, and Composition of Food

As mentioned earlier – and widely acknowledged – the presence of food slows the thermo-oxidative degradation of oils when compared to heating without food. However, the type and composition – including the shape and size – of the food can significantly affect the stability and performance of the frying oil. For instance, fatty food items containing highly unsaturated oils can compromise the integrity of the frying oil by increasing the overall level of the thermo-labile polyunsaturated fatty acids (PUFAs) in it through mass exchange (Dobarganes et al. 2000). Proteins, peptides, and amino acids in foods can react with lipid-free radicals or lipid oxidation products, primarily carbonyl-containing oxidation products, to inhibit thermo-oxidation during frying (Pokorn et al. 2003). Starch, on the other hand, reportedly increases the thermo-oxidative degradation of frying oils (Aladedunye 2015). According to Artz et al. (2005), transition metals such as iron in meat products can progressively accumulate in the oil during frying, accelerating thermo-oxidative degradation. Iron bonds in myoglobin or haemoglobin are easily released during the frying of meat and fish (Pokorn et al. 2003). A direct positive correlation has been reported between the moisture content of the food being fried and the thermo-hydrolytic and thermo-oxidative degradation of the frying oil (Bhattacharya et al. 2008). The oxidative status of the food is also important, as leaching of pre-formed hydroperoxides in the food (again, this is more important for fish or PUFA-fortified food products) will cause accelerated degradation of the frying oil. Regardless of its type and composition, an increased cross-sectional area of the frying food (e.g., thin-cuts strips versus thick-cut; cracks and rough surfaces versus a smooth surface) will often lead to increased thermo-oxidative degradation of the frying oil. This may be because the increased surface area removes more of the oil, increasing oxygen pressure in the fryer. Further, the increased surface area also increases the opportunity and rate of mass exchange between the food (moisture, pro-oxidant metals, colouring matters, etc.) and the frying oil.

One way of moderating unwanted mass exchange between food and frying oil is to use edible coatings. Various materials have been used for the development of edible coatings, including hydrocolloids based on proteins of animal or plant sources (e.g. whey, soy, corn, legumes), polysaccharides (e.g. cellulose derivatives, alginates, or starches), and even (semi)synthetic polymers (e.g. polyvinyl acetate, hydroxypropylmethylcellulose) (Porta et al. 2012). However, interactions between these often multicomponent coatings and the frying oil cannot be precluded, resulting in either enhancement or depression of the oil's frying performance. For instance, according to Holownia et al. (2000), a decrease in the degradation of peanut oil was observed when the fried food (marinated chicken strips) was coated with food-grade hydroxypropylmethylcellulose prior to frying. This protection was attributed to the ability of the coating to inhibit moisture migration from the food into the frying oil, thereby limiting the thermal hydrolytic reaction and, by extension, the thermo-oxidative deterioration of the frying oil. On the other hand, Lazarick et al. (2014) observed that using whey protein as a base for battering resulted in significant colour changes and thermo-oxidative deterioration of frying oil, with the changes being more pronounced in the presence of pre-formed lipid hydroperoxides. Glucose and glycine are two minor ingredients that also contribute to colour formation in oil. Breading materials are prone to cause a more significant amount of oil deterioration when compared to battering ingredients, most likely due to excess loose breading particles falling into the oil during frying. In a study by Guerra-Hernandez et al. (2013), the battering of zucchini with rice- or corn flour-based batter significantly increased the amount of acrylamide formed during frying in sunflower oil at 190 °C, compared to wheat and chickpea flours. Breaded products showed lower amounts of acrylamide than battered products during the frying process. According to Xue and Ngadi (2007), the type of batter system can significantly influence the thermal properties of food products; for instance, rice- and corn-based batters require more energy for gelatinization, indicating that food coated with batters based on these items may require relatively longer frying periods, with negative influence on the stability of the frying oil.

Considering the interesting interactions between food coatings and frying oil, the development of functional coatings to enhance the thermo-oxidative stability of such oil – for instance, through the consistent batchwise release of fresh antioxidative compounds (incorporated into the coating) following the introduction of food into the oil – may be of practical importance. Unfortunately, while the literature is replete with studies, including critical reviews, on the influence of battering and breading ingredients on the oil uptake and sensory attributes of fried foods (Dogan et al. 2005;

Chen et al. 2009; Brannan et al. 2014), corresponding data on their effects on frying oils are rather scarce (Holownia et al. 2000; Lazarick et al. 2014).

All things being equal, for a particular food product, temperature, oxygen, and time are the three optimizable cornerstones upon which the stability and performance of frying oil are based, as far as external factors are concerned (Figure 2.5). Upward deviation from the optimum for these parameters tapers off the oil stability and weakens its performance. On the other hand, working below the optimum conditions in a bid to 'slow down' oil deterioration (dotted lines on Figure 2.5) will result in fried products with compromised quality (Aladedunye 2015).

Other relatively minor factors, namely, surface-to-volume ratio, oil replenishment, oil filtration (passive), and food geometry and size (Lin et al. 1998; Rossell 2001; Gertz 2004), are really subsets of the three mega-parameters: temperature, oxygen, and time. For instance: frying with an optimized surface-to-volume ratio is related to controlling the oxygen concentration in the frying medium; replenishing used oil with fresh oil is related to both maintaining a good surface-to-volume ratio (oxygen concentration) and minimizing the effect of the extended frying period of the institutional and industrial deep fat frying operations; and passive filtering of oil is related to reducing or limiting unwarranted contact time between the frying oil and the food (debris, in this case), and thus preventing undue generation and leaching of colour-impacting pro-oxidative components from the debris into the frying oil.

2.4.2 Internal Factors

Unlike the external factors affecting frying stability of the oil, internal factors arise from the inherent composition of the oil itself. Edible oils are composed of TAGs (>96%) and endogenous minor components. It is generally agreed that the inherent composition of

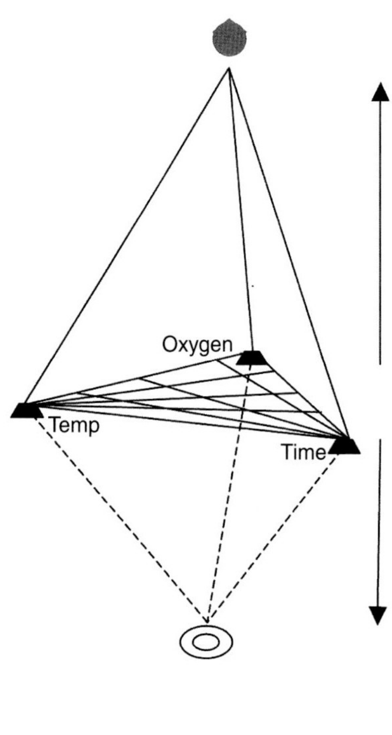

Figure 2.5 Trigonal bipyramid model showing the three basic optimizable parameters of frying. The base (striped) represents the optimized region for the highest stability of the frying oil and the best-quality fried product (operator's target). Higher values (bold lines) taper off stability, while lower values (dotted lines) decrease food quality. *Source:* Reproduced with kind permission of John Wiley & Sons (Aladedunye 2015).

edible oils exerts significant influence on their frying stability (Gertz et al. 2000; Normand et al. 2001; Shahidi and Zhong 2010; Aladedunye and Przybylski 2012).

2.4.2.1 Fatty Acid Composition and Distribution

The influence of the FA composition of oils on their stability has been variously reported (Neff et al. 1992; Zeb 2011). In general, oils that are more unsaturated oxidize more readily than those that are less unsaturated (Parker et al. 2003). In other words, as the number of double bonds in a FA increases, both the rate of formation and the amount of primary oxidation compounds accumulated at the end of the induction period also increase. This observation correlates with the relative rate of the FA alkyl radical formation (Zeb 2015). On the basis of oxygen uptake, linoleate was 40 times more reactive than oleate; linolenate was 2.4 times more reactive than linoleate; and arachidonate was 2 times more reactive than linolenate (Holman and Elmer 1947; Kamal-Eldin 2003). The oxidizability of linoleic (18:2), linolenic (18:3), arachidonic (20:4), and cervonic (22:6) acid was linearly correlated to the number of bis–allylic bonds in the FA ester (Cosgrove et al. 1987).

Based on this, the search for stable oils has led to several modifications to the FA composition of edible oils, usually by decreasing the contribution of linoleic and linolenic acids and increasing that of oleic acid (Zeb and Murkovic 2013). Almost all conventional oils now have counterparts with modified FA compositions. Various high oleic sunflower oils containing from 75 to 90% oleic acid are now available in the market. The amount of linolenic acid in conventional canola and soybean oils has been significantly reduced in low linolenic canola and soybean oils. Soybeans containing increased levels of oleic acid with reduced linoleic and linolenic acid levels have been developed or are being developed for commercialization (Wilkes 2008).

Notable methods for lowering unsaturation in oils include fractionation, hydrogenation, inter-esterification (chemical and enzymatic), conventional seed breeding, and genetic engineering. Several reviews covering some aspects of these methods are available (Dijkstra 2006; Holm and Cowan 2008; Clemente and Cahoon 2009; Kontkanen et al. 2011). The review by Dijkstra (2006) covers available modifications to the hydrogenation process that are geared towards reducing the formation of *trans* fat. Martin et al. (2010) reviewed the oxidative stability of vegetable oils with a modified FA composition by inter-esterification, technically called structured lipids, as compared to conventional oils.

Blending of polyunsaturated oils with more saturated or monounsaturated oil is also being used as a cost-effective way of reducing the amounts of linoleic and linolenic acids in frying oils (Gupta et al. 2001; Su and White 2004; Farhoosh and Esmaeilzadeh 2009). In a study by Khan et al. (2011), the contribution of linoleic acid to total FA was 50% lower in a blend of sesame and coconut oil compared to pure sesame oil. Panghal et al. (2010) prepared blends of soybean oil and palm oil with a linolenic acid content of 3.1%, compared to the 9.8% in soybean oil. In the same study, a blend of sunflower and palm oil contained 32.2% linoleic acid, lowered from 73.6% in pure sunflower oil. This technique is gaining wide acceptance, and the literature on the subject is increasing, with applications extending to many other nontraditional edible oils (Rehab and El Anany 2012; Durmaz and Talpur 2015).

Although the frying stability of vegetable oil blends is generally reported to be higher than that of the individual oils used in the blending (Su and White 2004; Farhoosh and

Esmaeilzadeh 2009; Alireza et al. 2010; Serjouie et al. 2010), it is inaccurate to attribute the increase in stability entirely to changes in FA composition, since the endogenous minor components of the constituent oils and other subtle factors may play significant roles. Thus, for the technique to be truly successful, a superior knowledge of the constituent oils is of prime importance in providing a better choice of the blending ratio, since: (i) depending on the composition, destructive interaction between endogenous minor components of the individual oils could impair the stability of the blend, or some of the endogenous minor compounds could exceed the optimum threshold above which they become pro-oxidative; and (ii) depending on the composition and the blending ratio, a significant change in the TAG composition of the blend could occur. For instance, according to Mezouari and Eichner (2006), raising the amount of rice bran oil from 10 to 50% in a sunflower/rice bran oil blend will increase OLO, OLP, PPL, OOO, OPO, oleic, and palmitic acids, and decrease LLL, LLO, and linoleic acid contents to a level that can significantly influence thermo-oxidative stability. Changes in the TAG composition have been reported to exert a strong influence on the stability of frying oils (Neff et al. 1992; Neff and El-Agaimy 1996; Zeb and Ahmad 2017). Indeed, according to Kim et al. (1988), the influence of changes in TAG composition on the foaming of oils during frying exceeded that of the FA composition. Hoshina et al. (2004) reported that PPL and the mixture of PPP/PLL (1 : 1, v/v) exhibited higher thermo-oxidative stability than PPP/LLL (2 : 1, v/v), as measured by the amounts of total polar compounds, polymeric compounds, carbonyl compounds, and FFAs formed during heating at 150 and 180 °C for 12 hours. Similarly, PPO was more stable than the mixture of PPP/OOO (2.1, v/v).

Some of the methods currently used to modify the FA compositions of oils are also known to negatively affect the nutrition, functionality, and stability of the oils (Lichtenstein et al. 2006). Hamam and Shahidi (2006) reported a significant reduction in tocopherol content during the synthesis of structured lipids by inter-esterification. Consequently, the modified oils were less stable than their unmodified counterparts, despite an increase in the degree of saturation in the products. In a study by Wang et al. (2010), sea blubber and menhaden oils were modified through chemical inter-esterification via sodium methoxide. The oxidative stability of the oils during accelerated storage at 60 °C for 4 days was assessed by measuring conjugated dienes and thiobarbituric acid value. The authors reported a significant reduction in oxidative stability, attributable to the loss of endogenous minor components during the modification process. Their observations agree with several other studies (Turan et al. 2007; Lee et al. 2008; Jennings and Akoh 2009).

The use of genetic engineering to modify edible oil is still plagued by consumer suspicion towards genetically modified organism (GMO) products. Therefore, a continued search for methods of improving the stability of frying oils without compromising their nutritional and functional qualities is warranted (Aladedunye 2015).

2.4.2.2 Antioxidative Minor Components

The minor components referred to as unsaponifiable matter are the non-TAG constituents of an oil; they constitute up to 5% of the total lipid composition (Abuzaytoun and Shahidi 2006). The composition of edible oil minor components is of great importance to oil stability both during storage and at frying. For instance, a statistical study showed that the phenolic compounds in olive oil contributed approximately 51% to its storage stability, as compared to a 24% contribution arising from its FA composition (Aparicio et al. 1999). Zambiazi (1998) estimated that only about half of the storage stability of oil can be adequately explained by its FA composition.

The influence of antioxidative minor components on the stability and performance of frying oils has been thoroughly reviewed (Yanishlieva and Marinova 2001; Márquez-Ruiz et al. 2014) and will not be reiterated here. Suffice it to say that: (i) these endogenous antioxidative minor components (tocochromanols, phytosterols, γ-oryzanol and related compounds, sesamins and related compounds, hydroxytyrosol and related compounds, carotenoids, and the phenolics) do exert significant influence on the stability of frying oils; (ii) there are optimal concentrations above which some of these antioxidant compounds become pro-oxidative; (iii) depending on the oil and the relative concentration, there is the possibility of both synergistic and antagonistic interactions among these antioxidants and other minor components in the frying oils; (iv) while some of these components are very effective antioxidants under oxidative and storage conditions, they are much less effective during frying, due to the prevailing stringent conditions and the diverse, not-completely-understood mechanisms of thermo-oxidative alterations of frying oils; (v) there is a need for a fortification of the endogenous antioxidants through the addition of potent exogenous antioxidants; (vi) synthetic antioxidants are less appealing to consumers and are less effective under frying conditions, principally due to their volatility and thermal inactivation; and (vii) there is a continued search for potent natural or semi-synthetic exogenous antioxidants specifically designed for frying applications, with recorded successes.

Extractives from vegetal sources, especially from common spices and herbs (rosemary, thyme, oregano, sage, turmeric, coriander, summer savoury, and turmeric), as well as other food-processing and agricultural wastes and byproducts (berries, citrus peel, olive leaf, hazelnut leaf, grape seed, and tea leaf), are continually being evaluated for their potential to curb the thermo-oxidative degradation of frying oils (Yanishlieva and Marinova 2001). Although much success has been reported, there are many challenges, including: possible inconsistency in activity arising from varying, unstandardized extraction and preparation methods or from genetic, climatic, and geographical variations in the active constituents of the extracts; a possible negative impact of the extracts' inherent colour and flavour on the sensory attributes of the frying oils and food; possible co-extraction/contamination with some toxic or allergenic compounds; the poor lipophilic nature of the active polyphenolic constituents of the extracts; and the very competitive market prices of available synthetic alternatives.

2.5 Types of Frying

There are four major types of frying: deep frying, pan or shallow frying, vacuum frying, and air frying. All can be highly effective, depending on the need and the type of food. Pan and deep frying may sometimes be a mixed process, depending on the food type.

2.5.1 Pan Frying

Pan frying is characterized by the use of small amount of oil or fat in a frying pan and usually a small amount of food in home kitchens. It is also called shallow frying.

Figure 2.6 Coated and noncoated stainless steel frying pans.

A frying pan is a small utensil made of stainless steel or aluminium. Traditionally, it was made of pure copper. The surface is sometimes coated in polytetrafluoroethylene (PTFE, or Teflon), to prevent sticking (Figure 2.6). The frying time and temperature must be carefully monitored to avoid burning the food. Pan frying can maintain the same level of moisture as other frying technique and is carried out under atmospheric pressure.

The pan frying process is very simple. A specific amount of frying oil or fat is heated to a specific temperature or to boiling. Food is immersed or kept in it and stirred continuously until fried. Foods may be covered with different types of coating in order to obtain specific tastes, textures, and qualities. Pan frying can be carried out using shallow or deep frying, depending upon the amount of oil added. It is commonly used for the frying of fish, meat, and a variety of vegetables. Heat transfer in pan frying is very quick as compared to other frying techniques. Turning the food during frying reduces the frying time and increases crust formation (Ikediala et al. 1996).

Mass transfer in pan frying occurs in two discrete periods: the frying period and the cooling period. During the frying period, the temperature of the product is raised from the ambient temperature to the boiling point of water (100 °C). The capillary pressure is negligible during this period, so almost no oil is absorbed (Moreira 2014). During the cooling period, the surface oil is absorbed through the pores on the food surface. This occurs due to the difference in pressure (decrease in temperature and pressure in the pore spaces) caused by the increase in capillary pressure (Moreira and Barrufet 1998). Moreira et al. (1997) showed that 20% of the total final oil content was absorbed by tortilla chips during frying, and 64% during cooling, leaving only 36% at the chips' surfaces.

2.5.1.1 Quality Characteristics in Pan Frying

Earlier experiments showed that pan frying produces a high amount of peroxides when the amount of oil is very small (i.e. a thin layer of oil) (Usuki et al. 1980). The type of oil may also affect the FA composition of the food, as well as lipid oxidation. The degrees of these deteriorative reactions are highly correlated with frying temperature and time, the unsaturation of frying fats, minor constituents, and water content. Soupas et al. (2004) showed that the oxidation of phytosterol occurs during pan frying. The highest phytosterol oxide content was reported in saturated butter oil enriched with free phytosterols, whereas the highest matrix dimer and polymer contents were found in free phytosterol-enriched liquid margarine, which contained 20% water. Sioen et al. (2006) showed that pan frying increased the fat content (0.55–4.15 and 0.55–2.30 g/100 g before and after pan frying with margarine and olive oil, respectively) in cod fillets (lean) and decreased that (13.91 to 10.57 g/100 g and 15.35 to 12.95 g/100 g before and after frying with margarine and olive oil, respectively) in salmon fillets (fat).

Andrikopoulos et al. (2002) showed that domestic pan frying reduces natural antioxidant species in edible oils. Thus, supplementation of frying medium with antioxidants is a good choice to increase the oxidative stability of fat and improve the fate of micronutrients present in foods. Chiou et al. (2009) supplemented palm oil, olive oil, and sunflower oil with an extract rich in polyphenols obtained from olive tree (*Olea europaea*) leaves at levels of 120 and 240 mg total polyphenols per kilogram of oil. The authors reported the polyphenolic content of raw potatoes and pan-fried French fries, as shown in Table 2.1. Chlorogenic acid was found as the main polyphenol in raw potatoes, making up 65% of their total individual polyphenol content. All pan-fried French fries contained polyphenols. Before supplementation, 13 polyphenols were detected in French fries fried in olive oil and 11 in those fried in sunflower oil or palm oil. Hydroxytyrosol, initially absent in raw potatoes, was detected in potatoes fried in nonsupplemented olive oil. French fries pan fried in the enriched oils had a significantly higher polyphenol content than those fried in the same oils before supplementation (e.g. 2.06 mg/100 g > 0.22 mg/100 g; Table 2.1).

Eighteen polyphenols were detected in French fries fried in the enriched olive oils and sunflower oils, and 16–17 in those fried in the enriched palm oils. In most cases, concentration increases and/or detection of polyphenols initially absent in vegetables fried in nonsupplemented oils were observed when enriched oils were used for frying. The differences in chlorogenic acid concentrations among French fries prepared in the enriched oils and the respective nonsupplemented oils may be due to the raw potatoes' storage time, necessary for the frying experiments to be completed, which may have resulted in a light-induced chlorogenic acid accumulation in the tubers. Oleuropein concentrations ranged from 0.62 to 3.39 mg/100 g, being affected to some extent by the degree of oil enrichment. Based on the results of this study, polyphenols (especially oleuropein) originating in olive leaf extract can be incorporated into fried food, where they may provide beneficial health effects.

2.5.2 Vacuum Frying

Vacuum frying is a process that is carried out under pressures well below atmospheric levels, preferably below 50 Torr (6.65 kPa). It is an efficient method of producing fruit and vegetable snacks with the necessary degree of dehydration without excessive darkening or scorching (Moreira 2014). Additionally, vacuum-fried products have higher retention of nutritional quality (phytochemicals), enhanced colour (less oxidation) (Shyi-Liang et al. 2005; Fan et al. 2005a; Da Silva and Moreira 2008), and reduced oil degradation compared with products of atmospheric frying. However, a de-oiling mechanism is necessary to remove the excessive oil absorption at the product surface (Moreira et al. 2009).

Low oil absorption is not the only health benefit of vacuum-frying systems. Lower operating temperatures reduce the formation of acrylamide in potato chips by 97%

Table 2.1 Polyphenol species (mg/100 g potato) identified and quantified in raw potatoes and in French fries pan-fried in olive oil, sunflower oil, and palm oil before and after their enrichment with a polyphenol-rich olive leaf extract.

No.	Phenolic compounds	Raw potatoes	French fries fried in olive oil OLE addition (mg TPP / kg oil)			French fries fried in sunflower oil OLE addition (mg TPP / kg oil)			French fries fried in palm oil OLE addition (mg TPP / kg oil)		
			0	120	240	0	120	240	0	120	240
1	Oleuropein	ND	ND	0.85 (0.10)[a,b]	1.48 (0.36)[b]	ND	1.32 (0.31)[b]	3.39 (0.38)[c]	ND	0.62 (0.13)[a,b]	0.78[ab]
2	HO-Tyrosol	ND	0.02 [a,b]	0.13 (0.02)[c,d]	0.15 (0.04)[d]	ND	0.17 (0.03)[d]	0.45 (0.14)[e]	ND	0.08 (0.01)[b,f]	0.11 (0.02)[f,g]
3	Quercetin	0.01	ND	0.07 (0.01)	0.08 (0.01)	ND	0.11	0.21 (0.08)	ND	0.07 (0.01)	0.07
4	Vanillin	0.01	0.02	0.05	0.04	0.01 (0.01)	0.04	0.05	0.02	0.05	0.03 (0.02)
5	Cinnamic acid	TR	TR	0.01	0.01	TR	0.01	0.01	TR	0.01 (0.01)	0.01
6	Tyrosol	TR	0.03	0.08 (0.01)	0.09 (0.01)	0.01	0.02	0.02	0.01	0.02	0.02
7	*p*-HO-Benzoic acid	TR	TR	0.01	0.01	ND	0.01	0.01	TR	0.01	0.01
8	*p*-HO-Phenylacetic acid	TR	0.01	0.02	0.01	0.01	0.02	0.02 (0.01)	0.01	0.01	0.02
9	Vanillic acid	TR	TR	0.01	0.01	TR	TR	TR	ND	TR	TR
10	Protocatechuic acid	ND	ND	0.01	0.01	ND	0.01	0.02 (0.01)	ND	0.01	0.01
11	Vanillethanediol	ND	ND	0.01	0.02	ND	0.01	0.02	TR	0.02	0.02
12	Syringic acid	TR	TR	0.01	0.01	TR	0.02	0.02 (0.01)	TR	0.01	0.01
13	*p*-Coumaric acid	TR	TR	0.02	0.01	0.01	0.02	0.03	TR	0.01	0.02
14	Ferulic acid	0.02 (0.02)	TR	0.01	0.02 (0.01)	TR	0.03	0.05 (0.02)	TR	TR	TR
15	Caffeic acid	TR	TR	0.01	0.01	TR	0.01	0.02 (0.01)	TR	0.01	0.01
16	Sinapic acid	TR	ND	TR	0.01	TR	0.01	0.02 (0.01)	ND	ND	TR

(*Continued*)

Table 2.1 (Continued)

No.	Phenolic compounds	Raw potatoes	French fries fried in olive oil			French fries fried in sunflower oil			French fries fried in palm oil		
			OLE addition (mg TPP / kg oil)			OLE addition (mg TPP / kg oil)			OLE addition (mg TPP / kg oil)		
			0	120	240	0	120	240	0	120	240
17	Kaempherol	TR	TR	0.02	0.02	ND	0.02	0.02	ND	ND	ND
18	Chlorogenic acid	0.13 (0.08)	0.12 (0.06)	0.81	0.53 (0.08)	0.17	2.51 (0.40)	2.16 (0.65)	0.20 (0.02)	0.57 (0.06)	0.67 (0.14)
—	Sum (chlorogenic acid excluded)	0.07 (0.03)^{a}	0.10 (0.01)^{a}	1.31 (0.15)^{b}	2.02 (0.44)^{b}	0.04 (0.01)^{a}	1.83 (0.36)^{b}	4.46 (0.61)^{c}	0.05^{a}	0.93 (0.14)^{a,b}	1.14^{a,b}
—	Sum	0.20 (0.07)^{a}	0.22^{a}	2.06^{b}	2.57 (0.52)^{c}	0.21^{a}	4.34 (0.58)^{d}	6.64^{e}	0.25 (0.03)^{a}	1.50 (0. 16)^{f}	1.86^{b,f}

Source: Reproduced with kind permission of John Wiley & Sons (Chiou et al. 2009).

Numbers in parentheses indicate standard deviation (SD). SD is not indicated in cases where SD < 10%. Mean values are the average of two determinations in two independent experiments.

In the case of oleuropein, hydroxytyrosol, and the sum of polyphenol, values in the same row not sharing superscript lowercase letters are significantly different at confidence level 95%.

OLE, olive leaf extract; TPP, total polyphenols by Folin–Ciocalteau; ND, not detected (for detection limits, see Materials); TR, traces (<0.005 mg /100 g); HO, hydroxy.

(Granda et al. 2004). Acrylamide is a potential carcinogen found in fried snacks that is formed by the Maillard reaction, with asparagine as the amino acid.

In recent years, vacuum-fried fruits and vegetables have become very popular in South East Asia. A couple of companies there have developed a system for processing fruits (apple, pineapple, grapes, banana, guava, mango, peach, etc.) and vegetables (sweet potato, potato, pumpkin, carrots, etc.) into chips and for vacuum frying fish and shellfish (octopus and cattle fish) (Moreira 2014).

2.5.2.1 Vacuum Frying Systems

A vacuum fryer consists of four components: (i) the frying vessel; (ii) the vacuum pump; (iii) the de-oiling mechanisms; and (iv) the condenser. Figure 2.7 shows a lab-scale vacuum fryer used to produce snacks in the Food Engineering Laboratory of the Biological and Agricultural Engineering Department at Texas A&M University. The vessel consists of a heating element, a thermostat, a basket where the product is placed, and a lid where the vacuum gauge is located. It has capacity for 5 L of oil.

The de-oiling mechanism is a centrifuge, in which a motor connected to the basket's rod spins the product at a constant rpm for a given period. The condenser is a spiral-type heat exchanger that condenses any vapour coming from the fryer before it reaches the pump. This is necessary to protect the pump from the water vapour, which would damage its mechanism and cause cavitation.

Compared to atmospheric frying, the quality and colour of fruits and vegetables fried under vacuum are superior. Products like apples and mangoes, for example, are not suitable for frying at high temperatures, because their structure will collapse (Da Silva and Moreira 2008; Nunes and Moreira 2009). Osmotic dehydration (OD) is a pretreatment used to improve the texture of these products prior to frying. Their high sugar

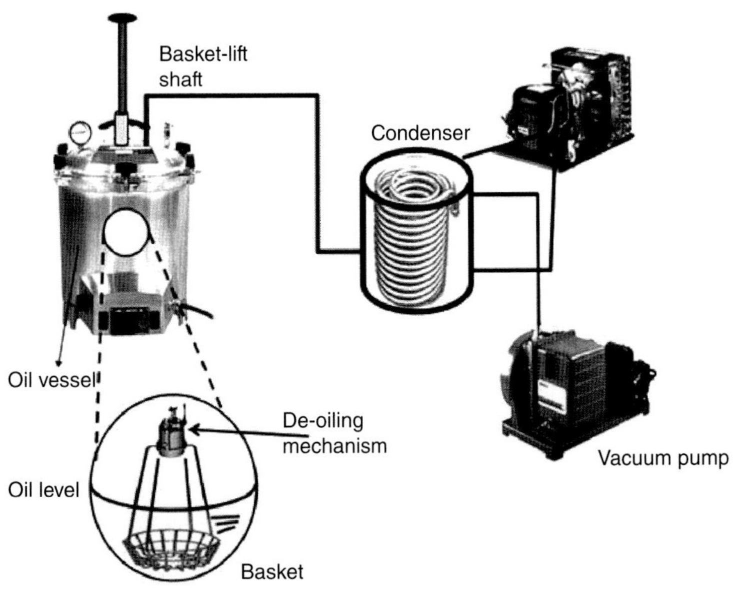

Figure 2.7 Laboratory-scale vacuum fryer used to produce snacks in the Food Engineering Laboratory of the Biological and Agricultural Engineering Department at Texas A&M University.

content produces a dark colour and a scorched taste when they are fried under atmospheric conditions. Therefore, vacuum frying is the only way to produce high-quality fried snacks from fresh fruits and vegetables.

A vacuum-frying operation consists of six steps: (i) heating the oil; (ii) loading the product in the basket and closing the lid; (iii) evacuating the vessel to about 10 Torr (1.33 kPa) and submerging the products into the oil; (iv) frying the product until it reaches the desired moisture level; (v) lifting the basket from the oil and de-oiling the samples by centrifugation or other means; and (vi) pressurizing the vessel and opening the lid.

The amount of oil in a fried product will depend on the centrifuge operation parameters (speed, residence time, and position of the chips in the basket). Too much oil is undesirable, but a product with low oil content does not have the taste and mouthfeel characteristics of a high-quality fried product (Nunes and Moreira 2009).

Many companies in Asia and Europe have designed commercial-size vacuum-frying systems. These systems can be of batch or continuous type. A commercial batch vacuum-frying system can fry 50–100 kg potato chips hour^{-1} and 20–50 kg fruit hour^{-1} and operates under a vacuum of 10.7 Torr (1.42 kPa) (I-Tung Machinery Industry, Taiwan). Apple & Snack Company (Hirosaki-shi, Aomori-ken, Japan) manufactures vacuum-frying equipment for the processing of different types of fruits and vegetables (Moreira 2014). Florigo (BMA, The Netherlands) has a pilot-scale batch system that can be used to test the effects of oil temperature, frying time, vacuum level, and the de-oiling process on the quality of the final product. Several US food companies are now using vacuum-frying systems for their European snack products. A major Japanese snack food company, Calbee Foods Co., uses vacuum fryers to produce Jagabee potato sticks, a healthy snack product that was introduced in the market with great success in 2006.

Because the frying process occurs in a vacuum tube, we can say that vacuum frying essentially is a batch process, with limited throughput. Even the largest units, which can measure 12 m long and 3 m wide, can only produce about 350 kg of product per hour. Such systems are also expensive.

The Florigo vacuum-frying system operates in a continuous mode. The vacuum frying pan is installed in a stainless-steel vacuum tube, into which the raw product is introduced through a rotary device. A conveyor takes the finished product out of the fryer towards the outlet. A closed chamber located at the exit of the vacuum tube prevents air from entering the vacuum zone. A special conveyor system takes the product from one zone to another. The vacuum is created by vacuum pumps. The frying system contains a frying oil circulation system and a filtration system that has been adapted to the special conditions in the installation. Units are equipped with centrifuges to carry out defatting after frying. The centrifuges are installed in a special vacuum dome attached to the vacuum fryer. A very important feature is the easy-to-use programmable logic controller (PLC) control system. Touch-screen monitors serve to survey and control the production line. The capacity of the unit is about 350 kg potato chips hour^{-1}. Certain applications require a two-stage frying process. In that case, the product is pre-fried in an atmospheric fryer and then subjected to vacuum frying until the final moisture content is reached. For two-stage frying, the capacity is about 1500 kg potato chips hour^{-1} (Moreira 2014).

In the United States, vacuum frying has traditionally been a curiosity in the market. However, with the rise of obesity and other health concerns, snack food manufacturers have begun to modify product formulations to deliver healthier products with reduced sodium, sugar, and oil. New equipment has been developed for use in the traditional batch fryers, like the heated de-oiling centrifuge developed by Heat and Control, Inc. (CA, USA). The centrifuge can remove 25–30% of the surface oil residing on a chip as it exits the fryer (Moreira 2014).

Ekezie et al. (2017) reviewed the efficiency of microwave-assisted food processing, showing that this technology is highly energy efficient and produces fried foods of the highest quality. The microwave-assisted vacuum frying (MVF) of potato chips shows enhanced moisture evaporation kinetics (Su et al. 2017). The use of ultrasound in combination with microwave-assisted frying is highly efficient in enhancing moisture evaporation kinetics, and retains important bioactive substances in sweet potato (Su et al. 2018a). Su et al. (2018b) studied the ultrasonic microwave-assisted vacuum frying (USMVF) of potato chips and found that USMVF markedly increased the moisture evaporation kinetics and that moisture diffusivity was effective. The drying time was reduced 20–28%, especially at lower frying temperatures. The oil uptake was also reduced at low temperatures. Texture properties, such as crispness and colour, were greatly improved by the combination of ultrasound and MVF.

2.5.2.2 Fried Product Characteristics

During vacuum frying, the mass transfer mechanisms can be divided into five periods: depressurization, immersion frying, de-oiling, pressurization, and cooling (Yagua and Moreira 2011). During the depressurization period, the product is placed in the headspace of the frying vessel until the pressure value falls, starting the second period. In immersion frying, heat and mass transfer phenomena occur; heat is convected from the oil to the surface of the product and then conducted from the surface to the centre, water is evaporated from the product, and a small amount of oil is absorbed by the food. Next, the fried products are taken out of the frying medium and held up in the vessel headspace, where the de-oiling process (centrifuging) is performed and most of the oil is removed from the product's surface. In the pressurization period, the vacuum is broken and the system recovers up to atmospheric pressure; an initial cooling of the product also occurs in this stage. Once the system is back to atmospheric pressure, the lid is opened and the product is removed from the frying vessel and allowed to cool to ambient temperature. As the product cools, during the cooling step, the pressure inside the pores changes (is reduced), creating a differential in pressure, ΔP, between the surface and the centre. This pressure difference generates a driving force, allowing the oil at the surface to penetrate the pores. However, the ΔP during the cooling period is much smaller than that during the pressurization period.

As soon as the product is removed from the oil bath, a layer of oil adheres to its surface. This oil will drain off as long as the product remains under vacuum. Oil will not flow into the product because the pressure inside the pores obstructs its passage. A centrifuge is generally used to accelerate the de-oiling period. If the vacuum is broken before the de-oiling process is carried out, all the surface oil will be pulled into the pores, followed by air (i.e. air will enter as soon as the surface oil is depleted and the pore entrance is free). During the pressurization period, there is a combination of pressure increase and temperature decrease in the environment surrounding the product.

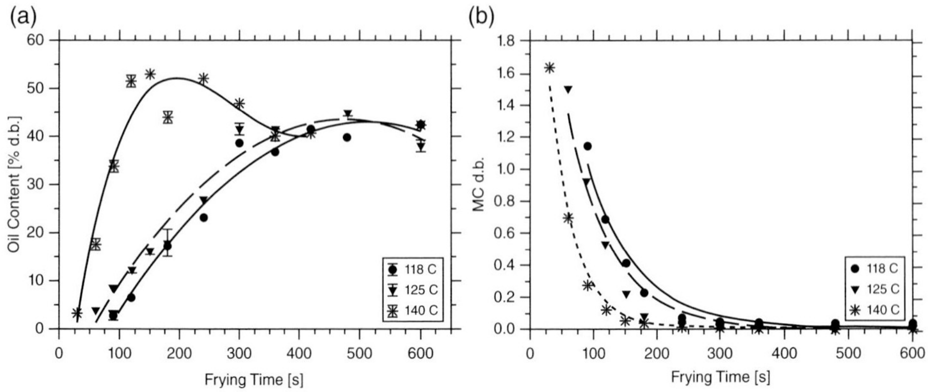

Figure 2.8 Oil absorption rate of potato chips fried at different frying oil temperatures, (a) vacuum frying at 1.33 kPa; (b) traditional frying at 101.33 kPa. *Source:* Reproduced with kind permission of John Wiley & Sons (Moreira 2014).

2.5.2.3 Frying Temperature and Final Oil Content

In general, the oil temperature does not affect the final oil content of chips during frying. The same trend was observed when comparing atmospheric and vacuum frying of potato chips (Figure 2.8). During vacuum frying, oil absorption kinetics show different behaviours at higher temperatures (140 °C at p = 1.33 kPa), as shown in Figure 2.8b (Moreira 2014). The chip's oil content reaches its maximum value early in the frying process and then decreases as frying time increases (i.e. the oil absorption rate increased as temperature increased). The highest oil content during vacuum frying at 140 °C coincides with the period in which water evaporates from the potato slices at the fastest rate (Figure 2.8a).

So, the higher the oil temperature, the higher the oil uptake by the potato chips during the first 150 seconds of frying. As the frying temperature decreases, the peak shifts to the right, in accordance with the period with the fastest drying rate at the given temperature. For example, at 125 °C, the oil content increased very rapidly during the first 360 seconds of frying (29.3% w.b.), reached a maximum at about 480 seconds (31.9% w.b.), and decreased again to 27.5% w.b. after 600 seconds. This indicates that the final oil content of the potato chips at constant pressure is not a function of temperature, but rather a function of time and remaining moisture, which increases with decreasing temperature (Garayo and Moreira 2002).

Figure 2.9 shows the influence of pressure on the final oil content of potato chips at 140 °C (vacuum) and 165 °C (atmosphere). Vacuum frying at different temperatures resulted in very similar final oil contents. However, at 140 °C, the frying time was reduced (compared to frying at 118 or 125 °C to the same final moisture content) to about 420 seconds, and chips with a nice light colour and good texture were obtained. The final oil content for the chips fried under vacuum at 140 °C (28.8% w.b.) was significantly lower ($p < 0.05$) than that of chips fried under atmospheric conditions at 165 °C (37.1% w.b.).

Yagua and Moreira (2011) studied the effect of temperature on the oil absorption of potato chips fried under vacuum (1.33 kPa). The oil content was divided into three types: (i) total oil content (TOC), defined as the oil content of the product after frying without any de-oiling process; (ii) internal oil content (IOC), defined as the oil content of the chips after the de-oiling process; and (iii) surface oil content (SOC), defined as TOC minus IOC. Yagua and Moreira (2011) noted that the IOC increased at a faster

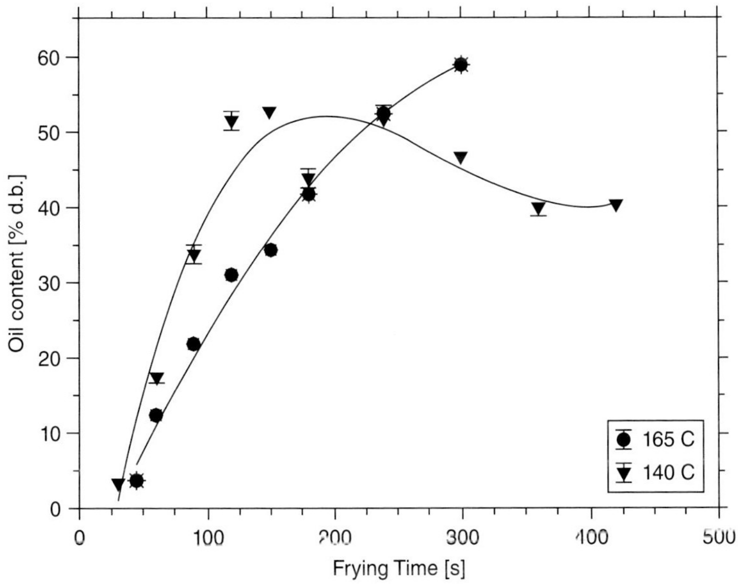

Figure 2.9 Comparison of oil absorption rates for potato chips fried under traditional (165 °C) and vacuum (10 Torr (1.33 kPa), 140 °C) frying. *Source:* Reproduced with kind permission of John Wiley & Sons (Moreira 2014).

rate first during the first seconds of frying (60–120 seconds) and then slowed as frying time increased (i.e. the surface oil was more easily removed from the chip surface by the de-oiling mechanism as frying time increased). Changes in product structure (e.g. pores, permeability (due to starch gelatinization), and interfacial tension) during frying may affect oil absorption during the pressurization period. The maximum oil content can be defined as a critical oil content, which coincides with the point at which most of the water has been removed from the chips and the structure becomes apparent, with a defined porous medium.

The final values of the TOC are comparable with those obtained in traditional frying, since oil is rapidly absorbed during the pressurization stage, which drives the surface oil into the product. Almost 75% of the TOC was absorbed in the first 70–80 seconds of frying. After that time interval, the TOC remained constant until the frying process was finished. The curves for TOC during vacuum frying at different temperatures show the same shape and behaviour – the only difference is that the equilibrium value is reached faster at higher temperatures.

The effect of the pressurization step can be diminished by the use of the de-oiling system, which removes the SOC. The final values of SOC follow the same behaviour as that of the TOC, quickly reaching the maximum value and then remaining constant for the rest of the process.

2.5.2.4 Vacuum Pressure and Final Oil Content

Vacuum pressure level does not affect the final oil content of potato chips fried at any temperature (Garayo and Moreira 2002). Higher vacuum levels (16.7, 9.9, and 3.1 kPa) lead to faster development of a crust, and thus to a higher oil absorption rate (due to loss of the characteristic hydrophilicity of raw potatoes). Therefore, the oil absorption rate increases as the vacuum level increases. The initial oil pickup (around 150 seconds) by potato chips is higher at lower versus higher vacuum pressures. This indicates that more oil likely adheres to the chips' surfaces as P_{vac} decreases. This is because the T_{sat} of water decreases (the driving force, ΔT, increases), the moisture loss rate increases, and thus the crust develops quickly.

Fan et al. (2005b) showed the effect of vacuum pressure (40, 20, and 5 kPa) on the moisture and oil contents of carrot chips fried at 80 °C. Lower vacuum pressure increased the rate of water evaporation and oil absorption. The final moisture and oil contents of carrot chips were not affected by vacuum pressures.

2.5.2.5 Pretreatment and Final Oil Content

Another way of reducing the oil absorption at the surface of vacuum-fried products is to pretreat the product before frying. Several studies have shown that less oil is absorbed during the atmospheric frying process when pretreatment is used (Gamble and Rice 1987; Fan et al. 2005b; Moyano and Pedreschi 2006; Durán et al. 2007).

Delicate products like fruits need to be pretreated before frying to provide texture (firm structure) before they can be processed as chips. Often, OD is used in combination with vacuum frying to produce high-quality fruit snacks (Da Silva and Moreira 2008). Swi-Bea et al. (2004) described the process of processing raw mango into chips in three steps: (i) blanching of mango slices in hot water (95 °C), cooling, freezing (−20 °C), and thawing in a solution containing 30° Brix maltose syrup and 1% calcium chloride; (ii) vacuum frying at 100 °C and 70 Torr (9.32 kPa) to produce chips; and (iii) de-oiling by centrifugation and flavouring by adding seasoning. The OD has been used to reduce the initial moisture content, and conserve and retain the initial quality of processed fruits and vegetables (Heng et al. 1990; Torreggiani and Bertolo 2001). Because of the high sugar content of the product after OD, vacuum frying is an excellent way of producing high-quality (in terms of sensory and physical properties and oil content) deep fat-fried fruit chips.

Nunes and Moreira (2009) produced chips with the lowest oil content (22% w.b.) by vacuum frying pretreated (OD for 45 minutes) mango slices at 120 °C for 120 seconds. More than 45% of the oil content was reduced by de-oiling the samples through centrifugation at 225 rpm for 25 seconds before the pressurization step. Compared to the vacuum-fried chips, the chips fried under atmospheric frying were completely burnt (caramelized due to the sugar content), oily, and 'soggy'.

Mariscal and Bouchon (2008) showed that predried (air dried at 80 °C) vacuum-fried (15 kPa and 115 °C) apple slices absorbed less than 50% of the oil absorbed by atmospheric-fried chips. Yang et al. (2012) produced vacuum-fried (20 kPa and 90 °C) sweet potato chips with 50% less oil content than chips fried under atmospheric pressure. The sweet potato slices were pretreated by blanching them in boiling water for 1.5 minutes, soaking them in 40°Bx maltodextrin for 2 hours, and freezing them before frying. A de-oiling system was used to remove the surface oil before the vessel was pressurized in both studies.

Troncoso et al. (2009), on the other hand, observed that pretreated potato slices (blanched in hot water at 85 °C for 3.5 minutes and air dried at 60 °C for a final moisture content of 0.6 kg water kg dry solid $^{-1}$) absorbed more oil during vacuum frying (5.4 kPa and 120–140 °C) than during atmosphere frying. The same results were obtained by

Troncoso and Pedreschi (2009). The higher oil contents for vacuum frying obtained in these studies suggest an improved understanding of the pressurization period. In both studies, no de-oiling mechanisms were applied to the products before pressurization, which explains the high oil content of the chips. The high oil uptake during vacuum frying is to be expected, since oil that is adhered to the surface of a chip is drastically absorbed during the pressurization stage, when the pressure difference ($P_{surroundings} - P_{pore}$) drives the surface oil into the product. Yagua and Moreira (2011) showed that 83–87% of the TOC in vacuum-fried potato chips was SOC, which was easily removed by the de-oiling system. They observed that 75% of the surface oil was absorbed during the first 70–80 seconds of frying.

2.5.2.6 Pressurization Period and De-oiling Period

The pressurization process plays an important role in the oil absorption mechanism. It can increase or decrease oil absorption, depending on the amount of surface oil and free water present in the product (Garayo and Moreira 2002). Furthermore, Moreira et al. (2009) established that a de-oiling process must be used to remove surface oil under vacuum after the product is fried. They found that 14% of the TOC was located in the core (internal oil) and that the remaining 86% of the oil content was surface oil. The de-oiling mechanism (a centrifuging system) used in their study removed surface oil before the pressurization step and was able to reduce the TOC by about 80–90%.

In order to understand and decrease oil absorption during the pressurization step, Mir-Bel et al. (2009) investigated the influence of various parameters of the pressurization and cooling stages on the final oil content of fried potato using different geometries. They found that oil absorption during the cooling stage is greatly influenced by not only the difference in temperature but also the vacuum break conditions as the system recovers atmospheric pressure. The volume of oil absorbed by the product is inversely proportional to the pressurization velocity, meaning that lower velocities favour oil absorption (increase of 70% for potato chips compared to the oil content when the vacuum breaks abruptly).

Pandey and Moreira (2012) showed that potato chips would absorb less oil with a decrease in pressurization rate, regardless of the frying temperature. Drainage of the surface oil may have played an important role in reducing the oil content from these chips. The results showed that the faster the pressurization rate (ΔP), and the faster the cooling, the higher the chips' final oil content. During the pressurization step, the pressure in the pore space increases rapidly, causing the adhered oil at the chip surface to penetrate into the porous food, until the pressure at the pores equals the atmospheric pressure.

A number of different ways of de-oiling the product during the pressurization stage can be used. Dueik et al. (2010) showed that 50% of the surface oil could be removed from the surface of vacuum-fried (6.5 kPa and 118 °C) carrot chips when the basket was lifted out and left to stand for 3 minutes before the vessel was pressurized. Most of the surface oil drained off easily during that step, because the high temperature inside the vessel maintained a low oil viscosity. Paper blotting the chips immediately after the frying vessel is opened helps to remove most of the absorbed oil (oil drainage by capillary forces), as demonstrated by Garayo and Moreira (2002). Centrifuges have also been used by several researchers to remove excess surface oil and to minimize oil absorption during the vacuum frying of foods (Sothornvit 2011; Yamsaengsung et al. 2011).

Without any de-oiling process, the oil content of potato chips can be as high as 40% w.b. (Garayo and Moreira 2002; Troncoso et al. 2009). However, when a centrifuge is used to de-oil the product under vacuum, it can be reduced to as low as 6% w.b. (87% less) (Yagua and Moreira 2011). It is important to note that not only the centrifugal force but also the centrifuge time can affect the de-oiling process. In addition, the position of the chips (horizontal or vertical) in relation to the axis direction of the centrifugal forces can also affect the de-oiling process. The de-oiling of chips in the horizontal position can result in oil content around 18–28% w.b. However, when chips are de-oiled in the vertical position, oil content can be as low as 7% w.b. (Moreira 2014).

The pore size distribution in vacuum-fried potato chips is more uniform than that in chips fried using traditional frying (Yagua and Moreira 2011), which makes it easier to remove the surface oil during the de-oiling process. Ravli et al. (2013) showed that sweet potato chips prepared under vacuum frying had a dark orange colour and a surface with very few blisters, as compared to chips fried under atmospheric conditions, which were brownish and had a rough surface with many bubbles. The oil content of the atmospheric pressure-fried chips was 60% higher than that of the chips fried under vacuum. The same authors also showed that frying sweet potato slices under a two-stage process (1 minute under atmospheric pressure and 2 minutes under vacuum pressure) resulted in a product with better sensory quality (flavour and mouthfeel). The vacuum-fried chips were more compact than the two-stage-fried ones, which had more blisters. The two-stage-fried chips shrunk more in diameter and expanded more in thickness compared to chips fried under vacuum only. Likewise, the TOC of the two-stage-fried chips was higher than that of chips fried under vacuum. However, the IOC was lower for the two-stage-fried samples. This suggests that the structures of these chips were very different and that the de-oiling process could more easily remove the surface oil from the chips fried under the two-stage process.

2.5.2.7 Vacuum Frying and Product Quality Attributes

Although oil content is an important quality parameter for fried products, the overall quality characteristics have to include colour, texture, and flavour. Vacuum frying is the only frying technology that can be used to fry delicate, high-sugar-content products like fruits and vegetables. In general, vacuum-fried snacks retain more of their natural colours and flavours due to the lower oxidation and lower frying temperature (Moreira et al. 2009).

The most important textural attribute of chips is crispness, which denotes freshness and high quality. A crispy product should be firm and snap easily when bent, emitting a crunchy sound (Krokida et al. 2001). Frying of raw vegetables induces major changes in their microstructure, which affects their final physical and sensory properties. Monitoring the changes that occur in product texture during frying can help us understand the transport mechanisms involved. Dueik et al. (2010) observed the textural changes of carrot chips fried using atmospheric and vacuum frying by measuring the maximum force required to fracture the samples in a three-point bending test. During frying, the carrot slices showed an initial softening that was followed by hardening due to the progressive development of a dehydrated crust. Dueik et al. (2010) modelled the texture changes (normalized maximum force) during frying by means of two terms: a fast phase, which describes the softening of the tissue, and a slow phase, which defines the hardening-dependent component. The final maximum force of fried carrot chips was not affected by the frying technology or the frying temperature. However, results from different studies (Yang et al. 2012) have shown that sensory panellists find great differences between vacuum-fried and atmospheric-fried chips.

Carotenoids make corn yellow, carrots orange, and tomatoes red. More than 600 carotenoids have been found in plants. About half of the roughly 50 carotenoids in the human diet are absorbed into the bloodstream. Carotenoids are nearly insoluble in water and are best absorbed when associated with oils (Moreira 2014). Fried carrot slices fried under vacuum (6.5 kPa) at temperatures of 98–118 °C retained 90% of *trans* α-carotene and 86% of *trans* β-carotene, preserving the colour of the raw vegetable (Dueik et al. 2010). A recent review by Ayustaningwarno et al. (2018) reveals that vacuum frying is also highly suitable for the frying of fruit products.

Da Silva and Moreira (2008) compared the changes in product quality attributes (PQAs) such as colour, texture, phytochemicals, oil content, and sensory characteristic for different fruits and vegetables (sweet potato, blue potato, mango, and green beans) fried in vacuum and traditional fryers. Sensory panellists overwhelmingly preferred the vacuum-fried products for colour, texture, taste, and overall quality. Most of the products retained or accentuated their original colours when fried under vacuum. Anthocyanin and total carotenoid contents were significantly higher (20–50% higher) for the products fried in the vacuum fryer than for those fried in the traditional fryer. In general, products fried under atmospheric conditions are darker than those fried under vacuum. In addition, vacuum technology allows for the production of fried products with good texture even when using lower oil temperatures (<100 °C). This is not possible with atmospheric frying, which requires a higher temperature to produce the right crunchiness.

Perez-Tinoco et al. (2008) prepared vacuum fried (at 24 kPa) pineapple chips at 106–117 °C and noted that the total phenolic content and dehydro-ascorbic acid content increased with increasing frying time and temperature. The chips had a golden yellow colour, low oil content, and high residual content of vitamin C.

Vacuum frying is also effective in the mitigation of furan and acrylamide production in potato chips (Mariotti-Celis et al. 2017). It was found to reduce the content of both contaminants and to preserve quality attributes. Belkova et al. (2018) studied a 23-member panel's evaluation of potato crisps prepared using conventional and vacuum frying. The study showed that the flavour of the vacuum-fried crisps was considered tasty.

2.5.3 Air Frying

Air frying is another highly important frying method that is now widely used in home kitchens. The basic concept includes the heating of pre-oiled foods using a heating element and the blowing of hot air using an electric fan, which circulates the air around the food at relatively high speed. Figure 2.10 shows a schematic representation of an air fryer. The fryer consists of mechanical fan, a heating element, a heating chamber, a frying basket, and a heating overflow. The fan above the heater pulls the hot air from the chamber, which creates a strong vertical air flow. The hot air is then passed down around the cooking chamber. The hot air is guided by a special guidance system present below the chamber that distributes it evenly in the vertical direction. The heat can only enters at the bottom of the chamber, where the air-guiding system is present. The process is controlled by a computer system that regulates the temperature, frying time, and air flow.

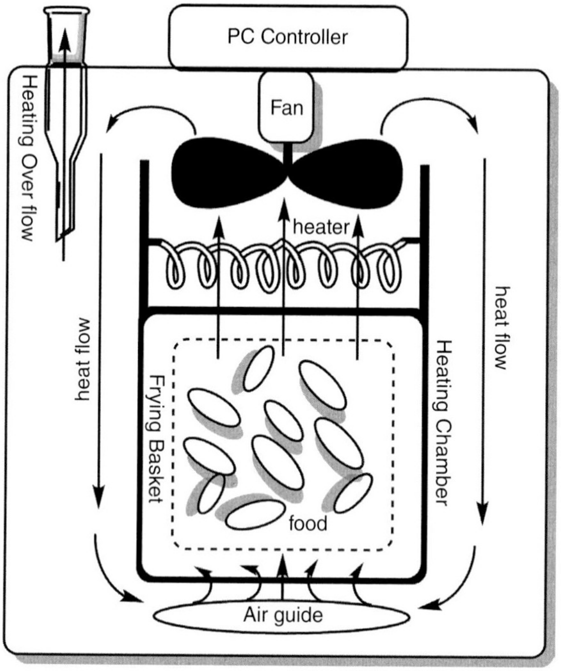

Figure 2.10 Diagrammatic representation of an air-frying system.

2.5.3.1 Heat and Mass Transfer

The heat transfer is very high in air frying as compared to conventional frying. This is due to the fact that air is circulated by a high-speed mechanical fan, which pulls it in the vertical direction. Andrés et al. (2013) showed that potato chips can be air fried with lower oil contents than could be obtained using conventional methods. This leads to a higher water loss, and consequently a higher volume loss. In comparison to deep frying, air frying shows similar moisture contents and colour characteristics, and lower oil uptake (Teruel Mdel et al. 2015). Teruel Mdel et al. (2015) compared the preparation time for French fries using deep versus air frying and found that air frying was more time-consuming (9 versus 21 minutes). Using differential scanning calorimetry (DSC) studies, they also found that air-fried products had lower starch gelatinization as compared to the products of deep frying.

2.5.3.2 Quality Characteristics in Air Frying

The effect of air-frying technology, in combination with pretreatment with citric acid, glycine, calcium lactate, sodium chloride, or nicotinic acid, on the generation of acrylamide was studied in fried potatoes (Sansano et al. 2015). The authors concluded that the rate of Maillard reaction decreased as the initial reducing sugar content increased in the raw material, and was lower for deep oil frying than for air frying regardless of the pretreatments applied. A significant reduction was observed in acrylamide content (90%) for air frying as compared to deep frying without a pretreatment.

Santos et al. (2017) compared deep frying and air frying using two commercial air-frying systems and four frying oils: sunflower, soybean, canola, and olive. The colour, moisture, lipid composition and degradation indicators, tocopherols, total ascorbic acid, β-carotene, antioxidant activity, acrylamide, and sensory attributes were compared. The results showed that fried potatoes obtained by air frying had an average of 70% less fat: a reduction of 45 kcal/100 g. Most chemical parameters were similar for both frying processes. The authors showed that all tested oils behaved similarly, and that they were mainly responsible for the enrichment of the potatoes in terms of tocopherols, phenolics, and β-carotene. However, lower lipid oxidation was observed with olive oil.

The effects of conventional frying and air frying on the microstructure, starch gelatinization, and digestibility of potato strips were reported by Tian et al. (2017). Lower starch gelatinization (86%) and slightly higher starch digestibility (59%) were observed in air-fried potato as compared to conventional frying. The authors observed that the microstructure of the air-fried potato showed a compact and well-organized structure, whereas the surface of the conventionally fried potato appeared to be oil-soaked and crusty (Figure 2.11). After *in vitro* digestion, the conventionally fried potato had a normal cell structure with some lamellar residuals, while the air-fried samples had a honeycomb-like structure.

2.5.4 Industrial Frying

In the early nineteenth century, a need emerged for the high-scale production of fried foods such as potato chips. Early fryers usually consisted of a large kettle of a given size and capacity, with heating provided by open gas burners. Over time, the size of the fryer

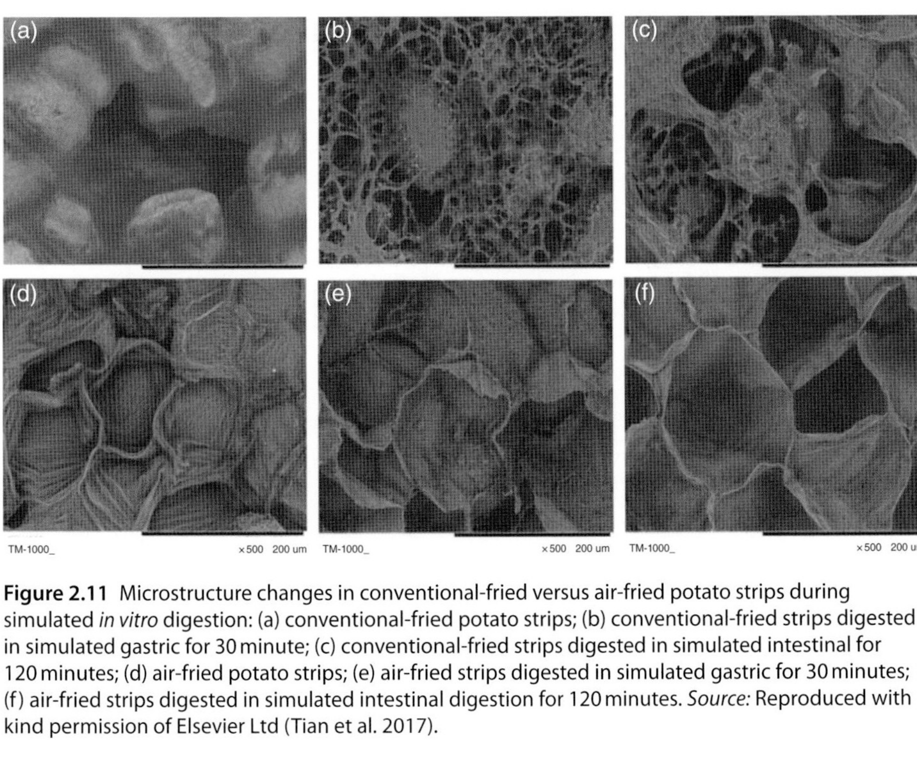

Figure 2.11 Microstructure changes in conventional-fried versus air-fried potato strips during simulated *in vitro* digestion: (a) conventional-fried potato strips; (b) conventional-fried strips digested in simulated gastric for 30 minute; (c) conventional-fried strips digested in simulated intestinal for 120 minutes; (d) air-fried potato strips; (e) air-fried strips digested in simulated gastric for 30 minutes; (f) air-fried strips digested in simulated intestinal digestion for 120 minutes. *Source:* Reproduced with kind permission of Elsevier Ltd (Tian et al. 2017).

and the burning efficiency were improved. A continuous fryer with a channel design was developed by J.D. Ferry Co. Chips were moved through the fryer with a rake system that submerged them for a few minutes before moving them on. This was a major development in large-scale fried food production.

Industrial frying is a type of deep frying in which large amounts of food are produced in a relatively small time using a high volume of oil and a continuous system. Industrial fryers are usually automatic and consume either electricity or gas (or both) (Banks 2007). The basic unit consists of heating elements, moving belts, oil stores and fillers, and food storage equipment. Usually, hot oil is circulated and passed through the food products before collecting below the fryer and being recirculated. Banks (2007) reports the details of the industrial frying process. This begins with washing of the raw products. The washed potatoes are then peeled, trimmed, and sliced. The sliced product is washed again and checked before going on to further processing, such as blanching and drying. After drying, the potato chips are deep fried in the fryer. This may be an immersion fryer or a filter fryer. In the latter case, the hot oil is poured on the chips and filtered, collected, and recirculated. The fried chips are de-oiled and the final products are sent for storage and packaging (Figure 2.12).

2.5.4.1 Advantages of Industrial Frying

Industrial frying has several advantages over other frying systems:

- It is a quick and mostly automatic process that produces a large amount of fried product in a very short amount of time.
- It includes computer-controlled probes for temperature, timing, and other processing.

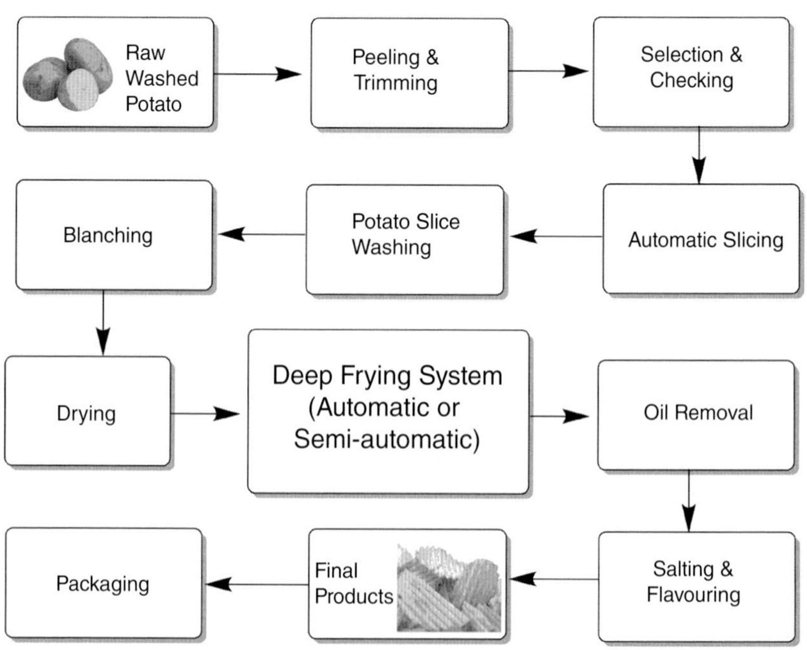

Figure 2.12 Schematic diagram of the steps involved in an industrial frying system.

- It produces foods with relatively uniform physical and chemical properties.
- It has very low variations in product quality.
- It has a limited labour requirement.
- It produces high-quality fried food.
- Loss of oil quality is limited as compared to manual deep frying.
- Its prodcuts have high consumer acceptability.
- Quality control is efficiently maintained.

2.5.4.2 Quality Characteristics in Industrial Frying

The ultimate aims of industrial frying are automation, quality, and consumer health (Stier 2013). Rytel et al. (2015) found that pretreatment of potatoes reduced their glycolalkaloid content to 74%, slicing to 29%, and frying to just 10%, relative to unpeeled potatoes. Sunflower oil is used as a frying medium in the industrial frying of potato chips. Dutta and Appelqvist (1996) reported that during industrial frying, several cholesterol and phytosterol oxidation products are formed. French fries prepared in a rapeseed/palm oil blend had the highest total sterol oxide level (191 ppm). Sitosterol and campesterol produce epoxides at a level of 172 ppm. French fries prepared using sunflower oil and high-oleic sunflower oil contained 7α-hydroxy-, 7β-hydroxy-, 7-keto-, and both epimers of epoxysitosterol in equal amounts. Thus, industrial frying requires highly stable oils such as high-oleic, high-palmitic sunflower oil (HOHPSO) (Marmesat et al. 2005). Palm oil has been found to be a good choice for industrial frying, producing good-quality fried products (Ferreira et al. 2014). Oil blends made of palm olein and sunflower oil show higher resistance towards oxidative and hydrolytic degradation as compared to other blends (Tarmizi and Ismail 2014).

Miranda et al. (2010) studied the changes in the proximate compositions and FA profiles of tuna pasties and ham nuggets induced by pre-frying in olive and sunflower oil, as well as domestic cooking methods. The pre-frying processes reduced the moisture and carbohydrate contents and increased the ash, fat, and protein contents in both products. During pre-frying and frying, the FA profiles of both became similar to those of the culinary fats used. Thus, the frying process reduced the proportion of saturated fatty acids (SFAs) and increased the proportion of monounsaturated fatty acids (MUFAs) in the foods when olive oil was used, and increased the proportion of PUFAs when sunflower oil was used.

2.6 Tips to Remember During Frying

Frying is a complex and laborious job. It is easy at industrial scale, but relatively hard in the home. There are several tips that should be remembered for a safe and tasty frying experience:

- ***Safety first:*** Your safety is one of the key elements in the production of good-quality fried foods. Frying is usually carried out at high temperatures – over 160 °C. A single drop of the hot oil can easily burn your skin, turning your cooking experience into a disaster. Thus, care is required to avoid accidents. With automatic fryers, the oil temperature is shown on the display screen. In manual frying, a thermometer should be used to monitor it. If you are burnt, baking soda should be poured on the affected area to denaturalize the damage. Keep children well away from the frying area. You should be aware of the approved guidelines for toxicity and the maximum permissible limit for frying oils.

- ***Your food:*** Not all foods should be fried, and foods should not all be fried under the same conditions. You should have some basic knowledge of the food you are going to fry.
- ***Choice of oil:*** As a basic component of the frying process, oil should be selected based on need and desired the cooking time. For longer times, a highly stable oil such as high oleic sunflower oil, palm oil, corn oil, or their blends should be selected. Do not use oils with a high smoke point. You should have some basic knowledge of the quality parameters of the oil you select.
- ***Fryer selection:*** In low-budget kitchens, a frying pan can be used for frying. However, several home fryers are available at relatively cheap prices. These semi-automatic fryers maintain a constant temperature and therefore allow you to reuse your oil. These fryers have specific baskets for your foods and work on the principal of immersion frying.
- ***Optimized conditions:*** All foods cannot be fried at the same temperature and for the same length of time. Specific recipes should be followed in order to prepare foods of the highest quality.
- ***Specific taste:*** In order to prepare foods with a specific taste, several ingredients (e.g. pepper, salt, lemon juice, and spices) can be added as a pre- or post-treatment. Again, a number of recipes can be followed.
- ***Coating:*** Foods can be prepared with a batter in order to impart several flavour and texture qualities. Batters are usually made of wheat flour, pea flour, or a mix of spices. For foods with slippery textures, such as sausages, the batter should be allowed to sit for a few minutes so that it sticks to the surface. Again, a number of recipes can be followed.
- ***Food overcrowding:*** The size of fryer limits how much food you can fry at a time. Trying to fry too much at once will make the food stick together and prevent you flipping it easily.
- ***De-oiling:*** After food is fried, the food basket should be drawn above the oils, and the oils allowed to run for a few minutes. A highly oiled food has lower stability and lower acceptability.
- ***Used oil:*** The oil left after frying can be reused if the amount of polar material remaining is above an approved level. The hot oil should be carefully carried to a cold and dark place for storage.

2.7 Choice of Frying Method

At its basic level, frying always involves cooking foods at high temperature in the presence of oils or fats. However, there are significant differences between the different types of frying procedure. Products fried using vacuum frying will have different physical and chemical characteristics to those fried using pan or deep frying. Likewise, the same food item fried in different edible oils will also have differences in its physical and chemical properties. Thus, the choice of frying medium plays a significant role in the frying procedure (Gupta 2004), and the search for a suitable frying medium is one of the hot topics of modern-day food science. Matthäus (2007) showed that palm oil and its products have similar frying performance to the high oleic oils and may thus be a good choice for frying. Several fractionations of palmolein are now available for specific types of frying.

Ghosh and Bhattacharjee (2013) reported that mustard oil is a better choice for repeated frying than sesame oil due to its lower changes in oxidation values and percentage radical scavenging activity. Mustard oil is very stable at frying temperature and contains several phenolic antioxidants (Zeb and Rahman 2017). Similarly, pumpkin oils with low smoke values have been found to be highly stable as compared to other frying oils and have been recommended for prolonged frying (Zeb and Ahmad 2017).

The choice of frying method depends on the type of food to be fried and the available equipment. In general, modern computer-controlled fryers give better products than manual pan frying. In all cases, a basic knowledge of frying chemistry, biochemistry, and safety is important when using any of the methods described in this chapter.

2.8 Key Concepts

- Deep frying is an important frying process, in which food is fried deep under the oil surface in order to obtain specific colour, taste, texture, and acceptability characteristics.
- Several automatic and semi-automatic fryers and accessories are available in the market.
- External factors influencing the stability and performance of a frying oil include the accessibility of oxygen, the temperature, the duration of frying, and the size, dimensions, and composition of the food.
- Internal factors influencing the stability and performance of a frying oil include its FA composition and minor components.
- Pan frying is a simple process involving a kitchen frying pan and shallow or deep oil. Despite its uncontrolled conditions, pan frying is one of the most widely used frying techniques.
- Vacuum frying involves frying under low pressure or vacuum. It has several advantages, including producing foods with a low oil content, a low content of acrylamide and other toxicants, and high consumer acceptability.
- Air frying is a modern home-kitchen frying process. It involves frying pre-oiled foods using hot air circulation. It produces foods with a low oil content compared to deep-frying process.
- Industrial frying is a highly complex and automatic process that involves a series of steps for the production of large amounts of fried foods per unit time. Significant development has been reported recently in the invention of new technologies that produce food of high quality and acceptability. Industrial frying has several advantages over other frying processes, including automation, consistent food quality, limited labour requirement, high acceptability, and a high standard of quality control.
- Several important tips should be remembered when engaging in frying procedures, such as ensuring your own safety, being knowledgeable about the food and oil you are using, selecting the most appropriate frying method, and maintaining optimal conditions.

References

Abuzaytoun, R. and Shahidi, F. (2006). Oxidative stability of algal oils as affected by their minor components. *Journal of Agricultural and Food Chemistry* 54: 8253–8260.

Adekunle, A.F. (2014). Natural antioxidants as stabilizers of frying oils. *European Journal of Lipid Science and Technology* 116: 688–706.

Aladedunye, F.A. (2015). Curbing thermo-oxidative degradation of frying oils: current knowledge and challenges. *European Journal of Lipid Science and Technology* 117: 1867–1881.

Aladedunye, F.A. and Przybylski, R. (2009a). Degradation and nutritional quality changes of oil during frying. *Journal of the American Oil Chemists' Society* 86: 149–156.

Aladedunye, F.A. and Przybylski, R. (2009b). Protecting oil during frying: a comparative study. *European Journal of Lipid Science and Technology* 111: 893–901.

Aladedunye, F.A. and Przybylski, R. (2012). Frying performance of canola oil triacylglycerides as affected by vegetable oils minor components. *Journal of the American Oil Chemists' Society* 89: 41–53.

Aladedunye, F., Przybylski, R., and Matthaus, B. (2017). Performance of antioxidative compounds under frying conditions: a review. *Critical Reviews in Food Science and Nutrition* 57: 1539–1561.

Alireza, S., Tan, C., Hamed, M., and Che MAN, Y. (2010). Effect of frying process on fatty acid composition and iodine value of selected vegetable oils and their blends. *International Food Research Journal* 17: 295–302.

Allam, S.S. and Mohamed, H.M.A. (2002). Thermal stability of some commercial natural and synthetic antioxidants and their mixtures. *Journal of Food Lipids* 9: 277–293.

Andrés, A., Arguelles, Á., Castelló, M.L., and Heredia, A. (2013). Mass transfer and volume changes in French fries during air frying. *Food and Bioprocess Technology* 6: 1917–1924.

Andrikopoulos, N.K., Dedoussis, G.V.Z., Falirea, A. et al. (2002). Deterioration of natural antioxidant species of vegetable edible oils during the domestic deep-frying and pan-frying of potatoes. *International Journal of Food Sciences and Nutrition* 53: 351–363.

Aparicio, R., Roda, L., Albi, M.A., and Gutiérrez, F. (1999). Effect of various compounds on virgin olive oil stability measured by Rancimat. *Journal of Agricultural and Food Chemistry* 47: 4150–4155.

Artz, W.E., Osidacz, P.C., and Coscione, A.R. (2005). Acceleration of the thermoxidation of oil by heme iron. *Journal of the American Oil Chemists' Society* 82: 579–584.

Ayustaningwarno, F., Dekker, M., Fogliano, V., and Verkerk, R. (2018). Effect of vacuum frying on quality attributes of fruits. *Food Engineering Reviews* 10: 154–164.

Banks, D. (2007). Industrial frying. In: *Deep Frying*, 2e (ed. M.D. Erickson). Champaign, IL: AOCS Press.

Bansal, G., Zhou, W., Barlow, P.J. et al. (2010). Evaluation of commercially available rapid test kits for the determination of oil quality in deep-frying operations. *Food Chemistry* 121: 621–626.

Belkova, B., Hradecky, J., Hurkova, K. et al. (2018). Impact of vacuum frying on quality of potato crisps and frying oil. *Food Chemistry* 241: 51–59.

Bhattacharya, A.B., Sajilata, M., and Singhal, R.S. (2008). Lipid profile of foods fried in thermally polymerized palm oil. *Food Chemistry* 109: 808–812.

Blumenthal, M. (1991). A new look at the chemistry and physics of deep-fat frying. *Food Technology (USA)* 45: 68–71.

Brannan, R.G., Mah, E., Schott, M. et al. (2014). Influence of ingredients that reduce oil absorption during immersion frying of battered and breaded foods. *European Journal of Lipid Science and Technology* 116: 240–254.

Bruscatto, M., Zambiazi, R., Sganzerla, M. et al. (2009). Degradation of tocopherols in rice bran oil submitted to heating at different temperatures. *Journal of Chromatographic Science* 47: 762–765.

Chen, S.-D., Chen, H.-H., Chao, Y.-C., and Lin, R.-S. (2009). Effect of batter formula on qualities of deep-fat and microwave fried fish nuggets. *Journal of Food Engineering* 95: 359–364.

Chiou, A., Kalogeropoulos, N., Salta, F.N. et al. (2009). Pan-frying of French fries in three different edible oils enriched with olive leaf extract: oxidative stability and fate of microconstituents. *LWT – Food Science and Technology* 42: 1090–1097.

Clemente, T.E. and Cahoon, E.B. (2009). Soybean oil: genetic approaches for modification of functionality and total content. *Plant Physiology* 151: 1030–1040.

Cosgrove, J.P., Church, D.F., and Pryor, W.A. (1987). The kinetics of the autoxidation of polyunsaturated fatty acids. *Lipids* 22: 299–304.

Da Silva, P.F. and Moreira, R.G. (2008). Vacuum frying of high-quality fruit and vegetable-based snacks. *LWT – Food Science and Technology* 41: 1758–1767.

DGF (2000). Recommendations for the 3rd International Symposium on Deep-Fat Frying–Optimal Operation. *European Journal of Lipid Science and Technology* 102: 594.

DGF (2013). *Proceedings of the 7th International Symposium of Deep-Fat Frying – Optimum Frying for Safe and Improved Quality Fried Food*. San Francisco, CA: German Society for Fat Research.

Dijkstra, A.J. (2006). Revisiting the formation of trans isomers during partial hydrogenation of triacylglycerol oils. *European Journal of Lipid Science and Technology* 108: 249–264.

Dobarganes, C., Márquez-Ruiz, G., and Velasco, J. (2000). Interactions between fat and food during deep-frying. *European Journal of Lipid Science and Technology* 102: 521–528.

Dogan, S.F., Sahin, S., and Sumnu, G. (2005). Effects of batters containing different protein types on the quality of deep-fat-fried chicken nuggets. *European Food Research and Technology* 220: 502–508.

Dueik, V., Robert, P., and Bouchon, P. (2010). Vacuum frying reduces oil uptake and improves the quality parameters of carrot crisps. *Food Chemistry* 119: 1143–1149.

Durán, M., Pedreschi, F., Moyano, P., and Troncoso, E. (2007). Oil partition in pre-treated potato slices during frying and cooling. *Journal of Food Engineering* 81: 257–265.

Durmaz, F. and Talpur, M.Y. (2015). Oxidation on the stability of canola oil blended with stinging nettle oil at frying temperature. *International Journal of Food Properties* 18: 1402–1415.

Dutta, P.C. and Appelqvist, L.-Å. (1996). Sterols and sterol oxides in the potato products, and sterols in the vegetable oils used for industrial frying operations. *Grasas y Aceites* 47: 1–2.

Ekezie, F.-G.C., Sun, D.-W., Han, Z., and Cheng, J.-H. (2017). Microwave-assisted food processing technologies for enhancing product quality and process efficiency: a review of recent developments. *Trends in Food Science & Technology* 67: 58–69.

Erickson, M.D. (2015). *Deep Frying: Chemistry, Nutrition, and Practical Applications*. New York, NY: Elsevier.

Fan, L.-P., Zhang, M., and Mujumdar, A.S. (2005a). Vacuum frying of carrot chips. *Drying Technology* 23: 645–656.

Fan, L.P., Zhang, M., Xiao, G.N. et al. (2005b). The optimization of vacuum frying to dehydrate carrot chips. *International Journal of Food Science and Technology* 40: 911–919.

Farhoosh, R. and Esmaeilzadeh, K.R. (2009). Anti-rancidity effects of sesame and rice bran oils on canola oil during deep frying. *Journal of the American Oil Chemists' Society* 86: 539–544.

Farhoosh, R. and Tavassoli-Kafrani, M.H. (2010). Polar compounds distribution of sunflower oil as affected by unsaponifiable matters of Bene hull oil (BHO) and tertiary-butylhydroquinone (TBHQ) during deep-frying. *Food Chemistry* 122: 381–385.

Ferreira, D.R., Beléia, A.D.P., and Silva, R.S.D.S.F. (2014). Alterations of palm oil (*Elaeis guineensis*) in the continuous industrial par frying of breaded chicken snacks. *Food Science and Technology (Campinas)* 34: 32–37.

Freeman, I., Padley, F., and Sheppard, W. (1973). Use of silicones in frying oils. *Journal of the American Oil Chemists Society* 50: 101–103.

Fujisaki, M., Endo, Y., and Fujimoto, K. (2002). Retardation of volatile aldehyde formation in the exhaust of frying oil by heating under low oxygen atmospheres. *Journal of the American Oil Chemists' Society* 79: 909–914.

Gamble, M.H. and Rice, P. (1987). Effect of pre-fry drying of oil uptake and distribution in potato crisp manufacture. *International Journal of Food Science and Technology* 22: 535–548.

Garayo, J. and Moreira, R. (2002). Vacuum frying of potato chips. *Journal of Food Engineering* 55: 181–191.

Gertz, C. (2004). Optimising the baking and frying process using oil-improving agents. *European Journal of Lipid Science and Technology* 106: 736–745.

Gertz, C. (2014). Fundamentals of the frying process. *European Journal of Lipid Science and Technology* 116: 669–674.

Gertz, C., Klostermann, S., and Kochhar, S.P. (2000). Testing and comparing oxidative stability of vegetable oils and fats at frying temperature. *European Journal of Lipid Science and Technology* 102: 543–551.

Ghidurus, M., Turtoi, M., Boskou, G. et al. (2010). Nutritional and health aspects related to frying (I). *Romanian Biotechnological Letters* 15: 5675–5682.

Ghosh, P. and Bhattacharjee, P. (2013). Alternative methods of frying and antioxidant stability in sesame and mustard oils. *Acta Alimentaria* 42: 109–123.

Granda, C., Moreira, R., and Tichy, S. (2004). Reduction of acrylamide formation in potato chips by low-temperature vacuum frying. *Journal of Food Science* 69: E405–E411.

Gueraud, F., Atalay, M., Bresgen, N. et al. (2010). Chemistry and biochemistry of lipid peroxidation products. *Free Radical Research* 44: 1098–1124.

Guerra-Hernandez, E., Plata-Guerrero, R., and Garcia-Villanova, B. (2013). Acrylamide formation in different batter and breading formulation during deep-frying of zucchini and exposure estimation. *Annals of Nutrition and Metabolism* 63: 1841–1842.

Gupta, M.K. (2004). Selection of frying oil. In: *Frying Technology and Practices* (ed. M.K. Gupta, K. Warner and P.J. White). Champaign, IL: AOCS Press.

Gupta, M.K., Warner, K., and White, P.J. (2004). *Frying Technology and Practices.* Champaign, IL: AOCS Press.

Gupta, N., Sinddhu, A., and Sundararaj, P. (2001). Oxidative stability of edible grade crude palm oil blends with sunflower oil or ground oil as deep frying medium. *International Journal of Oil Palm* 2: 11–18.

Hamam, F. and Shahidi, F. (2006). Acidolysis reactions lead to esterification of endogenous tocopherols and compromised oxidative stability of modified oils. *Journal of Agricultural and Food Chemistry* 54: 7319–7323.

Heng, K., Guilben, S., and Cuq, J. (1990). Osmotic dehydration of papaya: influence of process variables on the product quality. *Sciences des Aliments* 10: 831–848.

Holm, H.C. and Cowan, D. (2008). The evolution of enzymatic interesterification in the oils and fats industry. *European Journal of Lipid Science and Technology* 110: 679–691.

Holman, R.T. and Elmer, O.C. (1947). The rates of oxidation of unsaturated fatty acids and esters. *Journal of the American Oil Chemists Society* 24: 127–129.

Holownia, K., Chinnan, M., Erickson, M., and Mallikarjunan, P. (2000). Quality evaluation of edible film-coated chicken strips and frying oils. *Journal of Food Science* 65: 1087–1090.

Hoshina, R., Endo, Y., and Fujimoto, K. (2004). Effect of triacylglycerol structures on the thermal oxidative stability of edible oil. *Journal of the American Oil Chemists' Society* 81: 461–465.

Houhoula, D.P., Oreopoulou, V., and Tzia, C. (2003). The effect of process time and temperature on the accumulation of polar compounds in cottonseed oil during deep-fat frying. *Journal of the Science of Food and Agriculture* 83: 314–319.

Ikediala, J.N., Correia, L.R., Fenton, G.A., and Ben-Abdallah, N. (1996). Finite element modeling of heat transfer in meat patties during single-sided pan-frying. *Journal of Food Science* 61: 796–802.

Jennings, B.H. and Akoh, C.C. (2009). Effectiveness of natural versus synthetic antioxidants in a rice bran oil-based structured lipid. *Food Chemistry* 114: 1456–1461.

Kamal-Eldin, A. (2003). *Lipid Oxidation Pathways.* Champaign, IL: AOCS Press.

Khan, M.I., Asha, M.R., Bhat, K.K., and Khatoon, S. (2011). Studies on chemical and sensory parameters of coconut oil and its olein blends with sesame oil and palmolein during wheat flour-based product frying. *Journal of Food Science and Technology* 48: 175–182.

Kim, M.-A., Matoba, T., and Hasegawa, K. (1988). Relationship between the triacylglycerol composition and foaming of mixed coconut oil under deep-fat frying. *Agricultural and Biological Chemistry* 52: 693–699.

Kontkanen, H., Rokka, S., Kemppinen, A. et al. (2011). Enzymatic and physical modification of milk fat: a review. *International Dairy Journal* 21: 3–13.

Krokida, M.K., Oreopoulou, V., Maroulis, Z.B., and Marinos-Kouris, D. (2001). Effect of pre-treatment on viscoelastic behaviour of potato strips. *Journal of Food Engineering* 50: 11–17.

Lazarick, K., Aladedunye, F., and Przybylski, R. (2014). Effect of breading and battering ingredients on performance of frying oils. *European Journal of Lipid Science and Technology* 116: 763–770.

Lee, J.H., Akoh, C.C., and Lee, K.-T. (2008). Physical properties of trans-free bakery shortening produced by lipase-catalyzed interesterification. *Journal of the American Oil Chemists' Society* 85: 1–11.

Lichtenstein, A.H., Matthan, N.R., Jalbert, S.M. et al. (2006). Novel soybean oils with different fatty acid profiles alter cardiovascular disease risk factors in moderately hyperlipidemic subjects. *The American Journal of Clinical Nutrition* 84: 497–504.

Lin, S., Akoh, C.C., and Reynolds, A.E. (1998). The recovery of used frying oils with various adsorbents. *Journal of Food Lipids* 5: 1–16.

Mariotti-Celis, M.S., Cortés, P., Dueik, V. et al. (2017). Application of vacuum frying as a furan and acrylamide mitigation technology in potato chips. *Food and Bioprocess Technology* 10: 2092–2099.

Mariscal, M. and Bouchon, P. (2008). Comparison between atmospheric and vacuum frying of apple slices. *Food Chemistry* 107: 1561–1569.

Marmesat, S., Mancha, M., Ruiz-Méndez, M.V., and Dobarganes, M.C. (2005). Performance of sunflower oil with high levels of oleic and palmitic acids during industrial frying of almonds, peanuts, and sunflower seeds. *Journal of the American Oil Chemists' Society* 82: 505–510.

Marmesat, S., Morales, A., Velasco, J., and Dobarganes, M. (2010). Action and fate of natural and synthetic antioxidants during frying. *Grasas y Aceites* 61: 333–340.

Márquez-Ruiz, G., Velasco, J., and Dobarganes, M.C. (2004). Effectiveness of dimethylpolysiloxane during deep frying. *European Journal of Lipid Science and Technology* 106: 752–758.

Márquez-Ruiz, G., Ruiz-Méndez, M.V., and Velasco, J. (2014). Antioxidants in frying: analysis and evaluation of efficacy. *European Journal of Lipid Science and Technology* 116: 1441–1450.

Martin, D., Reglero, G., and Señoráns, F.J. (2010). Oxidative stability of structured lipids. *European Food Research and Technology* 231: 635–653.

Matthäus, B. (2007). Use of palm oil for frying in comparison with other high-stability oils. *European Journal of Lipid Science and Technology* 109: 400–409.

Mehta, U. and Swinburn, B. (2001). A review of factors affecting fat absorption in hot chips. *Critical Reviews in Food Science and Nutrition* 41: 133–154.

Mezouari, S. and Eichner, K. (2006). Effect of stirring on the thermooxidative stability of refined rice bran oil. *European Journal of Lipid Science and Technology* 108: 848–857.

Min, D.B. and Wen, J. (1983). Effects of dissolved free oxygen on the volatile compounds of oil. *Journal of Food Science* 48: 1429–1430.

Miranda, J.M., Martínez, B., Pérez, B. et al. (2010). The effects of industrial pre-frying and domestic cooking methods on the nutritional compositions and fatty acid profiles of two different frozen breaded foods. *LWT – Food Science and Technology* 43: 1271–1276.

Mir-Bel, J., Oria, R., and Salvador, M.L. (2009). Influence of the vacuum break conditions on oil uptake during potato post-frying cooling. *Journal of Food Engineering* 95: 416–422.

Mistry, B.S. and Min, D.B. (1987). Effects of fatty acids on the oxidative stability of soybean oil. *Journal of Food Science* 52: 831–832.

Moreira, R.G. (2014). Vacuum frying versus conventional frying – an overview. *European Journal of Lipid Science and Technology* 116: 723–734.

Moreira, R.G. and Barrufet, M.A. (1998). A new approach to describe oil absorption in fried foods: a simulation study. *Journal of Food Engineering* 35: 1–22.

Moreira, R.G., Sun, X., and Chen, Y. (1997). Factors affecting oil uptake in tortilla chips in deep-fat frying. *Journal of Food Engineering* 31: 485–498.

Moreira, R.G., Da Silva, P.F., and Gomes, C. (2009). The effect of a de-oiling mechanism on the production of high quality vacuum fried potato chips. *Journal of Food Engineering* 92: 297–304.

Moreno, M.M., Olivares, D.M., Lopez, F.A. et al. (1999). Determination of unsaturation grade and trans isomers generated during thermal oxidation of edible oils and fats by FTIR. *Journal of Molecular Structure* 482: 551–556.

Moyano, P.C. and Pedreschi, F. (2006). Kinetics of oil uptake during frying of potato slices: effect of pre-treatments. *LWT – Food Science and Technology* 39: 285–291.

Nawar, W.W. (1984). Chemical changes in lipids produced by thermal processing. *Journal of Chemical Education* 61: 299–302.

Neff, W.E. and El-Agaimy, M. (1996). Effect of linoleic acid position in triacylglycerols on their oxidative stability. *LWT – Food Science and Technology* 29: 772–775.

Neff, W., Selke, E., Mounts, T. et al. (1992). Effect of triacylglycerol composition and structures on oxidative stability of oils from selected soybean germplasm. *Journal of the American Oil Chemists' Society* 69: 111–118.

Normand, L., Eskin, N., and Przybylski, R. (2001). Effect of tocopherols on the frying stability of regular and modified canola oils. *Journal of the American Oil Chemists' Society* 78: 369–373.

Nunes, Y. and Moreira, R.G. (2009). Effect of osmotic dehydration and vacuum-frying parameters to produce high-quality mango chips. *Journal of Food Science* 74: E355–E362.

Ohta, S., Kusaka, H., Tokue, K., and Takaoka, K. (1987). Investigation of the surface conditions on the frying oils containing silicone oil by infrared ATR spectrometry and photoacoustic spectrometer. *Journal of Japan Oil Chemists' Society* 36: 480–486.

Pandey, A. and Moreira, R.G. (2012). Batch vacuum frying system analysis for potato chips. *Journal of Food Process Engineering* 35: 863–873.

Panghal, R., Jood, S., Khetarpaul, N., and Goyal, R. (2010). Chemical characteristics and fatty acid composition of cooking oils and their blends. *Acta Alimentaria* 39: 126–129.

Parker, T.D., Adams, D., Zhou, K. et al. (2003). Fatty acid composition and oxidative stability of cold-pressed edible seed oils. *Journal of Food Science* 68: 1240–1243.

Perez-Tinoco, M.R., Perez, A., Salgado-Cervantes, M. et al. (2008). Effect of vacuum frying on main physicochemical and nutritional quality parameters of pineapple chips. *Journal of the Science of Food and Agriculture* 88: 945–953.

Pokorn, J., Panek, J., and Trojakova, L. (2003). Effect of food component changes during frying on the nutrition value of fried food. *Forum of Nutrition* 56: 348–350.

Porta, R., Mariniello, L., Pierro, P. et al. (2012). Water barrier edible coatings of fried foods. *Journal of Biotechnology and Biomaterials* 2: e116.

Przybylski, R. and Eskin, N. (1988). A comparative study on the effectiveness of nitrogen or carbon dioxide flushing in preventing oxidation during the heating of oil. *Journal of the American Oil Chemists' Society* 65: 629.

Przybylski, R. and Eskin, N.M. (1995). Methods to measure volatile compounds and the flavor significance of volatile compounds. In: *Methods to assess quality and stability of oils and fat-containing foods* (ed. K. Warner and N.A.M. Eskin). Champaign, IL: AOCS Press.

Rani, A.K.S., Reddy, S.Y., and Chetana, R. (2010). Quality changes in trans and trans free fats/oils and products during frying. *European Food Research and Technology* 230: 803–811.

Ravli, Y., Da Silva, P., and Moreira, R.G. (2013). Two-stage frying process for high-quality sweet-potato chips. *Journal of Food Engineering* 118: 31–40.

Réblová, Z. (2006). The effect of temperature on the antioxidant activity of tocopherols. *European Journal of Lipid Science and Technology* 108: 858–863.

Rehab, F.A. and El Anany, A. (2012). Physicochemical studies on sunflower oil blended with cold pressed tiger nut oil during deep frying process. *Grasas y Aceites* 63: 455–465.

Rossell, J. (2001). *Frying: Improving Quality*. Sawston: Woodhead Publishing.
Rudzińska, M., Przybylski, R., and Wąsowicz, E. (2009). Products formed during thermo-oxidative degradation of phytosterols. *Journal of the American Oil Chemists' Society* 86: 651–662.
Rudzińska, M., Przybylski, R., Zhao, Y.Y., and Curtis, J.M. (2010). Sitosterol rhermo-oxidative degradation leads to the formation of dimers, trimers and oligomers: a study using combined size exclusion chromatography/mass spectrometry. *Lipids* 45: 549–558.
Rytel, E., Tajner-Czopek, A., Kita, A. et al. (2015). The influence of washing and selection processes on the contents of glycoalkaloid and other toxic compounds during industrial chip production. *International Journal of Food Science and Technology* 50: 1737–1742.
Sahin, S. and Sumnu, S.G. (2009). *Advances in Deep-Fat Frying of Foods*. Boca Raton, FL: CRC Press.
Sansano, M., Juan-Borras, M., Escriche, I. et al. (2015). Effect of pretreatments and air-frying, a novel technology, on acrylamide generation in fried potatoes. *Journal of Food Science* 80: T1120–T1128.
Santos, C.S.P., Cunha, S.C., and Casal, S. (2017). Deep or air frying? A comparative study with different vegetable oils. *European Journal of Lipid Science and Technology* 119: 1600375.
Sebastian, A., Ghazani, S.M., and Marangoni, A.G. (2014). Quality and safety of frying oils used in restaurants. *Food Research International* 64: 420–423.
Serjouie, A., Tan, C.P., Mirhosseini, H. et al. (2010). Effect of vegetable-based oil blends on physicochemical properties of oils during deep-fat frying. *American Journal of Food Technology* 5: 310–323.
Shahidi, F. and Zhong, Y. (2010). Lipid oxidation and improving the oxidative stability. *Chemical Society Reviews* 39: 4067–4079.
Shyi-Liang, S., Lung-Bin, H., and Sun, H.L. (2005). Effects of processing conditions on the quality of vacuum-fried carrot chips. *Journal of the Science of Food and Agriculture* 85: 1903–1908.
Shyu, S.-L., Hau, L.-B., and Hwang, L.S. (1998). Effect of vacuum frying on the oxidative stability of oils. *Journal of the American Oil Chemists' Society* 75: 1393–1398.
Sioen, I., Haak, L., Raes, K. et al. (2006). Effects of pan-frying in margarine and olive oil on the fatty acid composition of cod and salmon. *Food Chemistry* 98: 609–617.
Sionek, B., Krygier, K., Ukalski, K. et al. (2013). The influence of nitrogen and carbon dioxide on the oxidative stability of fully refined rapeseed oil. *European Journal of Lipid Science and Technology* 115: 1426–1433.
Sosińska, E., Przybylski, R., Aladedunye, F., and Hazendonk, P. (2014). Spectroscopic characterisation of dimeric oxidation products of phytosterols. *Food Chemistry* 151: 404–414.
Sothornvit, R. (2011). Edible coating and post-frying centrifuge step effect on quality of vacuum-fried banana chips. *Journal of Food Engineering* 107: 319–325.
Soupas, L., Juntunen, L., Lampi, A.M., and Piironen, V. (2004). Effects of sterol structure, temperature, and lipid medium on phytosterol oxidation. *Journal of Agricultural and Food Chemistry* 52: 6485–6491.
Stier, R.F. (2004). Frying as a science – an introduction. *European Journal of Lipid Science and Technology* 106: 715–721.
Stier, R.F. (2013). Ensuring the health and safety of fried foods. *European Journal of Lipid Science and Technology* 115: 956–964.

Su, C. and White, P. (2004). Flavor stability and quality of high-oleate and regular soybean oil blends during frying. *Journal of the American Oil Chemists' Society* 81: 853–859.

Su, Y., Zhang, M., Fang, Z., and Zhang, W. (2017). Analysis of dehydration kinetics, status of water and oil distribution of microwave-assisted vacuum frying potato chips combined with NMR and confocal laser scanning microscopy. *Food Research International* 101: 188–197.

Su, Y., Zhang, M., Bhandari, B., and Zhang, W. (2018a). Enhancement of water removing and the quality of fried purple-fleshed sweet potato in the vacuum frying by combined power ultrasound and microwave technology. *Ultrasonics Sonochemistry* 44: 368–379.

Su, Y., Zhang, M., Zhang, W. et al. (2018b). Ultrasonic microwave-assisted vacuum frying technique as a novel frying method for potato chips at low frying temperature. *Food and Bioproducts Processing* 108: 95–104.

Swi-Bea, J., Wu, M.C., and Wei, Y.P. (2004). Tropical fruits. In: *Processing Fruits: Science and Technology* (ed. D.M. Barrett, L.P. Somogyi and H.S. Ramaswamy). Boca Raton, FL: CRC Press.

Tarmizi, A.H. and Ismail, R. (2014). Use of pilot plant scale continuous fryer to simulate industrial production of potato chips: thermal properties of palm olein blends under continuous frying conditions. *Food Science & Nutrition* 2: 28–38.

Teruel Mdel, R., Gordon, M., Linares, M.B. et al. (2015). A comparative study of the characteristics of French fries produced by deep fat frying and air frying. *Journal of Food Science* 80: E349–E358.

Tian, J., Chen, S., Shi, J. et al. (2017). Microstructure and digestibility of potato strips produced by conventional frying and air-frying: an in vitro study. *Food Structure* 14: 30–35.

Torreggiani, D. and Bertolo, G. (2001). Osmotic pre-treatments in fruit processing: chemical, physical and structural effects. *Journal of Food Engineering* 49: 247–253.

Totani, N., Yawata, M., Mori, T., and Hammond, E.G. (2013). Oxygen content and oxidation in frying oil. *Journal of Oleo Science* 62: 989–995.

Troncoso, E. and Pedreschi, F. (2009). Modeling water loss and oil uptake during vacuum frying of pre-treated potato slices. *LWT – Food Science and Technology* 42: 1164–1173.

Troncoso, E., Pedreschi, F., and Zúñiga, R.N. (2009). Comparative study of physical and sensory properties of pre-treated potato slices during vacuum and atmospheric frying. *LWT – Food Science and Technology* 42: 187–195.

Tsuzuki, W., Matsuoka, A., and Ushida, K. (2010). Formation of trans fatty acids in edible oils during the frying and heating process. *Food Chemistry* 123: 976–982.

Turan, S., Karabulut, I., and Vural, H. (2007). Influence of sn-1, 3-lipase-catalysed interesterification on the oxidative stability of soybean oil-based structured lipids. *Journal of the Science of Food and Agriculture* 87: 90–97.

Tyagi, V. and Vasishtha, A. (1996). Changes in the characteristics and composition of oils during deep-fat frying. *Journal of the American Oil Chemists' Society* 73: 499–506.

Usuki, R., Fukui, H., Kamata, M., and Kaneda, T. (1980). Accumulation of peroxides in pan-frying oil. *Fette, Seifen, Anstrichmittel* 82: 494–497.

Wang, J., Reyes Suarez, E., Kralovec, J., and Shahidi, F. (2010). Effect of chemical randomization on positional distribution and stability of omega-3 oil triacylglycerols. *Journal of Agricultural and Food Chemistry* 58: 8842–8847.

Warner, K. (2004). Chemical and physical reactions in oil during frying. In: *Frying Technology and Practices* (ed. M.K. Gupta, K. Warner and P.J. White). Champaign, IL: AOCS Press.

Wilkes, R.S. (2008). Low linolenic soybeans and beyond. *Lipid Technology* 20: 277–279.

Wolff, R.L. (1993). Heat-induced geometrical isomerization of α-linolenic acid: effect of temperature and heating time on the appearance of individual isomers. *Journal of the American Oil Chemists' Society* 70: 425–430.

Xu, X.-Q., Tran, V.H., Palmer, M.V. et al. (2000). Chemical, physical and sensory properties of Monola oil, palm olein and their blends in deep frying trials. *Food Australia* 52: 77–82.

Xue, J. and Ngadi, M. (2007). Thermal properties of batter systems formulated by combinations of different flours. *LWT– Food Science and Technology* 40: 1459–1465.

Yagua, C.V. and Moreira, R.G. (2011). Physical and thermal properties of potato chips during vacuum frying. *Journal of Food Engineering* 104: 272–283.

Yamsaengsung, R., Ariyapuchai, T., and Prasertsit, K. (2011). Effects of vacuum frying on structural changes of bananas. *Journal of Food Engineering* 106: 298–305.

Yang, J.-H., Park, H.-Y., Kim, Y.-S. et al. (2012). Quality characteristics of vacuum-fried snacks prepared from various sweet potato cultivars. *Food Science and Biotechnology* 21: 525–530.

Yanishlieva, N.V. and Marinova, E.M. (2001). Stabilisation of edible oils with natural antioxidants. *European Journal of Lipid Science and Technology* 103: 752–767.

Zambiazi, R.C. (1998). Effect of endogenous minor components on the oxidative stability of vegetable oils. *Lipid Technology* 10: 58–62.

Zeb, A. (2011). Effects of β-carotene on the thermal oxidation of fatty acids. *African Journal of Biotechnology* 10: 15346–15352.

Zeb, A. (2015). Chemistry and liquid chromatography methods for the analyses of primary oxidation products of triacylglycerols. *Free Radical Research* 49: 549–564.

Zeb, A. and Ahmad, S. (2017). Changes in acylglycerols composition, quality characteristics and in vivo effects of dietary pumpkin seed oil upon thermal oxidation. *Frontiers in Chemistry* 5: 55.

Zeb, A. and Murkovic, M. (2013). Determination of thermal oxidation and oxidation products of β-carotene in corn oil triacylglycerols. *Food Research International* 50: 534–544.

Zeb, A. and Rahman, S.U. (2017). Protective effects of dietary glycine and glutamic acid toward the toxic effects of oxidized mustard oil in rabbits. *Food & Function* 8: 429–436.

3

Frying and Culture

3.1 Introduction

Generally, a cuisine is characterized by specific and distinctive ingredients, techniques, or dishes. The ingredients are the principal factor differentiating between different cultures. For example, Jewish dietary laws prohibit the ingestion of certain animals, such as pork, insects, molluscs, and crustaceans, while requiring that other animals be slaughtered according to a specific process known as *Shechita*. Other religions, such as Christianity and Islam, have their own restrictions. Several excellent books are available regarding the intersection of religions, cultures, and foods (Adamson 2004; Kittler et al. 2011; Brown 2018). Civitello (2011) discusses how different cultures adopted different cuisines across history, up to the present day.

Ingredients such as spices, plant extracts, and batters may be added before or after frying foods. These ingredients have a significant effect on the taste, colour, and quality of the final food. Salvador et al. (2005) studied the effects of wheat starch, modified corn starch, dextrin, dried egg, and gluten on the pick-up and flow properties, colour, and texture of fried coatings for squid rings. They found that gluten-added batters had the highest pick-up values, while there was no significant difference in colour due to the presence of the different ingredients. Dextrin showed high crispness, whereas gluten significantly lowered the oil content. Brannan et al. (2014) showed that batter ingredients have significant effects on oil absorption.

It is beyond the scope of this chapter to examine the frying process in every part of the world. Instead, we will look at just a few of the most famous fried foods on each continent. Knowledge of what fried foods are consumed in which parts of the world could provide us with a clue to the introduction of new technologies, safety methods, and lifestyle improvements.

3.2 The Common Point

Frying can be a simple or a complex method of preparing food in heated oil. Oil, food, and temperature are considered the common factors in the frying process, while others factors such as battering, additional ingredients, specific cultural

Food Frying: Chemistry, Biochemistry, and Safety, First Edition. Alam Zeb.

processes, and the technology used vary across cultures. The variable factors will be discussed in the coming chapters, while the basic frying concept has been presented in detail in Chapters 1 and 2.

There are some fried foods that are common in nearly all parts of the world. These include fried chicken, chicken nuggets, and fried fish. Kentucky Fried Chicken (KFC) is perhaps the most famous fried chicken brand worldwide. With little or no change in its process, KFC prepares fried chickens with high acceptability. This can be seen by the way the chain has spread across the globe; there is a significant correlation ($R^2 = 0.9973$) between the number of years in operation and the number of restaurants worldwide (Statista 2018b). The American Consumer Satisfaction Index (ACSI) shows that KFC had particularly high values (78) during the years 2016 and 2017.

The brand values of the top 10 fast food restaurants globally have increased significantly in recent times (Statista 2018a). However, with the growing popularity of fast food, there have been attendant negative effects on human health (Janssen et al. 2018). Targeted interventions to reduce the impact on public health are desperately needed. The Australian government has begun implementing a health star rating (HSR) system on fast food labels to increase consumer awareness of the health risks, although this system is currently voluntary (Dunford et al. 2017).

3.3 Frying in American Cuisines

Ray (2007) presents a detailed history of how different groups affected modern US cuisine. The Italians, Chinese, Germans, French, Japanese, and others all had a large effect on the American food landscape. Different parts of the country have their own local cuisines, with their own ingredients, compositions, and methods of preparation.

The United States is famous for a number of fried foods, including fried snacks, buffalo chicken wings, corn dogs, pickles, chicken fried steaks, green tomatoes, fried fish, and Twinkies. Southern states have their own fried foods and fast food chains. Among them, fried chicken is the most famous and most highly consumed.

Following are some famous American fried foods.

- ***Chicken Fries:*** Battered chicken strips, fried in high oleic edible oils. They come ready to eat and mixed with other ingredients, such as salad and lemon.
- ***Maryland Chicken:*** Lightly battered, fried, and mixed with bacon, corn, banana, chickpeas, cannelloni beans, fresh mint, and milk cream.
- ***Barberton Chicken:*** Introduced by Serbian immigrants. Usually consists of four pieces served with coleslaw, hot sauce, and potato fries.
- ***Chicken Waffles:*** Made of fried battered chicken and waffles, and served with peaches and milk cream or butter.
- ***Hot Chicken:*** Also known as Nashville hot chicken. Made of fried battered chicken parts, seasoned, floured, and covered with a spicy hot sauce.
- ***Po-Boy:*** Made of fried seafood such as shrimps and oysters, with lettuce, tomatoes, pickles, and mayonnaise, all piled in a French bread.
- ***Cheese Burgers:*** Consists of a cheese topping, fried meat patty, lettuce, tomatoes, onion, pickles, and ketchup.

- ***Fried Fish:*** Any fish, fried battered or nonbattered. Served with French fries, lemon, ketchup, and sauces.
- ***Doughnut:*** A dessert food made of fried dough, commonly sold in confectionery stores.

3.4 Frying in European Cuisines

European cuisines are very diverse, due to the number of different specialized cultures there. The original European cuisines were believed to be developed by royals and nobility; the use of cutlery while dining was considered a sign of nobility (Rebora 2010). Immigrants and invaders also played a significant role. For example, Arab invaders had a large influence on medieval European cuisines. The Renaissance led to many developments in the science of frying. In the late 1700s, French fries were invented by street vendors in Paris. This is still one of the most common fried foods in the world.

Following are some famous European fried foods.

- ***French Fries:*** Made of potatoes and served with fried fish and other ingredients, depending upon culture, needs, and personal choice. French fries are one of the most common fried foods worldwide, available in a large variety of styles.
- ***Fried Chicken:*** Fried either breaded or battered and served with different ingredients, depending upon taste. One of the most common fried foods in European countries.
- ***Bratwurst:*** German sausages made of beef, veal, or pork and fried with different ingredients.
- ***German fries:*** Sliced potatoes fried with onion, green pepper, and bacon. A very famous and traditional fried food in Germany.
- ***Fried Fish:*** Fried either breaded or battered.
- ***Fried Potatoes:*** Served with different fried vegetables and salad.
- ***Wiener schnitzel:*** A traditional Austrian dish made of breaded fried veal cutlets and served with lemons and garnished with parsley. Fried onions or mushrooms may also be added.
- ***Rice Polu:*** A common Eastern European food. Fried or cooked rice with fried meat pieces, tomatoes, and salad. Also called pilaf.
- ***Chebureki:*** Fried beef or lamb folded in dough and fried again. A popular fast food in Eastern European countries.
- ***Pirozhki:*** Different sizes of baked or fried buns stuffed with a variety of fillings. Originated in Russia and Ukraine. May be served alone or with French fries or salad.
- ***Fried pizza:*** Fried dough, cheese, and tomato, with a variety of vegetarian and nonvegetarian toppings.
- ***Fish and Chips:*** Fried fish and potato chips. Very common fast food in the United Kingdom.
- ***Doughnut:*** A dessert food made of fried dough, commonly sold in confectionery stores.

3.5 Frying in Asian Cuisines

The Asian continent is the largest and most populated in the world. Asian cuisines are very diversified, due to the number of different cultures and habits. By population, the largest are the Chinese, Indian, and Pakistani cuisines. Asian cuisine has also invaded the cuisines of other continents (Cwiertka and Walraven 2013). There are a large number of recipes (Trang 2003), many of which involve frying.

Following are some famous Asian fried foods.

- ***Fried Chicken:*** Fried either battered or not battered. Common as famous fast food and in the home.
- ***Fried Fish:*** The frying of coated, salted fish is common across the continent. Served with vinegar, lemon, and salad.
- ***Rice Polao:*** Fried rice with fried or roasted meat or chicken.
- ***Fried Kebab:*** Famous South Asian fried food made of meat, a mixture of spices, onion, tomatoes, and lemon.
- ***Samosa:*** A very famous fried food made of a mixture of savoury fillings, including spices, potatoes, tomatoes, peas, and meat or chicken pieces, and cooked in a trihedral shape. Served with sauces and pepper.
- ***Fried fritters:*** A variety of fritters are prepared in different Asian countries.
- ***Bakery Products:*** A variety of bakery products are prepared using frying. Can be obtained in bakery or confectionary stores.
- ***Jalebi:*** A fried sweet food made of plain flour and sugar, and fried in a circular manners. Famous in nearly all Asian countries.
- ***Halva:*** Stir-fried wheat, rice, or finely ground carrot. A very common and delicious fried food.
- ***Doughnut:*** A dessert food made of fried dough, commonly sold in confectionery stores.
- ***Laziji:*** A Chinese stir-fried food made of deep-fried chicken pieces, spiced beans, pepper, garlic, tomato, and ginger.
- ***Kung Pao Chicken:*** Fried chicken pieces with chillies, peanuts, and a spicy sauce.
- ***Crab Rangoon:*** Deep-fried wontons filled with crabmeat, cheese, and seasonings.
- ***Sesame Chicken:*** Deep-fried battered chicken pieces cooked in sesame oil and served with cooked broccoli.
- ***Ginger Beef:*** Ginger-marinated beef, fried twice and served with sweet-and-sour sauce.
- ***Lobster Cantonese:*** Stir-fried lobster served with savoury black bean sauce.
- ***Fried Mixed Vegetables:*** Different vegetables fried independently and served together with spices and sauces. Common in most Asian countries.
- ***Fried Vegetables:*** A single vegetable like ladyfinger or cauliflower, fried and served with tomato, onions, and salad.
- ***Szechuan Green Beans:*** Deep-fried beans served with potatoes and pepper.

3.6 Frying in African Cuisines

The African continent has a great number of cultures and cuisines, affected by geological variations in its animal and plant life (Harris 1998) and by its agricultural, cultural, and economic history (Mccann 2009).The invasion of Africa by European nations also played a role in its cuisine development.

Following are some famous African fried foods.

- ***Injera:*** A national cuisine in some African countries. Made of sour-dough risen bread with a spongy texture, with fried meat. The dough may be corn, barley, or wheat.
- ***Fried Fish:*** Highly popular in most African countries.
- ***Fried Chicken:*** Highly popular in most African countries.
- ***Yassa:*** A spicy food made of marinated fish or chicken and onions.
- ***Halva:*** A variety of halva is prepared through frying.
- ***Jollof rice:*** A form of fried rice made of tomatoes, palm oil, onions, salt, spices, chilli pepper, and fried meat, fish, or chicken.
- ***Acarajé:*** A deep-fried food made of peeled beans in the form of a ball.
- ***Koeksister:*** A fried confectionary food made of dough and immersed in honey or sugar syrup.
- ***Mandazi:*** A fried dough bread made of flour, yeast, milk, water, and sugar.

3.7 Frying in Middle Eastern Cuisines

In the Middle Eastern countries, meat and its products are very popular in a variety of cuisines. Significant diversity and homogeneity are present in cuisines, affected by different cultures and traditions. Olive and sesame oils are very popular as frying oils (Roden 2000). Saudi Arabia, Iran, and Iraq all played significant roles in the region's culinary variety. The area is largely influenced by Islamic cultural foods; however, the advent of US food companies has significantly increased culinary development. Basan (2007) provides recipes for several important dishes prepared in the Middle East.

Following are some famous Middle Eastern fried foods.

- ***Fried Chicken:*** Fried battered or nonbattered.
- ***Fried Fish:*** Served with cooked rice.
- ***Fried Rice:*** Served with tomatoes, onions, and fried chicken or fish.
- ***Couscous:*** Served with a variety of fried foods, including vegetables and chicken.
- ***Falafel:*** A deep-fried ball or doughnut made from chickpeas and fava beans. Served with salad
- ***Musakhan:*** Made of roasted chicken, onions, saffron, fried pine nuts, and spices spread over bread like a pizza.
- ***Maqluba:*** Made of a mixture of rice, meat, fried vegetables (tomatoes, cauliflower, eggplant), and fried chicken pieces.
- ***Dolma:*** Made from a variety of fried vegetables.
- ***Murtabak:*** A mixture of fried vegetables covered in bread.
- ***Turkey schnitzel:*** A modified form of Viennese schnitzel.
- ***Sabich:*** A bread sandwich consisting of pita stuffed with fried eggplant and boiled eggs.

3.8 Key Concepts

- Frying in different cuisines is largely affected by culture, food, and religion.
- A common point in different cuisines around the world is the presence of frying oil and fried foods. Some commercial restaurants are spreading a common style of fried food, such as fried chicken.

References

Adamson, M.W. (2004). *Food in Medieval Times*. Westport, CT: Greenwood.

Basan, G. (2007). *Middle Eastern Kitchen*. New York, NY: Hippocrene Books Inc.

Brannan, R.G., Mah, E., Schott, M. et al. (2014). Influence of ingredients that reduce oil absorption during immersion frying of battered and breaded foods. *European Journal of Lipid Science and Technology* 116: 240–254.

Brown, A.C. (2018). *Understanding Food: Principles and Preparation*. Boston, MA: Cengage Learning.

Civitello, L. (2011). *Cuisine and Culture: A History of Food and People*. Chichester: John Wiley & Sons.

Cwiertka, K.J. and Walraven, B.C. (2013). *Asian Food: The Global and the Local*. New York, NY: Routledge.

Dunford, E.K., Wu, J.H.Y., Wellard-Cole, L. et al. (2017). A comparison of the health star rating system when used for restaurant fast foods and packaged foods. *Appetite* 117: 1–8.

Harris, J.B. (1998). *The Africa Cookbook: Tastes of a Continent*. New York, NY: Simon and Schuster.

Janssen, H.G., Davies, I.G., Richardson, L.D., and Stevenson, L. (2018). Determinants of takeaway and fast food consumption: a narrative review. *Nutrition Research Reviews* 31: 16–34.

Kittler, P.G., Sucher, K.P., and Nelms, M. (2011). *Food and Culture*. Boston, MA: Cengage Learning.

Mccann, J.C. (2009). *Stirring the Pot: A History of African Cuisine*. Athens, OH: Ohio University Press.

Ray, K. (2007). Ethnic succession and the new American restaurant cuisine. In: *The Restaurants Book: Ethnographies of Where We Eat* (ed. D. Beriss and D.E. Sutton). Oxford: Berg.

Rebora, G. (2010). *Culture of the Fork: A Brief History of Everyday Food and Haute Cuisine in Europe*. New York, NY: Columbia University Press.

Roden, C. (2000). *The New book of Middle Eastern Food*. New York, NY: Alfred A. Knopf.

Salvador, A., Sanz, T., and Fiszman, S.M. (2005). Effect of the addition of different ingredients on the characteristics of a batter coating for fried seafood prepared without a pre-frying step. *Food Hydrocolloids* 19: 703–708.

Statista. 2018a. Brand value of the 10 most valuable fast food brands worldwide in 2017. Available at: https://www.statista.com/statistics/273057/value-of-the-most-valuable-fast-food-brands-worldwide [Accessed 14 September 2018].

Statista. 2018b. Number of Kentucky Fried Chicken restaurants worldwide from 2010 to 2017. Available at: https://www.statista.com/statistics/256793/kfc-restaurants-worldwide-by-geographic-region [Accessed 14 September 2018].

Trang, C. (2003). *Essentials of Asian Cuisine: Fundamentals and Favorite Recipes*. New York, NY: Simon and Schuster.

Part II

Chemistry of Food Frying

4

Chemistry of the Frying Medium

4.1 Frying Medium

The frying medium is an essential part of the frying process. Foods are partially or completely immersed and held in the frying medium at high temperature. The chemical nature of the frying medium is exclusively lipid. Lipids used as frying medium are classed as either fats or oils, depending on their physical properties. They are obtained from plant seeds known as oilseeds, nuts, and animal tissues, and are extracted using a variety of techniques. The overall process of extraction usually consists in several operational units working consecutively to produce the final product. These units include rendering or extraction, degumming, chemical and physical refining, bleaching, adsorption, hydrogenation, deodorization, and crystallization. One or more of these units may be avoided or repeated in order to obtain the frying lipid of choice, keeping in view the efficient and economic quality of the product.

The production of frying medium from animal tissues starts with the heating of the fatty tissues, in a process known as rendering. Heating solidifies the protein contents of the tissues and releases lipids. The rendering process can take two forms: dry and wet rendering. Dry rendering produces a product free of moisture, with a relatively dark colour and a cooked flavour. The end product may be used as such, or it can be processed further for physical or chemical refining. Wet rendering involves the cooking of tissues by high-pressure steam at relatively low temperatures. The product is usually lightly coloured and flavoured; its quality is based on the freshness and proper handling of the fatty tissues, especially in terms of preventing microbial and pest infection. The freshness of the frying medium can be assessed by checking the free fatty acid (FFA) content and the smell. High FFA values represent low quality; they can be reduced by physical adsorption or refining (Kalapathy and Proctor 2000). Butter, lard, and tallow are all examples of fats obtained from animal tissues or products.

The extraction of oils from plant seeds or tissues begins with the cleaning and crushing of the seeds, pretreatment, pressing or extraction, degumming, refining, neutralization, and deodorization. These processes may vary depending upon the type of product required or the nature of the raw materials. Palm fruits, for example, must first be heated or autoclaved, due to the presence of active lipase enzymes in their tissues. Lipases catalyse the production of FFAs, and consequently lower the quality of the

Food Frying: Chemistry, Biochemistry, and Safety, First Edition. Alam Zeb.

product. After separation of the kernel and pulp, the pulp is processed to give palm oil. Olive fruits, on the other hand, are processed without cooking. The oil obtained after pressing or centrifugation is used as such; it is widely known as "virgin olive oil." Extraction or pressing of oilseeds is carried out using mechanical or pressure extraction techniques, the details of which are reported by Erickson (1996). The qualities of the end product include good appearance, acceptable odour, and acceptable colour.

4.2 Classification and Choice of Frying Medium

Frying medium can be broadly classified into two main types: fats and oils. Each adds significant characteristic properties to the final fried food. Thus, the choice of frying medium is important to food manufacturers. It can vary from culture to culture and from country to country. Similarly, different fried foods have different requirements in terms of flavour, texture, appearance, keeping freshness, and quality. For example, Chapli kebab[1] is commonly fried in tallow instead of edible oils, due to its specific taste and high thermal stability (Zeb and Ali 2008). There are several important parameters that contribute to the choice of frying medium.

4.2.1 Appearance and Colour

The appearance of the frying medium plays a significant role in food frying. For example, red palm oil and sea buckthorn oils are rich in red colour pigments. The intensity of the colour is very high as compared to other edible oils, so consumers tend not to prefer them for frying. The appearance and colour of the final food fried also contribute to the selection of fats and oils.

4.2.2 Odour and Flavour

The odour and flavour of the frying medium contribute to the odour and flavour of the final fried food. Thus, they too are important parameters. The flavour intensity is one of the best indicators of overall quality in fresh potato chips. It has been found that a fatty acid (FA) composition of 16–42% oleic acid and 37–55% linoleic acid produces freshly fried chips with moderate fried food flavour intensity, good overall flavour quality, and low to moderate total polar compounds (Warner et al. 1997). Thus, a frying medium rich in oleic acid will give better frying results.

4.2.3 Texture of the Fried Food

The texture of the frying medium is usually liquid, oily, and fatty. However, the texture of the final fried food is of considerable importance for food fryers and consumers, and the frying medium plays a significant role in the development of this texture. The amount of lipid absorbed by potato crisps, and thus their texture, is dependent on

1 Chapli kebab is made of minced meat, grinded onion, tomato, garlic, salt, spices, wheat flour, and water. It is fried in beef tallow. It is a widely consumed South Asian street food.

the kind of oil used for frying. A strong correlation was observed between the oleic acid content of the frying medium and the texture of the final potato crisps (Kita et al. 2007). The fat content and texture of potato crisps were also influenced by frying temperatures. Thus, the crispness or hardness of fried foods is a function of the type of frying medium, the temperature, and any food additive used.

4.2.4 Taste

Edible oils and fats are mostly tasteless but contribute to the development of specific tastes in the final fried food. Olive contains a large amount of oleuropein, a polyphenol with significant biological properties, which imparts a bitter taste to olive oil (Gutiérrez-Rosales et al. 2003). Flaxseed oil, butter, coconut, and sesame oil have a low smoke point, and are thus not suitable for continuous or industrial frying. These frying mediums can develop a burning taste in fried foods and thus can be used for pan frying at lower temperatures.

4.2.5 Nutritional Value

A frying medium rich in polyphenols, natural antioxidants, and essential FAs is nutritionally significant. However, some of these nutrients may degrade or be destroyed during frying. For example, polyunsaturated fatty acid (PUFA)-rich oils oxidize faster than saturated oils. Most healthy oils contain a large amount of oleic acids, such as high oleic sunflower, canola, and safflower oils (Dubois et al. 2007). These oils do not require hydrogenation and are thus used as such. Natural antioxidants such as tocopherols play a synergic role in frying stability and the nutritional significance of the final fried food.

4.2.6 Availability and Economy

The frying medium must be easily available and economically acceptable so that as many customers as possible can afford to buy the fried food product. Increased production of high oleic sunflower oils was achieved with the help of sustainable agriculture and technologies. In 2016–17, vegetable oil production increased to 185.8 million metric tons worldwide – a 6.28% increase over the previous year. Thus, edible oils contribute significantly to the world economy.

4.3 Chemistry of the Frying Medium

4.3.1 Fatty Acids and Triacylglycerols

Foods fried in the frying medium (oil or fat) are composed largely (95–98% by weight) of triacylglycerols (TAGs). TAGs are the most important and abundant component of the natural lipids present in plants and animals. They possess significant nutritional value for humans. In plants, they are stored in the oilseeds, where they provide energy for growth. TAGs have high energy contents: about 37 kJ g^{-1}, compared with 17 kJ g^{-1} for protein and 16 kJ g^{-1} for carbohydrate, including glycogens – a form of energy storage in mammals (Brindley 1991).

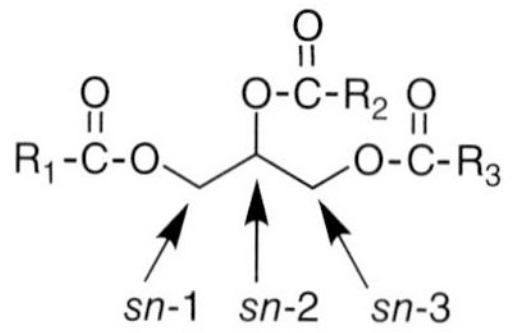

Figure 4.1 Chemical structure of triacylglycerol (TAG). R1, R2, and R3 represent fatty acids.

Chemically, TAG is made of glycerol, which is esterified with three FAs of one, two, or three different kinds. Each individual FA is located on a different carbon of the glycerol (Figure 4.1). The word 'tri-acyl-glycerol' echoes the presence of three esterified acyl groups attached to the glycerol backbone. The FA positions on the carbon 1, 2, and 3 are designated by *sn*-1, *sn*-2, and *sn*-3, where '*sn*' is the abbreviation of 'stereo-specific numbering' (Zeb 2015).

In plant oils, FAs are typically distributed in nonrandom fashion, and each position has a characteristic FA pattern. The FA at *sn*-2 is important in terms of TAG origin characterization and structure elucidation; those at *sn*-1 and *sn*-3 are digested in the first instance by the lipase enzymes, producing *sn*-2 mono-acylglycerols (Bernbäck et al. 1990). Thus, the TAG profile may reflect the nutritional and metabolic history of each cell as its anticipated energy storage requirement. The specific location of each acyl chain is important in terms of chemical and physical properties such as viscosity, pour point, melting point, fusion temperature, solubility, crystal structure, and polymorphism. The regio-isomer compositions (i.e. the positional distribution of FAs between the *sn*-2 and *sn*- 1/3 positions) have been shown to have biological effects on humans in such instances as infant nutrition, lipid absorption and metabolism, and atherogenesis.

FAs attached to glycerol are either saturated or unsaturated. The saturated FA family begins with methanoic acid, with a molecular formula of HCOOH. The hydrogen in HCOOH is replaced with a methyl group(s), forming short- to long-chain FAs. Methanoic, ethanoic, and propanoic acids are not present in natural fats. Along with butyric acid, they are water-soluble, and are often omitted from lipids. However, these acids are found in the non-esterified form in many food products. More than 1000 FAs are known, but only 20 or fewer are present in significant amounts in the oils and fats of commercial importance (Table 4.1). The most common acids are saturated and unsaturated C16 and C18. Anything below C16 is usually characterized as a short- or medium-chain FA, while anything above C18 is known as a long-chain acid. Most vegetable oils contain FAs with chain lengths between C16 and C22. The most dominant is saturated or unsaturated C18 FA, found in most plant oils. Palm kernel and coconut are the sources of medium-chain FAs and are referred to as lauric oils.

In contrast to plant oils, animal fats have a wider range of chain lengths, most of which are saturated. In most natural TAGs, a random distribution of FAs on the glycerol backbone is not common. Generally, in plant oils, unsaturated acids predominate at the *sn*-2 position, with more saturated acids at *sn*-1 and *sn*-3. The distributions of FAs at the *sn*-1 and *sn*-3 positions are often similar, although not identical. The FA composition in TAG may be either saturated or unsaturated, or a mixture of the two (Figure 4.2).

In the case of animal fats, the type of FA predominating at the *sn*-2 position is more variable; for example, palmitate may be selectively present in addition to unsaturated acids. Edible oils that are rich in one FA contain more monoacid TAG (e.g. olive, sunflower, corn, and rapeseed oils containing OOO and LLL). Table 4.2 shows the typical compositions of important TAGs in various edible oils and fats of frying significance. It has been observed that plant oils contain TAGs with PUFA moieties (i.e. LLnLn, LLLn,

Table 4.1 Typical composition (%) of some important fatty acids (FAs) in various frying media.

		Fatty acid range (%)				
Origin	**Frying Medium**	**C16:0**	**C18:0**	**C18:1**	**C18:2**	**C18:3**
Plant	Corn oil	10–17	1.6–3.3	25–42	39–61	0.7–1.3
	Olive oil	7.5–20	0.5–5	55–83	3.5–21	0–1.5
	Palm oil	40–46	4–7	36–41	9–12	10–13
	Rapeseed oil	3.3–6	1–2.5	52–67	16–25	6–14
	Sunflower oil	2.7–6	3–5	80–87	4–9	0–1.0
	Soybean oil	8–10	1–3	25–30	48–55	2–4
	Peanut oil	7–11	2–4	52–57	23–29	—
	Cottonseed oil	20–25	1–2	20–25	45–51	—
	Linseed oil	5–8	2–5	17–20	23–25	45–49
Animal	Beef tallow	25–32	12–18	45–52	—	—
	Butter	23–29	8–12	22–28	2–4	—
	Lard	25–30	9–13	45–50	3–6	—

The different range values shown here represent differences between varieties and studies. Only major FAs are shown; the values of others are excluded.
C16:0, palmitic acid; C18:0, stearic acid; C18:1, oleic acid; C18:2, linoleic acid; C18:3, linolenic acid.

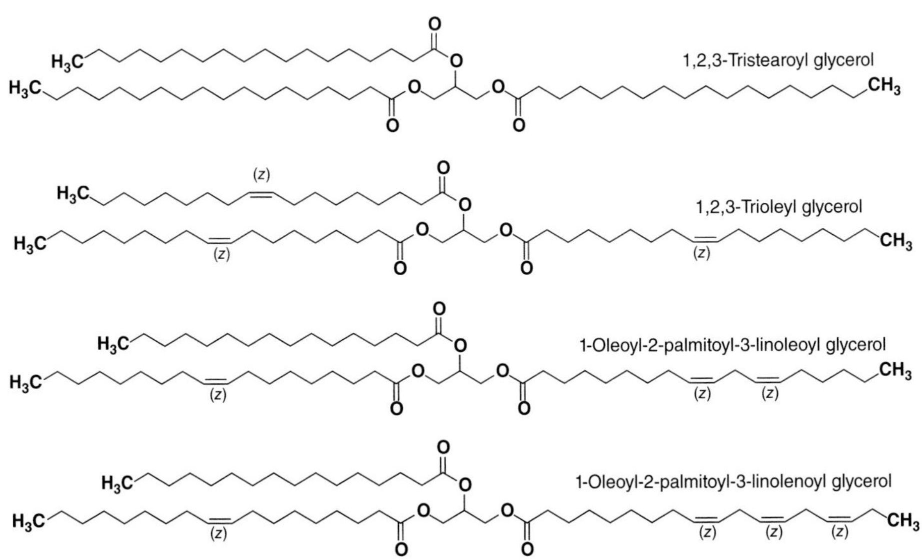

Figure 4.2 Structures of homo-TAGs and hetero-TAGs.

Table 4.2 Triacylglycerol (TAG) composition of some edible oils and fats used in frying.

	Relative Composition (%)						
TAG	**Sunflower oil (12)**	**Corn oil (12)**	**Olive oil (12)**	**Rapeseed oil (13)**	**Palm oil (13)**	**Lard (14)**	**Tallow (14)**
LLnLn	1.6	—	—	<0.5	—	—	—
LLLn	1.4	—	—	2.7	—	—	—
LLL	19.7	1.2	0.1	1.1	0.4	—	—
OLnL	—	—	—	5.7	—	—	—
OLL	27.8	14.6	0.9	8.0	2.2	2.58	—
PLL	9.9	0.6	0.3	1.1	—	2.18	—
OOL	18.0	18.1	11.6	23.5	—	4.97	—
PLO	12.4	13.3	2.2	5.4	—	10.03	3.8
PLP	0.2	—	—	0.7	8.3	2.47	—
OOO	4.2	17.0	60.7	16.8	3.8	5.67	3.4
POO	4.2	18.0	19.8	4.3	17.2	20.82	22.97
POP	0.2	1.3	0.6	0.8	11.2	8.0	10.79
GOO	—	5.3	—	1.3	—	—	—
OOS	0.1	8.1	3.2	1.2	1.7	4.61	11.39
SOP	—	0.9	0.2	—	1.9	15.54	14.85
AOO	—	0.4	—	—	0.09	—	—

The three letters represent the FA moieties in each TAG. Only TAGs of significance in terms of food frying and thermal stability are shown; the values of others are excluded.
Ln, linolenic acid; L, linoleic acid; O, oleic acid; P, palmitic acid; S, stearic acid; G, gadoleic acid; A, arachidic acid.

LLL, and OLnL) that are absent in animal fats such as lard and tallow. These TAGs are highly important for nutritional purposes but are one of the main reasons for the lower thermal stability of the oils during frying.

Studies have shown that oils and TAGs with high oleic acid contents are more resistant to thermal oxidation (Zeb 2012). Thus, high oleic sunflower and other edible oils are widely used in industrial and domestic frying.

TAGs rich in long-chain and saturated acids possess high melting points, while those rich in polyunsaturated acids have lower ones. However, the possibility that the FAs can be distributed in different molecular species with different melting points complicates the situation. Thus, oils with similar FA compositions may have different solid fat contents, polymorphic forms, and melting behaviours as a result of a different TAG composition. Mono- (MAG) and diacylglycerols (DAGs) are usually not significantly present in good-quality oils. High levels may be found in badly stored seeds, resulting from the activity of lipolytic enzymes. They can also be produced industrially by partial hydrolysis or glycerolysis of TAGs for use as food-grade emulsifiers.

4.3.2 Polyphenolic Compounds

Edible oils obtained from plant sources contain several important polyphenolic compounds. Among them, olive oil is rich in important phenolic compounds. Phenols and other related molecules that appear frequently in the literature contain mainly phenolic acids (hydroxybenzoic, hydroxyphenyl acetic, hydroxy cinnamic), simple phenols (tyrosol, hydroxytyrosol), derivatives of phenolic alcohols (hydroxytyrosol acetate), glycosides (oleuropein, ligstroside), aglyconic derivatives of glycosides (ligstroside aglycon, oleuropein aglycon), dialdehydic forms of decarboxymethyl elenolic acid linked to hydroxytyrosol (oleacein) and tyrosol (oleocanthal), lignans (acetoxypinoresinol, pinoresinol, 1-hydroxypinoresinol, syringaresinol), flavonoids (apigenin, luteolin), hydroxy-isochromans (1-phenyl-6,7-dihydroxy-isochroman, isochroman), and other phenols (vanillin, 4-hydroxy-3-methoxybenzaldeyde) (Boskou 2015) (Figure 4.3).

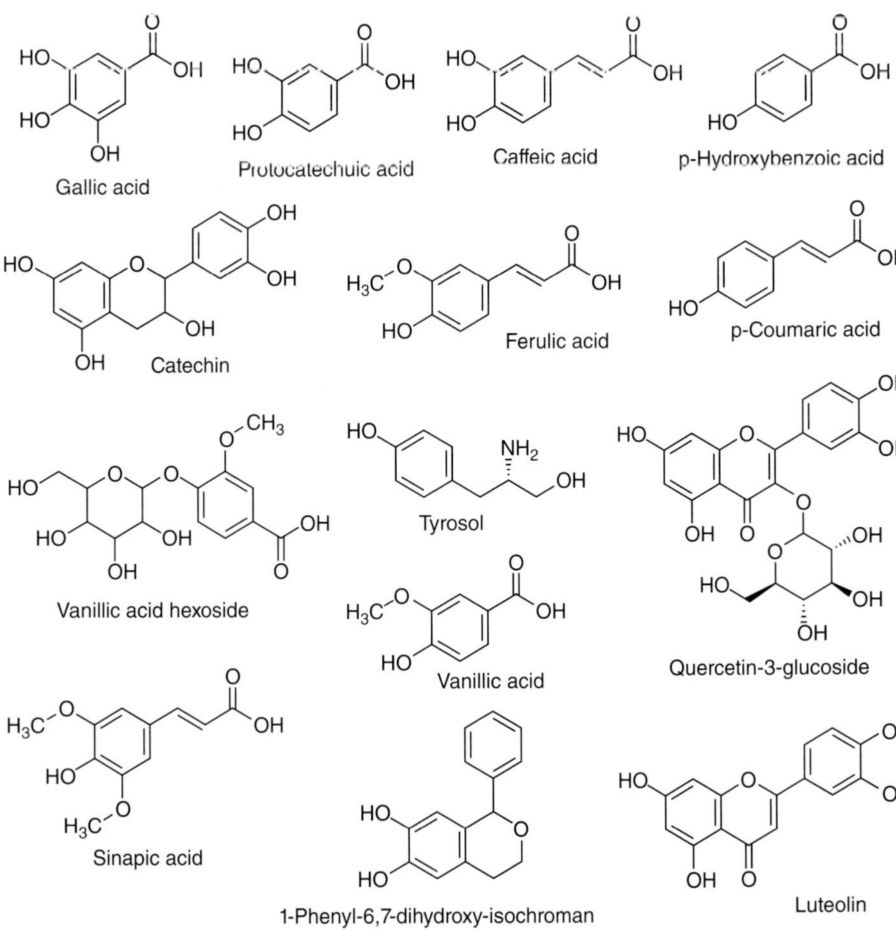

Figure 4.3 Chemical structures of important polyphenolic compounds in edible oils.

Similarly, pumpkin seed oil contains gallic acid, protocatechuic acid, p-hydroxybenzoic acid, vanillic acid, caffeic acid, p-coumaric acid, and ferulic acid in amounts less than 2 mg/100 g (Nyam et al. 2009). Fourteen phenolic compounds were identified and quantified in mustard oil using liquid chromatography (Figure 4.4); quercetin-3-glucoside, quercetin-3-feruloyl-sophoroside, catechin, quercetin-3-rutinoside, quercetin-3,7-diglucoside, sinapic acid, and vanillic acid hexoside were the major compounds (Zeb and Rahman 2017). These phenolic compounds provide strong antioxidant potential to the oils.

4.3.3 Hydrocarbons

Hydrocarbons form a minor component of edible oils. They include squalene and carotenes. Squalene is an unsaturated terpene that is widely distributed in nature. It is an important metabolite preceding sterol or carotene synthesis. Chemically, it is an all-*trans* 2,6,10,15,19,23-hexamethyl-2,6,10,14,18,22-tetracosahexaene (Figure 4.5). Its precursors in the cell are isoprenoid units. It forms an important part of the unsaponifiable matter of oils and fats, and constitutes more than 90% of the hydrocarbons fraction. The amount of squalene varies from 200 to 12 000 mg kg^{-1} in olive oils.

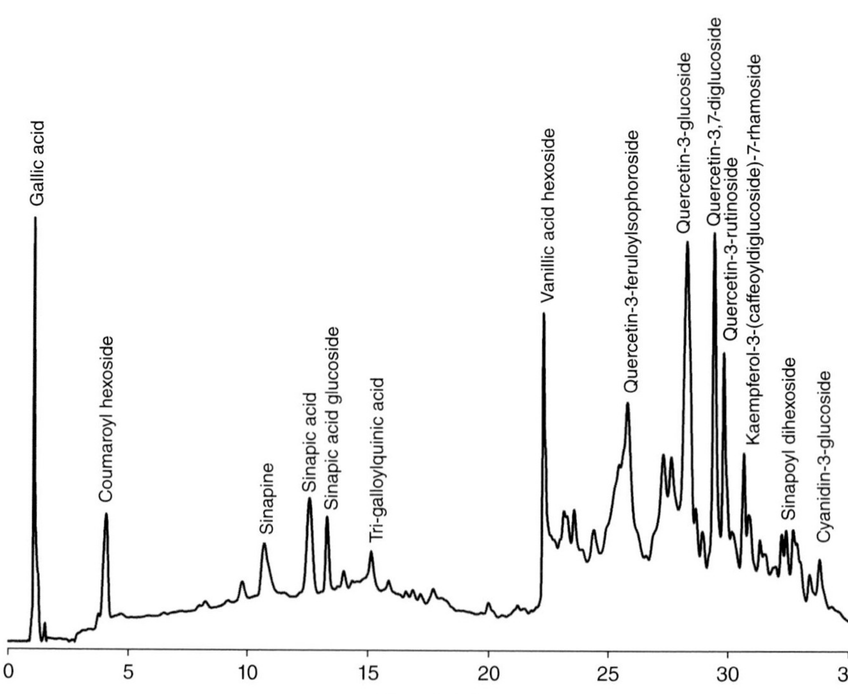

Figure 4.4 Reversed-phase HPLC-DAD chromatograms of mustard oil phenolic compounds at 320 nm. Fourteen phenolic compounds were identified and quantified as a total of 66.8 μg g^{-1}. *Source:* Reproduced with kind permission of Royal Society of Chemistry (Zeb and Rahman 2017).

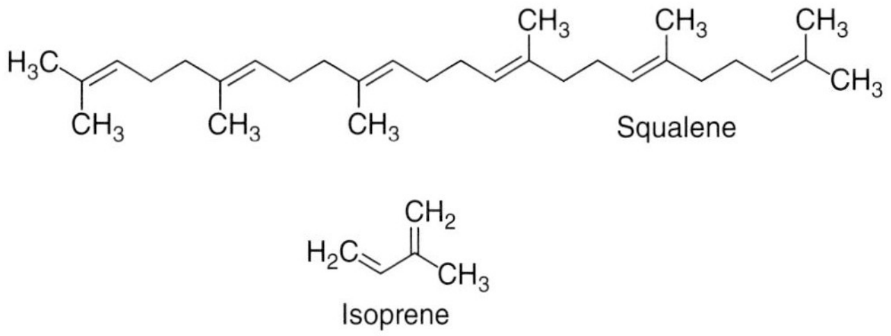

Figure 4.5 Structures of squalene and an isoprene unit.

4.3.4 Pigments

Pigments are usually present in plant oils and absent in animal fats. They come in two types: carotenoids and chlorophylls. Carotenoids are terpenoids made of 30 carbon compounds. They have two subclasses: carotenes, which are pure hydrocarbons, and xanthophylls, which are oxygenated carotenes. Among the carotenes, β-carotene is commonly present in edible oils, while among the xanthophylls, lutein is important (Figure 4.6). The amount of carotenoids in edible oils varies significantly with plant seed type. For example, sea buckthorn and red palm oils are rich in β-carotene, imparting a red colour. Chlorophylls (including pheophytins) are also present as a minor constituent of edible oils. The chlorophyll content of olive oil ranges from 5 to 30 mg kg^{-1} and may vary depending on variety, climate, and method of extraction and analysis.

4.3.5 Sterols

Sterols are minor hydrocarbons that are biochemically synthesized from isoprene units in plant seeds. Chemically, they are derivatives of cyclopenta hydroxy phenanthrene. In animal fats, sterols take the form of cholesterol or its derivatives. The edible oils from plant seeds are called phytosterols. Sterols are present either in free form or esterified with FAs or glycosides. In olive oil, four different types of sterols have been reported: 4α-desmethylsterols, 4α-methylsterols, 4,4-dimethylsterols, and triterpene dialcohols. The major components of the sterol fraction are β-sitosterol, Δ^5-avenasterol, and campesterol. Sterols present in trace amounts are stigmasterol, cholesterol, brassicasterol, ergosterol, sitostanol, campestanol, Δ^7-avenasterol, Δ^7-cholestenol, Δ^7-campestenol, Δ^7-stigmastenol,

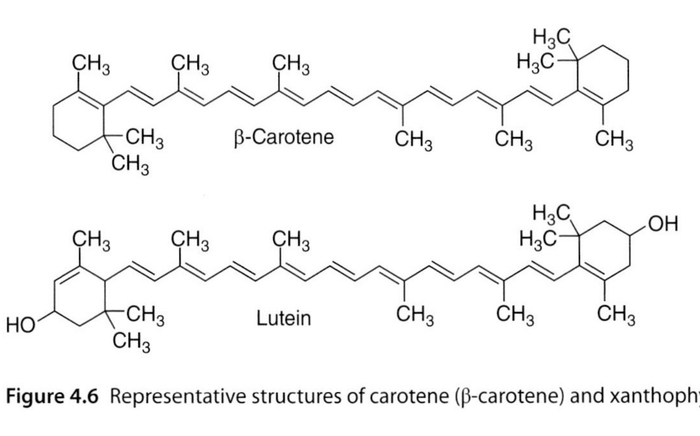

Figure 4.6 Representative structures of carotene (β-carotene) and xanthophyll (lutein).

$\Delta^{5,23}$-stigmastadienol, $\Delta^{5,24}$-stigmastadienol, $\Delta^{7,22}$-ergostadienol, $\Delta^{7,24}$-ergostadienol, 24-methylene-cholesterol, and 22,23-dihydrobrassicasterol (Figure 4.7).

4.3.6 Alcohols

Several free or bound alcohols are present as a minor component of edible oil. The key aliphatic alcohols found in edible oils (especially olive oil) are docosanol, tetracosanol, hexacosanol, and octacosanol (Figure 4.8). Fatty alcohols with odd carbon atoms (tricosanol, pentacosanol, heptacosanol) may also be present in trace quantities. Alkanols and alkenols having fewer than 10 carbon atoms in their molecule, and some aromatic alcohols, are constituents of the olive oil volatile fraction. Benzyl esters of hexacosanoic and octacosanoic acid have also been found.

4.3.7 Tocopherols and Tocotrienols

Tocopherol (vitamin E) is the generic name for a group of lipid-soluble compounds that includes four tocopherols and four tocotrienols. There are four isomers of each, for a total of eight homologues. In all of these isomers, the alpha-homologue makes up 90% of the total tocopherol content. Low amounts (~10–20 mg kg^{-1}) are reported for the

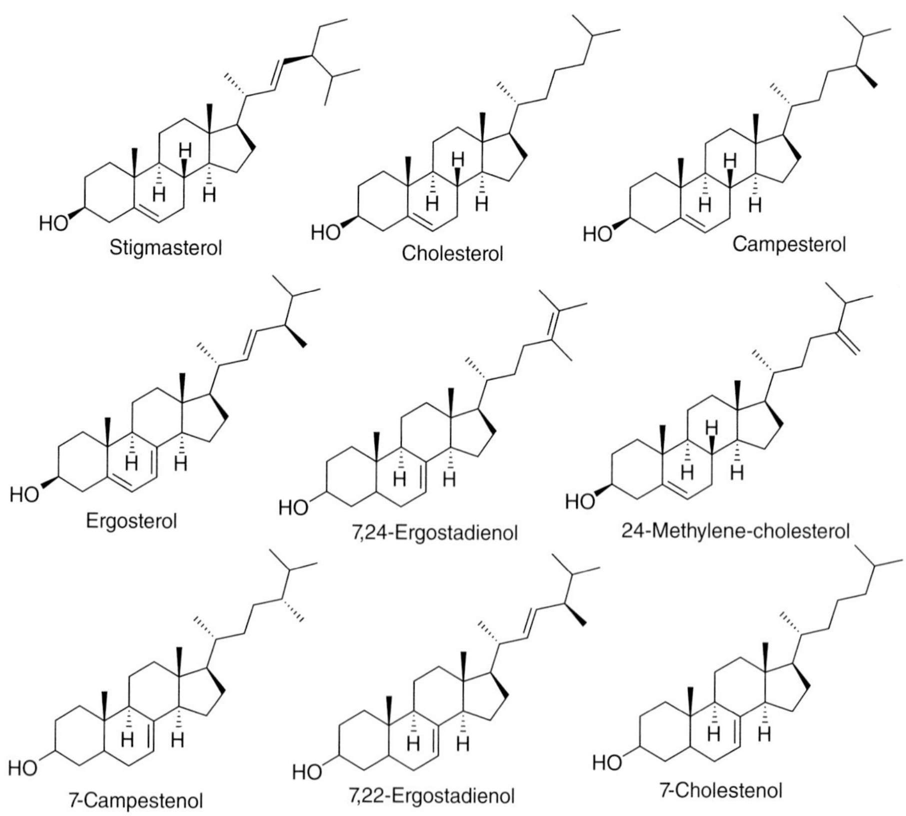

Figure 4.7 Chemical structures of cholesterol and phytosterol, with their derivatives.

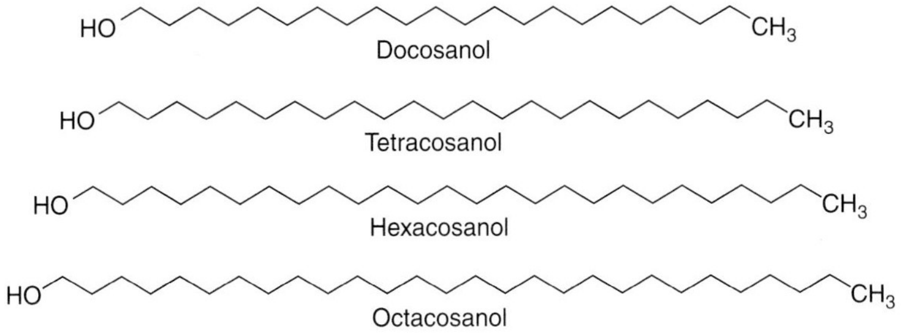

Figure 4.8 Chemical structures of medium- to long-chain fatty alcohols present in frying medium.

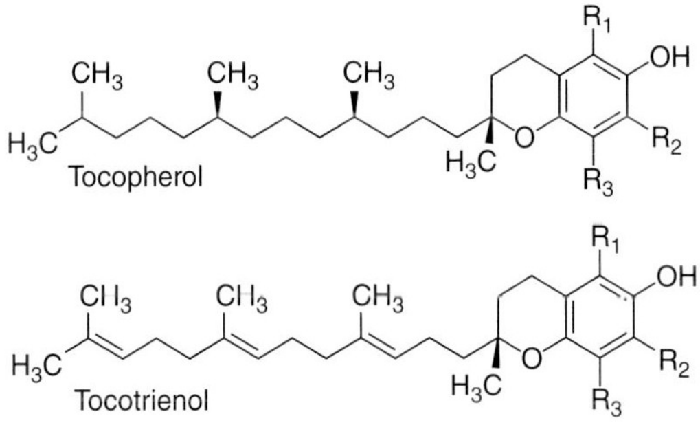

Figure 4.9 Representative structures of tocopherols and tocotrienols. The R groups are α-tocopherol (R_1, R_2 and $R_3 = CH_3$), β-tocopherol (R_1, $R_2 = CH_3$ and $R_3 = H$), γ-tocopherol (R_2, $R_3 = CH_3$ and $R_1 = H$), and δ-tocopherol (R_1, $R_2 = H$ and $R_3 = CH_3$); and α-tocotrienol (R_1, R_2 and $R_3 = CH_3$), β- tocotrienol (R_1, $R_2 = CH_3$ and $R_3 = H$), γ- tocotrienol (R_2, $R_3 = CH_3$ and $R_1 = H$), and δ- tocotrienol (R_1, $R_2 = H$ and $R_3 = CH_3$).

homologues β-tocopherol (~10 mg kg^{-1}), γ-tocopherol, and δ-tocopherol (Figure 4.9). The variations in the amounts have a strong correlation with the type of cultivar, method of extraction, and storage conditions. The high amount of α-tocopherol content indicates a wide range that depends on the cultivar potential and technological factors. Italian and Spanish oils were shown to vary from 55 to 370 mg kg^{-1} oil in terms of α-tocopherol content (Boskou 2015). Greek oils had a very high level of α-tocopherol, ranging from 98 to 370 mg kg^{-1} (Psomiadou et al. 2000). Values ranging from 93 to 260 mg kg^{-1} have been reported for Portuguese olive oil samples (Cunha et al. 2006). High levels of tocopherols (above 600 mg kg^{-1}) have been reported for Coratina, Arbequina, and Koroneiki cultivars grown in Egypt. Refining or hydrogenation causes loss of tocopherols.

Schwartz et al. (2008) determined the tocopherol, tocotrienol (i.e. tocol), and plant sterol contents of 14 vegetables and 9 industrial fats and oils available on the Finnish market in 2005 using normal phase high-performance liquid chromatographic (HPLC) with fluorescence detection and gas chromatography with flame ionization detection (GC-FID). They reported that the best sources of α-tocopherol were wheat germ (192 mg/100 g) and sunflower oil (59 mg/100 g). The oils richest in γ-tocopherol were camelina (72 mg/100 g), linseed (52 mg/100 g), and organic rapeseed oil (51 mg/100 g). Total tocotrienol contents were between 4.2 mg/100 g (coconut fat) and 268 mg/100 g in terms of wheat germ oil.

4.3.8 Phospholipids

Phospholipids are phosphate-containing DAGs. They are a highly important component of the cell membrane. They possess strong antioxidant potential in edible oils, where they are present in a concentration range of 1–6 mg kg^{-1}. Examples include phosphatidic acid, lysophosphatidic acid, and lysophosphatidylinositol (Figure 4.10).

4.3.9 Volatile and Aroma Compounds

Several small-chain hydrocarbons with volatile or aromatic properties are present in oils and fats. The aroma is commonly used for characterization or authentication purposes. Important aromatic constituents are C-6 aldehydes and alcohols, which are formed from PUFAs through a sequence of lipoxygenase- and transferase-catalysed reactions. The most important classes of volatiles are hydrocarbons, alcohols, aldehydes, ketones, esters, acids, phenols, phenol derivatives, oxygenated terpenes, furan derivatives, and thiols. Roughly 280 compounds have been identified in the volatile fraction of virgin olive oils. Some are odourless, while others present in small concentrations contribute to the aroma of oil. Important aroma compounds include hexanal, heptanal, Hexyl acetate, octenal, and nonenal (Figure 4.11).

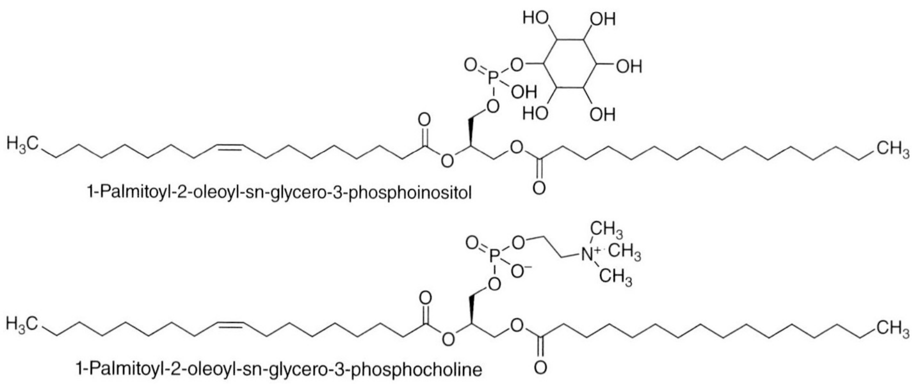

Figure 4.10 Chemical structures of important phospholipids (phosphatidyl inositol and phosphatidyl choline).

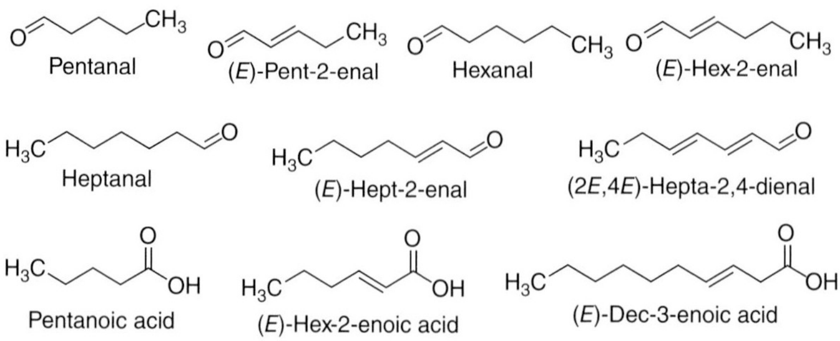

Figure 4.11 Chemical structures of important volatile compounds produced during frying.

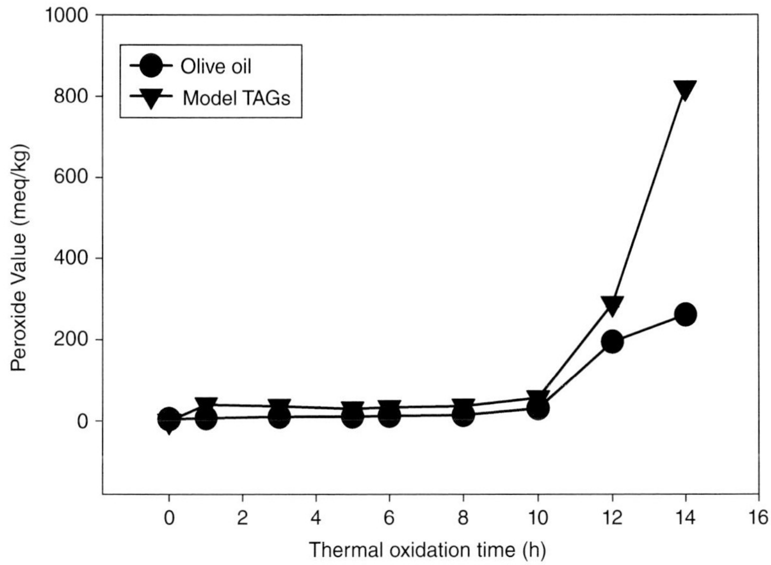

Figure 4.12 Effects of thermal oxidation on the formation of hydroperoxides in model TAGs and olive oils of similar composition. Model TAGs produced more peroxides than olive oils, showing the significance of minor components of oils. *Source:* Redrawn with permission from John Wiley & Sons Ltd (Zeb and Murkovic 2010) and Elsevier Ltd (Zeb and Murkovic 2011).

4.3.10 Importance of Minor Components in Frying

As already noted, TAGs form the major component of frying oils and fats. However, the minor components also play a significant role in frying. For example, more hydroperoxides were produced in the presence of β-carotene in model TAGs than in olive oil containing β-carotene under similar conditions (Figure 4.12). The β-carotene was stable and protected during the thermal treatment of the oil, which was attributed to the presence of other antioxidant compounds that are not found in model TAGs of similar composition (Zeb and Murkovic 2010, 2011). Further, phenolic compounds present in olive oil, such as hydroxyl-tyrosol acetate and lignin aglycan, are resistant to thermal treatment (180 °C). The behaviour of lignans is outstanding in this regard, as they belong to a family that is highly resistant to thermal treatment (Carrasco-Pancorbo et al. 2007). Thus, the contribution of minor components cannot be underestimated during frying.

4.4 Chemistry of Lipid Oxidation During Frying

When foods are fried in edible oils or fats, lipid oxidation occurs. Since the major part of the frying medium is composed of TAGs, this chapter will look at recent knowledge in this regard. Some well-known classical lipid oxidation mechanisms may not apply here. Complex lipid systems – even if they are formed only from TAGs – can have some significant chemistry. Lipid oxidation in real food frying is a complex subject, with several competing theories of oxidation. In this section, we will look at free radical oxidation induced by thermal stress. The role of FA moieties in TAGs, and their subsequent oxidation, will be discussed in order to clarify the basic principle of lipid or TAG oxidation during frying. These reactions may involve oxidation, hydrogen abstraction, electron transfer, isomerization, cyclisation, scission, and addition reactions (Figure 4.13), which are discussed in detail elsewhere (Kamal-Eldin et al. 2003; Schaich 2013; Barden and Decker 2016).

4.4.1 Fatty Acid Hydroperoxides

The FA moieties in TAGs are usually oxidized by a well-known mechanism known as free radical oxidation. This involves an attack of oxygen at the allylic position and the subsequent formation of unsaturated hydroperoxides. These hydroperoxides also take part in auto-oxidation, initiating a chain reaction (Zeb 2015). Thus, it is established that auto-oxidation of PUFAs occurs as a chain reaction, which involves three phases:

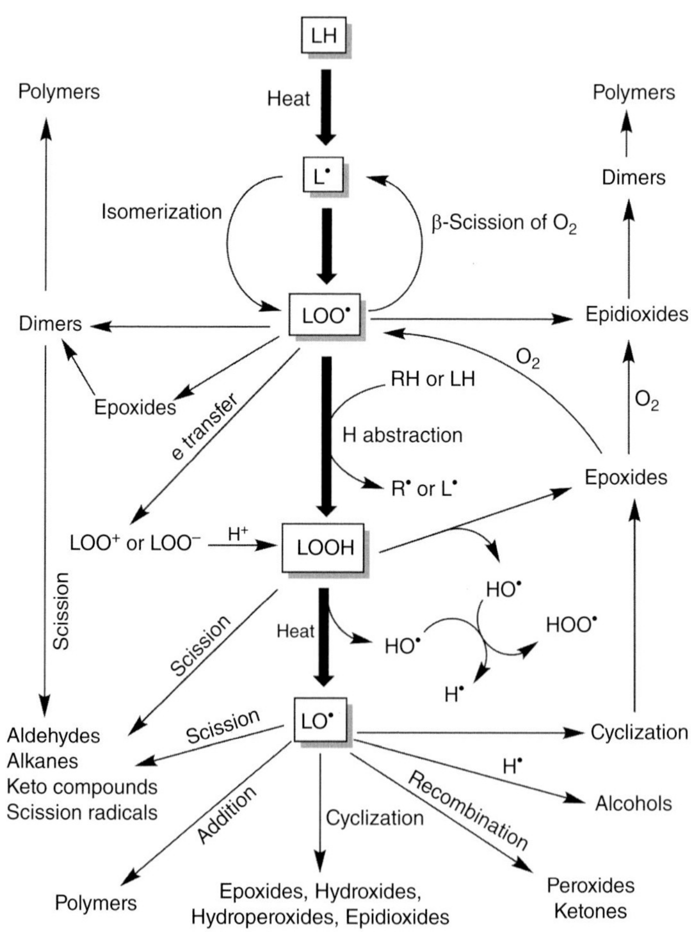

Figure 4.13 Representative sequence of reactions during frying. These reactions may not be as simple as represented here. R*, alkyl; L*, lipid radical; LH, lipid molecule; LOOH, lipid hydroperoxide; HO*, hydroxy radical; *OOH, hydroperoxy radical; LOO*, peroxy radical; LO*, alkoxy radicals respectively.

$$\left.\begin{array}{lcl} LH & \longrightarrow & L^* + H^* \\ LOOH & \longrightarrow & LO^* + OH^* \end{array}\right\} \text{(Initiation)}$$

$$\left.\begin{array}{lcl} L^* + O_2 & \rightleftharpoons & LOO^* \\ LOO^* + LH & \longrightarrow & LOOH + L^* \end{array}\right\} \text{(Propagation)}$$

$$\left.\begin{array}{lcl} 2LOO^* & \longrightarrow & \text{Non-Radical Products} \\ LOO^* + L^* & \longrightarrow & \text{Non-Radical Products} \\ 2L^* & \longrightarrow & \text{Non-Radical Products} \end{array}\right\} \text{(Termination)}$$

Figure 4.14 Representative free radical reaction sequence, showing the steps initiation, propagation, and termination steps. H^*, hydrogen radical; L^*, lipid radical; LH, lipid molecule; LOOH, lipid hydroperoxide; HO^*, hydroxy radical; LOO^*, hydroperoxy radical; LO^*, peroxy radical.

initiation, propagation, and termination (Figure 4.14). By the action of initiators, unsaturated FAs form a carbon-centred alkyl radical (L*) and a peroxy radical (LOO*) (Kamal-Eldin et al. 2003).

These radicals then propagate to form hydroperoxides (LOOH) as their primary oxidation product (Choe and Min 2006). The newly formed alkyl radical thus starts another chain reaction, which continues until all lipid molecules are consumed. Different FAs show different susceptibilities to auto-oxidation based on the dissociation energies of C—H bonds present in them.

Heavy metals in trace amounts have been found to significantly contribute to the acceleration of lipid oxidation. Lipid radical is generated either by electron transfer or by the decomposition of hydroperoxides present in the oils or fats. The decomposition of hydroperoxides involves the formation of metal-oxygen transition complexes ($M{-}O_2$) or metal hydroperoxide complexes (M–LOOH), which catalyse autoxidation and decomposition through one-electron redox reactions, as shown in Figure 4.15. Thus, food frying utensils also play a role in thermal oxidation, and therefore cannot be overlooked.

As a general rule, the removal of hydrogen is much easier in the presence of a double bond in the FA than a single bond (saturated FA). Experiments show that the oxidation

$$M^n + LH \longrightarrow M^{n-1} + L^* + H^+ \quad \text{(electron transfer)}$$

$$M^n + ROOH \longrightarrow M^{n+1} + RO^* + {}^*OH \quad \text{(decomposition)}$$

$$M^{n+1} + ROOH \longrightarrow M^n + ROO^* + H^* \quad \text{(decomposition)}$$

Figure 4.15 The role of heavy metal ions in the initiation reactions of lipid oxidation. Metal ion involves electron transfer and decomposition reactions. M^n, metal ion; LH, lipid molecule; L^*, lipid radical; H^*, hydrogen radical; ROOH, lipid hydroperoxide; RO^*, alkoxy radical; ${}^*OH$, hydroxide radical; ROO^*, peroxy radical.

of PUFAs is dependent on the number of allylic methylene groups present in the FAs (Wagner et al. 1994). Thus, the oxidation of monounsaturated fatty acids (MUFAs) like oleic acid is less pronounced than that of PUFAs (linoleic, linolenic acids, etc.). This was confirmed by Gardner (1989), who showed that the bond dissociation energy of a C—H bond of the bis-allylic methylene (75 kcal mol^{-1}) is less than that of the mono-allylic methylene (88 kcal mol^{-1}).

Figure 4.16 shows the radical oxidation of methyl linolenate. It has been observed that the hydrogen is abstracted by radical liberating initiators such as heat, light, metals, and irradiation from the two allylic methylene groups on carbon-11 and 14 to form two pentadienyl radicals. The addition of oxygen resulted in the formation of four peroxy radicals, which led to the formation of the corresponding conjugated dienoic hydroperoxides, such as 9-, 12-, 13-, and 16-hydroperoxides. These hydroperoxides may be present in *Z* or *E* forms. The probabilities of formation of external 9- and 16-hydroperoxides are significantly higher than those of the formation of 12- and 13-hydroperoxides (Frankel 1991).

About 95% of the primary oxidation of linoleate occurs due to the abstraction of the bis-allylic hydrogen atom and the resulting isomerization of one of the double bonds. Minor amounts of nonconjugated 8-, 10-, 12-, and 14-hydroperoxides can also be obtained, due to the abstraction of allylic hydrogen. Smaller amounts of the bis-allylic 11-hydroperoxide are also formed in the presence of α-tocopherol, as was shown by Brash (2000). Chan and Levett (1977) isolated four conjugated dienes hydroperoxides from the auto-oxidation of methyl linoleate. They showed that these hydroperoxides isomers are formed from both positional and geometrical isomerization of double bonds, which requires the opening of these double bonds and rotation

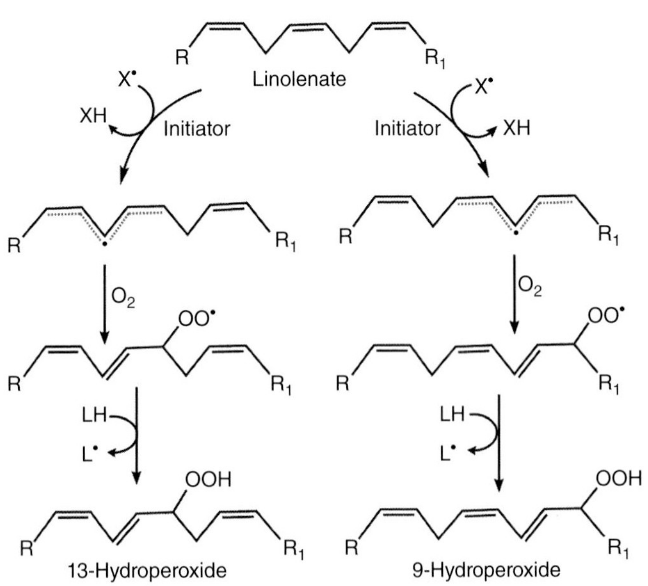

Figure 4.16 Free radical auto-oxidation of linolenate. R, CH_3CH_2–; R_1, $-(CH_2)_7$ COOH. *E*/*Z* isomers are not shown, for the purpose of simplicity. *Source:* Reproduced with permission from Taylor & Francis (Zeb 2015).

of the resulting single bonds. The β-fragmentation of oxygen from the *E/Z* hydroperoxy radicals followed by bond rotation leads to the formation of *E/E* hydroperoxides isomers. The addition of oxygen then leads to peroxyl radicals, which give 9- and 13-*E*, *E*-hydroperoxides, which are the thermodynamic products.

The linolenate peroxyl radical of 12- and 13-hydroperoxides also undergoes rapid cyclization to form hydroperoxy epidioxides by intramolecular rearrangement, as shown in Figure 4.17. It has been observed that this reaction is the major contributor of the smaller quantity of the 12- and 13-hydroperoxides as compared to the 9- and 16-hydroperoxides. Antioxidants donate hydrogen atoms to the free radicals, resulting in nonradical products. Generally, a compound which has a reduction potential lower than that of the free radical can only donate hydrogen, keeping in mind the kinetic favourability. Single-electron reduction potentials of 1600, 1000, and 600 mV were observed for the FAs alkoxy, peroxy, and alkyl radicals, respectively. The reduction potential of the antioxidants is 500 mV or less, so it is energetically favourable that antioxidants will react with these free radicals to form neutral species. Antioxidants such as α-tocopherol have been found to strongly inhibit the formation of hydroperoxy epidioxides. Peers et al. (1981) showed that 5% of α-tocopherol in methyl linolenate completely inhibits this cyclization.

During frying, linoleic acid is oxidized by the identical free radical mechanism, as shown Figure 4.18. Two auto-oxidation products are formed, in relatively equal amounts: 9- and 13-hydroperoxide. Other hydroperoxides formed from linoleic acid include 11- and 12-hydroperoxides. The use of modern techniques such as HPLC and nuclear magnetic resonance (NMR) reveals that these hydroperoxides are produced as both *cis* and *trans* isomers; a low temperature during frying favours the formation of *cis* isomers, while a higher temperature for a prolonged time results in *trans* isomers (Frankel et al. 1990).

In oleic acid oxidation (Figure 4.19), the mechanism involves hydrogen atom abstraction at the carbon atoms 8 and 11 to give two allylic radicals. The oxygenation of these allylic radicals gives rise to four peroxyl radicals: 11-*Z*, 9-*E*, 8-*Z*, and 10-*E*. It is important to note that the relative distribution of different hydroperoxide isomers is dependent on the hydrogen-donating ability of these carbons on the FA. It has been found that the formation of 11-*Z*, 9-*E*, 8-*Z*, and 10-*E* hydroperoxides is highly favoured in the presence of good hydrogen atom donors.

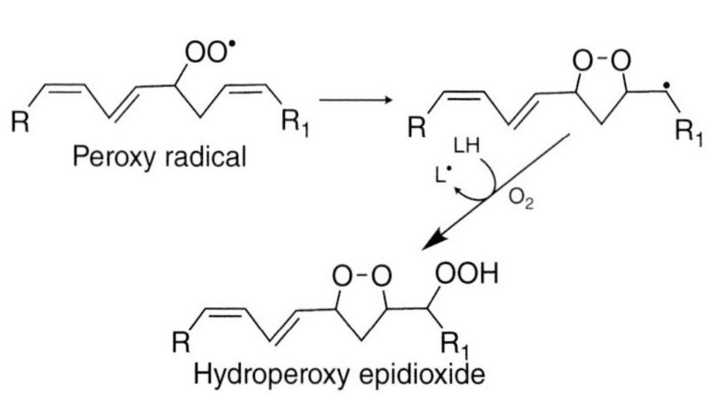

Figure 4.17 Formation of hydroperoxy epidioxide from linolenate peroxy radical. R, CH_3CH_2–; R_1, $-(CH_2)_7$ COOH. *E/Z* isomers are not shown, for the purpose of simplicity. *Source:* Reproduced with permission from Taylor & Francis (Zeb 2015).

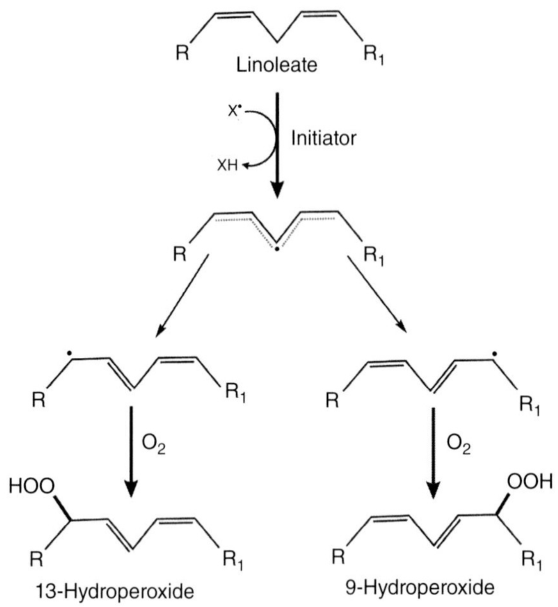

Figure 4.18 Formation of hydroperoxide in the auto-oxidation of linoleic acid. R, CH_3 $(CH_2)_{4}$–; R_1, $-(CH_2)_7$ COOH. *E/Z* isomers are not shown, for the purpose of simplicity. *Source:* Reproduced with permission from Taylor & Francis (Zeb 2015).

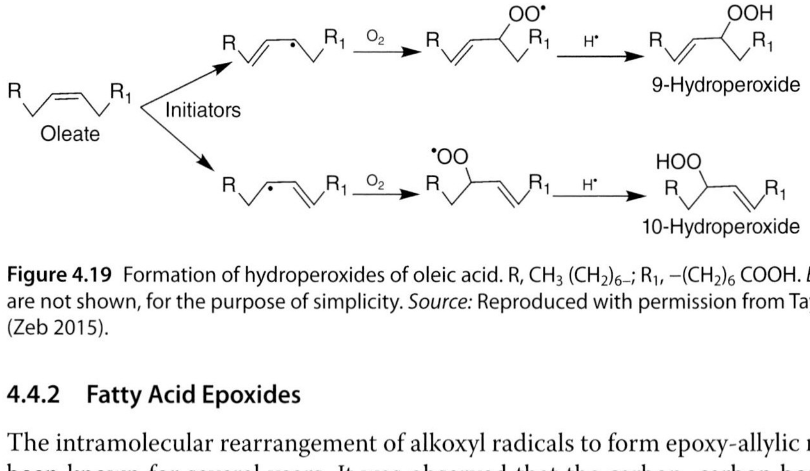

Figure 4.19 Formation of hydroperoxides of oleic acid. R, CH_3 $(CH_2)_{6}$–; R_1, $-(CH_2)_6$ COOH. *E/Z* isomers are not shown, for the purpose of simplicity. *Source:* Reproduced with permission from Taylor & Francis (Zeb 2015).

4.4.2 Fatty Acid Epoxides

The intramolecular rearrangement of alkoxyl radicals to form epoxy-allylic radicals has been known for several years. It was observed that the carbon–carbon bond rotation between the hydroperoxide groups and the double bond provides the least steric hindrance of the *E*-rotamer, so that *E*-isomers are predominant. Epoxides have been identified as products of the auto-oxidation of methyl oleate (Lercker et al. 2003) and methyl linolenate (Neff et al. 1981). The epoxide formation was attributed to the rearrangement of alkoxyl radicals of these FAs (Figure 4.20). The alkoxy radicals from either stable epoxides or epoxy-allylic radicals, which combined with other radicals form epoxy-hydroperoxides. The epoxy-allylic radicals can react with hydroxyl radicals to form epoxy-hydroxides, which are also stable products. The epoxy-allylic radicals can also react with peroxyl radicals to form epoxy-ketones or epoxy-hydroxides. It is worth mentioning

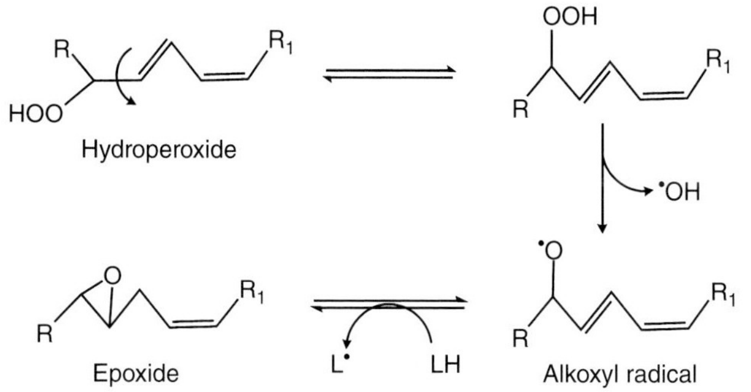

Figure 4.20 Formation of epoxides from hydroperoxides of linoleic acid. R, CH_3CH_2–; R_1, –$(CH_2)_7$ COOH. *E/Z* isomers are not shown, for the purpose of simplicity. *Source:* Reproduced with permission from Taylor & Francis (Zeb 2015).

that the reactivity of epoxy allylic radicals depends on many factors, such as the catalyst, light, heat, and the presence and absence of other free radicals and antioxidants.

4.4.3 Formation of Acrolein

Acrolein is known chemically as 2-propenal. It is an important end product of FA or lipid oxidation. Among the lipids, acrolein has been found to be synthesized from several reactants, including glycerol and FAs. Endo et al. (2013) showed that linolenic acid is the main source of acrolein formation during the frying or heating of vegetable oils. Due to heat stress, linolenic acid free radical is formed by the abstraction of a hydrogen atom at carbon 11. After rearrangement, the radical is shifted to carbon 13, where it reacts with molecular oxygen to form a peroxyl radical. The peroxyl radical reacts with another lipid molecule to form 13-hydroperoxy linolenic acid and another lipid free radical. Alkoxyl radical is formed from 13-hydroperoxide by cleavage of hydroperoxide moiety (Figure 4.21). Cleavage of alkoxyl radical produces a stable 13-oxotrideca-9,11-dienoic acid and an alkyl radical. The rearrangement of the radical and consequent oxidation produces 3-hydroperoxypent-1-ene. The cleavage of the hydroperoxide radical forms an alkoxy radical and a hydroxide radical. Upon cleavage, the alkoxy radical produce acrolein as an end product, as well as an alkyl radical.

In edible oils, a large amount of TAGs are present, which on hydrolysis and oxidation also produce acrolein. Glycerol produces 2-hydroxyacrylaldehyde upon dehydration, and consequently acrolein. Recent isotope labelling studies by Ewert et al. (2014) showed that the FA backbones of TAG in edible oils are the key precursors of acrolein and glycerol.

4.4.4 Oxidation of TAGs

The thermal oxidation of TAGs follows the free radical reaction mechanism. However, steric hindrance, the level of unsaturation, the presence of a pro-oxidant, and the thermal exposure time all play a critical role in this process. Model TAGs like trilinolein and trilinolenin were oxidized to determine the oxidative stability (OS) of TAGs in vegetable oil (Neff et al. 1981). The oxidation products, which are mono-, bis-, and tris-hydroperoxides, were formed by the sequential addition of oxygen to the trilinolein. It was found

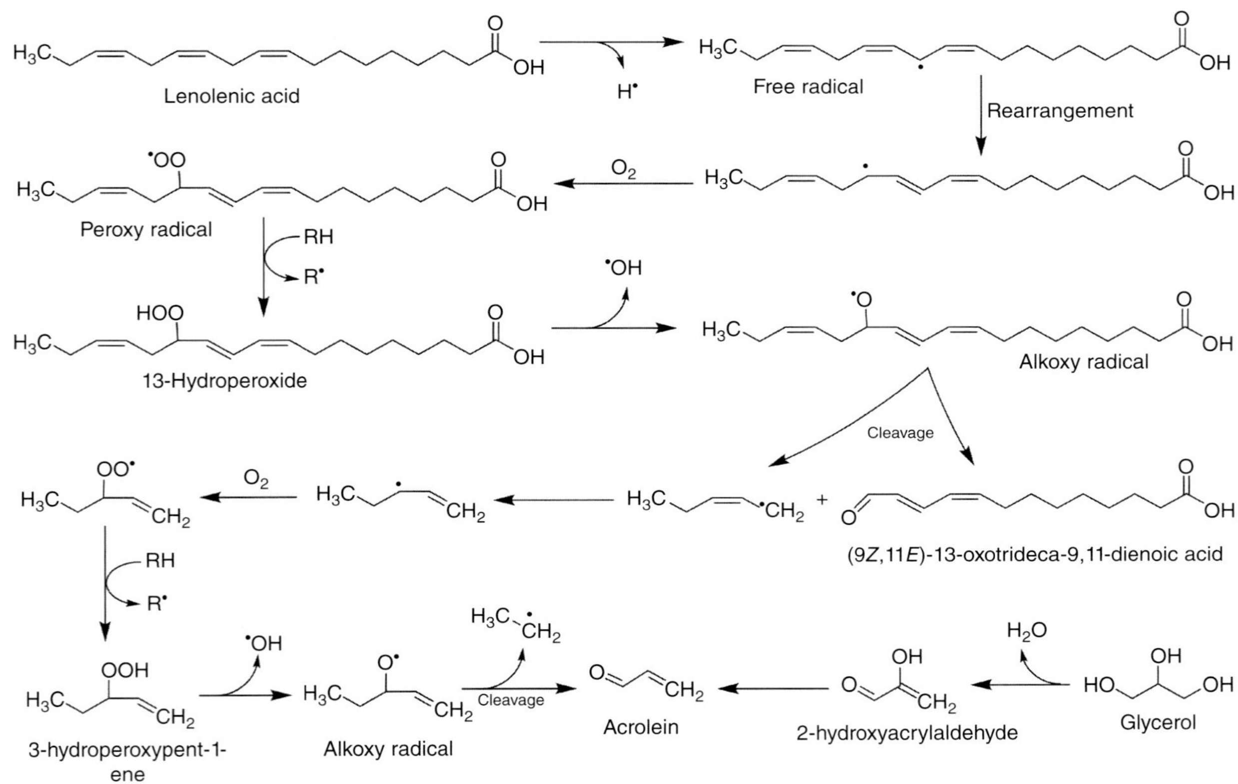

Figure 4.21 Mechanism of formation of acrolein from linolenic acid and glycerol.

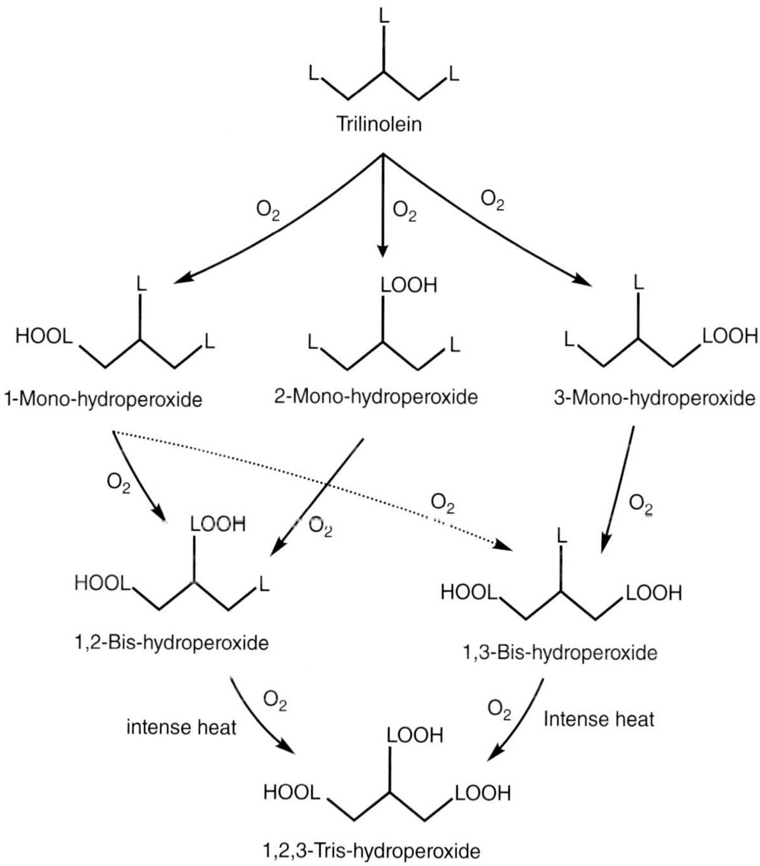

Figure 4.22 Mechanism of trilinolein auto-oxidation and the formation of hydroperoxides. L, linoleic acid moiety; LOOH, hydroperoxide of the linoleic acid moiety. *Source:* Reproduced with permission from Taylor & Francis (Zeb 2015).

that the initial products of oxidation were mono-hydroperoxides, which could further oxidize to form a mixture of 1,2-bis-hydroperoxides and 1,3-bis-hydroperoxides. These bis-hydroperoxides later formed tris-hydroperoxides, as shown in Figure 4.22. All these hydroperoxides were composed of their respective *Z-E* and *E-E* isomers. During the oxidation of trilinolein, no positional preference has been found between each of the *sn* positions (Frankel et al. 1990). Similarly, pure trilinolenin upon oxidation produces end products such as 1, 2, and 3-mono-hydroperoxides, 1,2 and 1,3-bis-hydroperoxides, and tris-hydroperoxides (Zeb 2017).

Other oxidation products include hydroperoxy epidioxides, formed by cyclization, as shown in Figure 4.17. These cyclic hydroperoxides were mixtures of 9 and 16-bis-hydroperoxy-epidioxides. HPLC separation revealed that the ratio of the *E-Z*-16-linolenate hydroperoxides on *sn*-1,3 was higher than that of the corresponding *E-Z*-9-linolenate hydroperoxides on *sn*-2. This suggests that an FA at the *sn*-2 position of the TAG is relatively more resistant to oxidation than one at *sn*-1 or *sn*-3. Oxidation of synthetic TAGs containing linolenic acid and linoleic acid at different positions on the glycerol has been observed to form mono-hydroperoxides and hydroperoxy-epidioxides as the main products (Miyashita et al. 1990).

Thermal oxidation of a pure TAG mixture at 110 °C resulted in the formation of oxidized species of TAGs, such as mono-hydroperoxides, bis-hydroperoxides, epoxy-epidioxides, and epoxides (Zeb and Murkovic 2010). Different TAGs showed different activity toward thermal treatment. The LLL was less stable, OLL and OLO were stable up to 10 and 12 hours, respectively, while POO, OOO, and OSO were stable up to 14 hours. Replacing linoleic acid with oleic acid was found to increase the stability of the corresponding TAG by 2 hours. The mechanism of formation of epoxy-epidioxide from triolein is shown in Figure 4.23. It was reported that the consumption of oxygen resulted in the formation of primary and di-hydroperoxides (Velasco et al. 2004). The attack of two lipid radicals – abstract hydrogen and the hydroxyl group of the di-hydroperoxides – led to the formation of alkoxy di-radical. The final termination reaction then produced epoxy epidioxide, which is a stable product that can be identified and quantified.

When refined olive oil was thermally oxidized under the previously stated conditions, a new class of oxidized compound, triolein hydroxyl epidioxide, was formed (Zeb and Murkovic 2011). Similarly, when corn oil was oxidized under the same conditions, two new classes of oxidized TAG, epidioxy bis-hydroperoxides and hydroxy bis-hydroperoxides, were identified (Zeb and Murkovic 2013). Other oxidation products of TAGs include mono-epoxides, mono-hydroperoxides, and epoxy hydroperoxides. The structure of the oxidized TAGs produced during thermal treatment is shown in Figure 4.24.

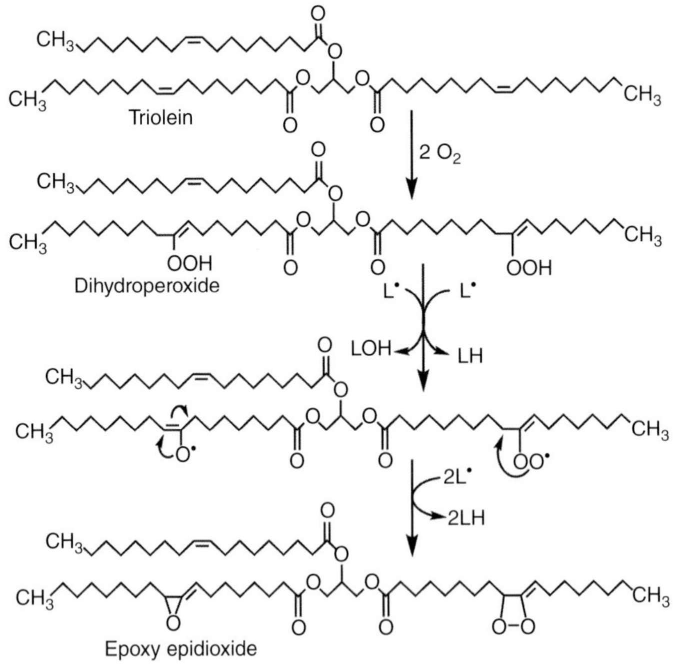

Figure 4.23 Formation of triolein epoxy epidioxides. *Source:* Reproduced with permission from John Wiley & Sons Limited (Zeb and Murkovic 2010).

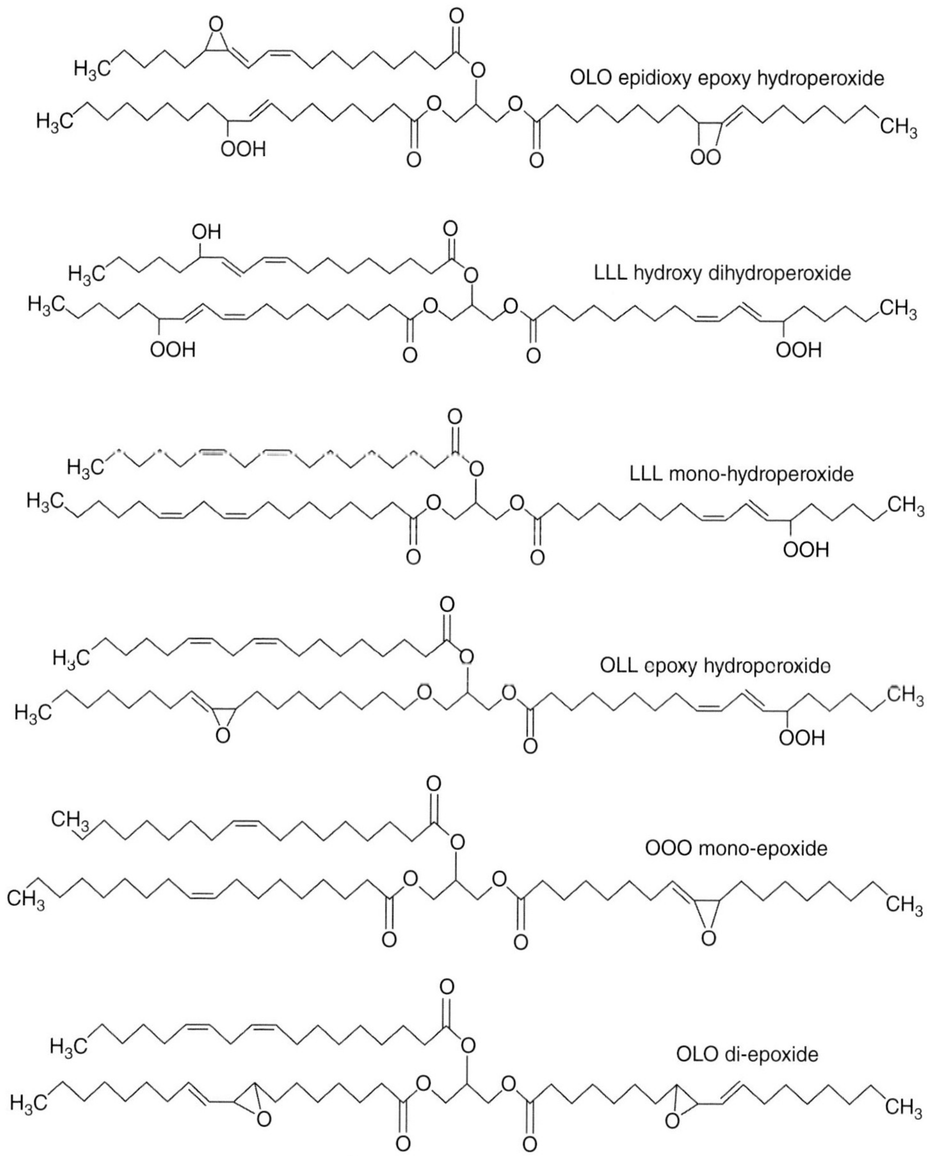

Figure 4.24 Chemical structures of oxidized TAGs.

4.4.5 Polymerization

Oxidized or un-oxidized TAGs join together to form high-molecular-weight dimers, trimers, tetramers, or polymers, depending upon the condition of thermal treatment. There are two types of polymerization: un-oxygenated and oxygenated. Un-oxygenated polymers are mostly formed by joining together saturated TAGs at relatively high temperatures. Oxygenated polymers are usually formed from oxidized TAGs. In earlier times, it was not possible precisely to elucidate these two classes. However, the advent of modern analytical techniques helps us to determine their exact structure. Byrdwell and Neff (2004) heated triolein, which is the major component of olive oil, canola oil, and other dietary oils, at frying temperature, and studied its polymerization with the help of reversed phase high-performance liquid chromatography/electrospray ionization mass spectrometry (HPLC-ESI-MS). Molecular masses corresponding to ammonium adducts of intact carbon-linked dimers having m/z 1783–1787, trimers having m/z 2666–2672, and tetramers having m/z 3547–3557 of triolein, with and without additional sites of unsaturation, were reported. These authors also reported dimers, trimers, and tetramers containing one, two, or three additional oxygens, also with and without additional sites of unsaturation. Based on the formation of certain types of triolein dimers (Figure 4.25), they suggested that tristearin might also form dimers, even though it has no freely oxidizable sites of unsaturation. Several oxidized tristearin monomers, tristearin dimers, chain addition products, and chain-shortened products are reported.

Przybylski et al. (2013) reported that modified canola oils, with low levels of linolenic acid, accumulated significantly lower amounts of polar components compared to other oils. They showed that canola oils generally display lower amounts of oligomers in their polar fraction. Thus, canola oil can be a good choice for long frying times at high temperatures. Recently, Cao et al. (2017) measured the triacylglycerol polymer (TGP) profiles for 12 different types of vegetable oil commonly seen in the Chinese market and 44 samples of used cooking oils, using high-performance size-exclusion chromatography (HPSEC) analysis. The TGP contents of the vegetable oils ranged from 0.10 to 1.67%, while those of the used cooking oils were between 3.18 and 18.75%. The same authors used the TGP contents to differentiate vegetable and waste cooking oils, showing that TGP quantification has promise in the quality evaluation of vegetable oils, especially those used in food frying.

4.5 Formation of Volatile Products

Volatile products are relatively small molecules formed from complex thermal reactions. Generally, these reactions follow two main pathways: hydrolysis and oxidation. The hydrolysis of TAGs occurs in the presence of water and heat stress and releases FFAs and glycerol. The sequence may involve the formation of one FFA and one DAG molecule at the first instance. The DAG molecule is then hydrolysed by the release of a mono-acylglycerol and a second FFA. Finally, the mono-acylglycerol releases a third FFA and a glycerol molecule (Figure 4.26). The FFAs thus formed are usually measured as a marker of oxidation. These FAs are very reactive and possess flavour characteristics. Most FFAs formed by hydrolytic cleavage become volatile before or after further oxidation or breakdown, and are consequently released from the oil as steam during frying. However, under the controlled frying of potatoes with a very high water content, the hydrolytic products were only minor components among the degradation compounds reported (Dobarganes et al. 1993).

Volatile products are also produced from the oxidation and degradation of primary oxidation products, such as hydroperoxides, hydroxides, epoxide, and epoxy

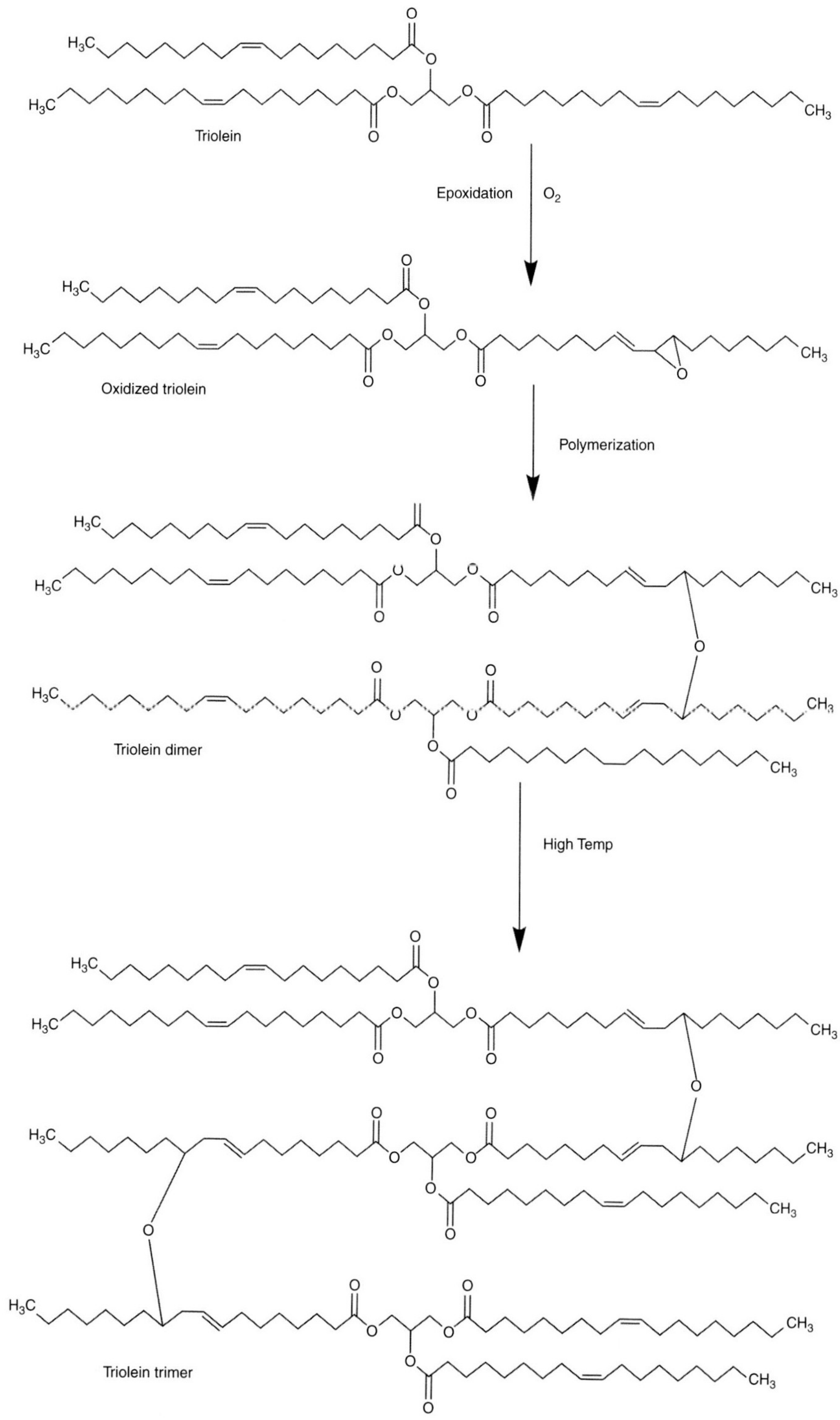

Figure 4.25 Representative sequence of chemical reactions involved in the formation of dimers and trimers of triolein during heating. *Source:* Redrawn with permission of AOCS Press (Byrdwell and Neff 2004).

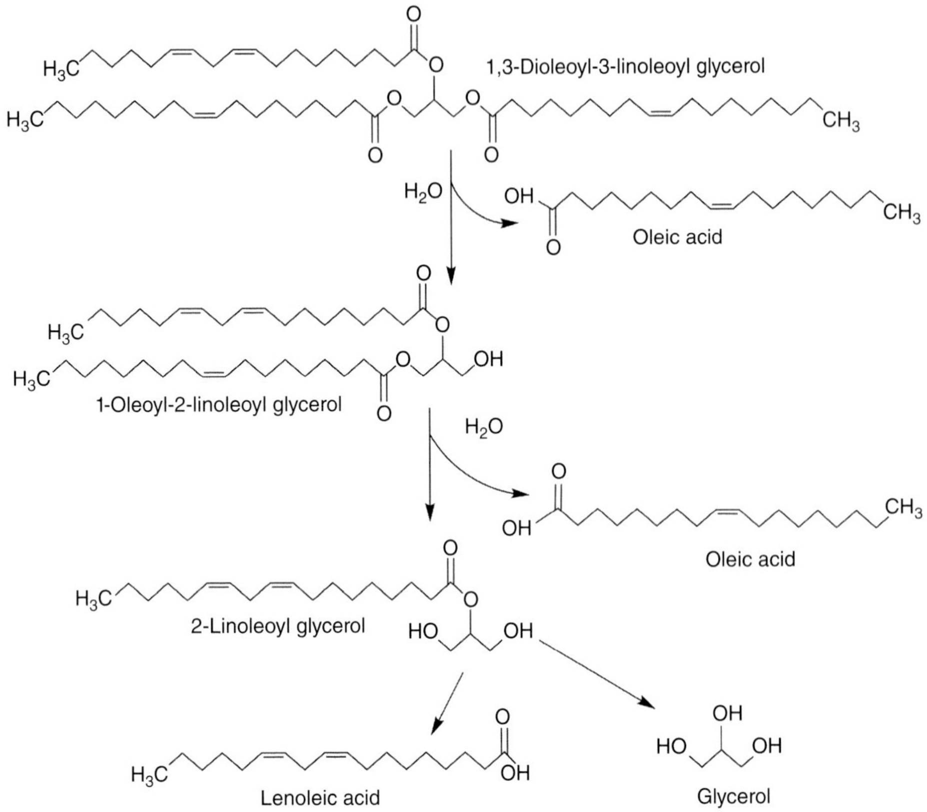

Figure 4.26 Hydrolytic reactions of 1,3-dioleoyl-2-linoleoyl glycerol. The reactions are catalysed by heat or lipase enzyme.

hydroperoxides. A variety of compounds consisting of a homologous series of alkanes, alkenes, alkanals, mono and di-unsaturated alkenals, ketones, and methyl ketones form. Figure 4.27 shows the decomposition of oleic acid hydroperoxides to form an alkoxy radical of oleic acid. Heat stress then further decomposes the radical to form two more alkoxy radicals, such as decyl or octyl. Attack of the oxygen radical by free radical hydrogen then produces an alcohol (dec-2-en-1-ol, or 2-decenol). On rearrangement, the free radical can also produce a carboxylic radical, which upon termination reaction produces an aldehyde (2-decenal). The octyl carboxylic acid radical may produce octanoic acid upon condensation with hydrogen radical, or it may produce 8-hydroxyoctanoic acid. The latter may also be produced by the action of hydroxyl radical. Thus, a wide variety of oxidation compounds with a volatile nature can be produced, depending on the position of the hydroxyl radical on the carbon chain of the FA.

The hydroperoxy epidioxides formed from linoleate (see earlier) are also important precursors of several volatile compounds. They are similar to those formed from the corresponding mono-hydroperoxides. Thus, a single oxidized compound such as 9-hydroperoxy-10,12-epidioxy-13-enoic acid yields hexanal, hept-2-enal, and 9-oxononanoate as major products (Figure 4.28). Minor volatile products include two common

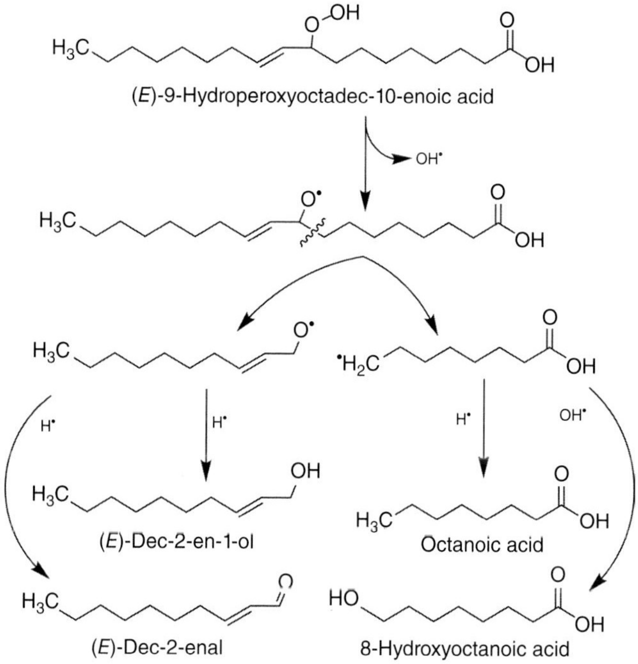

Figure 4.27 Formation of volatile compounds from the decomposition of oleic acid hydroperoxide. Only four products are shown, representing a carboxylic acid, hydroxy carboxylic acid, aldehyde, and alcohol.

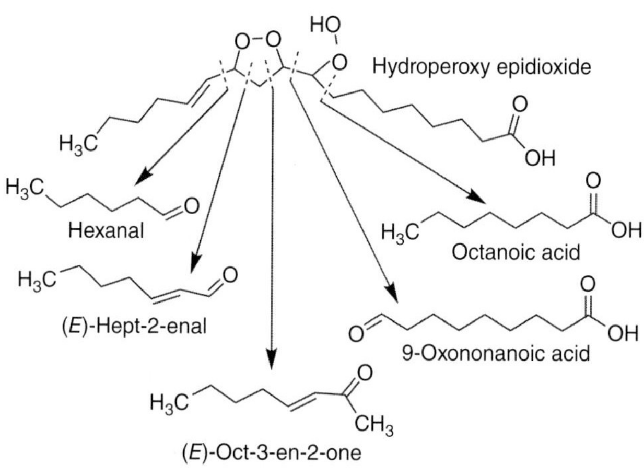

Figure 4.28 Mechanism of formation of volatile compounds from hydroxy epidioxide. The products shown are aldehydes, ketone, and carboxylic acids.

oxidation products of the mono-hydroperoxides – pentane and octanoate – and two distinctive compounds – 2-heptanone and oct-3-en-2-one. Thus, hydroperoxy epidioxides, epoxides, hydroxides, and so on may produce complex products, some of which may or may not be identical.

Neff et al. (2000) thermally oxidized model TAGs (triolein and trilinolein) at 190 °C and identified the odour significance of the volatile compounds produced using purge-and-trap gas chromatography ion trap mass spectrometry (GC-ITMS) with olfactometry (Table 4.3). The major odours of heated triolein at 1, 3, and 6 hours were fruity and plastic. Other negative odours were acrid and grassy. The epoxides were identified by reversed-phase HPLC coupled with atmospheric pressure chemical ionization mass spectrometry (APCI-MS) (Tables 4.4 and 4.5). The results revealed that two mechanisms were responsible for the formation of the epoxides. The first was hydroperoxide cyclization (Byrdwell and Neff 1999; Neff et al. 2000), which involves the loss of a hydroxy radical from a hydroperoxide. Through cyclization with an adjacent carbon, the alkoxy radical then forms an epoxide (see Figure 4.20). If the adjacent carbon is part of a double-bond system, then the double bond is usually shifted one carbon to form an epoxy group adjacent to a double bond. The cyclization mechanism produces triglyceride epoxides, such that the epoxy group is distributed among oleic acid carbons 7 through 12 (Neff et al. 2000). The second mechanism of epoxidation is hydroperoxidation, which is involved in the reaction of the unsaturated FA with an FA hydroperoxide to produce an epoxy group across the original oleic double bond. This reaction produces only one oleic epoxide, with the epoxide group across carbons 9 and 10. Trilinolein has been found to produce similar epoxide isomers. The first mechanism (cyclization)

Table 4.3 Composition of volatile compounds produced by thermal oxidation at 190 °C of triolein (OOO) and trilinolein (LLL) at 1, 3, and 6 hours. The amount of compound is presented as ppm using GC-MS with olfactometry.

	Heating Time (h)					
	1		3		6	
Volatile	**OOO**	**LLL**	**OOO**	**LLL**	**OOO**	**LLL**
Pentanal	15.9	42.5	24.4	28.5	48.9	50.8
Hexanal	17.2	58.0	22.3	50.4	45.4	82.5
2-Heptanol	2.1	—	1.8	—	4.0	—
2-Pentylfuran		28.2	—	33.0	—	40.7
4-Octen-3-one	0.4	1.4	1.3	1.1	0.8	4.4
E-2-Heptenal (heptanol)	(19.2)	98.6	(7.4)	80.0	(27.5)	127.3
E-2-Nonenal	9.0	1.3	14.7	2.0	44.7	5.2
E-2-Hexenal	—	11.3	1.9	6.9	1.9	13.5
Octanal	42.2	—	38.5	—	53.8	—
1-Octanol	103.4	—	28.1	—	168.4	—
Nonanal	485.1	—	348.0	—	403.0	—
E-2-Decenal	418.5	—	152.0	—	197.8	—
E-2-Undecenal	624.3	—	233.5	—	737.2	—

Source: Reproduced with kind permission of AOCS Press (Neff et al. 2000).
Values measure at n = 2. *E*, *cis* isomer of volatile compound.

Table 4.4 Mechanisms of volatile formation in heated triolein epoxides.

Epoxide	Double carbon location	Mechanism[a]	Precursor[b]	Volatile formed[c]
9,8	10	OOH cyclization	8-OOH	2-Decenal
7,8	9	OOH cyclization	8-OOH	2-Undecenal
9,10	11	OOH cyclization	9-OOH	2-Nonenal
8,9	10	OOH cyclization	10-OOH	2-Decenal
9,10	7	OOH cyclization	10-OOH	Nonanal, 2-Decenal
10,11	8	OOH cyclization	10-OOH	Octanal, 2-Nonenal
11,12	9	OOH cyclization	11-OOH	Heptanal, 2-Octenal
10,11	8	OOH cyclization	11-OOH	Octanal, 2-Nonenal
9,10	Saturated	Epoxidation	Oleic Acid	Nonanal, 2-Decenal

Source: Reproduced with kind permission of AOCS Press (Neff et al. 2000).
[a] Reaction to form the epoxide from the mono-hydroperoxide.
[b] Identification of triolein mono-hydroperoxide by hydroperoxy carbon location.
[c] Volatiles produced by thermal decomposition of the triglyceride epoxide via β-scission.

Table 4.5 Mechanisms of volatile oxidation product formation in heated trilinolein epoxides.

Epoxide	Double carbon location	Mechanism[a]	Precursor[b]	Volatile formed[c]
9, 10	11, 13	OOH cyclization	9-OOH	2,4-Decadienal, 2,4-Nonadienal
8, 9	10, 12	OOH cyclization	9-OOH	2,4-Decadienal, 2,4-Undecadienal
10, 12	8, 12	OOH cyclization	10-OOH	2,4-Nonadienal, 2-Octenal
9, 10	8, 11	OOH cyclization	10-OOH	2,4-Decadienal, 2-Nonenal
12, 13	9, 14	OOH cyclization	12-OOH	2,4-Heptadienal, 2-Hexenal
11, 12	9, 13	OOH cyclization	12-OOH	2,4-Octadienal, 2-heptenal
12, 13	8, 10	OOH cyclization	13-OOH	Hexanal, 2-Heptenal
13, 14	9, 11	OOH cyclization	13-OOH	Pentanal, 2-Hexenal
9, 10	12	Epoxidation	Linoleic	2,4-Decadienal, 3-Nonenal
12, 13	9	Epoxidation	Linoleic	Hexanal, 2-Heptenal

Source: Reproduced with kind permission of AOCS Press (Neff et al. 2000).
[a] Reaction to form the epoxide from the mono-hydroperoxide.
[b] Identification of trilinolein mono-hydroperoxide by hydroperoxy carbon location.
[c] Volatiles produced by thermal decomposition of the triglyceride epoxide via β-scission.

produces the epoxy group, which is distributed among linoleic carbons 8 through 14. In the second (hydroperoxidation), the epoxy group formed across the original linoleic double bonds produces two isomers: one with the epoxide group across carbons 9 and 10 and the other across carbons 12 and 13 (Neff et al. 2000).

The previously mentioned studies were based either on FA or on pure TAG. Thus, it was necessary to study a pure TAG mixture, capable of mimicking real edible oils and fats. Therefore, a mixture of TAGs with similar composition to olive oil was thermally oxidized for 1–14 hours under controlled conditions of temperature and air (Zeb and Murkovic 2010). The oxidized TAGs were identified and quantified using HPLC-ESI-MS. Among the oxidized species of TAGs, mono-hydroperoxides, bis-hydroperoxides, epoxy-epidioxide, and epoxides were the major compounds identified. Another study showed that oxidized TAGs species, such as mono- and bis-hydroperoxides, epoxy hydroperoxides, hydroxy hydroperoxides, hydroxy epidioxides, epoxy epidioxides, and mono-epoxides, were the major compounds produced by olive oil (Zeb and Murkovic 2011). In corn oil, similar thermal treatment produced epidioxy bis-hydroperoxides and hydroxy bis-hydroperoxides as new compounds (Zeb and Murkovic 2013); other oxidation products included mono-epoxides, mono-hydroperoxides, and epoxy hydroperoxides. These studies could be elaborated upon with the use of gas chromatography mass spectrometry (GC-MS), which would clarify our understanding of the formation of volatile compounds in a complex oil matrix.

4.5.1 Significance of Volatile Products in Frying

Volatile oxidation products of edible oils and fats have characteristic flavours. Therefore, in selecting a frying medium, it is necessary to determine its volatile profile. Chyau and Mau (2001) fried shallot[2] slices in soybean, corn oils, lard, and medium-chain triglycerides (MCT) at 150–160 °C for 5 minutes and studied the volatiles produced. They found that the flavouring of MCT-fried shallot contained 17 compounds. This was clearly different from the other shallot flavourings, which each revealed 28 compounds. The volatile composition was 70.2, 56.8, 43.6, and 23.3 $mg\,kg^{-1}$ for lard, corn oil, MCT, and soybean oil shallot flavourings, respectively. It was observed that volatiles containing sulfur and nitrogen were the characteristic flavour compounds of deep-fried shallot flavourings. These compounds were unusually high in MCT shallot flavouring. The lard shallot flavouring was rich in oxygen-containing compounds. The MCT shallot flavouring produced abundant lighter-aroma compounds, as it contained lower amounts of oxygen-containing compounds and was thus less rancid. This study suggests that the use of MCT as a frying medium provides a shallot flavouring product with a better acceptability.

The levels of different volatile products produced during frying can be a good indicator of quality. Toxic compounds such as 2,4-decadienal and acrolein are also produced. Extra virgin olive oil shows fewer unsaturated aldehydes, including toxic ones, and fewer degradation indicators, such as polar compounds (Molina-Garcia et al. 2017). Zhang et al. (2018) investigated the non-aldehyde volatile profile produced by the deep fat frying of soybean oil, wheat dough, and chicken breast. They identified alkanes, alkenes, alkynes, alcohols, ketones, nitrogen-containing volatiles (NCVs), and other volatiles. The NCVs were detected only in chicken breast oil samples. The amounts of 2-pentylfuran and 2-pentylpyridine increased with frying, while 4-methyl-1,4-heptadiene and 7-methyl-3,4-octadiene first increased and then decreased with increasing frying

---

2 Shallot is a variety of onion, also called the small onion. It is commonly used in the preparation of a variety of Asian foods. The characteristics flavour produced by frying shallot is used as a flavoring agent. It is often served with porridge.

time. Reduced quantities and concentrations of volatiles were observed in the different samples, perhaps due to the intensified reactions induced by the food components. The detection of some harmful volatiles in considerable amounts suggests attention should be paid to the safety of deep fat frying.

4.6 Sterol Oxidation

Cholesterol is present in animal fats and products of animal origin, but absent or very rare in plant oils. Frying in animal fats thus oxidizes cholesterol, forming a range of products known as cholesterol oxidation products. Rodriguez-Estrada et al. (2014) provide an excellent review of the cholesterol oxidation compounds in foods. They observe that 7-keto-cholesterol is a marker for cholesterol oxidation. Cholesterol has been found to degrade faster when heated alone rather than in the presence of TAGs (Ansorena et al. 2013). Likewise, the more unsaturated the matrix, the slower the degradation of cholesterol. As a general rule, the first products of cholesterol oxidation are hydroperoxides, which mainly occupy positions C-5 to C-7. Dismutation generates 7-α-hydroxycholesterol (7α-HC) and 7-β-hydroxycholesterol (7β-HC), along with 7-ketocholesterol (7-KC). The epoxy derivatives, such as 5-α,6α-epoxycholesterol (α-EC) and 5-β,6-β-epoxycholesterol (β-EC), are formed through a bimolecular reaction mechanism that involves the interaction of a hydroperoxy radical and cholesterol (Rodriguez-Estrada et al. 2014). Figure 4.29 shows the formation of 7-hydroperoxy cholesterol, 7-hydroxy cholesterol, and 7-keto cholesterol.

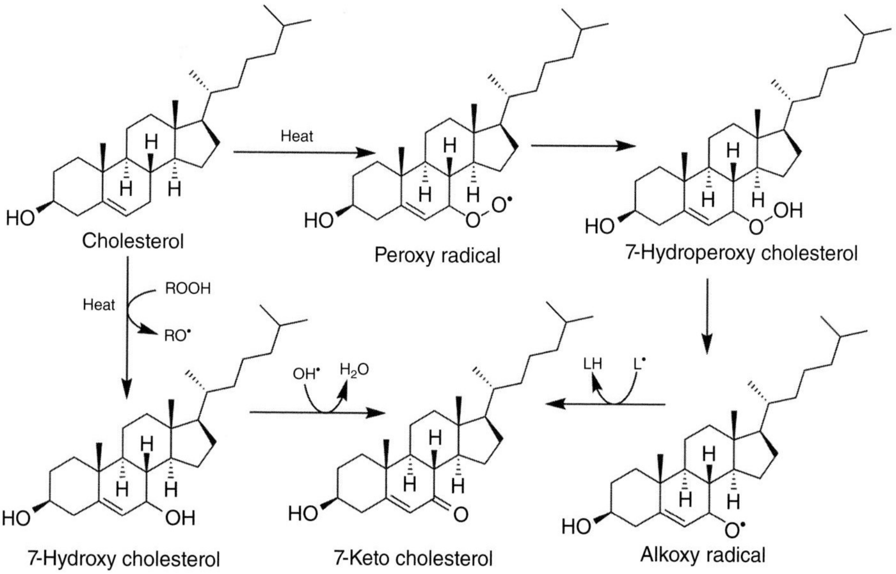

Figure 4.29 Formation of cholesterol oxidation products during frying. The cholesterol produces a peroxy radical, which ultimately produces 7-hydroperoxy cholesterol, a precursor of 7-keto cholesterol.

Sterols present in plant oils are also oxidized to form different oxidation products during frying. These products are called phytosterol oxidation products (POPs). Ramadan (2015) studied the oxidation of β-sitosterol (24α-ethyl-5-cholesten-3β-ol) and campesterol (24α-methyl-5-cholesten-3β-ol) in commercial sunflower oil (SFO) during deep and pan frying of French fries for 30, 60, 120, and 240 minutes. The total plant sterol content in SFO was 4732 $\mu g\,g^{-1}$, with the major one present being β-sitosterol and campesterol (Table 4.6). The total plant sterol content decreased with frying, with the highest decrease occurring after 240 minutes. The authors also evaluated the OS of different sitosterols and campesterols under both frying methods.

In general, pan frying resulted in greater plant sterol oxidation than deep frying. The β-sitosterol oxides predominated, while campesterol oxides were formed to a lesser extent. 7-Ketositosterol, followed by 7β-hydroxysitosterol, 5,6-epoxy derivatives, and 7α-hydroxysitosterol, was the main POP induced during frying. The proportion of 7-keto derivatives decreased during frying, while the proportion of 7β-hydroxy derivatives increased (Table 4.6). The formation of POP may be a limiting factor in frying in SFO for long periods.

Guardiola et al. (2004) reviewed the plant sterol oxidation products found in foods. Recently, Lin et al. (2016) studied the POP contents of 19 different foods prepared by typical household baking and cooking methods using margarine as a frying medium. The authors showed that median POP contents per portion size of cooked foods were 0.57 mg (range 0.05–1.11 mg) with control margarine versus 1.42 mg (range 0.08–20.5 mg) with plant sterol margarine. The oxidation rate of plant sterol (ORP) was 0.50% (median) with the plant sterol margarine and 3.66% with the control margarine. Shallow-fried potatoes had the highest POP content (20.5 mg per portion). Median POP contents in cookies, muffins, banana bread, and sponge cake baked with control and plant sterol margarine were 0.12 mg (range 0.11–0.21 mg) and 0.24 mg (range 0.19–0.60 mg) per portion, respectively, with a corresponding ORP of 1.38 and 0.06%. The authors showed that wide variation occurred in the distribution of individual POPs among different foods, with 7-keto phytosterol and 5,6-epoxy phytosterol being the major oxidation products.

4.7 Tocopherol Oxidation

Tocopherols and tocotrienols are strong natural antioxidants under oxidative stress (Kamal-Eldin and Appelqvist 1996). However, under the action of extreme heat stress (i.e. frying), they are oxidized to form tocopherol oxidation products (TOPs). The tocopherols are also oxidized through free radical reaction, just like any other lipid. The product is either epoxy, hydroperoxy, hydroxy, or epoxy hydroperoxides. Figure 4.30 shows two oxidation products of tocopherol (i.e. epoxy hydroperoxides).

Verleyen et al. (2001) assessed the effect of high-temperature treatment on the stability of α-tocopherol in triolein under a reduced atmospheric pressure (4–40 mbar) simulating the deodorization step of the refining of vegetable oils. They reported marked degradation of tocopherol, which showed increases in rate with increasing temperature (180–260 °C) and heating time (20–80 minutes). The degradation of tocopherol in triolein at 240 °C was inhibited by the addition of a synthetic antioxidant such as tertiary butyl hydroquinone (TBHQ) or by heating under a nitrogen atmosphere, indicating oxidative degradation (Tables 4.7 and 4.8). The oxidation products were identified as α-tocopherol quinone, 4,5-epoxy-α-tocopherol quinone, and 7,8-epoxy-α-tocopherol quinone.

Table 4.6 Phytosterol/stanol profile (μg g^{-1}) in commercial sunflower oil (SFO) during deep and pan frying.

	Deep-fried SFO				Pan-fried SFO			
Time (min)	Total plant sterols	β-Sitosterol	Campesterol	Total sitosterol and campesterol	Total POP	β-Sitosterol	Campesterol	Total sitosterol and campesterol
0	4732	2209 ± 0.3	296 ± 0.2	2505	4732	2209 ± 0.3	296 ± 0.1	2505
30	4268	1990 ± 0.2	230 ± 0.2	2220	4259	1960 ± 0.4	224 ± 0.3	2184
60	4096	1740 ± 0.2	215 ± 0.1	1955	3952	1730 ± 0.2	211 ± 0.2	1941
120	3878	1700 ± 0.2	205 ± 0.1	1905	3492	1380 ± 0.3	200 ± 0.2	1580
240	3547	1650 ± 0.1	200 ± 0.2	1850	3096	1280 ± 0.3	150 ± 0.1	1430

Source: Reproduced with kind permission of Springer UK (Ramadan 2015).
Data presented are mean ± standard deviation, measured using GC-MS.

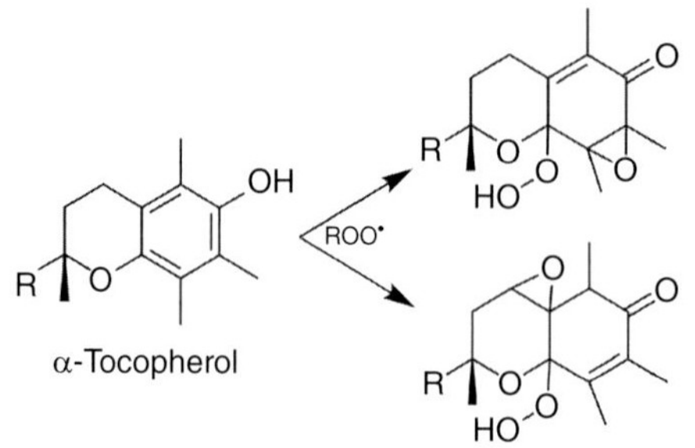

Figure 4.30 Formation of epoxy hydroperoxides of α-tocopherol in the presence of lipid free radical during frying.

Table 4.7 Distribution of major β-sitosterol and campesterol oxidation products ($\mu g\,g^{-1}$) in SFO during deep and pan frying.

	Deep fried Sunflower oil			Pan fried Sunflower oil		
Time (min)	**β-Sitosterol oxides**	**Campesterol oxides**	**Total oxides**	**β-Sitosterol oxides**	**Campesterol oxides**	**Total oxides**
0	23.8 ± 0.3	5.5 ± 0.3	29.3	23.8 ± 0.15	5.5 ± 0.09	29.5
30	34.7 ± 0.3	6.1 ± 0.3	40.8	38.0 ± 0.17	6.9 ± 0.10	44.9
60	39.5 ± 0.3	6.7 ± 0.3	46.2	46.8 ± 0.21	18.9 ± 0.21	65.7
120	45 ± 0.3	7.3 ± 0.3	52.3	79.6 ± 0.23	21.5 ± 0.27	101.1
240	63.5 ± 0.3	9.0 ± 0.3	72.5	137.5 ± 0.24	27.2 ± 0.25	164.7

Source: Reproduced with kind permission Springer UK (Ramadan 2015).
Data presented here is mean with standard deviation measured using GC–MS.

Table 4.8 Effects of heating on tocopherol degradation.

Temperature (°C)	Time (min)	Pressure (mbar)	Tocopherol (ppm)	Tocopherol Loss (%)
180	60	40	2070	5.7
200	20	40	2157	1.7
200	40	40	2118	3.5
200	60	40	2086	5.0
240	60	40	1996	9.1
240	60	45	2017	8.1
260	80	40	1895	13.7
260	60	45	1957	10.8

Source: Reproduced with kind permission of American Chemical Society (Verleyen et al. 2001).
The initial concentration of the tocopherol in all samples was 2195 ppm.

In a real matrix of food frying, it has been verified that, except in the lowest proportion, α-tocopherol not only acts as a pro-oxidant in soybean oil (Martin-Rubio et al. 2018) but also alters the lipid oxidation pathway by changing the generation rate, nature, relative proportions, and concentrations of various oxidation products. Its pro-oxidant role has largely been attributed to its high reactivity as a free radical scavenger (Elisia et al. 2013). Thus, it is important to consider the amount and types of tocopherol present in edible oils when selecting a frying medium: that with the high amount of tocopherols will provide the best thermal stability.

4.8 Formation of *Trans* Fatty Acids

In nature, all double bonds present in FAs are in *cis* or *E*- form. In free form or as an ester of glycerol, they are converted to *trans* or *Z*- form under heat stress. During frying, the heat is responsible for rotation at the double position. Due to the significant health effects of *trans* fatty acids (TFAs), this is a hot issue for food scientists (Bhardwaj et al. 2011; Akmar et al. 2013). The World Health Organization (WHO) has taken a special interest in updating UN member states regarding the issue (Nishida and Uauy 2009; Uauy et al. 2009). Figure 4.31 shows the representative chemical structures of some important TFAs.

Tsuzuki et al. (2010) evaluated the formations of TFAs during frying and heating in a model system. Sliced raw potatoes (10% of the frying oil (w/w)) were fried in commercially available canola oil at 160, 180, and 200 °C for 10 frying cycles. By the 10th frying operation, the TFA content was 0.99–1.05 g 100 g lipids^{-1}. Thus, when 100 g of potatoes fried in this process are consumed, the TFA intake is estimated at less than 0.1 g. After 10 frying operations, there was an increase in the *trans* 18:2 FAs and *trans* 18:3 FA content of the used frying oil as compared with the fresh oil. The authors also elucidated TFA accumulation in various edible oils during frying, showing small changes after 4 hours of heating. The results of this study suggest that an ordinary frying process using unhydrogenated edible oils has little impact on TFA intake. Similarly, during the deep frying of palm oil, TFA formation was shown to be directly proportional to the frying temperature and time (Chen et al. 2014). Thus, TFA content should be considered when selecting a frying medium.

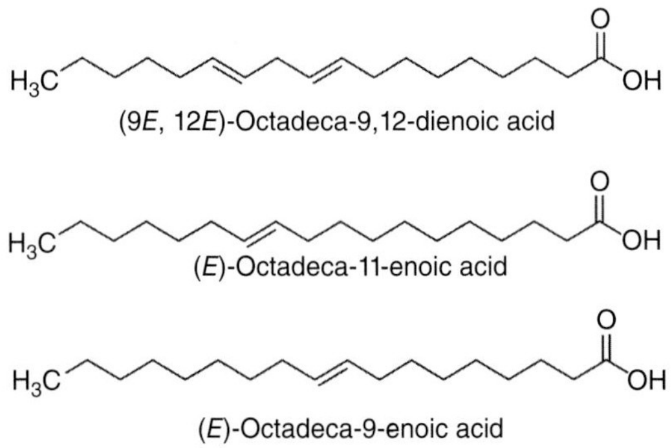

Figure 4.31 Chemical structures of important *trans* fatty acids (TFAs).

4.9 Techniques for Measuring Lipid Oxidation

Knowing the rate of lipid oxidation and the extent to which a frying medium can be reused is one of the key factors in maintaining the quality and safety of fried foods. Research laboratories usually employ chemical or analytical techniques to measure the lipid oxidation of a frying medium, but the equipment is expensive and not suitable for use in a restaurant or household kitchen. Therefore, food researchers have developed some simple techniques that can be used to determine for how long a frying medium can be used. These techniques are called rapid test methods (RTMs), some of which are given below:

4.9.1 Portable Oil Tester

An oil tester is a portable testing machine that measures the total polar materials (TPM) in a frying medium. Oil testers are manufactured by several companies, but their principal working mechanism is more or less the same. The tester tip is inserted into the frying oil for 30 seconds to 1 minute, and the TPM is displayed on a digital screen. These testers are robust, simple, affordable, and convenient for use in kitchens and restaurants. Their accuracy is usually in the range of ±1–2%. They ensure food safety and allow the cost-effective production of fried foods.

4.9.2 Viscometer

During frying, the amount of oxidized matter or polymerized material increases significantly, increasing the viscosity of the oil. With the help of a manual or digital viscometer, the level of polymerized oxidized materials (POMs) can easily be measured. The maximum acceptable level is 15% (Stier 2001). Small-scale viscometers are produced by a number of manufacturers, such as Leatherhead Food Research (United Kingdom) and PCE Holding GmbH (Germany). The viscosity measurements by viscometers are highly correlated with the polar contents of the degraded oils.

4.9.3 Dielectric Constant

Specific sensors are used to monitor changes in the dielectric constant of degrading frying fats. A sensitive electric circuit known as a food oil sensor (FOS) detects a very small dielectric change in the oil. This was one of the first oil-quality measurers used in food inspection.

4.9.4 Colour Testing

The colour of the frying oil can be evaluated by eye or using small-scale equipment such as the Oxifrit test (Merck, Germany). This test is based on chemical changes in the oil. It can be used in restaurant kitchens, but requires the handling of chemicals, with all its attendant safety issues. Other chemical tests for measuring changes in colour or polar materials are outside the scope of this chapter.

4.9.5 FFA Testing

FFAs can be used as a measure of the quality of oils and fats. For this purpose, 3M Corporation (United States) developed the shortening monitor. This monitor consists of a white paper strip measuring 1–10 cm with four blue bands across it, which can be dipped in oil to measure accumulated FFAs. The shortening monitor provides its users, especially those in the fast-food industry, with an inexpensive means of objectively measuring the quality of a cooking oil (Stier 2001)

4.9.6 Flavour and Sensory Testing

Flavour is one of the characteristic parameters that help us to monitor lipid oxidation in a frying medium. In 2000, the European Federation for the Science and Technology of Lipids (Euro Fed Lipid) concluded a deep-frying symposium by stating that 'The main index for deep fat frying should be sensory parameters of the food being fried' (Stier 2000). Thus, the best test for measuring oil oxidation is human sensory testing. Deteriorations in smell can be sensed by the human nose or by chemical equipment.

4.10 Key Concepts

- Foods are fried in a variety of frying media, which are selected on the basis of their appearance, colour, odour, flavour, texture, taste, nutritional value, and economically affordability.
- Frying fats and oils are composed largely of FAs and TAGs, with phenolic compounds, hydrocarbons, pigments, sterols, alcohols, tocopherols, tocotrienols, phospholipids, and volatile compounds forming their minor constituents.
- Edible oils and fats are oxidized by thermal stress to form FA hydroperoxides, epoxides, hydroxides, and epidioxides as their primary oxidation compounds.
- Primary oxidized compounds are degraded further to form volatile oxidation products, with significant effects on flavouring.
- Minor compounds are also oxidized to form primary and secondary oxidation products, which contribute to the increase in oxidation of the frying medium during frying.
- Several physical techniques can be used to determine the effect on the quality of the frying medium of thermal oxidation.

References

Akmar, Z.D., Norhaizan, M.E., Azimah, R. et al. (2013). The trans fatty acids content of selected foods in Malaysia. *Malaysian Journal of Nutrition* 19: 87–98.

Ansorena, D., Barriuso, B., Cardenia, V. et al. (2013). Thermo-oxidation of cholesterol: effect of the unsaturation degree of the lipid matrix. *Food Chemistry* 141: 2757–2764.

Barden, L. and Decker, E.A. (2016). Lipid oxidation in low-moisture food: a review. *Critical Reviews in Food Science and Nutrition* 56: 2467–2482.

Bernbäck, S., Bläckberg, L., and Hernell, O. (1990). The complete digestion of human milk triacylglycerol in vitro requires gastric lipase, pancreatic colipase-dependent lipase, and bile salt-stimulated lipase. *Journal of Clinical Investigation* 85: 1221.

Bhardwaj, S., Passi, S.J., and Misra, A. (2011). Overview of trans fatty acids: biochemistry and health effects. *Diabetes and Metabolic Syndrome: Clinical Research and Reviews* 5: 161–164.

Boskou, D. (2015). Olive oil: properties and processing for use in food. In: *Specialty Oils and Fats in Food and Nutrition: Properties, Processing and Applications* (ed. G. Talbot). Cambridge: Woodhead Publishing.

Brash, A.R. (2000). Autoxidation of methyl linoleate: identification of the bis-allylic 11-hydroperoxide. *Lipids* 35: 947–952.

Brindley, D.N. (1991). Metabolism of triacylglycerols. *New Comprehensive Biochemistry* 20: 171–203.

Byrdwell, W.C. and Neff, W.E. (1999). Non-volatile products of triolein produced at frying temperatures characterized using liquid chromatography with online mass spectrometric detection. *Journal of Chromatography A* 852: 417–432.

Byrdwell, W.C. and Neff, W.E. (2004). Electrospray ionization MS of high M.W. TAG oligomers. *Journal of the American Oil Chemists' Society* 81: 13–26.

Cao, H., Xue, B., Jiang, Y. et al. (2017). Application of triacylglycerol polymer determination in the quality evaluation of vegetable oil. *LWT – Food Science and Technology* 82: 243–247.

Carrasco-Pancorbo, A., Cerretani, L., Bendini, A. et al. (2007). Evaluation of the influence of thermal oxidation on the phenolic composition and on the antioxidant activity of extra-virgin olive oils. *Journal of Agricultural and Food Chemistry* 55: 4771–4780.

Chan, H.W.-S. and Levett, G. (1977). Autoxidation of methyl linolenate: analysis of methyl hydroxylinolenate isomers by high performance liquid chromatography. *Lipids* 12: 837–840.

Chen, Y., Yang, Y., Nie, S. et al. (2014). The analysis of trans fatty acid profiles in deep frying palm oil and chicken fillets with an improved gas chromatography method. *Food Control* 44: 191–197.

Choe, E. and Min, D.B. (2006). Mechanisms and factors for edible oil oxidation. *Comprehensive Reviews in Food Science and Food Safety* 5: 169–186.

Chyau, C.-C. and Mau, J.-L. (2001). Effects of various oils on volatile compounds of deep-fried shallot flavouring. *Food Chemistry* 74: 41–46.

Cunha, S.C., Amaral, J.S., Fernandes, J.O., and Oliveira, M.B.P.P. (2006). Quantification of tocopherols and tocotrienols in Portuguese olive oils using HPLC with three different detection systems. *Journal of Agricultural and Food Chemistry* 54: 3351–3356.

Dobarganes, M.C., Marquez-Ruiz, G., and Perez-Camino, M.C. (1993). Thermal stability and frying performance of genetically modified sunflower seed (*Helianthus annuus* L.) oils. *Journal of Agricultural and Food Chemistry* 41: 678–681.

Dubois, V., Breton, S., Linder, M. et al. (2007). Fatty acid profiles of 80 vegetable oils with regard to their nutritional potential. *European Journal of Lipid Science and Technology* 109: 710–732.

Elisia, I., Young, J.W., Yuan, Y.V., and Kitts, D.D. (2013). Association between tocopherol isoform composition and lipid oxidation in selected multiple edible oils. *Food Research International* 52: 508–514.

Endo, Y., Hayashi, C., Yamanaka, T. et al. (2013). Linolenic acid as the main source of acrolein formed during heating of vegetable oils. *Journal of the American Oil Chemists' Society* 90: 959–964.

Erickson, D.R. (1996). Production and composition of frying fats. In: *Deep-Frying Chemistry, Nutrition and Practical Applications* (ed. N. Weber). Chichester: John Wiley & Sons.

Ewert, A., Granvogl, M., and Schieberle, P. (2014). Isotope-labeling studies on the formation pathway of acrolein during heat processing of oils. *Journal of Agricultural and Food Chemistry* 62: 8524–8529.

Frankel, E.N. (1991). Recent advances in lipid oxidation. *Journal of the Science of Food and Agriculture* 54: 495–511.

Frankel, E.N., Neff, W.E., and Weisleder, D. (1990). Determination of methyl linoleate hydroperoxides by 13C nuclear magnetic resonance spectroscopy. *Methods in Enzymology* 186: 380–387.

Gardner, H.W. (1989). Oxygen radical chemistry of polyunsaturated fatty acids. *Free Radical Biology and Medicine* 7: 65–86.

Guardiola, F., Bou, R., Boatella, J., and Codony, R. (2004). Analysis of sterol oxidation products in foods. *Journal of AOAC International* 87: 441–466.

Gutiérrez-Rosales, F., Rios, J., and Gomez-Rey, M.L. (2003). Main polyphenols in the bitter taste of virgin olive oil. Structural confirmation by on-line high-performance liquid chromatography electrospray ionization mass spectrometry. *Journal of Agricultural and Food Chemistry* 51: 6021–6025.

Kalapathy, U. and Proctor, A. (2000). A new method for free fatty acid reduction in frying oil using silicate films produced from rice hull ash. *Journal of the American Oil Chemists' Society* 77: 593–598.

Kamal-Eldin, A. and Appelqvist, L. Å. (1996). The chemistry and antioxidant properties of tocopherols and tocotrienols. *Lipids* 31: 671–701.

Kamal-Eldin, A., Mäkinen, M., and Lampi, A. (2003). The challenging contribution of hydroperoxides to the lipid oxidation mechanism. In: *Lipid Oxidation Pathways* (ed. A. Kamal-Eldin). Champaign, IL: AOCS Press.

Kita, A., Lisińska, G., and Golubowska, G. (2007). The effects of oils and frying temperatures on the texture and fat content of potato crisps. *Food Chemistry* 102: 1–5.

Lercker, G., Rodriguez-Estrada, M.T., and Bonoli, M. (2003). Analysis of the oxidation products of cis- and trans-octadecenoate methyl esters by capillary gas chromatography–ion-trap mass spectrometry: I. Epoxide and dimeric compounds. *Journal of Chromatography A* 985: 333–342.

Lin, Y., Knol, D., Menéndez-Carreño, M. et al. (2016). Formation of plant sterol oxidation products in foods during baking and cooking using margarine without and with added plant sterol esters. *Journal of Agricultural and Food Chemistry* 64: 653–662.

Martin-Rubio, A.S., Sopelana, P., Ibargoitia, M.L., and Guillén, M.D. (2018). Prooxidant effect of α-tocopherol on soybean oil. Global monitoring of its oxidation process under accelerated storage conditions by 1H nuclear magnetic resonance. *Food Chemistry* 245: 312–323.

Miyashita, K., Frankel, E., Neff, W., and Awl, R. (1990). Autoxidation of polyunsaturated triacylglycerols. III. Synthetic triacylglycerols containing linoleate and linolenate. *Lipids* 25: 48–53.

Molina-Garcia, L., Santos, C.S.P., Cunha, S.C. et al. (2017). Comparative fingerprint changes of toxic volatiles in low PUFA vegetable oils under deep-frying. *Journal of the American Oil Chemists' Society* 94: 271–284.

Neff, W., Frankel, E., and Weisleder, D. (1981). High pressure liquid chromatography of autoxidized lipids: II. Hydroperoxy-cyclic peroxides and other secondary products from methyl linolenate. *Lipids* 16: 439–448.

Neff, W., Warner, K., and Byrdwell, W. (2000). Odor significance of undesirable degradation compounds in heated triolein and trilinolein. *Journal of the American Oil Chemists' Society* 77: 1303–1314.

Nishida, C. and Uauy, R. (2009). WHO scientific update on health consequences of trans fatty acids: introduction. *European Journal of Clinical Nutrition* 63 (Suppl. 2): S1–S4.

Nyam, K.L., Tan, C.P., Lai, O.M. et al. (2009). Physicochemical properties and bioactive compounds of selected seed oils. *LWT – Food Science and Technology* 42: 1396–1403.

Peers, K.E., Coxon, D.T., and Chan, H.W.S. (1981). Autoxidation of methyl linolenate and methyl linoleate: the effect of α-tocopherol. *Journal of the Science of Food and Agriculture* 32: 898–904.

Przybylski, R., Gruczynska, E., and Aladedunye, F. (2013). Performance of regular and modified canola and soybean oils in rotational frying. *Journal of the American Oil Chemists' Society* 90: 1271–1280.

Psomiadou, E., Tsimidou, M., and Boskou, D. (2000). α-Tocopherol content of Greek virgin olive oils. *Journal of Agricultural and Food Chemistry* 48: 1770–1775.

Ramadan, M.F. (2015). Oxidation of β-sitosterol and campesterol in sunflower oil upon deep- and pan-frying of French fries. *Journal of Food Science and Technology* 52: 6301–6311.

Rodriguez-Estrada, M.T., Garcia-Llatas, G., and Lagarda, M.J. (2014). 7-Ketocholesterol as marker of cholesterol oxidation in model and food systems: when and how. *Biochemical and Biophysical Research Communications* 446: 792–797.

Schaich, K. (2013). Challenges in elucidating lipid oxidation mechanisms: when, where, and how do products arise? In: *Lipid Oxidation Challenges in Food Systems* (ed. A. Logan, U. Nienaber and X. Pan). Champaign, IL: AOCS Press.

Schwartz, H., Ollilainen, V., Piironen, V., and Lampi, A.-M. (2008). Tocopherol, tocotrienol and plant sterol contents of vegetable oils and industrial fats. *Journal of Food Composition and Analysis* 21: 152–161.

Stier, R.F. (2000). Chemistry of frying and optimization of deep-fat fried food flavour – an introductory review. *European Journal of Lipid Science and Technology* 102: 507–514.

Stier, R. (2001). The measurement of frying oil quality and authenticity. In: *Frying Improving Quality* (ed. J.B. Rossell). Cambridge: Woodhead.

Tsuzuki, W., Matsuoka, A., and Ushida, K. (2010). Formation of trans fatty acids in edible oils during the frying and heating process. *Food Chemistry* 123: 976–982.

Uauy, R., Aro, A., Clarke, R. et al. (2009). WHO scientific update on trans fatty acids: summary and conclusions. *European Journal of Clinical Nutrition* 63: S68.

Velasco, J., Marmesat, S., Márquez-Ruiz, G., and Dobarganes, M.C. (2004). Formation of short-chain glycerol-bound oxidation products and oxidised monomeric triacylglycerols during deep-frying and occurrence in used frying fats. *European Journal of Lipid Science and Technology* 106: 728–735.

Verleyen, T., Verhe, R., Huyghebaert, A. et al. (2001). Identification of α-Tocopherol oxidation products in triolein at elevated temperatures. *Journal of Agricultural and Food Chemistry* 49: 1508–1511.

Wagner, B.A., Buettner, G.R., and Burns, C.P. (1994). Free radical-mediated lipid peroxidation in cells: oxidizability is a function of cell lipid bis-allylic hydrogen content. *Biochemistry* 33: 4449–4453.

Warner, K., Orr, P., and Glynn, M. (1997). Effect of fatty acid composition of oils on flavor and stability of fried foods. *Journal of the American Oil Chemists' Society* 74: 347–356.

Zeb, A. (2012). Triacylglycerols composition, oxidation and oxidation compounds in camellia oil using liquid chromatography-mass spectrometry. *Chemistry and Physics of Lipids* 165: 608–614.

Zeb, A. (2015). Chemistry and liquid chromatography methods for the analyses of primary oxidation products of triacylglycerols. *Free Radical Research* 49: 549–564.

Zeb, A. (2017). Chemistry of lipid oxidation in edible oils. In: *Lipid Peroxidation, Inhibition, Effects and Mechanism* (ed. A. Catala). Hauppauge, NY: Nova Science Publishers.

Zeb, A. and Ali, M. (2008). Thermal stability of animal tallow used in kebab preparation. *Journal of the Chemical Society of Pakistan* 30: 750–755.

Zeb, A. and Murkovic, M. (2010). Characterization of the effects of β-carotene on the thermal oxidation of triacylglycerols using HPLC-ESI-MS. *European Journal of Lipid Science and Technology* 112: 1218–1228.

Zeb, A. and Murkovic, M. (2011). Carotenoids and triacylglycerols interactions during thermal oxidation of refined olive oil. *Food Chemistry* 127: 1584–1593.

Zeb, A. and Murkovic, M. (2013). Determination of thermal oxidation and oxidation products of β-carotene in corn oil triacylglycerols. *Food Research International* 50: 534–544.

Zeb, A. and Rahman, S.U. (2017). Protective effects of dietary glycine and glutamic acid toward the toxic effects of oxidized mustard oil in rabbits. *Food & Function* 8: 429–436.

Zhang, Q., Wan, C., Wang, C. et al. (2018). Evaluation of the non-aldehyde volatile compounds formed during deep-fat frying process. *Food Chemistry* 243: 151–161.

5

Chemistry of Fried Foods

5.1 Introduction

Foods are made up of several chemical components, which are highly important to food manufacturers. The chemical composition of a food determines all aspects of the final product, including sensory, nutritional, and quality characteristics. Thus, it is important for a food scientist to have a fundamental knowledge of food composition and the various properties it can impart. Several peer-reviewed journals cover the different aspects of food chemistry, but they are all at an advanced level. Similarly, several textbooks are available on the subject, such as *The Chemistry of Food* by Velisek (2014).

It is important to have some essential knowledge regarding food biomolecules. These are classified as either macromolecules or micromolecules. Macromolecules are larger in size and are present in high numbers in biological systems. They include carbohydrates, proteins, and lipids. Micromolecules are small molecules with minute structures and are present in relatively low amounts. Examples are vitamins, minerals, pigments, and phenolic compounds.

5.2 Carbohydrates

Carbohydrates are the most abundant class of organic biomolecule in nature. Chemically, they are polyhydroxy aldehydes or ketones, or their derivatives. Carbohydrates are synthesized by plants during photosynthesis. The final product of photosynthetic reactions is glucose, which forms the principal component of dry matter in plant foods. Carbohydrates are classified into three groups based on their structural components: monosaccharides, oligosaccharides, and polysaccharides.

5.2.1 Monosaccharides

Monosaccharides, or simple sugars, are carbohydrates that cannot hydrolyze further into more simple compounds. They are the basic units of carbohydrates; therefore, they are often referred to by the generic term 'saccharide'. They are classed as either aldose or ketose, based on their functional group: aldehyde or ketone (Figure 5.1). These

Food Frying: Chemistry, Biochemistry, and Safety, First Edition. Alam Zeb.

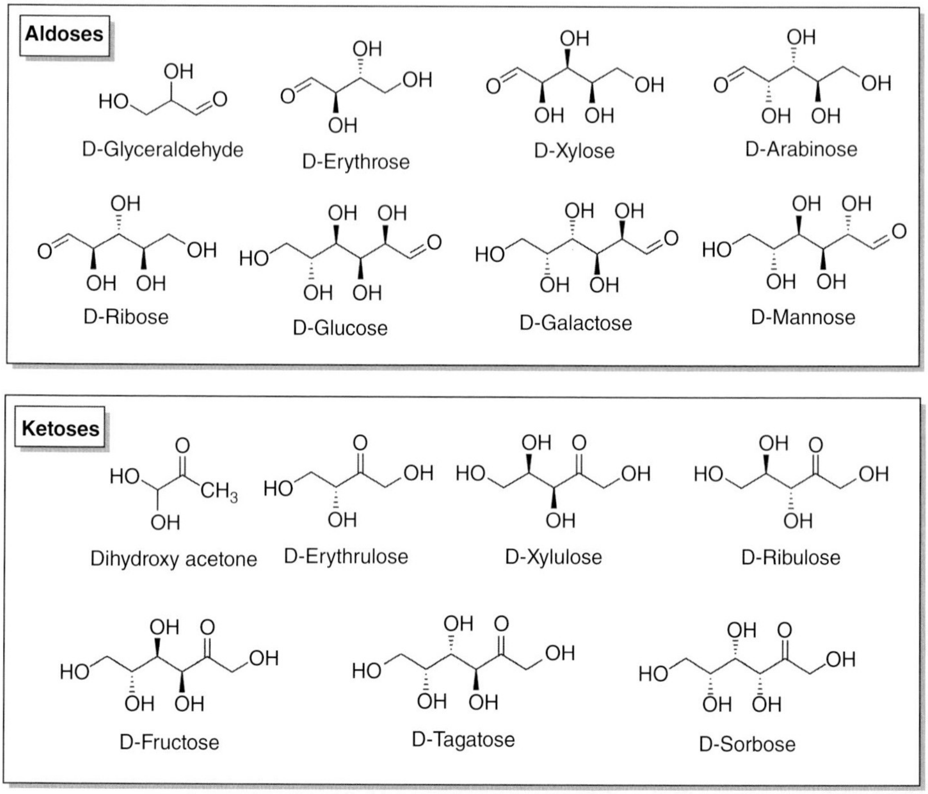

Figure 5.1 Structures of monosaccharide sugars. Aldoses contain aldehyde as a functional group, while ketoses contain ketone.

functional groups are important when considering chemical reactions in foods. Monosaccharides may contain three, four, five, or six carbon atoms. One with three carbons is called a triose, one with four is a tetrose, one with five is a pentose, and one with six is a hexose. Monosaccharides are a highly reactive form of carbohydrate, and their transformation during food preparation and manufacture creates several other compounds.

In nature, monosaccharides occur in two distinct structures: ring and cyclic. The ring structure is formed by the reaction of alcohol and an aldehyde or ketone moiety to form a hemiacetal or a hemiketal, respectively. This intramolecular reaction stabilizes the structure to form either a five- or a six-member ring – furan and pyran, respectively. The corresponding sugar is termed furanose or pyranose (Figure 5.2). The acyclic form exists in equilibrium with its corresponding cyclic structure. The structures shown in Figure 5.2 represent a zigzag conformation, which is now widely accepted. However, other representations of monosaccharides have been proposed, including the Howarth projection formula and the Fischer projection formula.

Pentoses and hexoses are the most common monosaccharides in foods. They are present in high amounts in fruits and vegetables, and in free or bound form in plant foods. They may be added to food products to enhance their flavour and texture.

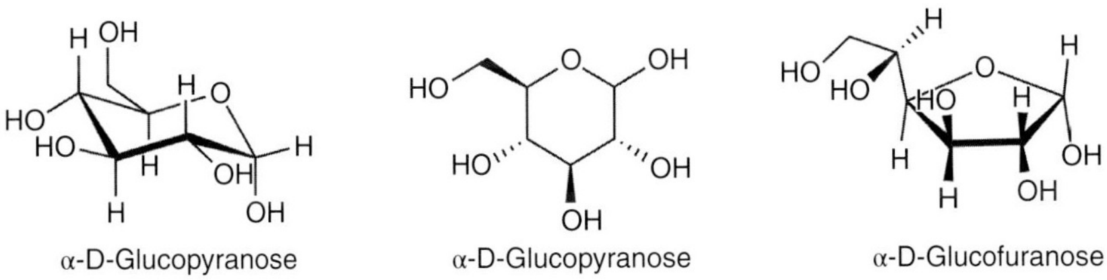

Figure 5.2 Ring structures of monosaccharides. Pyranose is a six-member ring and furanose is a five-member ring.

5.2.2 Oligosaccharides

Oligosaccharides are made of 2–10 monosaccharides units joined by a special bond called the glycosidic bond. The saccharide units usually take the furanose or the pyranose form. They are formed by the condensation of the α- or β-anomeric hydroxyl group of a monosaccharide with any other hydroxyl group of another monosaccharide. The condensation of two semiacetal hydroxyls yields a disaccharide, which is classed as either reducing or nonreducing depending on whether or not it contains a free anomeric hydroxyl group. Oligosaccharides are classified on the basis of the number of monosaccharide units they contain: disaccharide, trisaccharide, tetrasaccharide, and so forth. The most common example of an oligosaccharide is sucrose. Sucrose is a disaccharide made of a fructofuranose and a glucopyranose joined by a 1–1 glycosidic linkage. It is also called beet sugar, cane sugar, or table sugar – a common example of a nonreducing sugar. Sucrose is present in larger amounts in leaves and stems (Velisek 2014). Sugar cane is the richest source of sucrose, with a concentration of 12–26%, followed by common fruits (e.g. apples, oranges, apricots, peaches, pineapples, and papayas), with a concentration of about 8%.

Other disaccharides include maltose, which is made of two glucopyranose units joined by 1–4 linkage. It is a reducing sugar formed as a result of the hydrolysis of starch (a polysaccharide). Maltose is present at relatively high amounts in honey, with a concentration of 2.7–16%. It is also found in cereals, bread dough, and fruits. Foods such as rice, wheat, barley, and maize are rich in starch, and are ultimately good sources of maltose. Lactose is made of galactopyranose and glucopyranose units joined by 1–4 glycosidic linkage. It is also called milk sugar, and is unsurprisingly present in most dairy products. Cellobiose is made of two glucopyranose units linked by a 1–4 glycosidic bond. It is also a reducing sugar. Cellobiose is formed from the hydrolysis of cellulose (a polysaccharide) and is therefore rich in several plant foods, including honey. Another example of a nonreducing sugar is trehalose, which is made of two glucopyranose units, as shown in Figure 5.3. It is present in yeasts, mushrooms, and honey. Gentiobiose is made of two glucose units joined by 1–6 glycosidic linkage. It is present in honey and as glycosides in several plant foods. During food frying, gentiobiose is formed be the heat-induced caramelization of glucose.

Maltotriose is a trisaccharide made of three glucose units. It is one of the hydrolytic products of starch *in vitro* (food frying or cooking) and *in vivo* (digestion by human or animals). Another trisaccharide is cellotriose, which is made of three units of glucopyranose. It is produced by the hydrolysis of cellulose during food preparation and digestion.

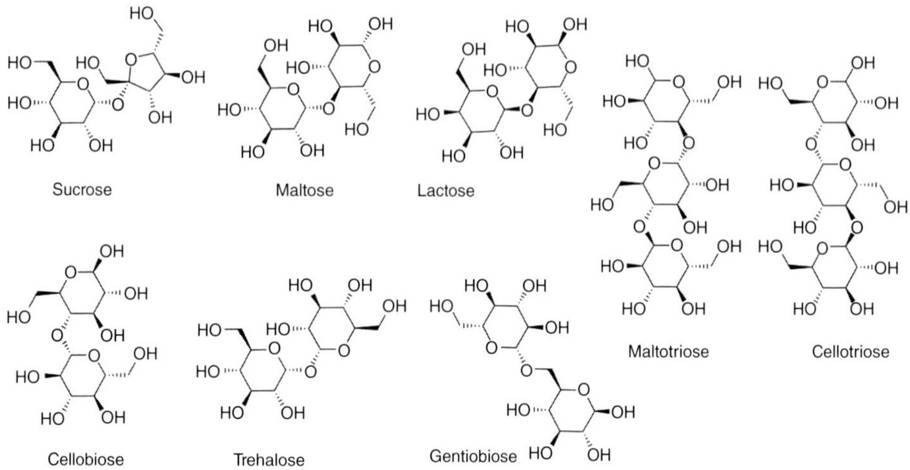

Figure 5.3 Chemical structures of important disaccharides and trisaccharides.

5.2.3 Polysaccharides

Polysaccharides, or glycans, contain more than 10 monosaccharide units joined by several types of glycosidic bond. They constitute a wide range of natural materials with significant importance in foods. Polysaccharides may contain a linear, cyclic, branched, or substituted chain of saccharide units. They are classified as either homopolysaccharides or heteropolysaccharides, based on the type of unit they contain. Starch, glycogen, and cellulose are common examples of homopolysaccharides, while chondroitin sulfate, hyaluronic acid, heparin, pectin, and proteoglycans are heteropolysaccharides. They can also be classified as storage, structural, and other types of functional polysaccharide, depending on their functional role.

Starch is a polymer of glucopyranose units joined by α 1–4 glycosidic linkage. It is a mixture of two homopolymers, amylose and amylopectin, in a 1 : 3 ratio by weight. Amylose is a linear polymer – normally, a repeating unit of maltose – while amylopectin is a highly branched polymer, with a branch every 25 units on average (Figure 5.4). The branch occurs every 25 units on average, making starch a highly branched polymer. Starch is present in large amounts in several foods, such as potato, wheat, rye, barley, oat, maize, and rice.

Cellulose is one of most abundant organic materials in nature. It occurs as a basic component of the cell wall of plant cells, contributing to the major components of the nonstarchy plants. It is a structural polysaccharide. It is a linear homopolymer of glucopyranose units joined by β 1–4 glycosidic linkage. Fruits and vegetables contain around 1–2% cellulose; cereals and legumes contain 2–4%; while wheat flour contains 0.2–3.0% (Velisek 2014).

Glycogen is a homopolysaccharide made of glucopyranose units joined by α 1–4 glycosidic linkage. It has a similar structure to that of amylopectin, but it is highly branched (every 8–12 units). It is one of the most important storage polysaccharides in animals.

Pectin is a group of very complex, widely dispersed, acidic polysaccharides with varying composition. With the exception of monocots, pectins occur in all tissues of higher

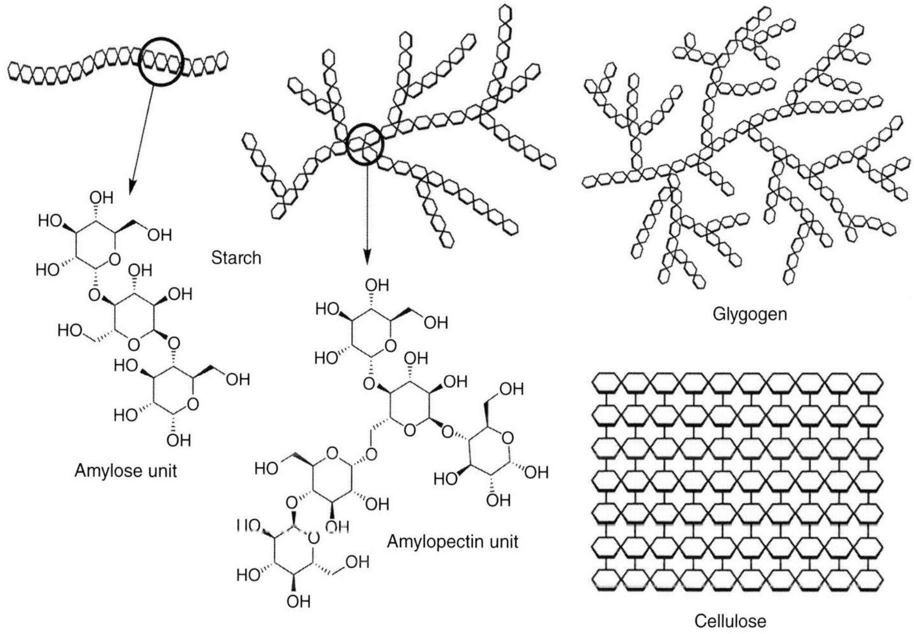

Figure 5.4 Structures of important polysaccharides present in all foods.

plants, forming the primary cell walls and intercellular spaces. They are formed and stored mainly in the early stages of growth, when the area of the cell walls increases. The presence of pectins and changes occurring in them during growth, maturation, storage, and processing have a major impact on the textures of fruits and vegetables.

5.3 Proteins and Amino Acids

Protein is made of a polymer of amino acids. An amino acid is made up of a carboxylic group and an amino group. There are 20 kinds of amino acid involved in the biological synthesis of proteins, which are known as proteogenic amino acids. They are joined by a carbonyl-amino group called an amide bond or peptide bond. Details on the reactions and occurrence of amino acids can be found in Velisek (2014), Nakai and Modler (2000), Brulé and Croguennec (2016), Almeida (2016), and Parry and Squire (2017). It is, however, necessary to have a general concept of their structure and chemical properties.

Amino acids occur in both free and bound form in proteins or with other biomolecules. The important amino acids in foods are alanine, leucine, valine, cysteine, serine, tyrosine, aspartic acid, asparagine, histidine, phenylalanine, and glutamic acid (Figure 5.5). They contribute to the specific properties of the proteins they form.

Aspartame is a dipeptide made of aspartic acid and phenylalanine. It is 200 times sweeter than sucrose, and thus is used as a sweetener in foods. Carnosine is another dipeptide, made of β-alanine and histidine. It has strong antioxidant properties and is an antiglycating agent, reducing the formation of the end products of advanced glycation. It is present in meat and meat products.

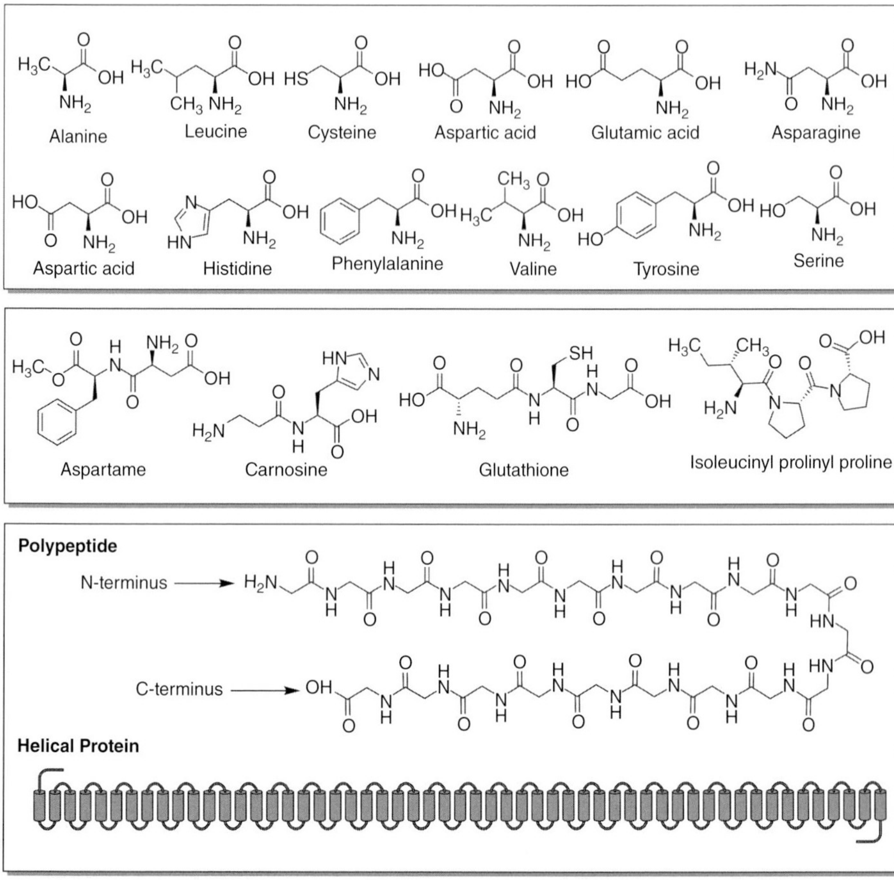

Figure 5.5 Chemical structures of amino acids and their derivatives, peptides and polypeptides. The helical structure of the proteins represents their simplest structural configuration.

Glutathione is a tripeptide made of γ-glutamate, cysteine, and glycine. It is a natural antioxidant and is involved in the prevention of several free radical reactions. It is present in both plants and animals. Another tripeptide is isoleucinyl prolinyl proline (IPP), which is found in milk and dairy products. It is a very good anticholinesterase inhibitor.

A polypeptide is a long chain of amino acids forming a number of proteins. One end of the chain contains the carboxylic acid group, while the other contains the amino group; they are represented as the C-terminus and N-terminus, respectively. Peptides are distinguished from proteins on the basis of their net molecular size. Peptide chains in proteins are arranged in specific three-dimensional conformations. The chains are held together to form secondary, tertiary, or quaternary structures. The specific protein structure is stabilized by hydrogen bonding, covalent bonding, and electrostatic and van der Waal forces. Helical and β-sheets are common secondary structures, which form the tertiary and quaternary structures of the proteins.

The richest sources of food proteins are animal products, such as meat, chicken, fish, egg, milk, and cheese. Cereals, nuts, and mushrooms are also the good sources (Figure 5.6).

Figure 5.6 Important daily sources of food proteins.

Meat and meat products contain 9–28% proteins, poultry products 11–29%, and milk and dairy products 0.5–41.0%. Similarly, cereals and cereal products contain 3–13% proteins, legumes and nuts 21–45%, vegetables 0.7–4%, fruits 0.3–4%, and mushrooms 2.6% (Velisek 2014).

5.4 Lipids

Lipids are organic and fatty biomolecules. They are soluble in organic solvent, and form esters of fatty acids (FAs) with alcohols. They are one of the major components of foods. Lipids are classified into three main classes: simple lipids, complex or compound lipids, and derived lipids. Simple lipids include neutral lipids such as triacylglycerols (TAGs). and waxes. TAGs were discussed in Section 4.3. Waxes are esters of FAs or their derivatives and higher alcohols. They are widely distributed in nature, in both plants and animals. Animal waxes are found in marine animals such as whale and bees. Plant waxes are found in plants of the brassica family.

Complex or compound lipids include phospholipids, glycolipids, and lipoproteins. Phospholipids include both glycerophospholipids and sphingophospholipids. Glycerophospholipids are made of glycerol (alcohol), FAs, and phosphoric acid, while sphingophospholipids contain sphingosine as their alcohol. Figure 5.7 shows the

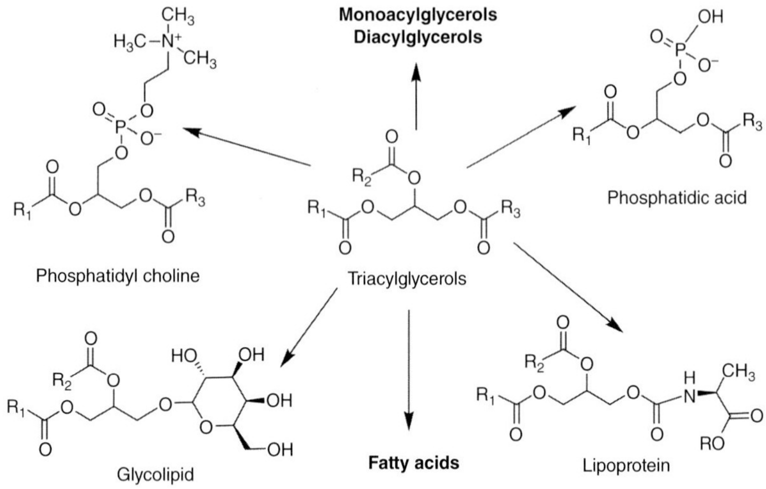

Figure 5.7 Contribution of triacylglycerols (TAGs) to the formation of other lipids.

contribution of TAGs to the formation of other lipids. Glycolipids contain one or more moieties of monosaccharides, while lipoproteins contain amino acids or protein moieties, together with a lipid part. Derived lipids include FAs (see Section 4.3), glycerol, sterols, and fat-soluble vitamins.

5.5 Micromolecules

Micromolecules include vitamins, minerals, pigments, and phenolic compounds. Vitamins are classed on the basis of their solubility as either fat-soluble (vitamins A, D, E, and K) or water-soluble (B-complex and ascorbic acid/vitamin C). Water-soluble vitamins are usually degraded during frying, while fat-soluble vitamins are rather stable. Minerals and vitamins are discussed in detail in Velisek (2014), while phenolic compounds are examined in Shahidi and Ho (2005). Pigments such as chlorophylls and carotenoids are also present in all foods. Carotenoids play a significant role in the colouration of food products, and are usually added as an ingredient during preparation.

5.6 Frying of Carbohydrate-Rich Foods

Foods of plants origin are rich in carbohydrates. Only a limited number of plant foods are fried, while the rest are either cooked or used as such. Potato (*Solanum tuberosum* L.) is a carbohydrate-rich food that is commonly fried. According to the Food and Agriculture Organization (FAO), worldwide potato production in 2015 was more than 376 million metric tonnes. Potato is a herb that yields tubers as food. Starch constitutes the major part of the potato tuber, containing amylose (22–30%) and amylopectin

(70–80%), and a large amount of water. Potatoes products such as whole potato, slices, crisps, and chips are usually fried in edible oils. Santos et al. (2018) fried whole potatoes at 175 °C and studied their freshness in terms of vitamin C contents; their results are shown in Table 5.1. Similarly, potato products such as crisps, chips, strips, and slices have also been studied during frying at temperatures of 100–210 °C (Lu et al. 2016; Rahimi et al. 2017; Su et al. 2018; Belkova et al. 2018) (Table 5.1).

Frying of other carbohydrate-rich foods includes maize starch (Chen et al. 2018), rich flour dough and its products (Shanthilal and Bhattacharya 2017), rice snacks (Lim and Han 2016), corn chips (Yuksel et al. 2017), wheat flour slices (Mariotti-Celis et al. 2017b), cassava chips (Oyedeji et al. 2017), and various cereals, legumes, and roots (Galani et al. 2017) (Table 5.1). Changes in parameters such as oil content, texture, solid content, colour, polar compounds, furan content, sensory characteristics, aroma compounds, and acrylamide content are indicators of the different chemical reactions that occur during frying.

Table 5.1 Recent studies on the frying of common carbohydrate-rich foods.

Food	Temperature (°C)	Frying Time (min)	Studied parameters	References
Potatoes	175	1680	Vitamin C	Santos et al. (2018)
Potato crisps	125 and 165	1–3	Sensory characteristics	Belkova et al. (2018)
Potato chips	100	22	Oil uptake, colour, texture, sensory characteristics	Su et al. (2018)
Potato strips	180	3.5	Moisture, texture, oil contents	Rahimi et al. (2017)
Potato slices	210	12	Aroma, acrylamide, vitamin C	Lu et al. (2016)
Maize starch	180	20	Oil contents	Chen et al. (2018)
Rice flour dough strands	180	2–3	Texture, sensory characteristics	Shanthilal and Bhattacharya (2017)
Rice snacks	160	1	Texture, sensory characteristics	Lim and Han (2016)
Corn chips	170–190	1	Texture, sensory characteristics	Yuksel et al. (2017)
Wheat flour slices	160–190	1.5–13	Solid contents, oil contents, polar compounds, furan contents	Mariotti-Celis et al. (2017a)
Cereals, legumes, roots	160	20	Acrylamide	Galani et al. (2017)
Cassava chips	160–180	2–12	Colour, texture	Oyedeji et al. (2017)

5.6.1 Formation of Furans

Carbohydrates in food are converted into several compounds, some of which are important indicators of food oxidation and flavour. Furan and furfural are synthesized from monosaccharides. The formation of furan and its derivative compounds from different carbohydrates has been extensively reviewed by several researchers (Kowalski et al. 2013; Van Putten et al. 2013). These reviews mostly discuss both the mechanistic features of dehydration reactions from previous studies and current progress in 5-hydroxymethyl-2-furfural (HMF) synthesis. Van Putten et al. (2013) places special importance on the effects of numerous solvent systems in the formation of HMF, and on the different formation pathways reported in the previous literature.

HMF is formed during caramelization and through the Maillard reaction. Caramelization occurs during the frying of carbohydrate-rich foods at specific temperatures. For example, fructose begins caramelization at 110 °C, while hexoses begin to caramelize at 160 °C. Maillard reactions, however, occur at relatively low temperature. The formation of HMF may follow either reaction mechanism, although in certain foods, such as coffee, HMF is formed by both caramelization and Maillard reactions. HMF has been found to be produced in other food preparation procedures as well, such as baking, smoking, and heating. Thus, a large set of foods contain HMF, and humans are continuously exposed to it.

The mechanism of formation of HMF may start directly with hexoses, such as fructose and glucose, or with their sources, such as starch, cellulose, sucrose, inulin, and oligo- and polysaccharides. These processes mainly involve dehydration reactions. However, no definite proof of the kinetics of dehydration reactions has been revealed to date. Several mechanisms have been proposed for the direct formation of HMF by dehydration of three water molecules, as shown in Figure 5.8. These reactions start from glucose and/or fructose. The formation of HMF from fructose involves cyclic intermediates. Fructose is dehydrated to form fructofuranosyl oxo cation. The elimination of a proton from this cation results in the formation of hydroxymethylene tetrahydrofuran diol. The next step is the second dehydration at C-4 to form a double bond in the furan ring. The resultant compound is dihydrofuran-2-carbaldehyde. The third dehydration reaction at C-3 of the furan ring produces HMF as the final product. Alternative insight into the mechanism of HMF production can be found in the different reactivity of fructose compared to that of glucose. Generally, fructose is more reactive and selective towards HMF than glucose (Van Putten et al. 2013). This selectivity may be due to the saturated furan ring present in fructose. Thus, pentoses may also follow the direct route, as already discussed.

The formation of HMF from glucose may follow a cyclic mechanism. Glucose and fructose are interconvertible via enediol formation. The enediol thus formed is dehydrated to release tetrahydroxy hex-2-enal. This compound has been found to isomerize to trihydroxy-2-oxohexanal. The second dehydration yields dihydroxy-2-oxohex-3-enal. This compound has been found to be less stable, and cyclizes to form 2-hydroxy HMF. The third and last dehydration yields HMF. All these reactions are catalyzed by the low pH provided by the different acids presents in plant foods.

HMF may be present as such or degraded further to form several products, such as alkoxy methyl furfural, furan dicarboxylic acid, hydroxymethylfuranoic acid, bishydroxymethylfuran, dimethylfuran, furan, furosine, levulinic acid, 1,6-hexandiol, and adipic acid, as shown in Figure 5.9. Most of these proposed pathways may or may not be

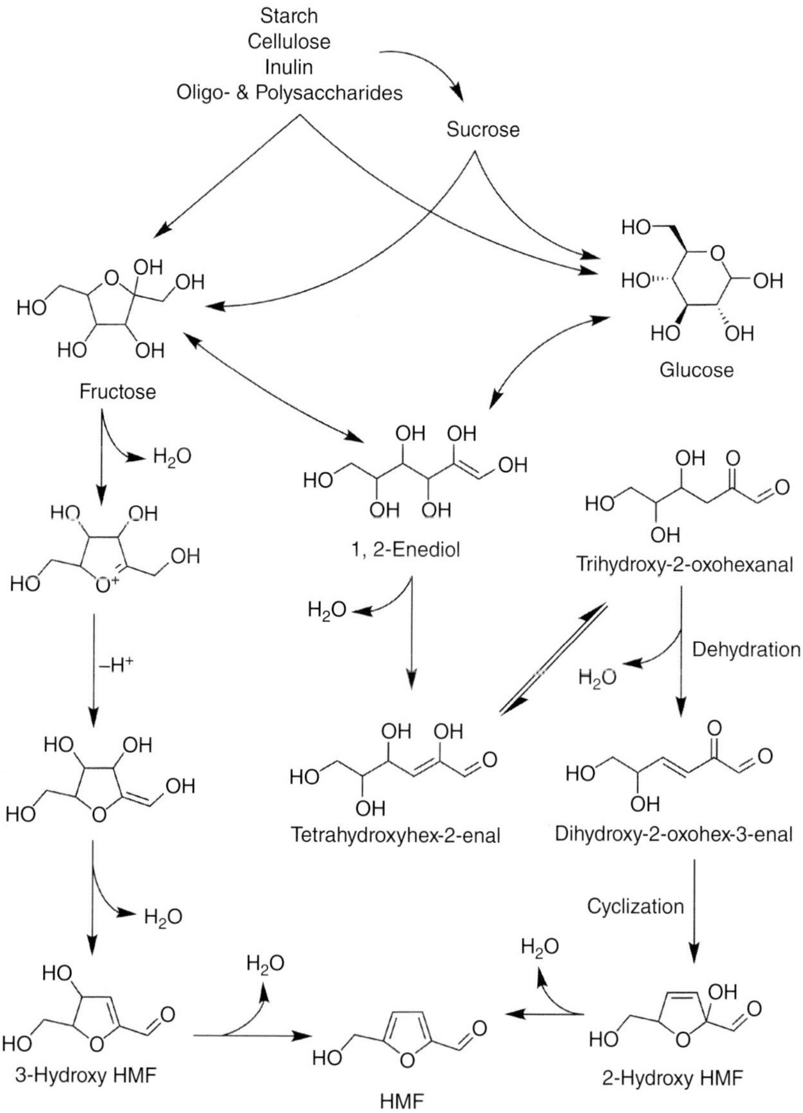

Figure 5.8 Mechanisms of formation of 5-hydroxymethyl-2-furfural (HMF) from carbohydrates.

followed in the real food matrix. Therefore, model food systems have been studied for the formation of HMF. However, most of these studies are limited to one or two compounds, and none includes mechanistic approaches to food frying.

Model wheat flour dough containing 25% glucose was prepared to assess the formation of HMF (Göncüoğlu and Gökmen 2013). The authors observed that HMF was also accumulated in frying oil in addition to the model food system, representing 5–6 mg L^{-1}. Researchers from Spanish universities and research centres have studied the formation of HMF in Spanish fried dough pastry. Morales and Arribas-Lorenzo (2008) showed

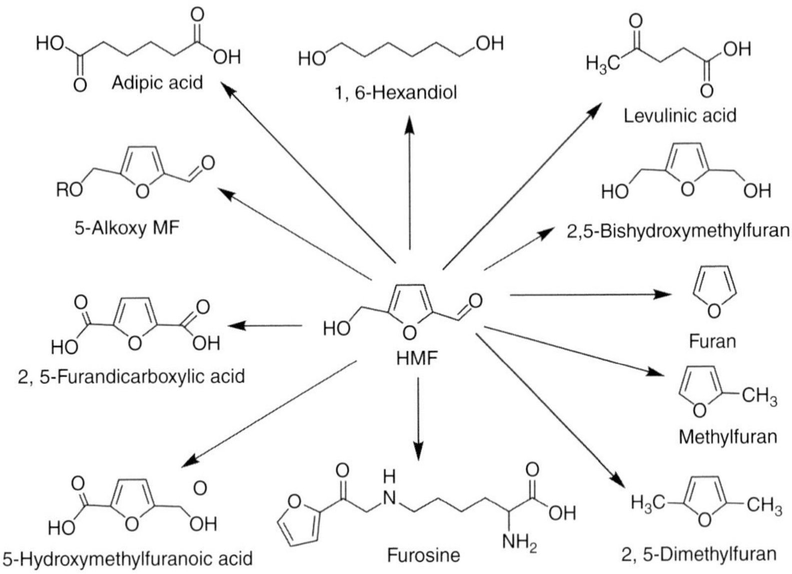

Figure 5.9 HMF is the precursor of several different organic compounds produced during the frying of foods.

that HMF ranged from 1.2 to 221.4 mg kg^{-1} in lab-scale experiments, while an average of 74.3 mg kg^{-1} was documented in commercial samples. They found that temperatures in the range 185–200 °C are commonly used to obtain churros with an acceptable texture. However, HMF formation was nearly twofold higher in the range 190–200 °C, which indicates that there should be a more precise control of frying temperatures. Similarly, authors from the same institution (Delgado-Andrade et al. 2010) determined the Maillard reaction products in several Spanish foods. The amount of furosine was 5.4–378.6 mg kg^{-1}, that of furfural was 0.1–1.2 mg kg^{-1}, and that of HMF was 0.1–37.9 mg kg^{-1} in lentil stew, paella, kid stew with garlic, anchovies, potato omelette, salmorejo, and churros. In another study, Spanish doughnuts were fried in sunflower oil at 100, 120, 140, and 180 °C (Mir-Bel et al. 2013). The authors compared the formation of HMF in a vacuum and an open fryer, and showed that vacuum frying resulted in much less HMF (< 0.05 mg kg^{-1}) and a better finished texture.

The formation of furan increases with increasing frying time. Mariotti-Celis et al. (2017b) studied the kinetics of furan formation in a wheat flour-based model system, and found that it followed a sigmoid trend with frying time at a temperature of 190 °C only. At lower temperatures, furan generation did not reach the asymptotic level. The authors showed that by keeping the frying temperature below 180 °C, reducing the frying time contributed to a reduction of furan formation in fried foods. Similarly, Mesias et al. (2016) reported the formation of HMF and furfural together with acrylamide during the frying of breadcrumbs. The breadcrumbs were prepared from wheat flour, salt, water, and yeast. Samples were deep fried in sunflower oil at 180 °C for 3 minutes using a domestic fryer. The temperature and time were adjusted according to industrial recommendations. The authors found that HMF and furan formation was influenced by sugar content and temperature, as shown in Figure 5.10.

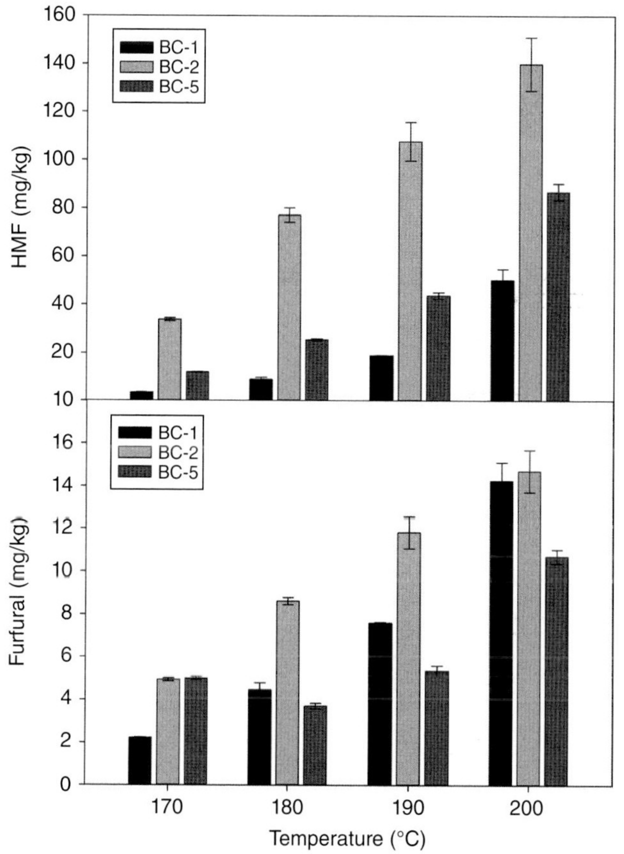

Figure 5.10 Effects of different temperatures on the formation of HMF and furfurals during the deep frying of breadcrumbs for 3 minutes. BC, breadcrumbs with a specific composition. *Source:* Redrawn with the kind permission of the Royal Society of Chemistry, UK (Mesias et al. 2016).

Mesias et al. (2016) showed that the formation of both HMF and furfural regularly increased with increasing frying temperature (Figure 5.11). The authors reported that HMF levels were 50.4, 140.0, and 86.9 mg kg^{-1} for BC-1, BC-2, and BC-5, respectively, when the frying temperature was 200 °C. Furfural levels were 14.3, 14.7, and 10.7 mg kg^{-1} under similar conditions. These experiments give rise to a basic question: What effect do proteins or amino acids present in the system have on furfural formation? Several authors (Kowalski et al. 2013; Van Putten et al. 2013) have reviewed this point. Proteins are usually degraded during frying to form their component amino acids.

Glucose shares some intermediates in the presence of an amino acid when forming HMF (Kowalski et al. 2013). It condenses with the amino acid under the heat stress during frying; the amino group of the amino acid reacts with the carbonyl carbon of the glucose to release a water molecule. The product is a Schiff base, as shown in Figure 5.12. Cyclization of the Schiff base produces a tautomeric N-substituted glucosyl amine. This compound is protonated to form a Schiff base cation. The proton may be initially attached to the oxygen atom of the amino acid moiety, which on rearrangement reaches the amino

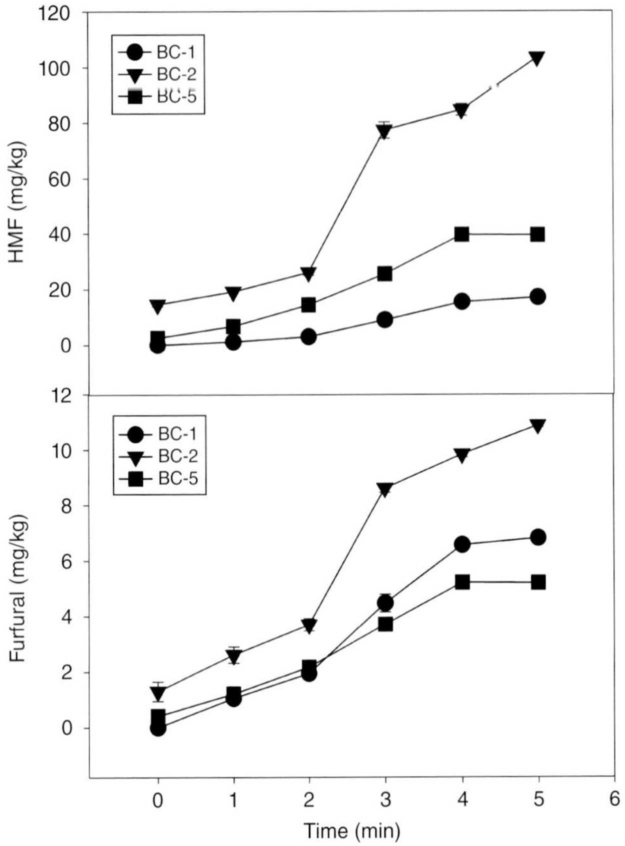

Figure 5.11 Effects of different frying times on the formation of HMF and furfurals during deep frying of breadcrumbs at 180 °C. BC, breadcrumbs with a specific composition. *Source:* Redrawn with the kind permission of the Royal Society of Chemistry, UK (Mesias et al. 2016).

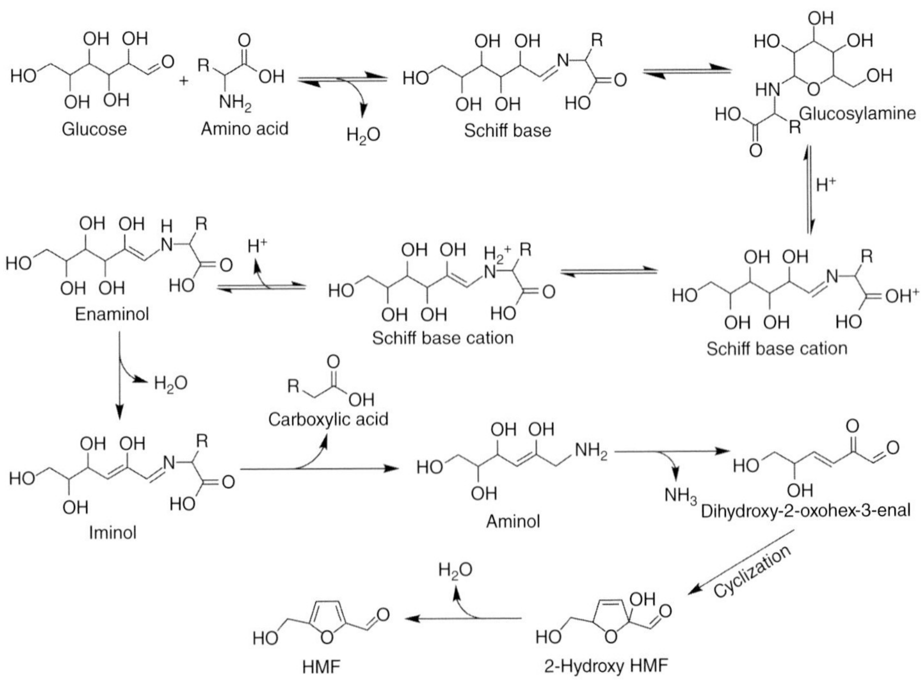

Figure 5.12 Mechanism of formation of HMF from glucose and an amino acid.

group. This amino cation of the Schiff base is more stable than the former one. The deprotonation may release a compound with enaminol as a base. The enaminol may form Amadori products or iminol. The latter is formed by the dehydration at C-3 of the glucose moiety. The iminol initially releases carboxylic acid, while the amino group remains with the glucose moiety, forming aminol. The dehydration of aminol and consequent oxidation produces oxo-hexenal and ammonia. The cyclization of oxo-hexenal produces 2-hydroxy HMF (Kowalski et al. 2013). The final step is the formation of HMF via dehydration. These pathways may also contribute to the formation of acrylamide and other toxic compounds.

5.6.2 Formation of Acrylamide

Acrylamide is a common name for prop-2-enamide, an important industrial chemical used in wastewater treatment, as a soil conditioner, in cosmetics, and as a laboratory chemical for electrophoresis. Although it has been found to be toxic (Kuperman 1958; Mccollister et al. 1964), its association with heated foods was determined by Tareke et al. (2002), and later by the World Health Organization (WHO/FAO, 2002). It was found that high levels of acrylamide were produced from heat-processed commercial foods and carbohydrate-rich foods cooked at high temperature. These studies sparked researchers around the world to investigate its presence in both unfried and unheated foods.

Stadler et al. (2002), Mottram et al. (2002), and Becalski et al. (2003) showed that acrylamide is produced from carbohydrates by the Maillard reaction. They found that reducing sugars with a free aldehyde group react with asparagine at above 100 °C to form an *N*-glycoside. This glucoside is then cleaved at the C—N bond to an intermediate that ultimately produces acrylamide. Yaylayan et al. (2003) proposed a pathway that leads *N*-glycoside to form acrylamide. It has been observed that 2-deoxyglucose, glyoxal, and glycerol also combine with asparagine to synthesize acrylamide. Two pathways have been proposed; one requires the participation of dicarbonyl moiety, while the other requires a monocarbonyl aldehyde or ketone.

The formation of acrylamide during frying can be explained as shown in Figure 5.13. The frying medium largely consists of TAGs, which upon hydrolytic cleavage produce FAs and glycerol. The FA oxidation product (e.g. acrolein, which is also formed from glycerol) is directly converted to acrylamide by the reaction of ammonia and acrylic acid. Similarly, asparagine is converted to 3-aminopropanamide, which upon deamination produces acrylamide. Glutamate and other similar amino acids produce propanamide or its derivatives, which upon deamination are transformed to acrylamide. This proposed mechanism is one of the best-explained pathways during frying.

Asparagine is considered a key participant in acrylamide formation. Foods are usually rich in asparagine. Among plant foods, asparagus, almond, potatoes, spinach, and wheat are all rich in asparagine (Friedman 2003), while animal foods contain relatively lesser amounts. Asparagine condenses with carbonyl carbon through the release of the water molecule to form a Schiff base. The decarboxylation of the Schiff base leads the double bond on the carbonyl side to be transferred to the asparagine side and back to the carbonyl carbon. Two possible pathways may produce acrylamide: one from 3-aminopropanamide and one from 3-oxopropanamide (Figure 5.14). Granvogl et al. (2004) showed that 3-aminopropanamide is a potent precursor in acrylamide formation in potatoes. In this reaction scheme, it clear that acrylamide is formed from asparagine, while a carbohydrate moiety helps release carbon dioxide and ammonia groups.

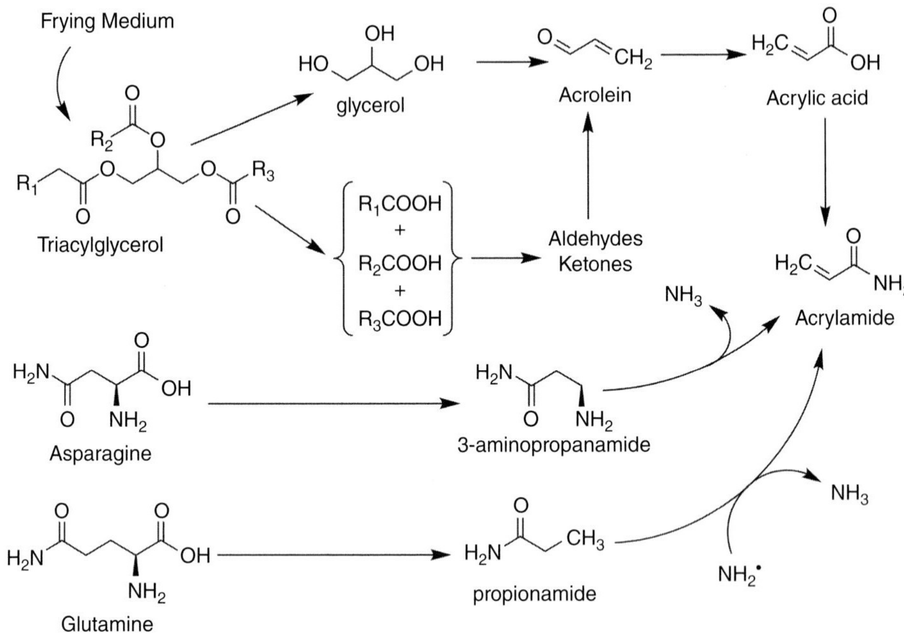

Figure 5.13 Formation of acrylamide from lipids and amino acids.

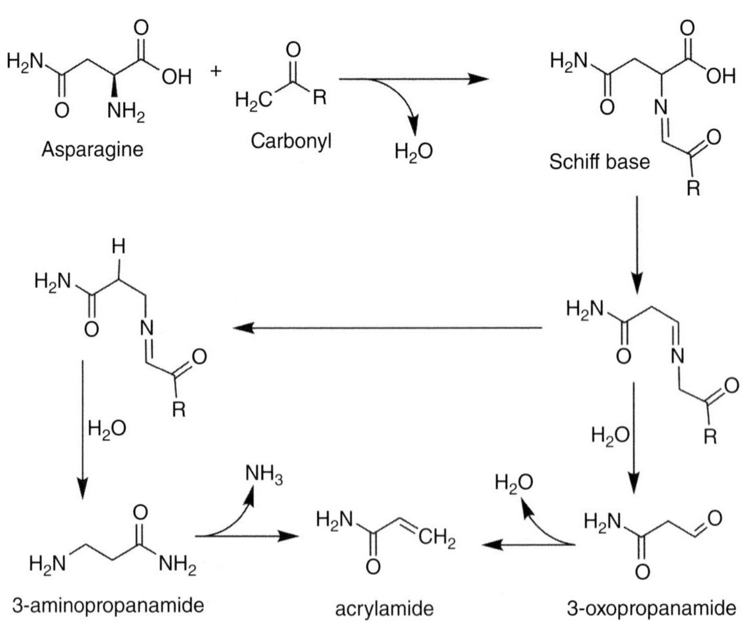

Figure 5.14 Mechanism of formation of acrylamide from asparagine.

Amrein et al. (2003) showed that fructose is a more efficient precursor for acrylamide formation than is glucose. Similarly, galactose-, lactose-, and sucrose-containing foods also produce acrylamide. Sucrose is a common ingredient in modern-day frying. It is believed to decompose into fructose and glucose, which are potent precursors of acrylamide during high-temperature frying or roasting (Kocadagli et al. 2012). Fried foods rich in carbohydrates contain a large amount of acrylamide, particularly French fries.

The formation of acrylamide increases when the frying medium is rich in polar materials or a diacylglycerols. Moisture escaping from the frying food also contributes significantly (Gertz et al. 2003). Matthäus et al. (2004) showed that the level of acrylamide in French fries increased with increasing temperature and frying time, especially at temperatures higher than 175 °C. Similarly, Mesias et al. (2016) showed that acrylamide formation during the deep frying of breadcrumbs was directly related to temperature and frying times (Figure 5.15).

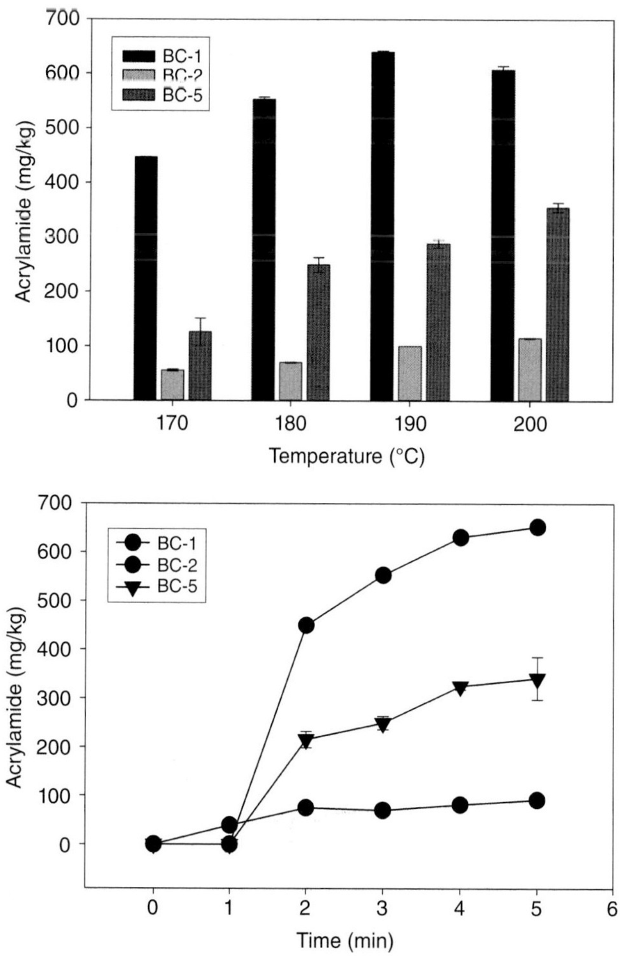

Figure 5.15 Effects of different temperatures and frying times on the formation of acrylamide during the deep frying of breadcrumbs. BC, breadcrumbs with a specific composition. *Source:* Redrawn with the kind permission of the Royal Society of Chemistry, UK (Mesias et al. 2016).

Mesias and Morales (2015) studied the acrylamide contents of potatoes products (crisps) available on the Spanish market during 2004–2014. The acrylamide content in all samples ranged from 108 to 2180 μg kg^{-1}, the average and median values of which were 630 and 556 μg kg^{-1}, respectively. The authors showed that there was a continuous decline from 2004 to 2014, with the average acrylamide content in 2014 being 14.6% lower than that in 2009 and 57.6% lower than that in 2004. Similarly, Powers et al. (2017) studied the acrylamide contents of potatoes crisps in 25 European countries during 2002–2016. Again, these showed a decline over time, due to the strict regulations of the European Food Safety Authority (EFSA).

The acrylamide content in select Colombian foods was studied by Pacetti et al. (2015). There was a wide variability among different foods and between different brands of the same food type, particularly for coffee powder, breakfast cereals, biscuits, and French fries. The highest amounts of acrylamide were found in bakery products, such as biscuit (1104 μg kg^{-1}) and wafers (1449 μg kg^{-1}), followed by potato chips (916 μg kg^{-1}). About 400 food samples in Hong Kong were analysed for acrylamide content by Leung et al. (2003). These authors found that deep frying at 170 °C had little effect on acrylamide formation, but that the effect increased as the temperature reached 210 °C. They proposed that the sequence of acrylamide formation begins on the food surface. The acrylamide then enters into the interior of the food matrix through heat transfer via radiation/conduction and diffusion of hot oil. In another study, 344 food products from the Netherlands were analysed for acrylamide content, which ranged from <30 to 3100 μg kg^{-1}. It was found that the foods with the highest mean acrylamide amounts were potato crisps (1249 μg kg^{-1}), deep-fried chips (351 μg kg^{-1}), cocktail snacks (1060 μg kg^{-1}), and gingerbread (890 μg kg^{-1}).

5.6.3 Formation of Acrolein

Heat induction during frying dehydrates glucose molecule to hydroxy ketone moiety. Hydroxy acetone is produced as a thermal cleave product from C-1–3 by retro-aldol cleavage. Hydroxy acetone is also formed via an Amadori product with the help of an amino acid (Stevens and Maier 2008; Shen et al. 2011). Formation of 1-deoxyglucosone is believed to isomerize to form a dione product. The cleavage between C-3 and C-4 produces hydroxy acetone and 3-hydroxy-2-oxopropanal (Figure 5.16). Upon

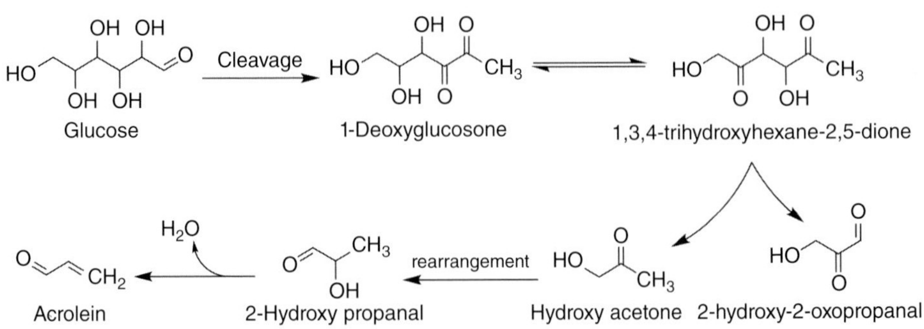

Figure 5.16 Mechanism of formation of acrolein from glucose.

rearrangement, the hydroxy acetone forms 2-hydroxy propanal, which upon dehydration produces acrolein. During frying, carbohydrate is exposed to FAs in the frying medium, which stimulates acrolein formation via glucose (Medina-Navarro et al. 2004). Auto-oxidation of glucose induces oxidation of free fatty acids (FFAs) and consequently increases the rate of acrolein formation. Thus, the formation of acrolein is one of the significant end products of oxidative reactions, an can involve both food and the frying medium.

5.7 Frying of Protein-Rich Foods

Frying of protein-rich foods is one of the oldest known methods of food preparation, beginning in prehistoric times. Thus, research on the composition and properties of protein-rich foods was a preliminary focus of food scientists. Bjeldanes et al. (1982) studied several of the most common protein-rich foods in the United States, including ground beef, beef steak, eggs, pork chops, fried chicken, pot-roasted beef, ham, roast beef, and bacon. These foods were prepared using normal household cooking methods, such as deep frying, griddle/pan frying, baking/roasting, broiling, stewing, braising, and boiling at 100–475 °C. The authors reported that well-fried ground beef, beef steak, ham, pork chops, and bacon contained significant numbers of mutagens. The high-temperature broiling of chicken and beef steak produced the most mutagenicity, followed by baking/roasting and frying. Other food preparation methods, such as stewing, braising, and deep frying, produced few mutagens. It was also found that eggs and egg products produced mutagens only after they were cooked at high temperatures, while commercially cooked hamburgers showed a wide range of mutagenic activity. This study concluded that mutagen formation during the cooking of protein-rich foods is a complex function of food type, cooking time, and cooking temperature. This work has opened doors to determining whether any human health hazard might result from the cooking of protein-rich foods.

Lyon and Mahoney (1988) were the first to report that fried foods may have an association with colon cancer. Several classes of organic compounds have been identified in protein-rich foods following frying and cooking, including heterocyclic aromatic amines (HAAs) (Felton and Knize 1990; Knize et al. 1994; Thiebaud et al. 1995; Persson et al. 2004; Hsu et al. 2006; Yao et al. 2013; Wang et al. 2017), acrylamide (Tareke et al. 2002; Friedman 2003), and polycyclic aromatic amines (Baghurst 1999; Aaslyng et al. 2013; Lu et al. 2017).

5.7.1 Formation of Heterocyclic Aromatic Amines

Aromatic hydrocarbons usually contain benzene. When such organic compounds contain rings of various kinds, they are referred to as heterocyclic aromatic compounds. They also contain one or more nitrogen, oxygen, or sulfur atoms. In foods, nitrogen-containing compounds are more common than sulfur-containing ones. Several HAAs have been identified in natural and processed foods. However, the majority are synthesized during thermal treatment. A list of some of the most important HAAs is given in Table 5.2. There are seven subclasses of HAAs, based on their chemical structure: quinolones, quinoxalines, pyridines, pyridoimidazoles and indoles, furopyridines, benzoxazines, and others

Table 5.2 Structures and sources of heterocyclic aromatic amines (HAAs) in protein-rich foods.

Name	Abbreviation	Structure
2-Amino-1,6-dimethylimidazo[4,5-b]-pyridine	DMIP	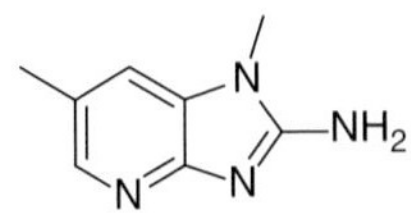
2-Amino-1,5,6-trimethylimidazo[4,5-b]-pyridine	1,5,6-TMIP	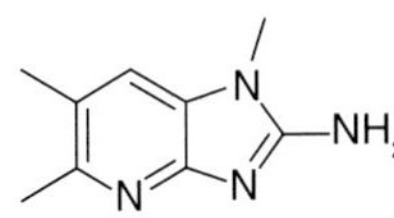
2-Amino-3,5,6-trimethylimidazo[4,5-b]-pyridine	3,5,6-TMIP	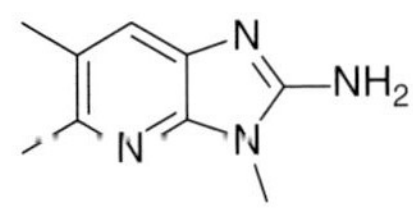
2-Amino-1-methyl-6-phenylimidazo[4,5-b]-pyridine	PhIP	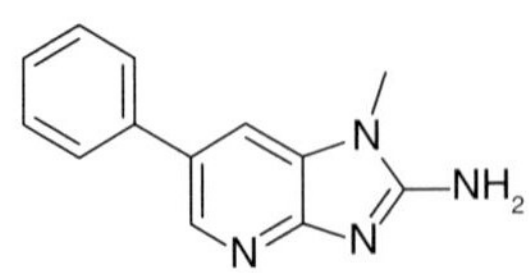
2-Amino-1-methyl-6-(4-hydroxyphenyl)-imidazo[4,5-b]-pyridine	4-OH-PhIP	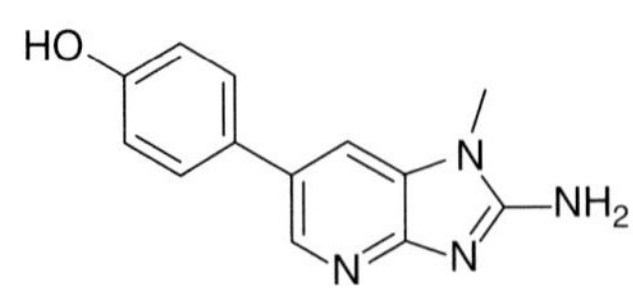
2-Amino-1,6-dimethyl-furo[3,2-e] imidazo[4,5-b]-pyridine	IFP	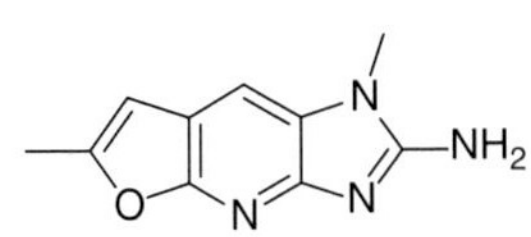
2-Amino-3-methyl-imidazo[4,5-f]-quinoline	IQ	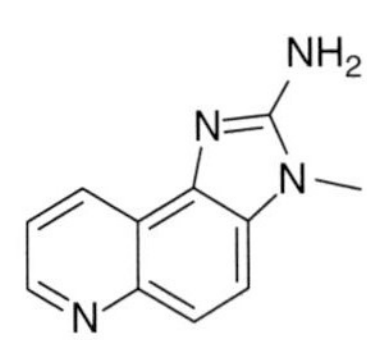
2-Amino-1-methyl-imidazo[4,5-f]-quinoline	Iso IQ	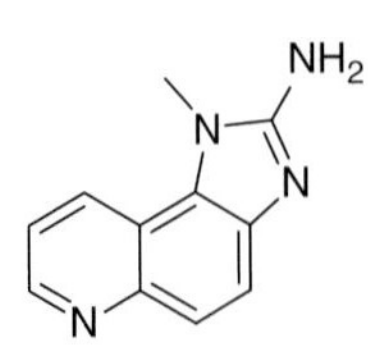
2-Amino-3,4-dimethyl-midazo[4,5-f]-quinolone	MeIQ	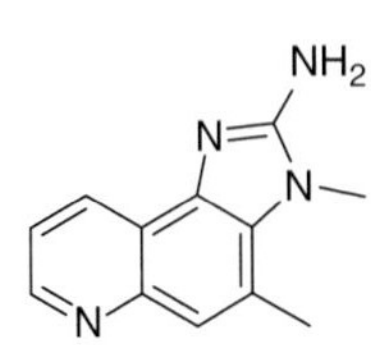

Table 5.2 (Continued)

Name	Abbreviation	Structure
2-Amino-3-methyl-imidazo[4,5-f]-quinoxaline	IQx	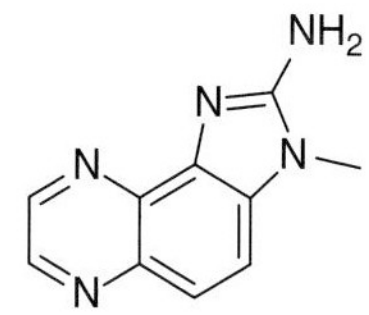
2-Amino-3,4-dimethyl-imidazo[4,5-f]-quinoxaline	4-MeIQx	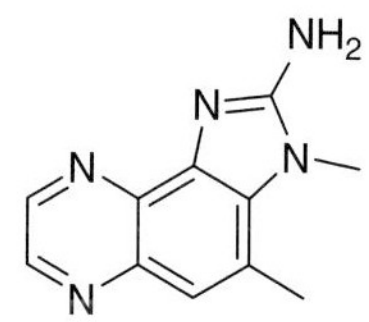
2-Amino-3,7,8-trimethyl-imidazo[4,5-f]-quinoxaline	7,8-DiMeIQx	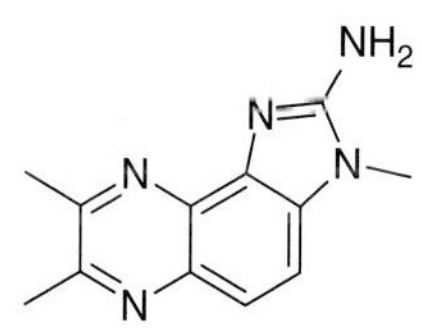
2-Amino-4-hydroxymethyl-3,8-dimethylimidazo[4,5-f]-quinoxaline	4-HMe-8-MeIQx	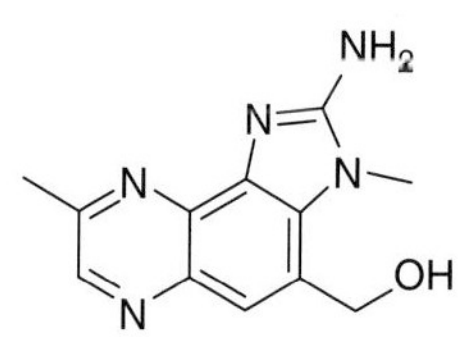
2-Amino-1-methyl-imidazo[4,5-g]-quinoxaline	IgQx	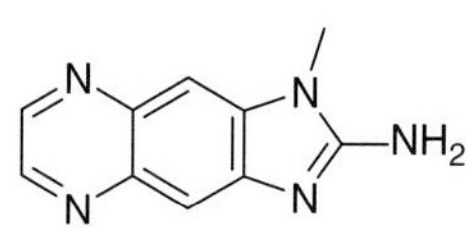
2-Amino-1,7-dimethyl-imidazo[4,5-g]-quinoxaline	7-MeIgQx	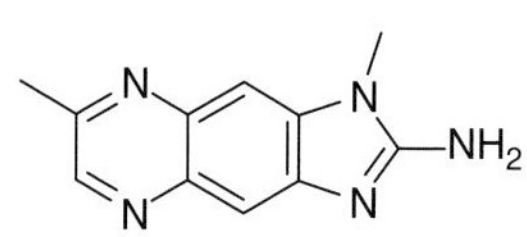
2-Amino-1,6,7-trimethyl-imidazo[4,5-g]-quinoxaline	6,7-DiMeIgQx	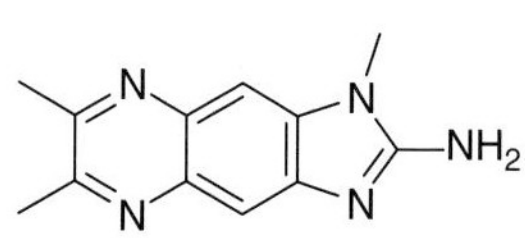
2-Amino-1,7,9-trimethyl-imidazo[4,5-g]-quinoxaline	7,9-DiMeIgQx	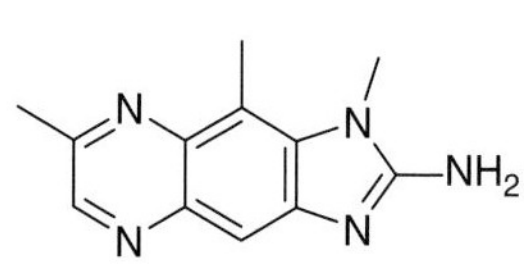
2-Amino-5-phenylpyridine	Phe-P-1	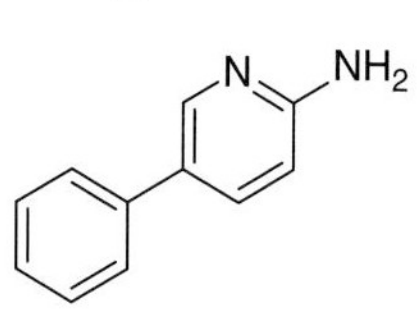

(*Continued*)

Table 5.2 (Continued)

Name	Abbreviation	Structure
2-Amino-9H-pyrido[2,3-b]indole	AαC	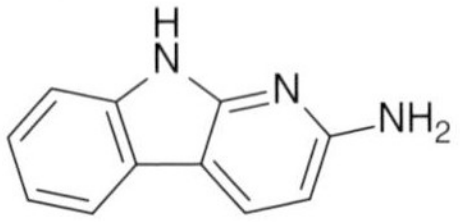
2-Amino-3-methyl-9H-pyrido[2,3-b]indole	MeAαC	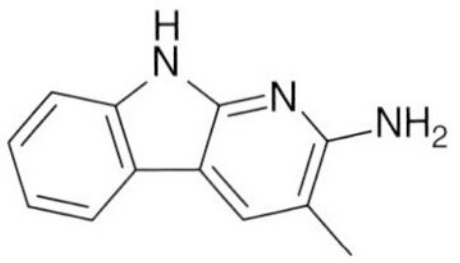
1-Methyl-9H-pyrido[3,4-b]indole	Harman	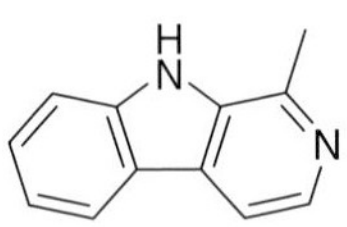
9H-pyrido[3,4-b]indole	Norharman	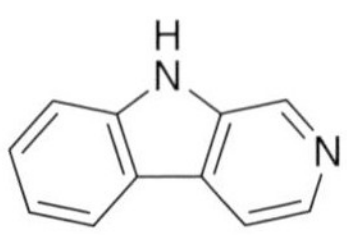
3-Amino-1-methyl-5H-pyrido[4,3-b]indole	Trp-P-2	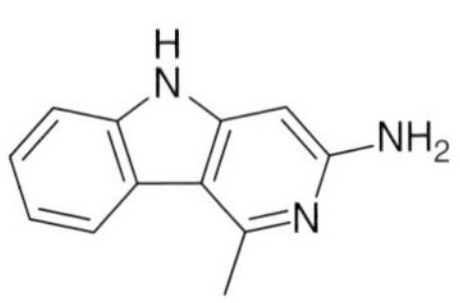
2-Aminodipyrido-[1,2-a:3_,2_-d]imidazole	Glu-P-2	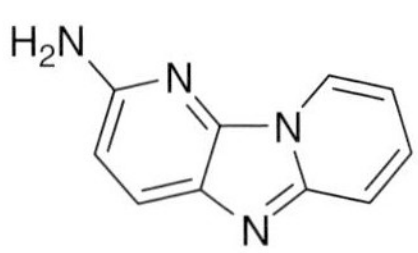
2-Amino-6-methyldipyrido-[1,2-a:3_,2_-d] imidazole	Glu-P-2	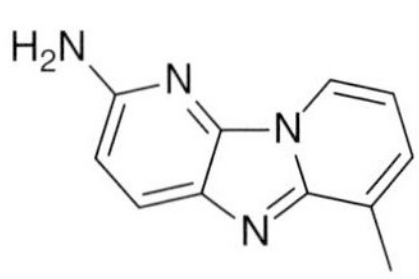
4-Amino-6-methyl-1H-2,5,10,10btetraazafluoranthene	Orn-P-1	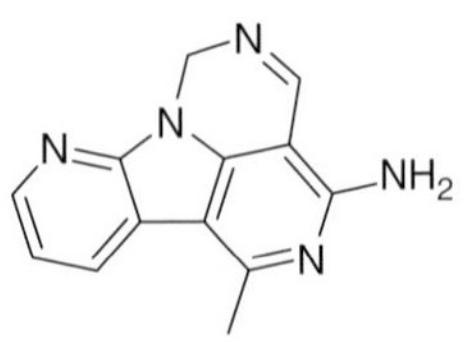
4-Amino-1,6-dimethyl-2-methylamino-1H,6H-pyrrolo-[3,4-f] benzimidazole-5,7-dione	Cre-P-1	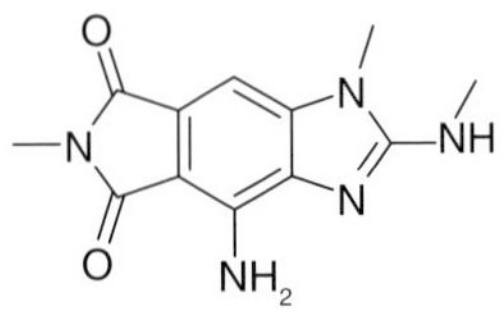

(Murkovic 2004; Alaejos and Afonso 2011). HAAs can also be classified on the basis of polarity and synthesis temperature. The latter classification gives two main types: thermic HAAs (formed below 300 °C; 2-amino-3-methyl-imidazo[4,5-f]-quinoline (IQ) and its derivatives) and pyrolytic HAAs (formed above 300 °C; non-IQ types).

It has been observed that most HAAs are synthesized from amino acids in the presence of heat. Protein-rich foods are thus a good reactant medium for HAAs formation (Table 5.3). Generally, the synthetic process involves nitrogenous compounds such as amino acids and creatine. The process of HAA formation is strictly dependent on temperature, heat transfer, the type of nitrogenous sources present, and heating conditions (Murkovic 2004). Pyridines and pyrazines are formed from hexose sugars and amino acids following Maillard reactions (Gibis 2016).

Phenylalanine and creatine are the reactants for the formation of 2-amino-1-methyl-6-phenylimidazo[4,5-b]-pyridine (PhIP). The sequence of the reactions was proposed by Zöchling and Murkovic (2002) and is shown in Figure 5.17. Thermal degradation of phenylalanine produces phenyl-acetaldehyde. This compound was isolated using solid phase microextraction and gas chromatography, as shown in Figure 5.18. The reaction was based on the model system of creatinine and phenylalanine. The authors also identified phenyl-acetaldehyde in fried meat samples, thus confirming the sequence of reaction. It is believed that during frying, creatinine is degraded to form formamide; alternatively, formamide may be a degradation product obtained from phenylalanine. During frying, oxidized lipid medium also plays a significant role in the formation of HAAs. Thus, other amino acids also compete for lipid against phenylalanine. There are several amino acid degradation products in which formamide may be important and reactive.

Creatinine and phenyl-acetaldehyde condense to form a stable and isolatable aldol condensation product. Under heat stress, this product is dehydrated to form an unsaturated condensation product. Two possible pathways may occur at this stage. One is the condensation of formamide with unsaturated aldol product and consequent cyclization to form PhIP. The other involves the addition reaction of ammonia, accompanied by the loss of water molecule to form an imino-imidazole compound. This compound may isomerize to transfer its double bond to the imidazole ring by forming a more stable product. Formaldehyde, a degradation product of lipid, carbohydrate, or amino acid, reacts with aminoimidazole compound to form an N-methylene aminoimidazole derivative. Next comes the cyclization reaction, followed by dehydration to form PhIP (Aeenehvand et al. 2016; Meurillon and Engel 2016).

Harman and norharman are synthesized by the condensation of tryptophan and glucose. The condensation product is an Amadori product, which upon oxidation and consequent cyclization produces norharman. Other HAAs are formed by the typical pyrolytic reactions of amino acids at above 300 °C, the details of which can be found in the literature (Baghurst 1999; Murkovic 2004; Randel et al. 2007; Alaejos and Afonso 2011; Gibis et al. 2015). It is imperative to consider the reactant as well as the conditions of frying when foods are being prepared.

Lu et al. (2017) evaluated the formation of HAAs and polycyclic aromatic hydrocarbons (PAHs) in pork patties with 40% fat replacement by olive, sunflower, or grapeseed oil, fried at 180 and 220 °C. They found that control patties contained the highest amounts of HAAs. Olive oil and sunflower oil replacement completely inhibited the formation of MeIQ, while grapeseed oil completely inhibited the formation of MeIQx,

Table 5.3 Precursors, sources, matrices, frying conditions, and concentrations of heterocyclic aromatic amines (HAAs) during the frying of protein-rich foods.

	Precursors					
HAA	Amino acid	Protein	Matrix	Temperature (°C)	Duration (min)	Concentration ($ng\,g^{-1}$)
AαC	Tryptophan	Albumin, casein, soybean, globulin	Ruminants	180–200	8	0.15–21
			Poultry	190	12	0.1–2.0
			Fish	200	6–24	0.1–1.0
IQ	Tryptophan	Animal or vegetable proteins	Meat	200	3	15.5–20.8
			Pork	180	Na	3.9
MeIQ	Tryptophan	Animal or vegetable proteins	Meat	200	3	10.7–18.8
			Pork	180	Na	18.3
MeIQx	Tryptophan	Animal or vegetable proteins	Meat	200	3	10.2–16.2
			Pork	180	Na	8.3
MeAαC	Tryptophan	Albumin, casein, soybean, globulin	Pork	150	10–30	0.01–0.5
			Poultry	190	12	Trace
Norharman	Tryptophan	Various proteins	Ruminants	176–191	4–33	0.1–41.0
			Pork	150–230	6–15	0.1–16.2
			Chicken	170–200	13	0.2–15.0
			Fish	200	6–24	2.0–184.0
Harman	Tryptophan	Various proteins	Ruminants	176–191	4–33	2.3–31.0
			Pork	150–230	6–15	5.0–25.0
			Poultry	140–220	12–34	5–300
			Fish	200	>15	3.0–60.1
Trp-P-1	Tryptophan	Animal or vegetable proteins	Ruminants	225	5	0.1–0.4
			Poultry	190	12	0.1–13.0
Trp-P-2	Tryptophan	Animal or vegetable proteins	Ruminants	225	5	0.2–14
			Poultry	190	12	0.2–3.1

Source: Alaejos and Afonso (2011), Meurillon and Engel (2016), and Aeenehvand et al. (2016).
AαC, 2-amino-9H-pyrido[2,3-*b*]indole; IQ, 2-amino-3-methyl-imidazo[4,5-f]-quinoline; MeIQ, 2-amino-3,4-dimethyl-imidazo[4,5-f]-Quinolone; MeIQx, 2-amino-3,4-dimethyl-imidazo[4,5-f]-Quinoxaline; MeAαC, 2-amino-3-methyl-9H-pyrido[2,3-*b*] indole; Norharman, 9H-pyrido[3,4-*b*]indole; Harman, 1-methyl-9H-pyrido [3, 4-*b*] indole; Trp-P-1, 3-amino-1,4-dimethyl-5H-pyrido[4,3-*b*] indole; Trp-P-2, 3-amino-1-methyl-5H-pyrido [4, 3-*b*] indole.

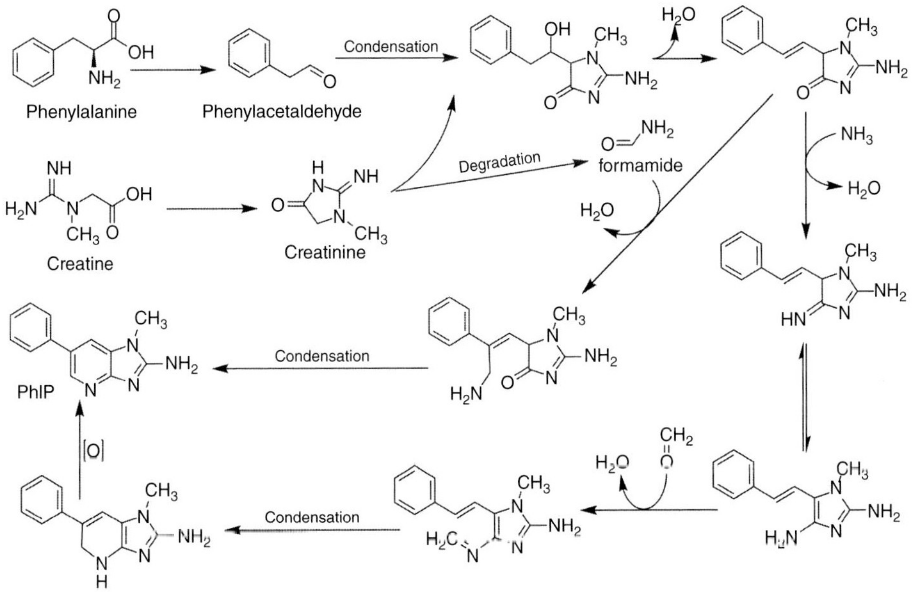

Figure 5.17 Mechanism of formation of 2-amino-1-methyl-6-phenylimidazo[4,5-b]-pyridine (PhIP) from phenylalanine and creatine. *Source:* Major parts of the sequence of reactions are reproduced with kind permission of Elsevier Science Ltd (Zöchling and Murkovic 2002).

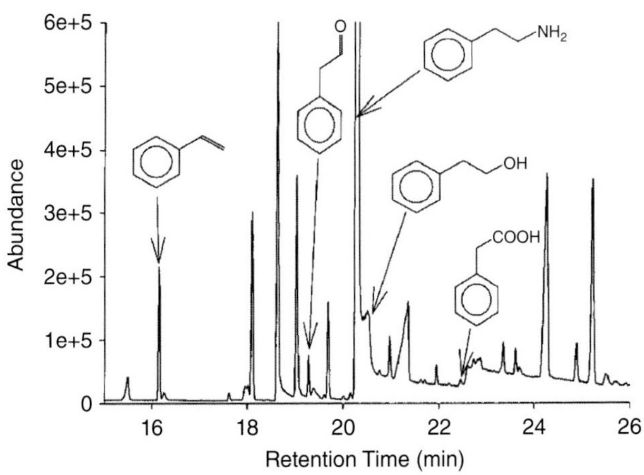

Figure 5.18 Headspace gas chromatography mass spectrometry (GC-MS) analysis of products formed after the heating of phenylalanine at 150 °C for 10 minutes. The identified structures (left to right) are styrol, phenylacetaldehyde, 2-phenylethylamine, 2-phenylethanol, and phenylacetic acid. *Source:* Reproduced with kind permission of Elsevier Science Ltd (Zöchling and Murkovic 2002).

4,8-DiMeIQx, and PhIP. Grapeseed oil had the greatest inhibitory effect, as compared with sunflower and olive oil. The authors concluded that fat replacement with sunflower, olive, or grapeseed oil in pork patties could reduce the formation of HAAs without compromising eating quality.

5.7.2 Polycyclic Aromatic Hydrocarbons

PAHs are organic compounds with several aromatic rings, as shown in Figure 5.19. They are hydrophobic, nonvolatile or with low volatility, nonbiodegradable, and highly toxic compounds. They are considered carcinogens. PAHs constitute the main component of the fumes originating during frying. Shuguang et al. (1994) extracted, chromatographed, and measured the PAHs in fumes from three commercial cooking oils and three catering shops. They found that all samples contained benzo (a) pyrene (BaP) and dibenzo (a,h) anthracene (DBahA). The high concentration of carcinogens such as BaP and DBahA in cooking oil fumes was correlated with a high incidence of pulmonary adenocarcinoma in Chinese women. Similarly, Zhu and Wang (2003) studied the PAHs in the air of six domestic kitchens and four commercial kitchens in China during 1999–2000. The total PAH content in the commercial kitchens (mainly three- and four-ring PAHs) was 17 $\mu g\,m^{-3}$, while that in the domestic kitchens (mainly two- and three-ring PAHs) was 7.6 $\mu g\,m^{-3}$. BaP levels were 0.0061–0.024 $\mu g\,m^{-3}$ in domestic kitchens and 0.15–0.44 $\mu g\,m^{-3}$ in commercial kitchens. These authors showed that conventional Chinese cooking methods are responsible for heavy PAH pollution. They also found that during fish frying, a large amount of PAHs are produced, as compared with grilling. Finally, increasing the frying temperature increased the levels of PAHs in foods, especially acenaphthene.

Chen and Chen (2003) studied the amount and variety of PAHs in fumes produced during the frying of chicken legs in edible oils. Chicken legs were fried in soybean oil, canola oil, or sunflower oil at 163 °C for 1–4 hours, and the PAHs produced were studied using GC-MS. The authors showed that most of the smoke (99%) was collected in the condensation apparatus, while the rest (1%) was adsorbed on to adsorption wool. Canola oil was found to generate the highest amounts of PAHs in smoke, followed by soybean and then sunflower oil. Perelló et al. (2009) reported that the highest concentration of PAHs was found in French fries and fish. They found 16 PAHs in fried products such as sardine, tuna, veal, hake, chicken, pork, lamb, and potatoes. Parameters such as the type of frying medium, frying time, temperature, and air properties play a significant role in the formation of PAHs in foods. An et al. (2017) reported a strong correlation between total polar compounds and PAHs during frying. Similarly, Zhao et al. (2018) investigated the polycyclic and oxygenated aromatic hydrocarbon (PAH and OPAH) concentrations in vegetable oils during storage for 270 days at 25 or 4 °C.

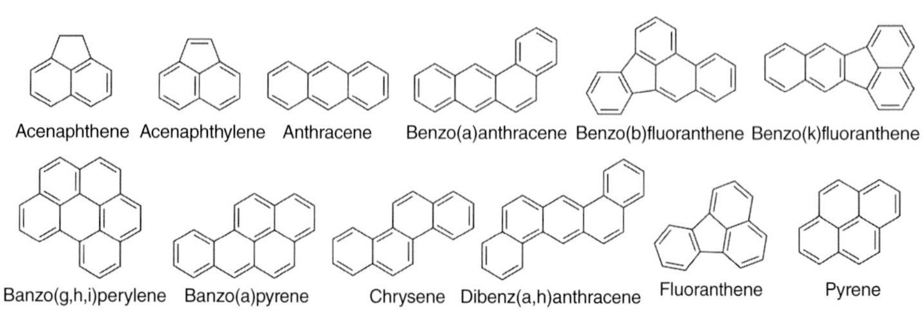

Figure 5.19 Chemical structures of important polycyclic aromatic hydrocarbons (PAHs) produced during the frying of foods.

They showed that the amounts of OPAHs and PAHs increased with storage time. Total PAHs increased from 33.11–36.77 to 45.12–58.04 μg kg^{-1} in 26 crude oil samples, while that in 26 refined oil samples increased from 16.52–20.02 μg kg^{-1} to 25.73–40.01 μg kg^{-1}. These results indicate that PAH and OPAH formation was inhibited in oils stored at a lower temperature.

Another study showed the effects of microwave oven frying on the production of PAHs (anthracene, fluorene, phenanthrene, and pyrene) in four tissues (heart, liver, neck, and wings) of Aseel chickens (Zahoor et al. 2015). The tissue samples were fried in a microwave oven and the PAHs produced were measured using high-performance liquid chromatographic (HPLC). The results indicated that the neck contained the maximum level (0.079 mg kg^{-1}) of anthracene, while the wings contained the lowest amounts (0.006 mg kg^{-1}). In the case of fluorene, the heart showed the greatest amount (225.04%), while the wings showed the least (16.18%). Phenanthrene and pyrene were maximal in wings, at 0.249 mg kg^{-1} and 0.094 mg kg^{-1}, respectively. Szterk (2015) showed that no acridine derivatives were produced during the frying of pork meat for 9 minutes at 180 °C.

Hao et al. (2016) reported the changes in 16 PAHs during the deep frying of potatoes and chicken nuggets in rapeseed, soybean, peanut, and olive oil. The HPLC results showed that all four fresh edible oils contained PAHs (189.9-2754.8 mg kg^{-1}), the amount of which increased with increasing frying time. The total mean values of four PAHs – benzo[a]anthracene, chrysene, benzo[b]fluoranthene, and benzo[a]pyrene – in the two deep-fried oils exceeded the maximum permitted level (10 mg kg^{-1}) set by the European Union by 9.2- and 6.8-fold, respectively, as shown in Table 5.4. These results indicate that more standards and regulations must be established concerning the presence of PAHs in edible oils, especially in countries where the deep-fried market is very high. They also show that rapeseed oils are best in regards to PAH formation in protein-rich foods.

Similarly, the effects of the frying medium on PAH formation in Turkish beef and lamb kokorec were reported by Babaoglu et al. (2017). The authors showed that the highest amounts were produced in beef kokorec and the tail fat group (Table 5.5).

Table 5.4 Total polycyclic aromatic hydrocarbon (PAH) compounds (benzo[a]anthracene, chrysene, benzo[b]fluoranthene, and benzo[a]pyrene) in edible oils during the frying of potato chips and chicken nuggets.

	Deep-fried potatoes			Deep-fried chicken nuggets		
Frying medium	**15 minutes**	**30 minutes**	**45 minutes**	**15 minutes**	**30 minutes**	**45 minutes**
Rapeseed oil	ND	ND	ND	ND	ND	ND
Soybean oil	ND	ND	29.66	ND	ND	ND
Peanut oil	15.1	43.0	121.3	55.9	67.1	84.0
Olive oil	23.1	55.3	71.7	20.3	54.7	88.4

Source: Redrawn with permission of Elsevier (Hao et al. 2016).
Frying conditions: 200 g oil and 150 g food at 200 °C. ND, not detected.

Table 5.5 Effects of animal fats on the formation of polycyclic aromatic hydrocarbons (PAHs) in beef and lamb kokorec samples.

		Concentration of PAHs ($\mu g\,kg^{-1}$)							
Kokoreç	**Animal fat treatment**	**BaA**	**Chry**	**BbF**	**BkF**	**BaP**	**DhA**	**BgP**	**IcP**
Beef	Tallow fat	1.35 ± 0.3 cd	1.63 ± 0.3bc	1.33 ± 0.3 cd	nq	1.84 ± 0.4c	0.46 ± 0.1c	1.80 ± 0.4c	0.71 ± 0.1b
	Subcutaneous fat	3.99 ± 0.2b	4.65 ± 0.8a	3.92 ± 0.2b	2.43 ± 0.2b	6.15 ± 0.1d	2.34 ± 0.07d	5.45 ± 0.5b	1.18 ± 0.2b
	Tail fat	4.86 ± 0.2a	5.60 ± 0.7a	5.37 ± 0.2a	2.83 ± 0.2a	7.73 ± 0.7ba	3.25 ± 0.5ba	6.81 ± 0.2a	3.66 ± 1.5a
Lamb	Tallow fat	1.77 ± 0.1c	2.15 ± 0.01b	1.77 ± 0.04c	0.91 ± 0.1c	2.44 ± 0.2c	0.83 ± 0.2c	1.44 ± 0.1 cd	0.40 ± 0.1b
	Subcutaneous fat	1.04 ± 0.08d	1.33 ± 0.09bc	1.13 ± 0.1d	nq	1.45 ± 0.3de	0.37 ± 0.1 cd	0.91 ± 0.1de	nq
	Tail fat	0.47 ± 0.02e	0.56 ± 0.03c	0.50 ± 0.04e	nq	0.63 ± 0.05e	Nq	0.47 ± 0.06e	Nd

Source: Redrawn with kind permission of Babaoglu et al. (2017).
Letters (a–e) within the column represent significant difference ($p < 0.05$).
nd, not detected; nq, not quantified; BkF, benzo[k]fluoranthene; BaP, dibenz[a,h]anthracene (DhA); BgP, benz[g,h,i]perylene; IcP, indo[1,2,3-cd]pyrene.

The interaction of lamb kokorec and the tail fat group had the lowest levels. These results suggest that the choice of frying medium is of significant value when considering PAH formation.

5.7.3 Formation of Acrolein

Acrolein is also synthesized as the end product of protein or amino acid oxidation. The principal amino acids forming acrolein are methionine, threonine, and asparagine. Methional is produced from methionine, which releases acrolein. Acrolein is also formed from threonine, through Strecker degradation. Stevens and Maier (2008) reviewed the details concerning the formation of acrolein from amino acids. Thus, frying of protein-rich food also produces acrolein with toxic properties.

5.8 Frying of Seafood

Seafood makes up 20% of all processed food consumed worldwide. In this book, we consider seafood to refer to fish, crustaceans, and molluscs. In fish, the main constituent of the flesh is water, accounting for about 70–80% by weight. The water molecules in fish muscle are tightly bound to the proteins in the structure. Protein is the second largest component by weight. Seafood proteins are classified as sarcoplasmic, myofibrillar, and stroma. Their properties in seafood are discussed in detail by Shahidi and Botta (2012). The crude protein content of seafood ranges from 17 to 22%. In crustaceans and molluscs, it may vary from 7 to 23%. Generally, the protein content of seafood increases in spring, when the food level is relatively high. Lipids are the third major constituent in fish muscle. The main manufacturers of marine lipids in the marine environment are microalgae. In fish muscle, TAGs and phospholipids containing long-chain FAs constitute the major components of lipids. The phospholipids present in the tissues account for 0.7–0.8% of all structural lipids. Tropical fish are rich in saturated phospholipids, while those in temperate waters contain unsaturated phospholipids (Shahidi and Botta 2012).

A small quantity of cholesterol is also present in seafood. Seafood such as bivalve molluscs contains phytosterol, which is obtained from microalgae and sediments. In higher fishes such as sharks, the liver stores nearly all of the lipids. These lipids may contain high-molecular-weight hydrocarbons, such as cholesterol and squalene. The lipid composition of seafood varies widely, even within the organs of the same species. The difference may be due to several factors, including the type of muscle and its location, age, sex, and maturation. Fish are often classified on the basis of their fat content as lean (<5% fat), fat (5–10% fat), and fatty (>10% fat) (Soares et al. 2015). Seafood is also rich in minerals, vitamins, enzymes, and carotenoids.

Changes in the lipid, protein or amino acid, and carbohydrate content of seafood during frying has been the main focus of research in this area. The quality of fresh seafood falls rapidly during frying, due to a variety of changes produced by chemical reactions, such as oxidation, reduction, hydrolysis, deamination, polymerization, and decarboxylation. The rate of deterioration depends on the type of species being fried, its compositional characteristics, the frying medium, the processing conditions, the packaging, and the storage conditions.

5.8.1 Changes in Carbohydrates

Seafood contains fewer carbohydrates than proteins. Carbohydrates – especially glucose – play a primary role in the metabolism of lipids and proteins. Hemre and Kahrs (1997) reported that radiolabelled glucose, when injected intraperitoneally into cod, was largely recovered as CO_2 within 24 hours. They also reported that the organs most active in glucose utilization in this species (activity per gram tissue) are, in descending order: gills > heart > red muscle > liver > white muscle. In the liver and muscle, small amounts of ingested carbohydrates are transformed into glycogen, while some fraction is converted into lipids (liver only) (Hemre et al. 2002). These results show that carbohydrates may be an important constituent during frying. However, changes in the carbohydrates content and its transformation into other compounds have received little research attention. Acrylamide and HMF have been reported as the metabolic products of carbohydrates in seafood.

The use of olive oil for deep-frying was found to promote higher levels of furanic compounds compared to sunflower oil (Pérez-Palacios et al. 2013). The amounts of furanic compounds produced decreased with decreasing temperature and frying time, as well as with an increasing delay after deep frying. The authors suggested that the generation of furanic compounds could be reduced by regulating the cooking method and conditions (e.g. using an electric oven and deep frying in sunflower oil at 160 °C for 4 minutes. Furanic compounds were also formed in frozen coated fish products during frying (Pérez-Palacios et al. 2014). Higher amounts of furfuryl alcohol (22.2 $\mu g\,g^{-1}$) and furan (11.8 $\mu g\,g^{-1}$), followed by 2-pentylfuran (3.76 $\mu g\,g^{-1}$) and 2-furfural (0.57 $\mu g\,g^{-1}$), were reported during frying in edible oil. The authors suggested that breaded crusts could contribute to the furfural content of fish during frying. Juániz et al. (2016b) studied the furan contents of bread-coated seafood in different frying oils (Figure 5.20).

When shrimp chips were deep fried or heated in a microwave oven, the acrylamide contents were higher in the microwaved ones (Chen et al. 2016). Chen et al. (2016) showed the effects of calcium supplements on the formation of acrylamide and found

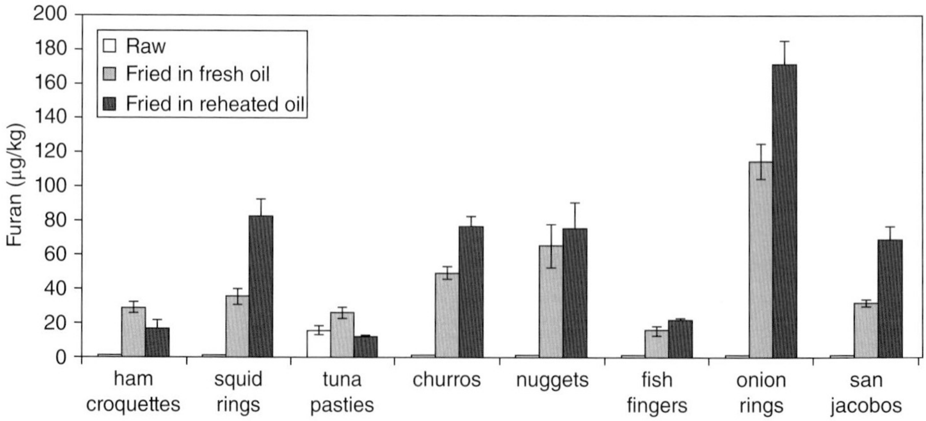

Figure 5.20 Furan (mg kg^{-1}) in frozen precooked bread-coated foods before and after deep frying with fresh and reheated olive oil. *Source:* Reproduced with kind permission of Elsevier Ltd (Juániz et al. 2016a).

the highest reduction in 0.1% calcium lactate in fried chips. Qin et al. (2017) reported acrylamide and HMF contents in different kinds of seafood using isotope dilution HPLC-MS/MS. They studied ready-to-eat seafood such as spicy pollack, spicy tuna, crispy croaker, and wild octopus, and found they contained 12.4, 21.2, 35.4, and 39.1 $\mu g\,kg^{-1}$, respectively. The HMF contents of pollack and tuna were 30.4 and 12.5 $\mu g\,kg^{-1}$, respectively. The authors showed that acrylamide and HMF contents were higher in fried ready-to-eat seafood than in boiled and roasted seafood.

5.8.2 Changes in Lipids

Seafood is rich in important lipids, such as long-chain polyunsaturated fatty acids (PUFAs), phospholipids, cholesterol, and TAGs. Yamsaengsung et al. (2017) studied the effects on tofu fish of atmospheric frying at 165 °C and vacuum frying at 120 °C. They found that high-temperature atmospheric frying caused an increase in temperature towards the centre of the fish and a destruction of the microstructure. High temperature and oxidation initiated a dark colouration and affected the chemical characteristics of the frying oil. Vacuum frying helped preserve the quality of the quality and maintained its light colour. These results suggest vacuum frying process as the method of choice for frying high-quality tofu fish and conserving oil quality. Similarly, Zeng et al. (2016) investigated the reduction of fat absorption during the deep frying of battered and breaded fish balls for 50 seconds at 170 °C, followed by 10 seconds at 190 °C. The addition of dietary fibre at 6% in the batter significantly improved the sensory quality of the balls. The fat content of the crust and the core decreased from 25.5 and 2.4% to 17.7 and 1.3%, respectively. These findings demonstrate that specific dietary fibre added to a batter can significantly reduce fat absorption by allowing the formation of low-fat fried fish products.

Ersoy (2011) showed the effects of microwave cooking and frying on the nutritional composition of European eels (*Anguilla anguilla*). An increase in the protein and fat content was observed on both cooking and frying. In cooked fish, the amount of saturated FAs and monounsaturated fatty acids (MUFAs) showed a decline, while both cooking and frying cause an increase in PUFAs. The amount of docosahexaenoic acid (DHA) increased during frying, which enhanced the eels' nutritional value. Similarly, king salmon (*Oncorhynchus tshawytscha*) from New Zealand was processed using common techniques such as poaching, steaming, microwaving, pan frying, oven baking, and deep frying in sunflower oil, and compared with the raw product (Larsen et al. 2010). The FA composition was examined to find the optimal preparation technique for both sensory and nutritional qualities (in particular, the levels of long-chain omega-3 PUFAs). Moisture and lipid losses were reported under various cooking methods. The FA composition indicated only slight changes in the different methods. There was an increase in PUFAs in the deep-fried salmon due to linoleic acid uptake from the frying oil. Omega-3 FAs were well preserved in all cooking methods except in deep frying.

Similarly, six species of finfish, common squid, and mussels freshly caught from several regions of the Mediterranean were analysed for proximate composition, energy content, cholesterol, squalene, and FA composition after frying in virgin olive oil (VOO), a traditional method of preparation in Greece. The results showed that pan frying of the selected seafood caused significant water loss and an increase in the total fat, crude protein, and energy contents. During frying, the amount of VOO absorbed ranged from

4.5 to 14.2% of fresh matter. The absorption of fat showed a nonlinear negative correlation with initial fat, fish length, and fish weight. The amount of cholesterol increased from 25–106 mg/100 g fresh weight (FW) to 33–130 mg/100 g FW after frying. The VOO absorbed during frying caused an increase in the squalene content of 20–78 times. There were significant changes in FA composition, especially in terms of MUFAs, which predominated in all fried samples (Table 5.6). The intakes of fat and major FA classes on consumption of the pan fried seafood were comparable with the respective average Greek values. Frying in VOO resulted in a two to three times reduction of the atherogenic index and a somewhat smaller reduction of the thrombogenic index. From these results, the authors concluded that pan frying in VOO provides a healthy lipid profile in terms of n-6/n-3 ratio, important FAs, and total fat content in fish and shellfish (Kalogeropoulos et al. 2007).

Weber et al. (2008) reported the effects of boiling, conventional and microwave baking, grilling, and deep frying in soybean, canola, or partially hydrogenated vegetable oil on the oxidation and proximate and FA composition of silver catfish (*Rhamdia quelen*) fillets. Frying in canola oil increased the n-3/n-6 ratio, while frying in soybean oil increased the PUFA content and frying in hydrogenated vegetable oil incorporated *trans* fatty acids (TFAs) into the fillets. Similarly, the long-chain omega-3 PUFA content of mackerel was not affected by deep frying (Sebedio et al. 1993). This study showed that small quantities of polar compounds, polymers, and geometrical isomers of the long-chain PUFAs were formed during deep frying. The FA composition of the frying medium, the fish skin, and the meat all showed evidence of extensive lipid exchange between the fish and the frying medium. The major isomers of eicosapentaenoic acid (EPA) were formed by thermal action, constituting less than 0.1% of the total C20:5 recovered from the used canola oil.

When pollock was deep fat fried without batter, extensive absorption of the frying oil occurred (Nawar et al. 1990). The batter coating was found to protect the fillets against loss of moisture, absorption of oil, and dilution or loss of flavour volatiles. Frying fish may accelerate or delay the oxidation of oil heated to frying temperatures, depending on the conditions of frying. Oxidation of cholesterol in squid occurred upon exposure to the sun and to ultraviolet (UV) and infrared (IR) lamps. Similarly, pan frying and deep fat frying of breaded and nonbreaded fillets of lake trout (*Salvelinus namacush*), white sucker (*Catastomus commersonni*), and bluegill (*Lepomis macrochirus*) were reported by Mai et al. (1978). These authors found that breading absorbed significant amounts of frying oil. Trout fillets did not absorb lipids, but sucker and bluegill absorbed significant amounts of oil. The FA composition of the trout showed slight changes, while the sucker and bluegill reflected the FA composition of the oil. The changes in the concentration of phospholipids varied with the method of cooking and the species of fish. These results suggest that the frying of seafood increases its nutritional quality, but also produces changes in its lipid composition; thus, quality should be the prime target of frying.

5.8.3 Changes in Proteins

Seafood is a rich source of important high-quality proteins, which provide large amounts of the essential FAs required by the human body. The absorption and digestibility of seafood are better than those of other food proteins. The nutritional value and net health benefits of proteins are essential to seafood's consumer acceptability. Proteins are degraded or denatured during heating and frying. Therefore, it is

Table 5.6 Changes in the FA composition ($g\,kg^{-1}$) of Mediterranean fish and molluscs fried in virgin olive oil.

FA	*Atherina boyeri*		*Engraulis encrasicholus*		*Boops boops*		*Mullus barbatus*		*Spicara smaris*		*Trachurus trachurus*		*Loligo vulgaris*		*Mytilus galloprovincialis*	
	Raw	Fried	Raw	Fried	Raw	Fried	Raw	Fried	Raw	Fried	Raw	Fried	Raw	Fried	Raw	Fried
12:0	3.9	1.0	1.8	0.5	1.9	0.5	4.6	1.2	1.4	0.5	4.0	1.4	0.5	—	0.8	—
14:0	94.3	14.3	102.1	17.9	97.7	20.0	57.8	18.2	76.0	15.4	65.0	20.4	25.7	7.8	72.5	16.6
15:0	13.6	1.9	13.7	2.3	24.0	5.1	20.3	6.0	14.2	3.5	8.4	2.3	6.1	1.8	30.6	13.1
16:0	220.0	167.4	235.2	151.0	255	151.0	248	179.7	268	178.5	202	167.6	228	155.8	191	162.8
17:0	8.2	1.2	9.8	1.7	11.7	5.3	13.3	3.8	75.3	2.7	7.0	2.6	7.0	2.3	10.1	3.9
18:0	49.8	31.3	33.7	21.6	55.5	30.0	44.0	29.1	31.4	31.4	59.2	35.1	37.2	27.5	34.8	31.0
20:0	—	2.7	4.0	1.9	4.0	1.5	1.8	2.2	—	—	—	—	—	—	1.0	2.4
24:0	2.0	0.8	4.0	0.7	2.8	1.4	8.1	4.5	2.6	1.3	2.6	1.1	1.3	0.9	1.6	1.3
14:1	18.3	2.4	4.3	0.6	1.9	—	5.3	0.6	2.4	—	2.9	0.6	2.5	0.7	3.2	0.6
16:1, n-9	26.5	6.3	4.1	3.0	5.8	3.4	10.9	4.8	4.6	2.8	4.9	3.5	3.5	1.0	2.6	2.7
16:1, n-6	155.8	37.9	101.6	29.3	63.9	23.8	139	47.4	95.9	28.2	127	55.7	9.1	9.8	151	67.4
17:1	6.4	0.6	10.4	1.2	2.3	—	3.9	1.1	2.6	1.2	7.0	3.2	3.8	—	10.6	1.5
18:1 n-9	119.7	508.0	71.3	562.6	155	511.2	182	505.6	153	482.2	175	459.2	25.3	514.6	35.8	417.6
18:1 n-6	49.4	29.5	28.3	14.0	28.2	23.0	51.2	17.8	31.2	25.1	46.3	28.6	11.9	23.2	3.4	24.7
20:1 n-9	7.6	9.3	50.0	9.8	22.9	10.4	14.3	5.7		—	12.7	13.0	16.1	6.7	44.0	18.3
22:1 n-9	2.3	2.8	6.9	6.0	6.5	2.8	3.1	2.7	1.0	3.1	3.5	1.9	3.2	0.6	4.6	1.7
24:1 n-9	2.1	0.9	4.2	0.5	2.9	0.8	1.0	—	3.3	1.2	1.0	0.5	5.0	1.5	0.9	—
18:2 n-6	44.4	72.2	16.0	64.4	17.9	68.6	13.4	64.9	18.1	64.9	9.2	52.2	11.0	64.2	18.5	71.2
18:3 n-6	1.6	1.3	1.2	1.0	1.3	1.3	1.6	0.5	—	—	1.0	0.7	—	—	2.1	0.8

(*Continued*)

Table 5.6 (Continued)

FA	*Atherina boyeri*		*Engraulis encrasicholus*		*Boops boops*		*Mullus barbatus*		*Spicara smaris*		*Trachurus trachurus*		*Loligo vulgaris*		*Mytilus galloprovincialis*	
	Raw	Fried	Raw	Fried	Raw	Fried	Raw	Fried	Raw	Fried	Raw	Fried	Raw	Fried	Raw	Fried
20:2 n-6	1.7	2.9	1.8	1.0	3.1	5.2	3.8	1.0	1.8	1.2	2.5	1.9	2.5	1.6	3.3	1.7
20:3 n-6	0.9	0.7	0.9	—	1.2	0.6	1.3	0.5	0.7	0.3	0.8	0.5	0.3	0.1	1.7	0.5
20:4 n-6	12.0	12.8	7.4	1.9	10.6	4.9	19.9	10.8	1.5	2.0	5.2	6.2	15.9	4.3	18.0	6.6
22:2 n-6	2.2	1.4	2.7	0.9	1.4	2.4	1.7	1.2	5.5	1.4	1.0	1.5	0.7	1.0	2.3	0.9
22:4 n-6	1.4	0.8	6.2	0.8	2.9	0.8	2.0	0.8	4.5	1.1	2.4	1.8	2.0	1.0	6.9	2.2
22:5 n-6	2.2	1.2	3.7	0.8	4.8	2.9	3.7	2.4	1.4	1.2	1.4	1.1	—	—	5.7	1.9
18:3 n-3	18.7	21.5	8.3	9.9	8.4	10.8	4.1	9.9	5.3	12.2	3.4	8.6	0.9	6.3	15.5	12.0
20:3 n-3	2.4	0.4	1.8	—	2.0	—	2.9	0.5	6.5	0.5	1.4	—		—	1.6	—
20:5 n-3	32.7	17.0	82.3	20.4	60.9	18.2	44.8	26.8	42.9	22.9	69.3	44.8	174	47.8	173	67.9
22:5 n-3	8.0	4.9	9.1	1.4	7.9	5.8	12.7	7.1	5.9	2.5	9.9	7.8	5.9	1.7	9.7	3.0
22:6 n-3	63.9	25.0	152.2	41.7	105	66.8	45.1	22.5	110	48.3	143	52.7	386	100.1	120	43.5

Source: Reproduced with permission (Kalogeropoulos et al. 2007).
Values are means of triplicate readings.

important to determine the quality of the seafood during frying. The total protein content of anchovy kamaboko (*Engraulis japonica*) was found to significantly decrease during frying (Ramos et al. 2012).

Different cooking methods were applied, such as grilling, frying, steaming, and microwaving to determine the changes in proximate composition, and protein quality of chub mackerel (*Scomber japonicus*). It was found that the protein and total essential amino acid contents increased in all cooked samples, from 47.21% in the raw sample to 63.87% in the grilled samples. *In vitro* protein digestibility increased from 79.4% in the raw sample to 86.43% in the fried samples. These results suggest that grilling and steaming have more advantageous effects on the protein quality of chub mackerel than does frying.

Oluwaniyi et al. (2010) showed the effects of different processing methods on the amino acid composition of four fishes commonly consumed in Nigeria: *Clupea harengus, Scomber scombrus, Trachurus trachurus,* and *Urophycis tenuis.* The samples were boiled, roasted, or fried in different oils. Total amino acids ranged from 76.72 to 98.01 g/100 g protein and total essential amino acids from 28.95 to 44.73 g/100 g protein in all samples. Reduced levels of total amino acids were observed during frying, with the lowest values found in palm oil frying. In contrast, boiling and roasting increased the total amino acid contents, with higher values in the roasted compared to the boiled samples. The raw, boiled, and roasted fish samples all gave amino acid scores greater than 100%; for the fried samples, the limiting amino acids were either leucine or threonine. The results show that roasting and boiling have more desirable effects on the amino acid content of fish samples than frying.

Erkan et al. (2010) showed that the amino acid contents of six raw fishes were significantly lower than those found in cooked fishes. The amounts of semi-essential and other amino acids in grilled Atlantic bonito and anchovy and in fried bluefish, red mullet, and hake were significantly higher than in cooked fishes. In another study, Ismail and Ikram (2004) compared the protein and amino acid contents of boiled and fried fishes: kembong (*Rastrelliger kanagurta*), sardine (*Sardina pilchardus*), red tilapia (*Oreochromis mossambicus*), and black tilapia (*Oreochromis mossambicus*). Seventeen amino acids were present in all the fishes, both essential (lysine, histidine, threonine, valine, methionine, leucine, isoleucine, and phenylalanine) and non-essential (arginine, aspartic acid, serine, glutamic acid, proline, glycine, alanine, cysteine, and tyrosine). These results, however, showed no significant difference in amino acid content among the boiled and fried fishes. Heat treatment in boiling water for 5 minutes at 100 °C and frying in palm oil for 3 minutes at 160 °, did not change the quality of proteins in any of the studied fishes.

5.8.4 Heterocyclic Aromatic Amines

The production of HAAs has also been reported during the frying of seafood, such as fishes. These compounds are toxic and possess mutagenic potential (Sugimura et al. 2004). Oz et al. (2010) showed that the highest total HAAs were present in fish samples processed via microwave, oven, hot plate, pan frying, and barbecuing, with values of 18.0, 4.28, 3.12, 6.98, and 5.22 ng g^{-1}, respectively.In another study, Oz et al. (2007) showed the effects of different cooking methods (deep fat frying, pan frying, grilling, and barbecuing) on the formation of HAAs in fillets of rainbow trout (*Oncorhynchus mykiss*) and brown trout (*Salmo trutta fario*). Barbecued brown trout contained 0.12 ng g^{-1} of IQ and 0.02 ng g^{-1} of 4, 8-DiMeIQx, while grilled rainbow trout contained

0.02 ng g^{-1} of 4,8-DiMeIQx. Thus, optimum conditions for obtaining minimum levels of heterocyclic aromatic compounds in seafood can be calculated in order to obtain better nutritional acceptability among consumers.

5.8.5 Cholesterol Oxidation

The cholesterol present in seafood is also thermally oxidized during frying. Such oxidized species are known as cholesterol oxidation products (COPs). The primary oxidation products formed are 7-hydroperoxycholesterol, 5-hydroperoxycholesterol, and 5,6-epoxycholesterol. The 7-hydroperoxycholesterol is degraded to form 7-hydroxycholesterol, which consequently produces 7-ketocholesterol. Rodriguez-Estrada et al. (2014) showed that 7-ketocholesterol is a marker for cholesterol oxidation in marine foods. On rearrangement, 5-hydroperoxycholesterol produces either 6-ketocholesterol or, more likely, 5,6-epoxycholesterol. The 5,6-epoxycholesterol is converted to 6,7-dihydroxycholesterol by reaction of a hydroxyl radical. The 6,7-dihydroxycholesterol then produces 6-hydroxy-7-ketocholesterol, which may be converted to 7-ketocholesterol or 6-ketocholesterol, as shown in Figure 5.21. Ohshima et al. (1993) strongly suggest that cholesterol oxidation in fish products proceeds in combination with the oxidative decomposition of the PUFAs in fish oils.

The amount of 7-ketocholesterol produced during the cooking and frying of salmon has been reported by Echarte et al. (2001, 2005). Al-Saghir et al. (2004) showed that salmon fillets produce several COPs on frying, such 7-ketocholesterol, 5,6-epoxycholesterol, 7-hydroxycholesterol, and their isomers. Freitas et al. (2015) showed that different cooking methods, including frying, for various fish species produce 7-ketocholesterol in relatively high amounts. Lira et al. (2017) showed that pan frying of king mackerel produced only 7-hydroxycholesterol. More work is required in other seafood, however.

5.9 Frying of Vegetables

Vegetables constitute the major part of our daily diet. They are obtained from plant sources. The widely accepted classification of vegetables is based on the consumable or edible part: bulbs, flowers, fruits, leaves, roots, seeds, stems, and tubers (Table 5.7). The chemistry differs by class. Significant variation in nutrient content can occur within a single family of vegetables, perhaps due to differences in soil composition, maturity, environmental conditions, temperature, and sunlight availability (Andersen et al. 2017; Marles 2017). It is not possible to discuss the chemistry of all vegetables in this chapter, but some of the most important bioactive components apart from lipids and carbohydrates will be discussed in this section; further details can be found elsewhere (Yalcin and Çapar 2017). It is worth mentioning that not all vegetables are fried. Those vegetables that are most commonly fried include potato, tomato, and leafy vegetables.

5.9.1 Phenolic Compounds

Vegetables are rich in a variety of phenolic compounds. The key dietary sources of phenolic compounds are fruits and beverages (i.e. fruit juices, tea, coffee). Fruits such as apple, grape, pear, cherry, and several berries contain up to 300 mg/100 g phenolic

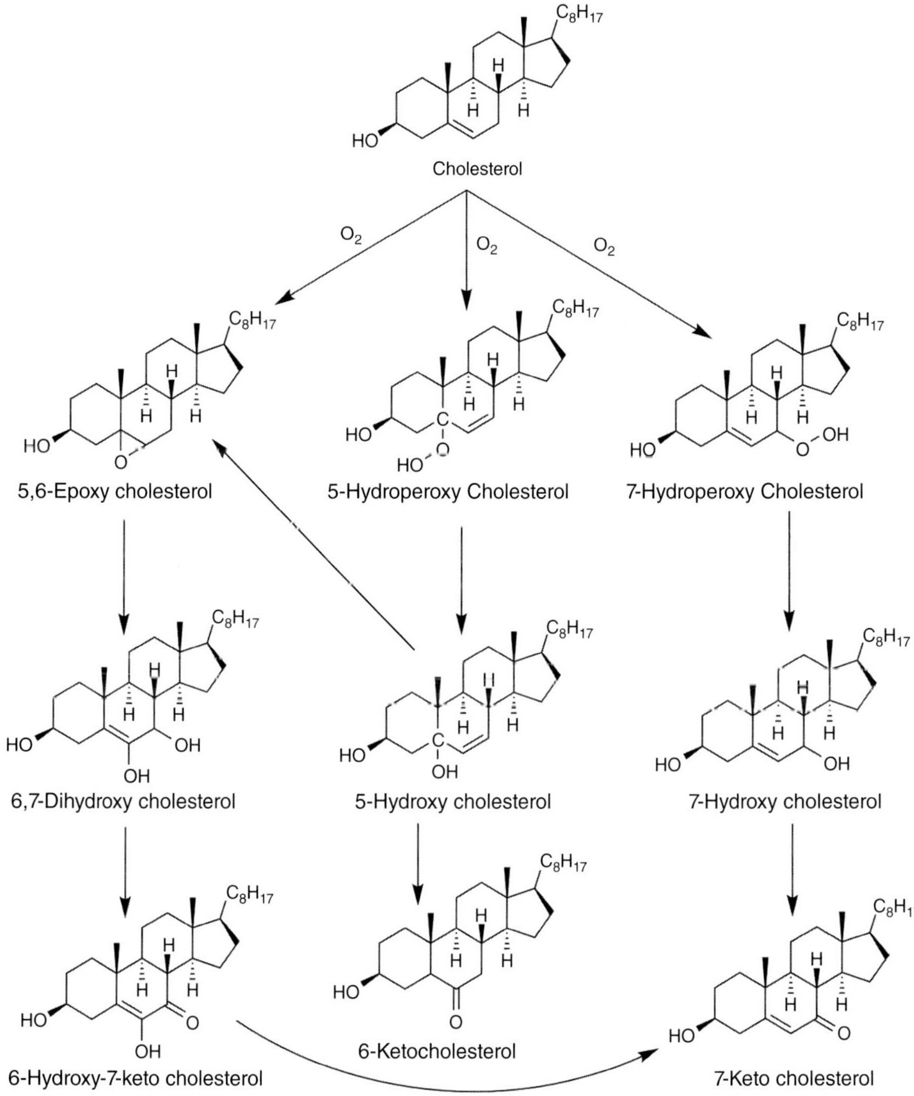

Figure 5.21 Mechanism of formation of different cholesterol oxidation products.

compounds FW. Strawberry and blackberry are considered among the best sources. Phenolic compounds from different vegetable sources can be used as ingredients to increase the stability, taste, and texture of frying foods. For example, onion, tomato, garlic, and spices are added to Chapli kebab. These ingredients provide several phenolic compounds, such as free gallic acid, caffeic acid, quercetin, ellagic acid, kaempferol, vanillic acid, luteolin, p-coumaric acid, catechin, rosmaric acid, apigenin, isorhamnetin, diosmetin, valonic acid, and caftaric acid or their derivatives (Zeb and Haq 2017). Complex or higher polyphenols degrade to simpler phenolic compounds during frying.

Table 5.7 Classification of common vegetables.

Class	Vegetables	Bioactive composition
Bulbs	Fennel, garlic, leek, onion, shallot, spring onion	Phenolic compounds
Flowers	Artichoke (globe), broccoflower, cauliflower, broccoli, choi sum, courgette, and other squash flowers	Phenolic compounds
Fruits	Bitter melon, capsicum, chilli, choko, courgette, cucumber, eggplant, fuzzy melon, Indian marrow, marrow, plantain, pumpkin and squash, scallopini, tindora, tomatillo, tomato, turia (ribbed gourd)	Phenolic compounds, carotenoids, vitamins
Leaves	Bok choy, Brussels sprout, cabbage, lettuce, radicchio, silverbeet, sorrel, spinach, watercress, Peking cabbage, and all leafy vegetables	Phenolic compounds, carotenoids, vitamins
Roots	Beetroot, carrot, celeriac, parsnip, radish, swede, turnip	Phenolic compounds, carotenoids, vitamins
Seeds	Bean (green, French, butter, snake), broad bean, pea, snow pea, sweet corn	Phenolic compounds, carotenoids, vitamins
Stems	Asparagus, celery, kohlrabi	Phenolic compounds
Tubers	Earth gem, Jerusalem artichoke, kumara, potato, yam	Phenolic compounds

Chinese cabbage and pakchoi were stir fried at 160–250 °C for 1–8 minutes (Nugrahedi et al. 2017). The results showed that aliphatic glucobrassicanapin is the most abundant glucosinolate in these vegetables, contributing 48 and 63% of their total glucosinolate content, respectively, followed by glucoiberin and gluconapin. Stir frying was found to retain glucosinolates at high temperature, which was attributed to the quick inactivation of myrosinase (glucosinolate's hydrolytic enzyme) during the first minutes of frying, as shown in Table 5.8. Similarly, Juániz et al. (2016a) studied the effect of frying in olive and sunflower oil on the polyphenolic composition of onion, green pepper, and cardoon. Their ultra-high-performance liquid chromatography with photodiode array detector and high-resolution mass spectrometry (UHPLC-PDA-HR-MS) results showed that the major compounds were quercetin and isorhamnetin derivatives in onion, quercetin and luteolin derivatives in green pepper, and chlorogenic acids in cardoon. Frying produced an increase in the number of phenolic compounds in all vegetables. The authors suggested that this increase may be due to a thermal destruction of the cell walls and subcellular compartments during frying that favours the release of these compounds.

In carrot, significant losses occurred in the chlorogenic acid content during frying, from 44.0 to 3.0 mg/100 g (Miglio et al. 2008). The amount of caffeic acid increased from 18.0 to 29.3 mg/100 g. The amount of coumaric acid decreased upon frying at 170 °C. Chiou et al. (2007) identified 12 polyphenolic compounds in frying oils and zucchini samples, among which tyrosol predominated; chlorogenic acid was the major compound in other vegetable samples. The triterpene acids maslinic, oleanolic, and ursolic were also reported in frying oils and fried vegetables. Similarly, Kalogeropoulos et al. (2007) fried potatoes, green peppers, zucchinis, and eggplants using Mediterranean

Table 5.8 Changes in the glucosinolate contents of Chinese cabbage and pakchoi during stir frying.

Sample	Frying Time (minutes)	Total glucosinolate (μmol g^{-1} DW)			
		160	200	225	250
Chinese cabbage	0	3.07 ± 0.11 (a)	3.07 ± 0.11 (a)	3.07 ± 0.11 (a)	3.07 ± 0.11 (a)
	1	4.91 ± 0.38 (bc)	4.71 ± 0.25 (b)	5.37 ± 1.00 (a)	6.60 ± 0.33 (b)
	2	4.96 ± 0.08 (bc)	4.99 ± 0.50 (b)	4.93 ± 0.70 (a)	6.35 ± 0.10 (b)
	4	4.19 ± 0.10 (ab)	5.01 ± 0.16 (b)	5.70 ± 1.17 (a)	6.37 ± 0.70 (b)
	8	6.14 ± 0.88 (c)	5.13 ± 0.11 (b)	5.20 ± 0.78 (a)	5.50 ± 0.54 (b)
Pakchoi	0	2.29 ± 0.01 (a)	2.29 ± 0.01 (a)	2.29 ± 0.01 (ac)	2.29 ± 0.01 (a)
	1	2.94 ± 0.41 (abd)	2.77 ± 0.44 (a)	2.31 ± 0.14 (ac)	2.56 ± 1.05 (a)
	2	3.74 ± 0.06 (bc)	2.94 ± 0.28 (a)	1.76 ± 0.12 (bc)	2.23 ± 0.62 (a)
	4	3.97 ± 0.10 (c)	2.67 ± 0.94 (a)	2.51 ± 0.20 (a)	2.63 ± 0.69 (a)
	8	2.29 ± 0.19 (ad)	2.39 ± 0.48 (a)	1.60 ± 0.23 (b)	1.61 ± 0.44 (a)

Source: Reproduced with Permission of Springer Nature (Nugrahedi et al. 2017).
Different letters within a column indicate significant differences ($p < 0.05$).

shallow frying in VOO and found that they provided an additional intake of α-tocopherol, terpenic acids, and polyphenols such as tyrosol and chlorogenic acid. These results indicate that significant changes occur during prolonged frying.

5.9.2 Carotenoids

Vegetables – particularly leafy vegetables – are also rich in several carotenoids, which provide their colour. Carotenoids include more than 700 more or less identical organic compounds, which are mainly classed as oxygenated or nonoxygenated. Lutein, β-carotene, lycopene, zeaxanthin, flavoxanthin, canthaxanthin, capsanthin, capsorubin, and β-cryptoxanthin are the best known in vegetables (Shahidi et al. 2017). Lutein, neoxanthin, violaxanthin, apo-carotenal, and β-carotene are present in high amounts in leafy vegetables of the brassica family and spinach varieties, as shown in Table 5.9 (Zeb 2017).

Spinach is a widely consumed leafy vegetable that provides significant nutritional and health benefits. The effects of high-temperature frying on the carotenoid, chlorophyll, and tocopherol contents of spinach leaves were reported by Zeb and Nisar (2017). They thermally processed the leaves in sunflower oil at 250 °C for 15, 30, 45, and 60 minutes. Reversed phase HPLC-DAD results showed a total of eight carotenoids, four chlorophylls, and one α-tocopherol in the fried leaves. Lutein, neoxanthin, violaxanthin, and β-carotene-5,6-epoxide were the major carotenoids observed. Amounts of β-carotene-5,6-epoxide, luteoxanthin, and lutein and its *Z*-isomers and chlorophyll *b*' isomer were increased. There was a significant decrease in the amounts of neoxanthin and violaxanthin with increasing frying time, as shown in Table 5.10.

Kao et al. (2012) studied the effects of boiling, stir frying, and deep frying on the composition of lutein, zeaxanthin, β-carotene and their *trans*/*cis* (*E*/*Z*) isomers in 25 different

Table 5.9 Carotenoid and pigment compositions of selected vegetables leaves.

	Composition (mg/100 g)					
Identity	***Bactris Compestris***	***Brassica rapa***	***Brassica juncea***	***Malva neglecta***	***Spinach***	***Spinach oriental***
α-Tocopherol	4.97 ± 0.12a	1.15 ± 0.02b	1.35 ± 0.01b	5.66 ± 0.02c	2.61 ± 0.01d	1.74 ± 0.11e
Chlorophyll *b*	0.46 ± 0.05a	3.59 ± 0.11b	8.1 ± 0.21c	3.96 ± 0.03d	0.18 ± 0.01e	2.96 ± 0.05f
All-*E*-anhydrolutein III	0.15 ± 0.02a	0.16 ± 0.01a	0.63 ± 0.11b	0.35 ± 0.01c	0.12 ± 0.01a	—
β-Carotene-5,6-epoxide	0.16 ± 0.01a	0.61 ± 0.01b	1.44 ± 0.22c	0.97 ± 0.01d	—	0.18 ± 0.02a
All-*E*-Neoxanthin	2.86 ± 0.03a	2.82 ± 0.04a	2.88 ± 0.01a	3.99 ± 0.02b	2.91 ± 0.05a	2.95 ± 0.04a
All-*E*-violaxanthin	7.15 ± 0.11a	7.66 ± 0.01b	7.82 ± 0.21b	11.2 ± 0.02c	10.4 ± 0.04d	10.9 ± 0.01d
13-*Z*-Zeaxanthin	0.34 ± 0.02a	0.36 ± 0.02a	0.29 ± 0.01a	0.65 ± 0.001b	0.94 ± 0.06c	1.12 ± 0.01c
9-*Z*-Zeaxanthin	0.56 ± 0.04a	0.34 ± 0.01b	0.48 ± 0.02a	0.52 ± 0.01a	0.91 ± 0.04c	0.77 ± 0.02d
All-*E*-Lutein	19.1 ± 0.08 a	18.9 ± 0.09a	18.4 ± 0.41a	24.1 ± 0.41b	20.6 ± 0.52a	18.7 ± 0.26a
9-*Z*-Lutein	0.59 ± 0.02a	0.69 ± 0.01b	0.68 ± 0.03b	1.11 ± 0.01c	1.72 ± 0.05d	1.75 ± 0.02d
9′-*Z*-Lutein	0.98 ± 0.11a	0.93 ± 0.02a	1.1 ± 0.01b	1.18 ± 0.03c	0.95 ± 0.01a	0.84 ± 0.01d
Apo-caroten-8-al	17.7 ± 0.15a	17.7 ± 1.30a	16.6 ± 0.06a	15.1 ± 0.02b	16.9 ± 0.21a	16.9 ± 0.13a
Chlorophyll *a*	26.1 ± 1.21a	25.3 ± 1.12a	22.2 ± 0.22b	20.5 ± 0.72b	24.1 ± 0.23c	22.8 ± 0.32b
β-Cryptoxanthin ester	0.46 ± 0.03a	0.62 ± 0.02b	0.59 ± 0.01b	0.16 ± 0.01c	0.18 ± 0.01c	0.18 ± 0.01c
All-*E*-β-carotene	15.3 ± 0.51a	15.1 ± 0.44a	12.4 ± 0.62b	6.52 ± 0.81c	12.4 ± 0.13b	11.7 ± 0.24b
13-*Z*-β-carotene	0.73 ± 0.02a	0.72 ± 0.11a	0.67 ± 0.01a	0.28 ± 0.01b	0.69 ± 0.02a	0.48 ± 0.001c

Source: Reproduced with kind permission of Springer Nature (Zeb 2017).
Values are expressed as mg 100 g^{-1} FW (n = 5). Different letters (a–f) in the same row represent significance at $p < 0.05$ (Tuky's test, n = 5).

Table 5.10 Effects on carotenoid, chlorophyll, and α-tocopherol contents of the frying of spinach leaves in sunflower oil at 250 °C.

			Carotenoids (mg/100 g)				
Peak	**Compound**	**p < value**	**Control**	**15**	**30**	**45**	**60**
1	α-Tocopherol	0.01	0.73 ± 0.05^{a}	1.91 ± 0.11^{b}	2.87 ± 0.16^{c}	3.18 ± 0.05^{d}	3.64 ± 0.12^{e}
2	β-Carotene-5,6-epoxide	0.01	1.44 ± 0.05^{a}	2.91 ± 0.13^{b}	1.33 ± 0.11^{a}	4.65 ± 0.08^{c}	6.65 ± 0.25^{d}
3	All-*E*-Neoxanthin	0.05	2.62 ± 0.15^{a}	4.78 ± 0.21^{b}	1.27 ± 0.09^{c}	1.53 ± 0.11^{c}	2.18 ± 0.15^{d}
4	All-*E*-Violaxanthin	0.01	5.15 ± 0.21^{a}	2.24 ± 0.12^{b}	1.79 ± 0.08^{c}	1.83 ± 0.06^{c}	0.78 ± 0.13^{d}
5	All-*E*-Luteoxanthin	0.01	0.42 ± 0.05^{a}	1.83 ± 0.03^{b}	1.52 ± 0.08^{c}	1.12 ± 0.11^{d}	2.13 ± 0.04^{e}
6	All-*E*-Mutatoxanthin	0.01	0.58 ± 0.04^{a}	2.71 ± 0.11^{b}	0.46 ± 0.02^{c}	0.63 ± 0.03^{a}	0.83 ± 0.02^{d}
7	All-*E*-Lutein	0.001	19.1 ± 1.13^{a}	37.5 ± 0.95^{b}	44.2 ± 1.21^{c}	47.8 ± 1.14^{d}	50.3 ± 1.31^{d}
8	9-*Z*-Lutein	0.01	10.1 ± 0.05^{a}	1.99 ± 0.16^{b}	4.68 ± 0.21^{c}	5.11 ± 0.11^{d}	7.49 ± 0.33^{e}
9	9′-*Z*-Lutein	0.01	2.84 ± 0.04^{a}	2.01 ± 0.02^{b}	2.45 ± 0.05^{c}	2.52 ± 0.5^{c}	1.37 ± 0.06^{d}
10	Chlorophyll *b*	0.01	2.11 ± 0.01^{a}	3.14 ± 0.03^{b}	2.01 ± 0.13^{a}	0.13 ± 0.01^{c}	0.11 ± 0.03^{c}
11	Chlorophyll *b*’	0.05	15.6 ± 0.91^{a}	13.9 ± 0.45^{b}	11.3 ± 0.45^{c}	9.93 ± 0.53^{d}	8.45 ± 0.31^{e}
12	Chlorophyll *b*’ isomer	0.05	2.61 ± 0.01^{a}	2.05 ± 0.01^{b}	2.74 ± 0.02^{c}	2.92 ± 0.04^{d}	3.61 ± 0.05^{e}
13	Chlorophyll *a*	0.01	19.3 ± 1.01^{a}	10.3 ± 0.72^{b}	5.11 ± 0.31^{c}	4.0 ± 0.31^{d}	2.82 ± 0.31^{e}

Source: (Zeb and Nisar 2017).

Values are expressed as mean ± SD of n = 3 readings. Different letter (a–e) in the same row represent significant at the given p values.

vegetables commonly consumed in Taiwan. They reported that cilantro, Thai basil leaves, sweet potato leaves, and choy sum had the highest total carotenoid contents (TCCs). There was no effect from boiling, which thus preserved the most carotenoids, while stir frying and deep frying considerably reduced the TCC and especially all-*E*-forms of the carotenoids. Similarly, Rajagopal et al. (2007) reported the effects of induction stir frying on carotenoid retention in carrot, green spring onion, red pepper, snow pea, white spring onion, and yellow pepper. The stir-fried vegetables retained more than 85% (w/w) β-carotene, lutein, zeaxanthin, α-carotene, β-cryptoxanthin, and lycopene. The majority of the consumer panellists reported liking extremely/liking very much or liking moderately/liking slightly the appearance, colour, flavour, texture, and overall acceptability of the stir-fried carrot, green spring onion, red pepper, snow pea, and yellow pepper. The authors thus showed that induction stir-fried vegetables had excellent carotenoid retention and good consumer acceptability.

Kidmose et al. (2006) analysed a number of leafy vegetables in order to study the variation in β-carotene and vitamin A in commonly consumed vegetables in Asia. The β-carotene content varied between 16 and 6630 mg/100 g FW in the selected vegetables. The retention of β-carotene and the formation of *cis* isomers were studied during stir frying. The results showed that retention of all-*E*-β-carotene varied between 73 and 98% in sweet bell pepper, sweet potato, and tomato. In sweet potato, 13-*Z*-β-carotene was the major *cis* isomer of β-carotene, while minor amounts of 15-*Z*- and 9-*Z*-β-carotene were also formed. The total amount of *Z*-isomers of β-carotene formed during frying depended on the frying time. In leafy vegetables, only 13-*Z*-β-carotene was detected during frying. Studies on the standard β-carotene oxidation in oils showed that β-carotene either produced its *Z*-isomers or oxidized directly to 5,6-epoxide or diones without isomerization (Zeb 2012). The epoxides likewise either transformed into the *Z*-isomers or formed *Z*-isomers of apo-carotenals, whereas the *Z*-isomer of β-carotene, upon oxidation, could produce *Z*-isomers of epoxides or apo-carotenals (Figure 5.22). Apo-carotenals can also be produced directly upon the oxidative breakdown of β-carotene. The dione produced may be isomerized to epoxides or cleaved form apo-keto-carotenal, which upon further cleavage produces short-chain species. This mechanism is in accordance with the scheme reported by Caris-Veyrat et al. (2001) for the model system and by Pénicaud et al. (2011) for the oxidation of β-carotene in fruits and vegetables. The final degradation products are short-chain products, which are volatile in nature and thus evaporate from the food oils during thermal treatment.

5.9.3 Vitamins

Vitamins are natural organic compounds that perform several biological roles. Fat-soluble vitamins such as vitamins A, E, D, and K are mostly stable during the frying of vegetables. However, water-soluble vitamins are mostly degraded or destroyed during frying and are therefore given less consideration (Fillion and Henry 1998). Maharaj et al. (2015) showed that the boiling and frying of several Fijian vegetables, such as drumstick leaves, taro leaves, bele leaves, amaranth leaves, fern/ota, okra, and French bean, affected their folate contents. Interestingly, boiling destroyed more folate than frying. Similarly, a large set of foods representing dietary patterns in Asia, Africa, Europe, the United States, and Latin America was cooked or fried and then analyzed in terms of its vitamins compositions (Agte et al. 2002). Cooking or frying

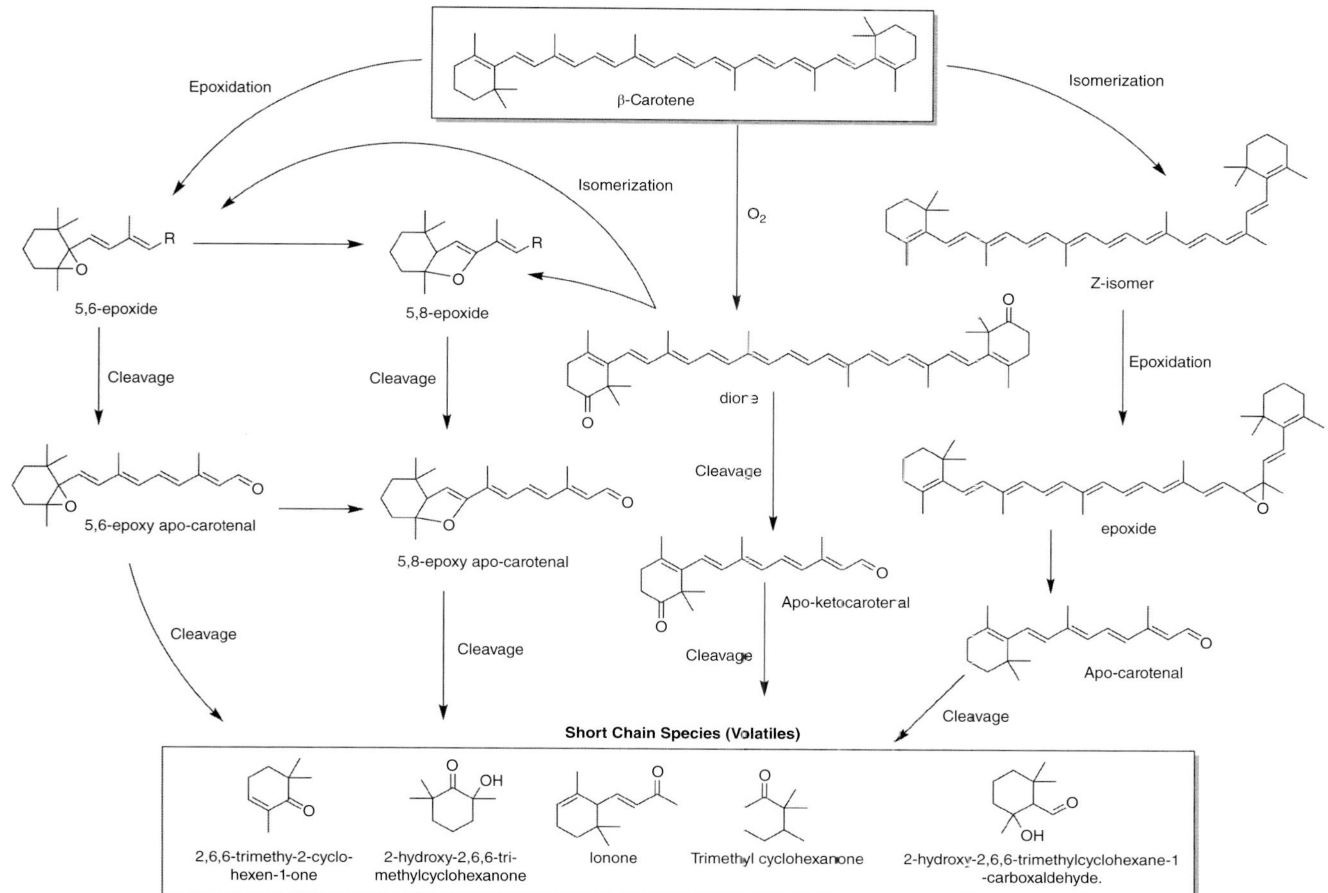

Figure 5.22 Proposed mechanism of degradation and oxidation of β-carotene during frying.

losses were 34.6, 30, 52.2, 45.9, and 32.2% in the case of ascorbic acid, thiamine, riboflavin, beta-carotene, and folic acid, respectively.

Tocopherols and their isomers are present in a variety of vegetables. They are strong antioxidants and protect the food matrix during frying and storage (Seppanen et al. 2010). Tocopherols have been widely studied during the frying process, especially in frying medium. They have been found to increase in number during frying, which has been attributed to their migration from the frying medium into the vegetable matrix (Chiou et al. 2012).

5.9.4 Phytosterols

Phytosterols are triterpene compounds with a similar structure to that of cholesterol. They are present in both free and bound form with glycosides, FAs, phenolic compounds, and so forth. In the diet, phytosterols are obtained from plant sources at 150–400 mg day^{-1} in the form of sitosterol, stigmasterol, and campesterol (Ryan et al. 2009). Frying of potato chips and other vegetables results in the formation of phytosterol oxidation products (POPs). These products may enter the vegetable food matrix and thus be consumed.

5.9.5 Maillard Reaction Products

Maillard reaction products such as Amadori compounds (ACs) are produced during food processing, especially cooking and frying. The ACs are much more reactive than their precursor compounds, such as fructose, glucose, and amino acids. The formation of Maillard reaction products depends on the frying temperature, frying time, and vegetable product (Wu et al. 2018). Under similar conditions, different vegetables or vegetables of the same family may produce significant variations in AC contents during frying. For example, bell pepper, red pepper, yellow onion, purple onion, tomato, and carrot were evaluated for quantitative AC composition by Yu et al. (2018). Bell pepper produced more deoxy fructosyl leucine than red pepper during frying. Similarly, yellow onion produced more deoxy fructosyl arginine than purple onion, as shown in Table 5.11. These results indicate that a number of different culinary methods used in vegetable frying may produce Maillard reaction products. The formation of ACs can be reduced by applying high temperatures and short frying times.

5.10 Physicochemical Characteristics of Fried Foods

Fried food can be analysed in terms of a number of physicochemical parameters, such as oil content, moisture content, colour, texture, and solids contents. These characteristics significantly affect the quality of the food, and hence consumer preference.

5.10.1 Oil Content

The frying of foods is carried out in frying media such as edible oils or fats. During frying, hot oils enter the food matrix and are absorbed. The oil content of a food may be measured as oil uptake or oil usage. The absorption of oil depends on the food type; for

Table 5.11 Effects of frying on the formation of Amadori compounds (ACs) (mg kg dry matter^{-1}).

Vegetable	Type	Fru-Arg	Fru-Phe	Fru-His	Fru-Met	Fru-Glu	Fru-Leu	Fru-Val	Fru-Ala
Bell pepper	Fresh	16.15 ± 0.8	10.10 ± 0.2	12.34 ± 0.1	ND	11.20 ± 0.4	4.90 ± 0.3	2.71 ± 0.1	7.53 ± 0.1
	Fried	132.73 ± 3.7	60.21 ± 1.6	84.17 ± 1.9	3.45 ± 0.1	42.75 ± 1.5	66.23 ± 1.4	29.52 ± 1.4	28.81 ± 0.5
Red pepper	Fresh	22.10 ± 0.7	11.05 ± 0.1	30.32 ± 0.9	ND	10.91 ± 0.4	11.22 ± 0.6	3.30 ± 0.1	5.92 ± 0.08
	Fried	137.72 ± 2.7	60.42 ± 1.5	91.25 ± 1.4	ND	31.87 ± 0.8	40.15 ± 1.8	18.23 ± 0.5	15.37 ± 0.3
Yellow Onion	Fresh	67.71 ± 1.4	21.56 ± 0.1	34.25 ± 0.1	ND	20.44 ± 0.4	17.72 ± 0.1	ND	ND
	Fried	307.54 ± 3.7	37.56 ± 0.5	47.12 ± 0.7	6.57 ± 0.2	37.81 ± 0.5	28.97 ± 0.3	12.63 ± 0.1	10.17 ± 0.1
Purple onion	Fresh	93.55 ± 1.7	24.77 ± 0.2	38.72 ± 0.7	ND	18.97 ± 0.2	16.65 ± 0.1	ND	ND
	Fried	418.31 ± 4.8	20.75 ± 0.2	32.42 ± 0.5	5.26 ± 0.09	20.25 ± 0.3	21.07 ± 0.5	7.82 ± 0.08	6.38 ± 0.1
Tomato	Fresh	22.24 ± 0.8	7.83 ± 0.1	12.07 ± 0.2	ND	48.83 ± 1.6	9.86 ± 0.3	10.66 ± 0.06	8.74 ± 0.1
	Fried	51.56 ± 1.3	29.45 ± 0.3	16.98 ± 0.4	8.50 ± 0.2	147.85 ± 2.7	30.57 ± 0.5	22.20 ± 0.2	31.04 ± 0.1
Carrot	Fresh	15.11 ± 0.2	7.53 ± 0.1	8.92 ± 0.1	ND	10.74 ± 0.2	5.56 ± 0.1	ND	ND
	Fried	26.83 ± 0.5	11.45 ± 0.7	7.67 ± 0.1	5.62 ± 0.1	17.86 ± 0.5	12.87 ± 0.2	9.32 ± 0.3	17.97 ± 0.7

Source: Reproduced with permission from John Wiley & Sons Limited (Yu et al. 2018).

Fru-Arg, fructosyl arginine; Fruc-Phe, fructosyl phenylalanine; Fru-Met, fructosyl methionine; Fru-Leu, fructosyl leucine; Fru-Val, fructosyl valine; Fru-Ala, fructosyl alanine; ND, not detected.

example, oil absorption will be high in foods with a porous structure and small size, and relatively low in foods with a nonporous structure and large size. Thus, potato chips will absorb more oil than whole potatoes. Oil absorption also depends on the type of frying. For example, sweet potatoes and green beans fried under vacuum showed less oil content (24 and 16% less, respectively) than the same vegetables fried in a traditional fryer (Da Silva and Moreira 2008). Mariotti-Celis et al. (2017a) showed that vacuum frying significantly ($p \leq 0.05$) decreased the oil content of potato chips, by 28%. These results are in accordance with results obtained in other studies of vegetables fried under vacuum conditions. The content may also vary within a fried food, with the maximum content on the surface crust and the minimum in the centre (Rahimi et al. 2017).

High oil content is usually not preferred by consumers and may reduce a product's stability during storage. Thus, processes or technologies capable of producing fried foods with low oil contents are in high demand.

5.10.2 Colour

The colour of fried material is of significant importance to consumer acceptance. Colour may be defined as the energy distribution of the light reflected by a fried food product. The reflected light is affected by the light source, the angle of viewing, and the background. Collectively, all these parameters affect colour perception. A change in the colour of a food is directly proportional to changes in its chemistry. The colour of fried foods is usually measured using the standard CIE scale[1] or the Hunter scale[2] of chromaticity parameters.[3] In food industries, colour is measured by a colourimeter.

Da Silva and Moreira (2008) studied fruit and vegetable snacks during vacuum frying at 120–130 °C. There noted significant differences in lightness (L*), greenness (a*), and blue-yellow chromaticity (b*) for most products, with the exception of sweet potato chips. Vacuum-fried mango chips and green beans had significantly higher L* values than potato chips fried under atmospheric conditions, as shown in Figure 5.23. Values of a* were significantly higher in atmospheric-fried mango chips and green beans compared to vacuum-fried, representing additional Maillard reactions in the atmospheric-frying samples. Values of b* were significantly higher in vacuum-fried mango chips and green beans compared to atmospheric-fried. Vacuum-fried sweet potato chips were lighter and more yellowish than their atmospheric-fried equivalents. The values of a* and b* for blue potato chips were significantly different. The vacuum-fried blue potatoes had higher b* values and lower a* values as compared to their atmospheric-fried counterparts.

In fried foods, redness is considered undesirable, as it results from the formation of a crust. Thus, an increase in redness is directly proportional to increasing frying temperature. The atmospheric frying of cassava chips resulted in more redness, while vacuum frying produced lower values (Oyedeji et al. 2017). The yellowness of the

1 The CIE scale was developed as tool for standard visual colour representation by the International Commission of Illumination, based in Vienna, Austria.

2 The Hunter scale is similar to the CIE scale. It was developed by Richard S. Hunter, an American colour scientist.

3 Saturation or colour intensity is called chromaticity. The parameters are L*, a*, and b* using the CIE scale, or L, a, and b using the Hunter scale, where L is lightness, a is degree of redness or greenness, and b is degree of yellowness or blueness. The value of L ranges from 0 to 100, representing black to white. These parameters represent accurate and standard human colour perceptions in terms of hue, lightness, and saturation.

Figure 5.23 Colour differences between products fried under vacuum and in traditional fryers. Top: (a) blue potato fried in a vacuum fryer; (b) blue potato fried in a traditional fryer; (c) sweet potato fried in a vacuum fryer; and (d) sweet potato fried in a traditional fryer. Bottom: (a) mango fried in a vacuum fryer; (b) mango fried in a traditional fryer; (c) green bean fried in a vacuum fryer; and (d) green bean fried in a traditional fryer. *Source:* Reproduced with kind permission of the Swiss Society of Food Science & Technology (Da Silva and Moreira 2008).

cassava chips decreased with frying, especially under vacuum. Oyedeji et al. (2017) showed that vacuum-fried cassava chips had the lowest overall colour decline as compared with their atmospheric-fried equivalents. This suggests that the colour deviation of the resulting chips from the fresh slices due to processing conditions was the least in vacuum-fried chips for all the frying times investigated. A similar trend was also reported by Mariscal and Bouchon (2008) for fried apple slices and by Sobukola et al. (2013) for wheat starch and gluten-based snacks.

5.10.3 Texture

Texture is a sensory characteristic that is perceived by touch, vision, and taste (Kilcast 2004). It may be considered as a combination of the structural elements of the food and the manner in which it is perceived by the physiological senses. The characteristics of texture may be classified into two main types: mechanical and geometrical. Mechanical texture involves hardness, cohesiveness, viscosity, elasticity, adhesiveness, brittleness, chewiness, and gumminess. Geometrical characteristics are usually characteristics of the mouthfeel, such as taste, oiliness, wateriness, flavour, and greasiness. Details of the texture of solid food can be found in the work of Kilcast (2004). The texture of fried foods may be measured through different techniques, such as specialized equipment or human sensory measurements.

Visual recording such as photomicrographs offer qualitative information about the surfaces of foods and cells. Several methods have been used to analyse the textures of images, such as fractional Brownian motion (FBMM), box counting (BCM), and fractal dimension (FD) estimation from frequency domain (FDM). Quevedo et al. (2002) used these methods to numerically define the surfaces of foods and the microstructure of potato cells. The surfaces of analysed foods had FDs ranging from 2.22 for chocolate to 2.44 for pumpkin shell.

Pedreschi and Moyano (2005) showed that frying temperature and pre-drying management had a significant effect on the final texture and oil content of fried potato chips. In another study, eight different kinds of vegetable oils (sunflower, rapeseed, soybean, olive, peanut, palm, partially hydrogenated rapeseed, and a mixture of hydrogenated rapeseed and palm oil) were used to fry potato crisps at 150, 170, and 190 °C (Kita et al. 2007). It was found that oil used affected the amount of fat absorbed by the crisps, and thus their texture. A strong relationship was also reported between the oleic acid content of the oil and the texture of the fried crisps. The fat content and texture of fried potatoes products are also influenced by frying temperature. Crisps absorbed less fat and had a lower hardness with increasing temperature (Kita 2014).

Confocal laser scanning microscopy (CLSM) was used to determine the microstructure of deep-fried chicken nuggets batter coated at 170, 180, and 190 °C (Adedeji et al. 2011). The porosity in the fried samples ranged between 4.97 and 32.7% and was significantly affected by frying temperature. Similarly, the pore size in the samples varied between 1.2 and 550.0 μm. It was found that frying under the selected conditions led to the formation of more micropores (pores <40 μm) and of larger pores (pore ≥ 216 μm). Crust formation is a significant factor in evaluating the crispness of fried foods. Van Koerten et al. (2015) studied the detailed structural and textural properties of crust in relation to crispness during frying as a function of process temperature and time. X-ray tomography showed a larger overall pore volume with higher frying times and a greater number of large pores with lower final moisture content. Texture analysis revealed that increasing porosity resulted in a more crispy behaviour after frying with oil up to 180 °C; at temperatures above 180 °C, the crispness decreased again, due to the increased plastic behaviour of the crust. The authors suggested that the decrease in crispness may be due to the increased sugar degradation at very high temperatures, which may be associated with the reduced glass transition temperature of the crust.

Frying of cassava chips using atmospheric and vacuum frying showed that a sharp reduction in the breaking force was obtained for all frying treatments after 8 minutes; this effect was lower in vacuum-fried chips (Oyedeji et al. 2017). First-order kinetics of the quality attributes showed that the rate constants k (min^{-1}) for vacuum frying were close to those of atmospheric frying, while activation energies for hardness and colour change were 53.30 and 467.11 $KJ\,mol^{-1}$, respectively. Similarly, textural values of fried potatoes increased with frying time and also followed a first-order kinetics model (Nourian and Ramaswamy 2003). These results indicate that vacuum frying produces a more acceptable texture in fried chips.

5.10.4 Moisture Content

Water is one of the main components of all foods, with the largest amount in plant foods. Water is evaporated during frying, and thus there is a decrease in the water content of fried foods according to the frying temperature, frying medium, and original water content. Frying of French fries at 140, 155, 170, and 185 °C for 60–240 seconds revealed that the porosity of the product increased with frying time (Ziaiifar et al. 2010). A kinetics study of moisture loss, oil uptake, and colour change during deep frying of Gethi (*Dioscorea kamoonensis kunth*) strips at a temperature of 120–180 °C was reported by Manjunatha et al. (2014). The results showed that the moisture loss and oil uptake followed first-order kinetics. The kinetic coefficients for moisture loss and oil uptake increased significantly with temperature. A clearer picture of the relationship between moisture loss and oil uptake can be seen in Figure 5.24.

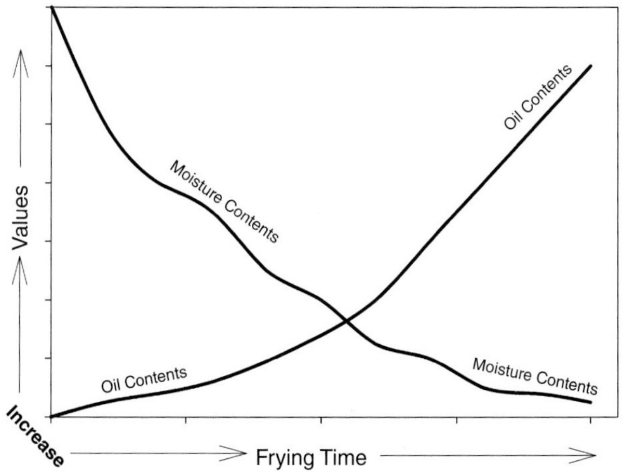

Figure 5.24 Correlation of moisture loss with oil uptake during the frying of food.

This figure shows that with an increase in frying time, there was a decrease in the moisture content, which was supplemented by an increase in oil content.

It was not clear until recently how initial moisture content affects oil absorption. Zhang et al. (2016) provided evidence suggesting that although the initial moisture content has no significant effect on either the final oil content or the oil fraction, the moisture content does greatly affect the rate of oil absorption. The authors analyzed different oil contents, such as total oil (TO), surface oil (SO), structural oil (STO), penetrated surface oil (PSO), and oil distribution during the frying of once-fried and twice-fried potato chips. They found that the SO fraction was lowest among three different fractions of TO, whereas PSO was the dominant section for TO. Their findings also suggested that the initial moisture content may have no effect on oil absorption. The twice-fried chips reached equilibrium at 180 °C more quickly than once-fried chips. This may be due to the absence of inner resistance to water evaporation in the twice-fried chips. CLSM analysis revealed that oil entered the core of twice-fried chips faster than it did that of once-fried chips. The initial pore size of the food may also affect oil absorption during frying (Li et al. 2016).

It has thus been established that initial moisture content and pore size may play a role in the frying process, and so predrying, preheating, and post-frying holding strategies have been introduced into many frying processes (Rahimi et al. 2017).

5.11 Improving Product Quality

The quality of fried foods is important to both food manufacturers and consumers. Thus, it is highly important for a food manufacturer to provide relevant knowledge to their staff involved in the frying of foods. The following main points should be borne in mind when using chemistry to improve the quality of fried foods; further details regarding the improvement of product quality are discussed in Chapter 13.

- Essential knowledge of the chemistry of foods (Velisek 2014).
- Chemical reactions occurring during the frying of foods (Chapter 4 and Perkins and Erickson 1996).
- Knowledge on how to improve specific product quality (Rossell 2001).
- Online and offline literature on food chemistry.
- Seminars and conferences on food quality improvement.

5.12 Key Concepts

- Carbohydrates, proteins, lipids, and minor components constitute most of the foods we usually fry.
- Frying of carbohydrate-rich foods such as potato, rice, wheat flour products, corn chips, and so on produces several toxic compounds.
- Furans such as HMF are produced by the action of carbohydrate alone or carbohydrate plus amino acid; this process is favoured with increasing frying time and temperature.

- HMF may be degraded into several other toxic organic compounds which are present in deep-fried foods.
- Acrylamide is produced from monosaccharide alone or monosaccharide plus an amino acid such as asparagine.
- Frying of protein-rich foods produces HAAs and PAHs.
- The presence of several HAAs and PAHs in a fried food is a marker for toxicity and quality changes.
- Seafood is rich in carbohydrates, proteins, and lipids. These components are transformed into several oxidized compounds, including COPs, POPs, acrylamide, furans, HAAs, and PAHs.
- Vegetables are rich in carbohydrates, polyphenolic compounds, carotenoids, and phytosterols. Frying of vegetables produces significant changes in the composition of these components.
- Frying alters the frying medium, while tocopherols, phenolic compounds, and carotenoids present in the food help to stabilize it.
- Frying of foods can also alter their physiochemical parameters, such as colour, texture, oil content, and moisture content. These parameters can be used as a standard marker along with other chemical parameters.
- Knowledge of chemistry is a fundamental tool in improving the quality of fried foods.

References

Aaslyng, M.D., Duedahl-Olesen, L., Jensen, K., and Meinert, L. (2013). Content of heterocyclic amines and polycyclic aromatic hydrocarbons in pork, beef and chicken barbecued at home by Danish consumers. *Meat Science* 93: 85–91.

Adedeji, A.A., Liu, L., and Ngadi, M.O. (2011). Microstructural evaluation of deep-fat fried chicken nugget batter coating using confocal laser scanning microscopy. *Journal of Food Engineering* 102: 49–57.

Aeenehvand, S., Toudehrousta, Z., Kamankesh, M. et al. (2016). Evaluation and application of microwave-assisted extraction and dispersive liquid–liquid microextraction followed by high-performance liquid chromatography for the determination of polar heterocyclic aromatic amines in hamburger patties. *Food Chemistry* 190: 429–435.

Agte, V., Tarwadi, K., Mengale, S. et al. (2002). Vitamin profile of cooked foods: how healthy is the practice of ready-to-eat foods? *International Journal of Food Sciences and Nutrition* 53: 197–208.

Alaejos, M.S. and Afonso, A.M. (2011). Factors that affect the content of heterocyclic aromatic amines in foods. *Comprehensive Reviews in Food Science and Food Safety* 10: 52–108.

Almeida, P. (2016). *Proteins: Concepts in Biochemistry*. Philadelphia, PA: Garland Science.

Al-Saghir, S., Thurner, K., Wagner, K.-H. et al. (2004). Effects of different cooking procedures on lipid quality and cholesterol oxidation of farmed Salmon fish (*Salmo salar*). *Journal of Agricultural and Food Chemistry* 52: 5290–5296.

Amrein, T.M., Bachmann, S., Noti, A. et al. (2003). Potential of acrylamide formation, sugars, and free asparagine in potatoes: a comparison of cultivars and farming systems. *Journal of Agricultural and Food Chemistry* 51: 5556–5560.

An, K.-J., Liu, Y.-L., and Liu, H.-L. (2017). Relationship between total polar components and polycyclic aromatic hydrocarbons in fried edible oil. *Food Additives & Contaminants. Part A, Chemistry, Analysis, Control, Exposure & Risk Assessment* 34: 1596–1605.

Andersen, C.J., Ragonesi, N., Cintron-Rivera, L. et al. (2017). Variation in food group availability and nutrient composition across food pantry inventories. *FASEB Journal* 31: lb379.

Babaoglu, A.S., Karakaya, M., and Öz, F. (2017). Formation of polycyclic aromatic hydrocarbons in beef and lamb kokorec: effects of different animal fats. *International Journal of Food Properties* 20: 1960–1970.

Baghurst, P.A. (1999). Polycyclic aromatic hydrocarbons and heterocyclic amines in the diet: the role of red meat. *European Journal of Cancer Prevention* 8: 193–199.

Becalski, A., Lau, B.P.Y., Lewis, D., and Seaman, S.W. (2003). Acrylamide in foods: occurrence, sources, and modeling. *Journal of Agricultural and Food Chemistry* 51: 802–808.

Belkova, B., Hradecky, J., Hurkova, K. et al. (2018). Impact of vacuum frying on quality of potato crisps and frying oil. *Food Chemistry* 241: 51–59.

Bjeldanes, L.F., Morris, M.M., Felton, J.S. et al. (1982). Mutagens from the cooking of food. II. Survey by Ames/Salmonella test of mutagen formation in the major protein-rich foods of the American diet. *Food and Chemical Toxicology* 20: 357–363.

Brulé, G. and Croguennec, T. (2016). Functional properties of ingredients. In: *Handbook of Food Science and Technology 3: Food Biochemistry and Technology* (ed. R. Jeantet, T. Croguennec, P. Schuck and G. Brulé). Chichester: John Wiley & Sons.

Caris-Veyrat, C., Amiot, M.-J., Ramasseul, R., and Marchon, J.-C. (2001). Mild oxidative cleavage of [small beta],[small beta]-carotene by dioxygen induced by a ruthenium porphyrin catalyst: characterization of products and of some possible intermediates. *New Journal of Chemistry* 25: 203–206.

Chen, Y.C. and Chen, B.H. (2003). Determination of polycyclic aromatic hydrocarbons in fumes from fried chicken legs. *Journal of Agricultural and Food Chemistry* 51: 4162–4167.

Chen, T.Y., Luo, H.M., Hsu, P.H., and Sung, W.C. (2016). Effects of calcium supplements on the quality and acrylamide content of puffed shrimp chips. *Journal of Food and Drug Analysis* 24: 164–172.

Chen, L., Tian, Y., Sun, B. et al. (2018). Measurement and characterization of external oil in the fried waxy maize starch granules using ATR-FTIR and XRD. *Food Chemistry* 242: 131–138.

Chiou, A., Salta, F.N., Kalogeropoulos, N. et al. (2007). Retention and distribution of polyphenols after pan-frying of French fries in oils enriched with olive leaf extract. *Journal of Food Science* 72: S574–S584.

Chiou, A., Kalogeropoulos, N., Boskou, G., and Salta, F.N. (2012). Migration of health promoting microconstituents from frying vegetable oils to French fries. *Food Chemistry* 133: 1255–1263.

Da Silva, P.F. and Moreira, R.G. (2008). Vacuum frying of high-quality fruit and vegetable-based snacks. *LWT – Food Science and Technology* 41: 1758–1767.

Delgado-Andrade, C., Morales, F.J., Seiquer, I., and Pilar Navarro, M. (2010). Maillard reaction products profile and intake from Spanish typical dishes. *Food Research International* 43: 1304–1311.

Echarte, M., Zulet, M.A., and Astiasaran, I. (2001). Oxidation process affecting fatty acids and cholesterol in fried and roasted Salmon. *Journal of Agricultural and Food Chemistry* 49: 5662–5667.

Echarte, M., Conchillo, A., Ansoren, D., and Astiasaran, I. (2005). Cholesterol oxidation products in fresh and frozen shrimps, raw and grilled. *Nutrición Hospitalaria* 20: 293–296.

Erkan, N., Ozden, O., and Selcuk, A. (2010). Effect of frying, grilling, and steaming on amino acid composition of marine fishes. *Journal of Medicinal Food* 13: 1524–1531.

Ersoy, B. (2011). Effects of cooking methods on the proximate, mineral and fatty acid composition of European eel (*Anguilla anguilla*). *International Journal of Food Science and Technology* 46: 522–527.

Felton, J.S. and Knize, M.G. (1990). Heterocyclic-amine mutagens/carcinogens in foods. In: *Chemical Carcinogenesis and Mutagenesis I* (ed. C.S. Cooper and P.L. Grover). Berlin, Heidelberg: Springer.

Fillion, L. and Henry, C.J. (1998). Nutrient losses and gains during frying: a review. *International Journal of Food Sciences and Nutrition* 49: 157–168.

Freitas, M.T., Amaral, C.A.A., Coutrim, M.X. et al. (2015). Effect of cooking method on the formation of 7-ketocholesterol in Atlantic hake (*Merluccius hubbsi*) and smooth weakfish (*Cynoscion leiarchus*) fillets. *LWT – Food Science and Technology* 62: 1141–1147.

Friedman, M. (2003). Chemistry, biochemistry, and safety of acrylamide. A review. *Journal of Agricultural and Food Chemistry* 51: 4504–4526.

Galani, J.H.Y., Patel, N.J., and Talati, J.G. (2017). Acrylamide-forming potential of cereals, legumes and roots and tubers analyzed by UPLC-UV. *Food and Chemical Toxicology* 108: 244–248.

Gertz, C., Klostermann, S., and Parkash Kochhar, S. (2003). Deep frying: the role of water from food being fried and acrylamide formation. *Oléagineux, Corps Gras, Lipides* 10: 297–303.

Gibis, M. (2016). Heterocyclic aromatic amines in cooked meat products: causes, formation, occurrence, and risk assessment. *Comprehensive Reviews in Food Science and Food Safety* 15: 269–302.

Gibis, M., Kruwinnus, M., and Weiss, J. (2015). Impact of different pan-frying conditions on the formation of heterocyclic aromatic amines and sensory quality in fried bacon. *Food Chemistry* 168: 383–389.

Göncüoğlu, N. and Gökmen, V. (2013). Accumulation of 5-hydroxymethylfurfural in oil during frying of model dough. *Journal of the American Oil Chemists' Society* 90: 413–417.

Granvogl, M., Jezussek, M., Koehler, P., and Schieberle, P. (2004). Quantitation of 3-aminopropionamide in potatoes-a minor but potent precursor in acrylamide formation. *Journal of Agricultural and Food Chemistry* 52: 4751–4757.

Hao, X., Li, J., and Yao, Z. (2016). Changes in PAHs levels in edible oils during deep-frying process. *Food Control* 66: 233–240.

Hemre, G.I. and Kahrs, F. (1997). 14C-glucose injection in Atlantic cod, *Gadus morhua*, metabolic responses and excretion via the gill membrane. *Aquaculture Nutrition* 3: 3–8.

Hemre, G.I., Mommsen, T.P., and Krogdahl, Å. (2002). Carbohydrates in fish nutrition: effects on growth, glucose metabolism and hepatic enzymes. *Aquaculture Nutrition* 8: 175–194.

Hsu, H.Y., Inbaraj, B.S., and Chen, B.H. (2006). Lack of formation of heterocyclic amines in fumes from frying French fries. *Journal of Food Protection* 69: 2230–2236.

Ismail, A. and Ikram, E.H.K. (2004). Effects of cooking practices (boiling and frying) on the protein and amino acids contents of four selected fishes. *Nutrition & Food Science* 34: 54–59.

Juániz, I., Ludwig, I.A., Huarte, E. et al. (2016a). Influence of heat treatment on antioxidant capacity and (poly) phenolic compounds of selected vegetables. *Food Chemistry* 197: 466–473.

Juániz, I., Zocco, C., Mouro, V. et al. (2016b). Effect of frying process on furan content in foods and assessment of furan exposure of Spanish population. *LWT – Food Science and Technology* 68: 549–555.

Kalogeropoulos, N., Mylona, A., Chiou, A. et al. (2007). Retention and distribution of natural antioxidants (α-tocopherol, polyphenols and terpenic acids) after shallow frying of vegetables in virgin olive oil. *LWT – Food Science and Technology* 40: 1008–1017.

Kao, F.J., Chiu, Y.S., Tsou, M.J., and Chiang, W.D. (2012). Effects of Chinese domestic cooking methods on the carotenoid composition of vegetables in Taiwan. *LWT – Food Science and Technology* 46: 485–492.

Kidmose, U., Yang, R.Y., Thilsted, S.H. et al. (2006). Content of carotenoids in commonly consumed Asian vegetables and stability and extractability during frying. *Journal of Food Composition and Analysis* 19: 562–571.

Kilcast, D. (2004). *Texture in Food: Solid Foods*. Cambridge: Woodhead Publishing.

Kita, A. (2014). The effect of frying on fat uptake and texture of fried potato products. *European Journal of Lipid Science and Technology* 116: 735–740.

Kita, A., Lisińska, G., and Gołubowska, G. (2007). The effects of oils and frying temperatures on the texture and fat content of potato crisps. *Food Chemistry* 102: 1–5.

Knize, M.G., Dolbeare, F.A., Carroll, K.L. et al. (1994). Effect of cooking time and temperature on the heterocyclic amine content of fried beef patties. *Food and Chemical Toxicology* 32: 595–603.

Kocadagli, T., Goncuoglu, N., Hamzalioglu, A., and Gokmen, V. (2012). In depth study of acrylamide formation in coffee during roasting: role of sucrose decomposition and lipid oxidation. *Food & Function* 3: 970–975.

Kowalski, S., Lukasiewicz, M., Duda-Chodak, A., and Zięć, G. (2013). 5-Hydroxymethyl-2-furfural (HMF) – heat-induced formation, occurrence in food and biotransformation – a Review. *Polish Journal of Food and Nutrition Sciences* 63: 207–225.

Kuperman, A.S. (1958). Effects of acrylamide on the central nervous system of the cat. *Journal of Pharmacology and Experimental Therapeutics* 123: 180–192.

Larsen, D., Quek, S.Y., and Eyres, L. (2010). Effect of cooking method on the fatty acid profile of New Zealand King Salmon (*Oncorhynchus tshawytscha*). *Food Chemistry* 119: 785–790.

Leung, K.S., Lin, A., Tsang, C.K., and Yeung, S.T. (2003). Acrylamide in Asian foods in Hong Kong. *Food Additives and Contaminants* 20: 1105–1113.

Li, J., Zhang, T., Liu, Y., and Fan, L. (2016). Effects of initial pore diameter on the oil absorption behavior of potato chips during frying process. *Journal of Oleo Science* 65: 303–310.

Lim, S.-T. and Han, J.-A. (2016). Improvement in antioxidant functionality and shelf life of yukwa (fried rice snack) by turmeric (*Curcuma longa* L.) powder addition. *Food Chemistry* 199: 590–596.

Lira, G.M., Cabral, C.C.V.Q., de Oliveira, Í.B.A. et al. (2017). Changes in the lipid fraction of king mackerel pan fried in coconut oil and cooked in coconut milk. *Food Research International* 101: 198–202.

Lu, F., Kuhnle, G.K., and Cheng, Q. (2017). Vegetable oil as fat replacer inhibits formation of heterocyclic amines and polycyclic aromatic hydrocarbons in reduced fat pork patties. *Food Control* 81: 113–125.

Lu, R., Yang, Z., Song, H. et al. (2016). The aroma-active compound, acrylamide and ascorbic acid contents of pan-fried potato slices cooked by different temperature and time. *Journal of Food Processing and Preservation* 40: 183–191.

Lyon, J.L. and Mahoney, A.W. (1988). Fried foods and the risk of colon cancer. *American Journal of Epidemiology* 128: 1000–1006.

Maharaj, P.P.P., Prasad, S., Devi, R., and Gopalan, R. (2015). Folate content and retention in commonly consumed vegetables in the South Pacific. *Food Chemistry* 182: 327–332.

Mai, J., Shimp, J., Weihrauch, J., and Kinsella, J.E. (1978). Lipids of fish fillets: changes following cooking by different methods. *Journal of Food Science* 43: 1669–1674.

Manjunatha, S.S., Ravi, N., Negi, P.S. et al. (2014). Kinetics of moisture loss and oil uptake during deep fat frying of Gethi (*Dioscorea kamoonensis* Kunth) strips. *Journal of Food Science and Technology* 51: 3061–3071.

Mariotti-Celis, M.S., Cortés, P., Dueik, V. et al. (2017a). Application of vacuum frying as a furan and acrylamide mitigation technology in potato chips. *Food and Bioprocess Technology* 10: 2092–2099.

Mariotti-Celis, M.S., Zuniga, R.N., Cortes, P., and Pedreschi, F. (2017b). A kinetic study of furan formation in wheat flour-based model systems during frying. *Journal of Food Science* 82: 232–239.

Mariscal, M. and Bouchon, P. (2008). Comparison between atmospheric and vacuum frying of apple slices. *Food Chemistry* 107: 1561–1569.

Marles, R.J. (2017). Mineral nutrient composition of vegetables, fruits and grains: the context of reports of apparent historical declines. *Journal of Food Composition and Analysis* 56: 93–103.

Matthäus, B., Haase, N.U., and Vosmann, K. (2004). Factors affecting the concentration of acrylamide during deep-fat frying of potatoes. *European Journal of Lipid Science and Technology* 106: 793–801.

Mccollister, D.D., Oyen, F., and Rowe, V.K. (1964). Toxicology of acrylamide. *Therapie der Gegenwart* 103: 172–181.

Medina-Navarro, R., Duran-Reyes, G., Diaz-Flores, M., and Hicks, J. (2004). Glucose-stimulated acrolein production from unsaturated fatty acids. *Human & Experimental Toxicology* 23: 101–105.

Mesias, M. and Morales, F.J. (2015). Acrylamide in commercial potato crisps from Spanish market: trends from 2004 to 2014 and assessment of the dietary exposure. *Food and Chemical Toxicology* 81: 104–110.

Mesias, M., Holgado, F., Marquez-Ruiz, G., and Morales, F.J. (2016). An investigation of process contaminants' formation during the deep frying of breadcrumbs using a bread coat model. *Food & Function* 7: 1645–1654.

Meurillon, M. and Engel, E. (2016). Mitigation strategies to reduce the impact of heterocyclic aromatic amines in proteinaceous foods. *Trends in Food Science & Technology* 50: 70–84.

Miglio, C., Chiavaro, E., Visconti, A. et al. (2008). Effects of different cooking methods on nutritional and physicochemical characteristics of selected vegetables. *Journal of Agricultural and Food Chemistry* 56: 139–147.

Mir-Bel, J., Oria, R., and Salvador, M.L. (2013). Reduction in hydroxymethylfurfural content in 'churros', a Spanish fried dough, by vacuum frying. *International Journal of Food Science and Technology* 48: 2042–2049.

Morales, F.J. and Arribas-Lorenzo, G. (2008). The formation of potentially harmful compounds in churros, a Spanish fried-dough pastry, as influenced by deep frying conditions. *Food Chemistry* 109: 421–425.

Mottram, D.S., Wedzicha, B.L., and Dodson, A.T. (2002). Acrylamide is formed in the Maillard reaction. *Nature* 419: 448–449.

Murkovic, M. (2004). Chemistry, formation and occurrence of genotoxic heterocyclic aromatic amines in fried products. *European Journal of Lipid Science and Technology* 106: 777–785.

Nakai, S. and Modler, H.W. (2000). *Food Proteins: Processing Applications*. New York, NY: Wiley-VCH.

Nawar, W., Hultin, H., Li, Y. et al. (1990). Lipid oxidation in seafoods under conventional conditions. *Food Reviews International* 6: 647–660.

Nourian, F. and Ramaswamy, H.S. (2003). Kinetics of quality change during cooking and frying of potatoes: part I. Texture. *Journal of Food Process Engineering* 26: 377–394.

Nugrahedi, P.Y., Oliviero, T., Heising, J.K. et al. (2017). Stir-frying of chinese cabbage and pakchoi retains health-promoting glucosinolates. *Plant Foods for Human Nutrition* 72: 439–444.

Ohshima, T., Li, N., and Koizumi, C. (1993). Oxidative decomposition of cholesterol in fish products. *Journal of the American Oil Chemists' Society* 70: 595–600.

Oluwaniyi, O.O., Dosumu, O.O., and Awolola, G.V. (2010). Effect of local processing methods (boiling, frying and roasting) on the amino acid composition of four marine fishes commonly consumed in Nigeria. *Food Chemistry* 123: 1000–1006.

Oyedeji, A., Sobukola, O., Henshaw, F., Adegunwa, M., Ijabadeniyi, A., Sanni, L. & Tomlins, K. 2017. Effect of frying treatments on texture and colour parameters of deep fat fried yellow fleshed cassava chips. *Journal of Food Quality*, 2017, art. 8373801.

Oz, F., Kaban, G., and Kaya, M. (2007). Effects of cooking methods on the formation of heterocyclic aromatic amines of two different species trout. *Food Chemistry* 104: 67–72.

Oz, F., Kaban, G., and Kaya, M. (2010). Effects of cooking methods and levels on formation of heterocyclic aromatic amines in chicken and fish with Oasis extraction method. *LWT – Food Science and Technology* 43: 1345–1350.

Pacetti, D., Gil, E., Frega, N.G. et al. (2015). Acrylamide levels in selected Colombian foods. *Food Additives & Contaminants. Part B, Surveillance* 8: 99–105.

Parry, D.A. and Squire, J. (2017). *Fibrous Proteins: Structures and Mechanisms*. Berlin: Springer.

Pedreschi, F. and Moyano, P. (2005). Effect of pre-drying on texture and oil uptake of potato chips. *LWT – Food Science and Technology* 38: 599–604.

Pénicaud, C., Achir, N., Dhuique-Mayer, C. et al. (2011). Degradation of β-carotene during fruit and vegetable processing or storage: reaction mechanisms and kinetic aspects: a review. *Fruits* 66: 417–440.

Perelló, G., Martí-Cid, R., Castell, V. et al. (2009). Concentrations of polybrominated diphenyl ethers, hexachlorobenzene and polycyclic aromatic hydrocarbons in various foodstuffs before and after cooking. *Food and Chemical Toxicology* 47: 709–715.

Pérez-Palacios, T., Petisca, C., Henriques, R., and Ferreira, I.M.P.L.V.O. (2013). Impact of cooking and handling conditions on furanic compounds in breaded fish products. *Food and Chemical Toxicology* 55: 222–228.

Pérez-Palacios, T., Petisca, C., Casal, S., and Ferreira, I.M. (2014). Changes in chemical composition of frozen coated fish products during deep-frying. *International Journal of Food Sciences and Nutrition* 65: 212–218.

Perkins, E.G. and Erickson, M.D. (1996). *Deep Frying: Chemistry, Nutrition, and Practical Applications*. Champaign, IL: AOCS Press.

Persson, E., Sjoholm, I., Nyman, M., and Skog, K. (2004). Addition of various carbohydrates to beef burgers affects the formation of heterocyclic amines during frying. *Journal of Agricultural and Food Chemistry* 52: 7561–7566.

Powers, S.J., Mottram, D.S., Curtis, A., and Halford, N.G. (2017). Acrylamide levels in potato crisps in Europe from 2002 to 2016. *Food Additives & Contaminants. Part A, Chemistry, Analysis, Control, Exposure & Risk Assessment* 34: 2085–2100.

Qin, L., Zhang, Y.Y., Xu, X.B. et al. (2017). Isotope dilution HPLC-MS/MS for simultaneous quantification of acrylamide and 5-hydroxymethylfurfural (HMF) in thermally processed seafood. *Food Chemistry* 232: 633–638.

Quevedo, R., Carlos, L.-G., Aguilera, J.M., and Cadoche, L. (2002). Description of food surfaces and microstructural changes using fractal image texture analysis. *Journal of Food Engineering* 53: 361–371.

Rahimi, J., Adewale, P., Ngadi, M. et al. (2017). Changes in the textural and thermal properties of batter coated fried potato strips during post frying holding. *Food and Bioproducts Processing* 102: 136–143.

Rajagopal, L., Giraud, D.W., Hamouz, F.L., and Driskell, J.A. (2007). Carotenoid retention and sensory characteristics of selected vegetables prepared by induction stir-frying. *Journal of Food Quality* 30: 703–717.

Ramos, L.R.O., Choi, N.-D., and Ryu, H.-S. (2012). Effects of processing conditions on the protein quality of fried anchovy kamaboko engraulis japônica. *Fisheries and Aquatic Sciences* 15: 265–273.

Randel, G., Balzer, M., Grupe, S. et al. (2007). Degradation of heterocyclic aromatic amines in oil under storage and frying conditions and reduction of their mutagenic potential. *Food and Chemical Toxicology* 45: 2245–2253.

Rodriguez-Estrada, M.T., Garcia-Llatas, G., and Lagarda, M.J. (2014). 7-Ketocholesterol as marker of cholesterol oxidation in model and food systems: when and how. *Biochemical and Biophysical Research Communications* 446: 792–797.

Rossell, J.B. (2001). *Frying: Improving Quality*. Cambridge: Woodhead Publishing.

Ryan, E., Mccarthy, F.O., Maguire, A.R., and O'brien, N.M. (2009). Phytosterol oxidation products: their formation, occurrence, and biological effects. *Food Reviews International* 25: 157–174.

Santos, C.S., Molina-Garcia, L., Cunha, S.C., and Casal, S. (2018). Fried potatoes: impact of prolonged frying in monounsaturated oils. *Food Chemistry* 243: 192–201.

Sebedio, J.L., Ratnayake, W.M.N., Ackman, R.G., and Prevost, J. (1993). Stability of polyunsaturated omega-3 fatty acids during deep fat frying of Atlantic mackerel (*Scomber scombrus* L.). *Food Research International* 26: 163–172.

Seppanen, C.M., Song, Q., and Saari Csallany, A. (2010). The antioxidant functions of tocopherol and tocotrienol homologues in oils, fats, and food systems. *Journal of the American Oil Chemists' Society* 87: 469–481.

Shahidi, F., Ambigaipalan, P., and Chandrasekara, A. (2017). Recent advances in phytochemicals in fruits and vegetables. In: *Fruit and Vegetable Phytochemicals: Chemistry and Human Health* (ed. E.M. Yahia). Chichester: John Wiley & Sons.

Shahidi, F. and Botta, J.R. (2012). *Seafoods: Chemistry, Processing Technology and Quality*. Berlin: Springer Science & Business Media.

Shahidi, F. and Ho, C.-T. (2005). *Phenolic Compounds in Foods and Natural Health Products*. Washington, DC: American Chemical Society.

Shanthilal, J. and Bhattacharya, S. (2017). Frying of rice flour dough strands containing gum Arabic: texture, sensory attributes and microstructure of products. *Journal of Food Science and Technology* 54: 1293–1303.

Shen, C., Zhang, I.Y., and Fu, Xu, G., X. (2011). Pyrolysis of D-glucose to acrolein. *Chinese Journal of Chemical Physics* 24: 249.

Shuguang, L., Dinhua, P., and Guoxiong, W. (1994). Analysis of polycyclic aromatic hydrocarbons in cooking oil fumes. *Archives of Environmental Health* 49: 119–122.

Soares, N.F., Vicente, A.N.A., and Martins, C.M. (2015). *Food Safety in the Seafood Industry: A Practical Guide for ISO 22000 and FSSC 22000 Implementation*. Chichester: John Wiley & Sons.

Sobukola, O.P., Dueik, V., Munoz, L., and Bouchon, P. (2013). Comparison of vacuum and atmospheric deep-fat frying of wheat starch and gluten based snacks. *Food Science and Biotechnology* 22: 177–182.

Stadler, R.H., Blank, I., Varga, N. et al. (2002). Acrylamide from Maillard reaction products. *Nature* 419: 449–450.

Stevens, J.F. and Maier, C.S. (2008). Acrolein: sources, metabolism, and biomolecular interactions relevant to human health and disease. *Molecular Nutrition & Food Research* 52: 7–25.

Su, Y., Zhang, M., Zhang, W. et al. (2018). Low oil content potato chips produced by infrared vacuum pre-drying and microwave-assisted vacuum frying. *Drying Technology* 36: 294–306.

Sugimura, T., Wakabayashi, K., Nakagama, H., and Nagao, M. (2004). Heterocyclic amines: Mutagens/carcinogens produced during cooking of meat and fish. *Cancer Science* 95: 290–299.

Szterk, A. (2015). Acridine derivatives (PANHs, azaarenes) in raw, fried or grilled pork from different origins, and PANH formation during pork thermal processing. *Journal of Food Composition and Analysis* 43: 18–24.

Tareke, E., Rydberg, P., Karlsson, P. et al. (2002). Analysis of acrylamide, a carcinogen formed in heated foodstuffs. *Journal of Agricultural and Food Chemistry* 50: 4998–5006.

Thiebaud, H.P., Knize, M.G., Kuzmicky, P.A. et al. (1995). Airborne mutagens produced by frying beef, pork and a soy-based food. *Food and Chemical Toxicology* 33: 821–828.

Van Koerten, K.N., Schutyser, M.A.I., Somsen, D., and Boom, R.M. (2015). Crust morphology and crispness development during deep-fat frying of potato. *Food Research International* 78: 336–342.

Van Putten, R.-J., Van der Waal, J.C., De Jong, E. et al. (2013). Hydroxymethylfurfural, a versatile platform chemical made from renewable resources. *Chemical Reviews* 113: 1499–1597.

Velisek, J. (2014). *The Chemistry of Food*. Chichester: John Wiley & Sons.

Wang, P., Hong, Y., Ke, W. et al. (2017). Formation of heterocyclic amines in Chinese marinated meat: effects of animal species and ingredients (rock candy, soy sauce and rice wine). *Journal of the Science of Food and Agriculture* 97: 3967–3978.

Weber, J., Bochi, V.C., Ribeiro, C.P. et al. (2008). Effect of different cooking methods on the oxidation, proximate and fatty acid composition of silver catfish (Rhamdia quelen) fillets. *Food Chemistry* 106: 140–146.

WHO/FAO. 2002. Health implications of acrylamide in food: report of a joint FAO/WHO consultation. WHO Headquarters, Geneva, Switzerland, 25–27 June 2002, Geneva, Switzerland. Food Safety Programme, World Health Organization.

Wu, J., Sabag-Daigle, A., Metz, T.O. et al. (2018). Measurement of fructose-asparagine concentrations in human and animal foods. *Journal of Agricultural and Food Chemistry* 66: 212–217.

Yalcin, H. and Çapar, T.D. (2017). Bioactive compounds of fruits and vegetables. In: *Minimally Processed Refrigerated Fruits and Vegetables* (ed. F. Yildiz and R.C. Wiley). Boston, MA: Springer US.

Yamsaengsung, R., Yaeed, S., and Ophithakorn, T. (2017). Vacuum frying of fish tofu and effect on oil quality usage life. *Journal of Food Process Engineering* 40: e12587.

Yao, Y., Peng, Z.Q., Shao, B. et al. (2013). Effects of frying and boiling on the formation of heterocyclic amines in braised chicken. *Poultry Science* 92: 3017–3025.

Yaylayan, V.A., Wnorowski, A., and Perez Locas, C. (2003). Why asparagine needs carbohydrates to generate acrylamide. *Journal of Agricultural and Food Chemistry* 51: 1753–1757.

Yu, J., Zhang, S., and Zhang, L. (2018). Evaluation of the extent of initial Maillard reaction during cooking some vegetables by direct measurement of the Amadori compounds. *Journal of the Science of Food and Agriculture* 98: 190–197.

Yuksel, F., Karaman, S., Gurbuz, M. et al. (2017). Production of deep-fried corn chips using stale bread powder: effect of frying time, temperature and concentration. *LWT – Food Science and Technology* 83: 235–242.

Zahoor, A.F., Yousaf, M., Awais, M. et al. (2015). Synthesis, extraction, and identification of light polycyclic aromatic hydrocarbons (PAH) in microwave oven fried Aseel chicken tissues. *Toxicological & Environmental Chemistry* 97: 145–154.

Zeb, A. (2012). Thermal degradation of β-carotene in food oils. In: *Vitamin A and Carotenoids: Chemistry* (ed. V.R. Preedy). London: The Royal Society of Chemistry.

Zeb, A. (2017). A simple, sensitive HPLC-DAD method for the simultaneous determination of carotenoids, chlorophylls & α-tocopherol composition of leafy vegetables. *Journal of Food Measurement and Characterization* 11: 979–986.

Zeb, A. and Haq, I. (2017). Polyphenolic composition, lipid peroxidation and antioxidant properties of chapli kebab during repeated frying process. *Journal of Food Measurement and Characterization* 12: 555–563.

Zeb, A. and Nisar, P. (2017). Effects of high temperature frying of spinach leaves in sunflower oil on carotenoids, chlorophylls, and tocopherol composition. *Frontiers in Chemistry* 5: 19.

Zeng, H., Chen, J., Zhai, J. et al. (2016). Reduction of the fat content of battered and breaded fish balls during deep-fat frying using fermented bamboo shoot dietary fiber. *LWT – Food Science and Technology* 73: 425–431.

Zhang, T., Li, J., Ding, Z., and Fan, L. (2016). Effects of initial moisture content on the oil absorption behavior of potato chips during frying process. *Food and Bioprocess Technology* 9: 331–340.

Zhao, X., Gong, G., and Wu, S. (2018). Effect of storage time and temperature on parent and oxygenated polycyclic aromatic hydrocarbons in crude and refined vegetable oils. *Food Chemistry* 239: 781–788.

Zhu, L. and Wang, J. (2003). Sources and patterns of polycyclic aromatic hydrocarbons pollution in kitchen air, China. *Chemosphere* 50: 611–618.

Ziaiifar, A.M., Courtois, F., and Trystram, G. (2010). Porosity development and its effect on oil uptake during frying process. *Journal of Food Process Engineering* 33: 191–212.

Zöchling, S. and Murkovic, M. (2002). Formation of the heterocyclic aromatic amine PhIP: identification of precursors and intermediates. *Food Chemistry* 79: 125–134.

6

Chemistry of Interactions in Frying

6.1 Introduction

Frying has two important components: the frying medium and the food. These components are classified further into several biomolecules, as described in Chapters 4 and 5 and elsewhere (Perkins and Erickson 1996; Stier 2000; Velisek 2014). These biomolecules interact with one another or transform alone to form new products. None is exempted from the interactions and transformation. Since temperature is an essential component of frying, it is not considered a limiting factor. Applied pressure and frying time also affect interactions during frying. However, all physical and chemical reactions are the result of either the food or the frying medium. The interactions may be either physical or chemical in nature (Figure 6.1). Thus, the interactions occurring during food frying may be classified on the basis of their components as follows:

1) ***Food–Food Interactions:*** Interactions producing chemical reactions within the food components are called food–food interactions or homogenous interactions. These include the formation of 5-hydroxymethyl-2-furfural (HMF), acrylamide, heterocyclic aromatic amines (HAAs), and polycyclic aromatic hydrocarbons (PAHs). Such interactions do not involve the frying medium, but are rather caused by the heat supplied.
2) ***Food–Oil Interactions:*** Interactions involving the transformation of food components or their interaction with the oil components are called food–oil interactions or heterogeneous interactions. They include the browning reactions caused by the lipid medium and the leaching of pigments or solid matter from the food and their consequent interactions or reactions in the frying medium.
3) ***Oil–Oil Interactions:*** Interactions involving the lipid oxidation processes in the frying medium are called oil–oil interactions. These include primary and secondary lipid peroxidation and the formation of cholesterol oxidation products, acrolein, HAAs, and PAHs.

Such reactions may be more prominent in either of the frying components. For example, browning or Maillard reactions are more pronounced in food–food interactions than in oil–oil interactions. Lipid peroxidation is higher in frying medium than in foods. Fillion and Henry (1998) reviewed the micro- and macronutrient losses during the frying of foods, while Dobarganes et al. (2000) reviewed the interactions of fat and food during frying. They showed that fat absorption and lipid exchanges are the main

Food Frying: Chemistry, Biochemistry, and Safety, First Edition. Alam Zeb.

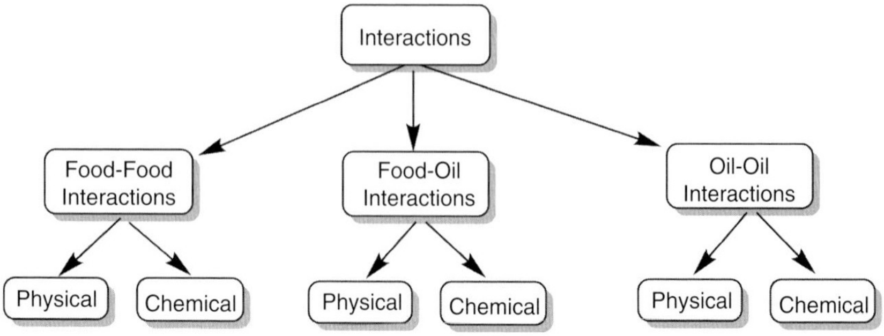

Figure 6.1 Classification of interactions during food frying.

physical changes in frying, while chemical interactions include the chemical reactions between foods and the frying medium. Most reviews show the physical interactions occurring in the frying medium alone (Vitrac et al. 2000; Ling et al. 2015), the physico-chemical interactions in the frying medium (Nayak et al. 2016), and the chemical interactions between the frying medium and the food (Zhang et al. 2012; Ling et al. 2015; Duda-Chodak et al. 2016). Figure 6.2 summarizes these interactions. When considering frying a particular food, the chemistry and possible interactions must be kept in mind in order to obtain an acceptable and healthy product. It is particularly important to know which micro- or macrocomponents are interacting and affecting the food and the frying medium.

6.2 Factors Affecting the Frying Medium

The frying of food is affected by several factors that can significantly alter the frying medium. These factors may be physical or chemical, depending upon the type of food fried. The composition of the frying medium plays a significant role in overall physical as well as chemical changes during frying. Several desirable and undesirable flavour compounds can affect the flavour stability and quality, colour, and texture of fried foods, and consequently their nutritional quality. Chemical reactions such as hydrolysis, oxidation, and polymerization of the frying medium produce volatile or nonvolatile compounds. Most of the volatile compounds evaporate into the atmosphere along with steam. Some of these compounds present in oil undergo further reactions or are absorbed by the fried food. The physical and chemical properties of the oil, as well as of the fried food, are significantly affected by the nonvolatile compounds. These compounds also affect flavour stability, and consequently the quality and texture of fried foods during storage. Deep fat frying decreases the unsaturation and moisture contents of the frying medium and increases the colour, viscosity, density, specific heat, and contents of free fatty acids (FFAs), polar materials, and polymeric compounds, as shown in Figure 6.3. These parameters are affected by several other factors present in the frying system.

The deterioration of frying fats is significantly affected by the different conditions used during frying. Gere (1983) studied several parameters in laboratory heating and frying experiments in order to get further information on each effect. Infrared (IR)

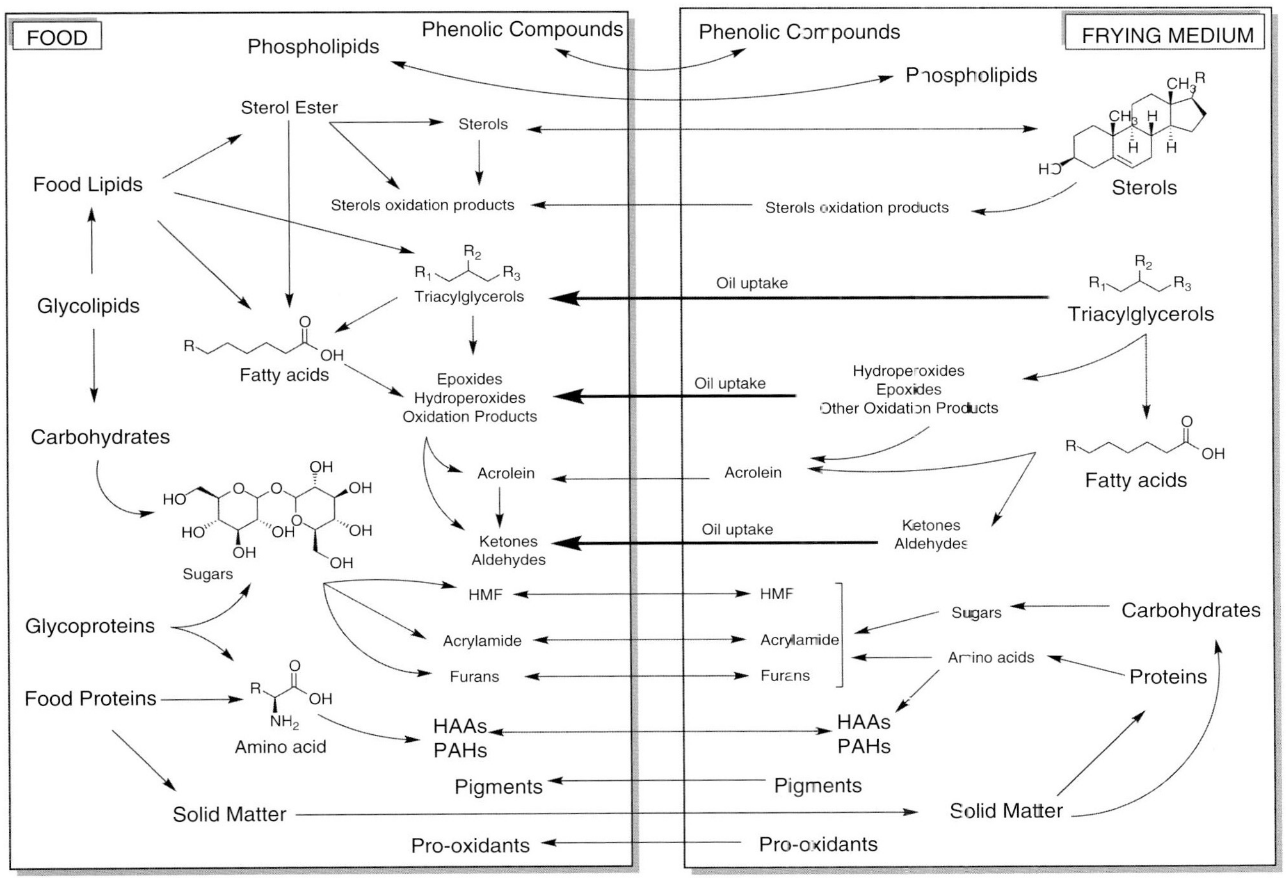

Figure 6.2 Schematic representation of several chemical interactions between food and frying medium. The bold arrow represents uptake of the oil by the food.

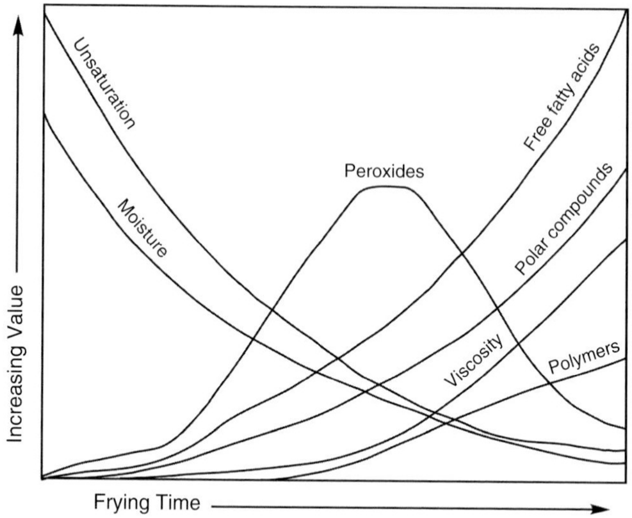

Figure 6.3 Changes in important physicochemical characteristics of the frying medium.

study of the chemical nature of total polar products showed that frying under severe conditions not only increased the rate of formation of decomposition products but also increased the deterioration of the polar fraction. This study provided numerical data showing the importance of the effects of temperature, intermittent use, surface-to-volume (S/V) ratio, and the use of fresh oil.

6.2.1 Temperature and Surface-to-Volume Ratio

Temperature is one of the key factors affecting the frying medium. High temperature may cause the formation of oxidized triacylglycerols (TAGs) (Zeb 2015), *trans* fatty acids (TFAs) (Bhardwaj et al. 2016), 3-chloropropane-1,2-diol fatty acid (FA) esters (Zhou et al. 2014), and glycidyl esters (Ben Hammouda et al. 2017), and affects the quality of refined olive-pomace oil blended with refined palm oil and several other frying oils. The physical and chemical interactions during frying are also severely affected by the S/V ratio of the oil, even if the temperature is kept constant.

Da Silva and Pereira (2008) reported the emission rates of certain carbonyl compounds formed by palm and soybean oils at 180 °C in the presence of air, through different time intervals and at different S/V ratios, in both continuous and intermittent processes. The authors collected and derivatized several carbonyl compounds on silica C-18 cartridges impregnated with an acid 2,4-dinitrophenylhydrazine solution. Acetaldehyde, acrolein, propanal, butanal, hexanal, 2-heptenal, and 2-octenal were the main emission products in both oils and at all S/V ratios, followed by hexanal and 2-heptenal (Figure 6.4). Greater emission rates of acrolein were observed for soybean oil than for palm oil. When different S/V ratios were used (e.g. 0.23, 0.43, and 0.77 cm^{-1}), saturated and unsaturated carbonyl compounds had different behaviours towards oxidation reactions. Similarly, there was an increasing trend in the emission rates of saturated aldehydes during intermittent heating, while there was a decreasing trend with unsaturated aldehydes. This was attributed to the reactivity of the double bond present in these compounds.

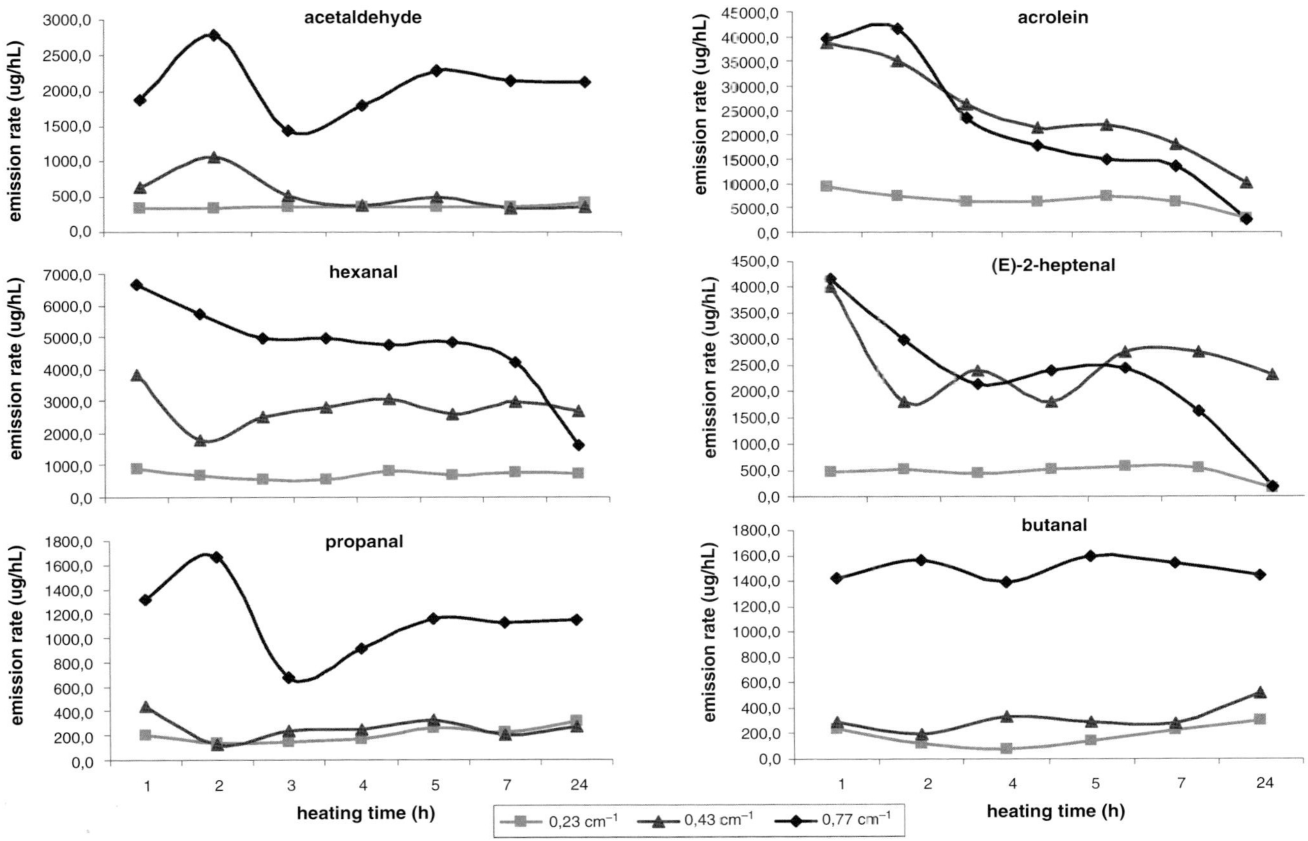

Figure 6.4 Emission rates of carbonyl compounds from palm oil during continuous heating at 180 °C under three surface-to-volume (S/V) ratios. *Source:* Reproduced with kind permission of the American Chemical Society (Da Silva and Pereira 2008).

6.2.2 Triacylglycerol Composition

TAGs constitute the major part of the frying medium, so their chemistry plays a significant role in frying interactions. Saturation and unsaturation of TAGs imparts stability to the frying medium. The saturated TAG composition of fats such as tallow, lard, and butter causes higher stability during frying, thus favouring food–oil interactions. Glycidyl esters in palm oil were found to decrease during frying, which was significant in unsaturated FAs (Aniołowska and Kita 2016) Unsaturation increases the chances of oxidation and thus favours oil–oil interactions. Thus, polyunsaturated oils will be prone to oxidation and may not be effective for prolonged frying.

6.2.3 Moisture Contents

Moisture is present in frying media in relatively small amounts, while higher amounts are present in plant foods. During the initial heating of oils or fats, the moisture contents of the food and frying medium is heated to form steam. Steam is evaporated with a bubbling action and thus lost. Moisture from the food enters the frying medium, which causes hydrolysis of the oil. Krokida et al. (2000) showed that water loss from food is a function of frying. The hydrolysis of TAGs produces FFAs, which causes a decrease in the quality of the oil.

On the other hand, the water released from foods during frying improves heat transfer and prevents oxidation (Saguy and Dana 2003). Thus, in order to improve oil quality, some simple integrated approaches such as water injection can be used during frying. Dana et al. (2003) used corn and canola oils to fry foods at 170 °C under different water injection flow rates. They found that water injection should be considered for prolonged frying of foods. Continuous water injection may be helpful in the food service and restaurants, where high-temperature frying oil is maintained idle for long periods of time (Peri and Saguy 2015).

6.2.4 Antioxidants

Antioxidants are those organic compounds which prevent the oxidation of a substrate or a free radical chain reaction. The basic mechanism of antioxidation involves the reaction of a hydrogen atom, which is provided by an antioxidant to stop free radical reactions. It has been observed that a compound that can easily provide a hydrogen atom (is loosely bonded) will be a stronger antioxidant than one with tightly bonded hydrogen. Food antioxidants present in the frying process include tocopherols, phenolic compounds, sterols, squalene, vitamin A, and carotenoids (Hudson 2012). They play an important role during the frying of foods (Márquez-Ruiz et al. 2014; Aladedunye et al. 2017).

The mechanisms of action of the natural food antioxidants are similar. For example, the hydroxyl group of tocopherol provides a hydrogen atom to a lipid free radical to perform a termination reaction. The tocopherol free radical is stabilized by the resonance hybridization of a radical on the benzene ring. This radical is converted to its active tocopherol antioxidant if another antioxidant such as ascorbic acid is present. Thus, tocopherol acts as an antioxidant during frying (Rossi et al. 2007). A phenolic compound such as gallic acid that contains more than one hydroxyl group is a more

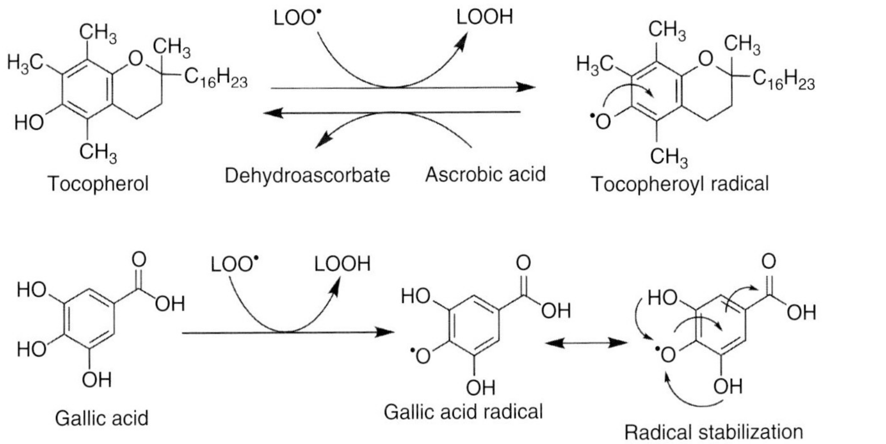

Figure 6.5 Antioxidant roles of tocopherol and gallic acid, showing the stabilization of antioxidant radical after hydrogen is donated.

efficient antioxidant than tocopherol. The para hydroxyl group is of low binding energy and thus can easily donate a hydrogen atom to scavenge free radical. The resultant gallic acid free radical is more stable than a tocopheroyl radical. Stabilization is provided by the resonance on the benzene ring, the interaction of adjacent hydroxyl groups, and the electron withdrawal of the carboxylic group, as shown in Figure 6.5. Thus, polyphenolic compounds present in the frying medium provide protection during frying (Chiou et al. 2007).

The effects of several antioxidants on the stability of rapeseed oil during the deep fat frying of potato chips (French fries) were evaluated by Gordon and Kourkimskå (1995). These authors measured the formation of polymers and the increase in the peroxide value, supported by the determination of the increase in mass during heating and changes in tocopherols during frying. The order of antioxidant activity during heating was: TBHQ > lecithin > ascorbyl palmitate > rosemary extract > BHT, BHA, and D-δ-tocopherol. Lampi and Kamal-Eldin (1998) showed that high oleic/high γ-tocopherol canola and soybean oils are better protected during frying than high oleic/high α-tocopherol sunflower oil.

Similarly, Normand et al. (2001) showed that the frying stability of canola oil was more affected by tocopherol degradation than by changes in the FA composition. In another study, Warner and Moser (2009) reported the effects of α-, γ-, and δ-tocopherols and TAGs in various combinations on stripped mid/high oleic sunflower oil used in the frying of tortilla chips. They studied the deterioration of frying oil in terms of total polar compounds (TPC) and the retention of tocopherols in the oil or the chips. The effects of tocopherols on chips were evaluated for hexanal as an indicator of oxidative stability and for odour characteristics. TPC was higher in the control oils, followed by the oils containing only α-tocopherol. These results showed that oil containing γ-tocopherol had the best fry life, as indicated by its low TPC. Nogala-Kalucka et al. (2005) studied the effects of deep frying on antioxidant stability and activity by adding α- and δ-tocopherol to a mixture of partially hydrogenated rapeseed and palm oil called Planta. The amounts added were 0.01, 0.05, and 0.1%, and peroxide value, anisidine

Table 6.1 Effects of the addition of α-tocopherol (α-T) or δ-tocopherol (δ-T) on the fatty acid (FA) pattern of Planta after 6 hours of frying at 160 °C.

FA	Initial	Control	0.05% δ-T	0.1% δ-T	0.2% δ-T	0.05% α-T	0.1% α-T	0.2% α-T
Total SFA	48.1 ± 0.5	47.5 ± 0.9	49.1 ± 1.2	47.4 ± 1.7	47.3 ± 1.6	48.3 ± 1.5	47.3 ± 1.8	47.4 ± 1.5
Total MUFA	39.1 ± 0.7	39.0 ± 0.8	37.8 ± 1.1	39.7 ± 0.9	40.2 ± 0.8	39.9 ± 0.7	40.6 ± 1.0	40.8 ± 0.9
Total PUFA	12.7 ± 0.1	12.5 ± 0.3	12.4 ± 0.5	12.5 ± 0.3	12.4 ± 0.4	11.6 ± 0.5	12.0 ± 0.4	11.8 ± 0.4
n-3/*n*-6	0.30	0.30	0.30	0.28	0.29	0.29	0.27	0.26

Source: Reproduced with kind permission of Springer Nature (Nogala-Kalucka et al. 2005).
SFA, saturated fatty acids; MUFA, monounsaturated fatty acids; PUFA, polyunsaturated fatty acids.

value, and the hexanal were used as a measure of oxidation at 160 °C (Table 6.1). The results showed that the addition of an antioxidant improved the stability of the frying fat, but no concentration-dependent effects were observed. The δ-tocopherol was shown to be more stable than α-tocopherol during high-temperature frying.

Carotenoids are also strong natural antioxidants that are found in edible oils but not in animal fats. During frying, the high temperature significantly affect the carotenoid content. In a model TAG system, when 300 μg of astaxanthin[1] was added per gram of oil and heated, the degradation of astaxanthin was significantly affected by increasing temperature, as shown in Figure 6.6. Carotenoids were found to be more stable at lower temperatures, with lower degradation. Higher degradation was observed at 140 °C.

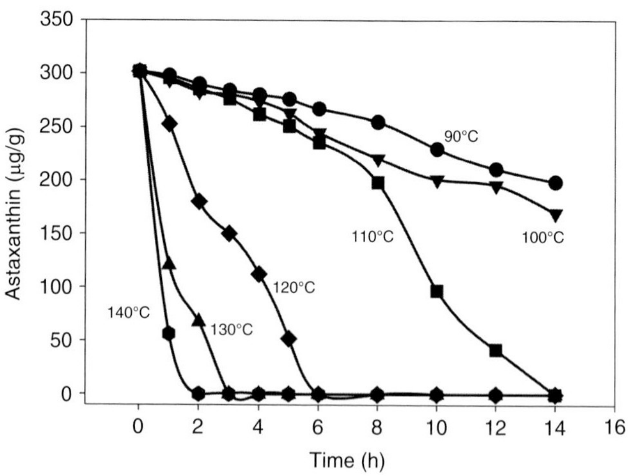

Figure 6.6 Degradation of astaxanthin at different temperatures.

1 Astaxanthin is a derivative of β-carotene with two ketone functional groups at the 4 position of the β-ring. It is an example of xanthophyll: oxygenated carotene. It is reddish in colour and lipid-soluble. Astaxanthin is commonly present in marine organisms such as microalgae, yeast, trout, shrimp, salmon, crayfish, and several of the crustaceans. It is a natural dietary component, commonly used as a food supplement or in food fortification as a colourant.

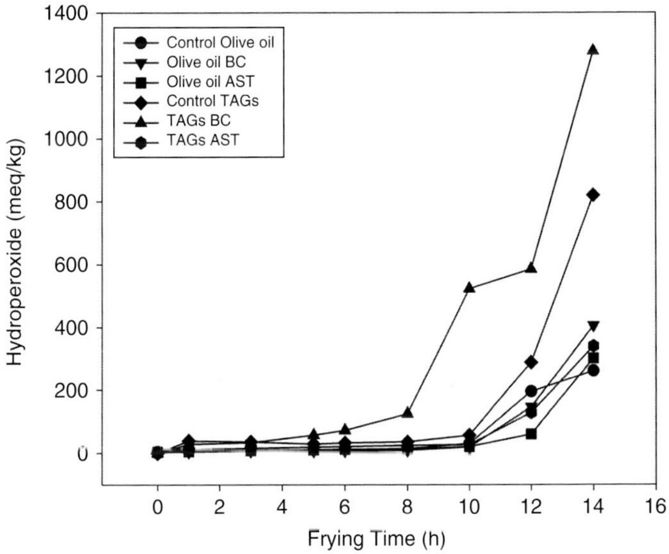

Figure 6.7 Effects of astaxanthin and β-carotene on the formation of hydroperoxides in model TAGs and olive oil at 110 °C. BC, β-carotene; AST, astaxanthin; TAG, model triacylglycerol.

Interestingly, the degradation of carotenoids plays a reverse role to that of the antioxidants, having pro-oxidant activity. When equal amounts of β-carotene and astaxanthin were added to a model frying medium with similar TAG composition as olive oil and actual olive oil, both carotenoids played a pro-oxidant role, but in different manners (Figure 6.7). The pro-oxidant activity was measured in terms of hydroperoxides formation. More than 300% as many hydroperoxides were formed in model TAGs as in olive oil in the presence of β-carotene. Similar, less significant observations were also reported in astaxanthin, although the amount of hydroperoxides was significantly lower than with β-carotene (Zeb and Murkovic 2010; Zeb and Murkovic 2011). These results show that other antioxidants, such as tocopherols and phenolic compounds, help in retarding hydroperoxide formation in edible oils.

Frying of foods containing carotenoids or phenolic compounds increases the stability of oil towards oxidation. Among the edible oils, the highest amount of carotenoids is found in red palm oil, followed by sea buckthorn oil (Zeb 2012). Frying of French fries in red palm oil showed the generation of lower volatility acrolein as compared with other edible oils, suggesting a positive effect of carotenoids on the formation of individual oxidation compounds (Andreu-Sevilla et al. 2009). Lee et al. (2003) fried flour dough containing carrot powder at 160 °C in soybean oil to evaluate lipid oxidation. The amount of carotenoid in the oil increased as frying was repeated, due to the dissolution of carotenoids from the dough. Frying of watercress leaves at 180 °C in sunflower oil showed an initial increase in the oxidation parameters (peroxide values, anisidine values, and FFAs), but the further increase was not significant, as shown in Figure 6.8 (Zeb and Habib 2018). Similar observations were also reported during the frying of spinach leaves at a relatively high temperature (Zeb and Nisar 2017). These results show that during the frying of vegetables, frying oils are relatively stable due to the leakage of natural antioxidants into the frying medium. Frying of animal foods, however, does not provide the same sort of stability.

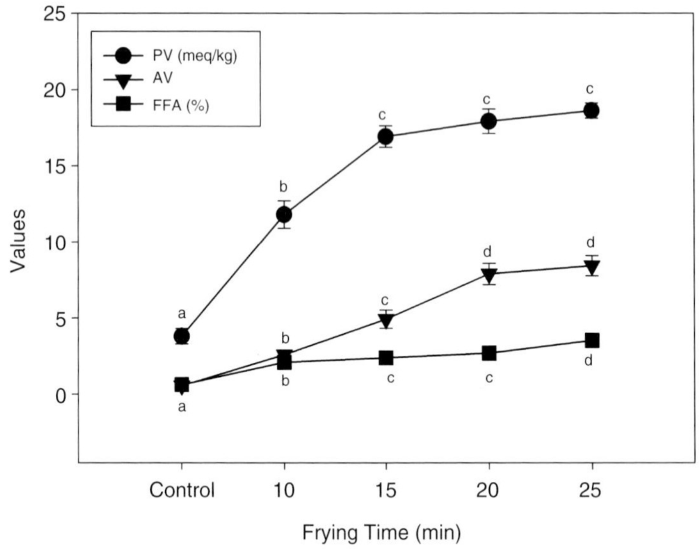

Figure 6.8 Effects of frying on the lipid oxidation parameters (peroxide values (PV), anisidine values (AV), and free fatty acids (FFA)) of sunflower oil and watercress leaves. Different letters (a–d) in the same parameter represent significance at $p < 0.05$ (PV), $p < 0.01$ (AV and FFA). *Source*: Reproduced with kind permission of Springer Nature (Zeb and Habib 2018).

6.3 Factors Affecting the Food

The frying of foods is affected by several factors, including temperature, air, frying apparatus, S/V ratio, and the type of food being fried. In the case of a direct heating system, the frying process is relatively fast, with more interactions occurring in less time as compared to indirect heating. As indicated in Figure 6.9, moisture losses from food will be higher in direct heating than in indirect heating. Thus, the frying system or type of frying plays an important role in determining different interactions.

6.3.1 Physical Parameters

Physical parameters such as temperature, time, and S/V ratio are of considerable importance during frying, significantly affecting various chemical interactions. For example, Gökmen and Palazoğlu (2009) studied changes in the amount of acrylamide in potatoes during frying, using water vaporization. Water vapour was condensed by a cold trap and then analysed for acrylamide content. To assess the thermal effect on acrylamide formation, the surface and internal temperatures were measured at different frying temperatures (150, 170, and 180 °C), times (1, 3, and 4 minutes), and S/V ratios (strips, cubes, and discs). The percentage ratio of the amount of acrylamide lost by evaporation ranged between 0.7 and 1.8 during frying at temperatures ranging between 150 and 180 °C. The authors concluded that an insignificant fraction of the acrylamide formed during frying is removed with vaporizing water.

Similarly, Su et al. (2016) showed that temperature is one of the main factors affecting quality in the microwave-assisted vacuum frying of potato chips. The authors suggested

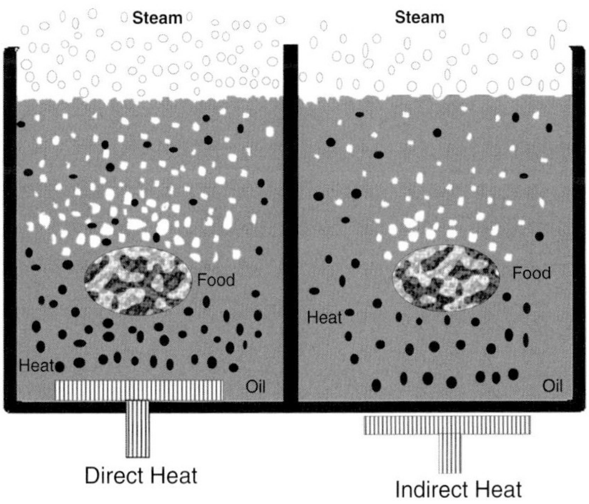

Figure 6.9 Deep frying of food. Heat transfer and steam are higher in direct heating than in indirect heating.

that potato chip manufacturers need to find suitable temperature conditions to obtain a product of acceptable quality and safety. Kavousi et al. (2015) showed that both pH and heating time significantly affected the formation of HMF in model systems.

Saeleaw and Schleining (2011) produced cassava crackers by deep frying cassava starch and water in oil at 140, 150, and 160 °C. They evaluated physical properties such as frying time, moisture content, bulk density, linear expansion, penetration forces, and sound emission. With increasing frying time and temperature, there was a significant increase in the moisture content, linear expansion, number of forces, and sound peaks. There was a significant decrease in bulk density and maximum and mean force peaks with increasing frying time and temperature. Significant correlations were reported between physical and textural properties and sound emission. A negative correlation was found between moisture content, bulk density, maximum sound peak, maximum force and mean force with linear expansion, number of sound peaks, and number of force peaks.

Different types of frying also produce significant changes in the physical parameters of foods. Dueik et al. (2010) and Dueik and Bouchon (2011) showed that vacuum frying can reduce oil content to nearly half while preserving carotenoids such as all-*E*-α-carotene (by 90%) and all-*E*-β-carotene (by 86%). Yagua and Moreira (2011) showed that vacuum frying significantly improved the physical properties of fried potato chips.

Coating of foods prevents loss of water and nutrients, and thus may alter the normal interactions seen in the same foods without coatings. For example, water barrier coatings of fried doughnuts and French fries were prepared from whey protein/pectin edible films in the presence of transglutaminase (Rossi Marquez et al. 2014). Textural characteristics such as hardness and chewiness showed no significant difference between uncoated and whey protein/pectin/transglutaminase-coated samples, while a small increase in both hardness and chewiness was observed in soy protein/whey protein-coated samples only (Table 6.2). The authors proved that coating with an edible whey protein/pectin film prepared in the presence of transglutaminase enzyme was a very useful way of decreasing significantly the oil content in widely consumed deep-fried foods and of preventing water absorption by baked foods over time.

Table 6.2 Texture profile analysis of uncoated and differently coated doughnuts and French fries.

	Doughnuts		French fries	
	Hardness	Chewiness	Hardness	Chewiness
Sample	(*N*)	(*N* × mm)	(*N*)	(*N* × mm)
Uncoated	15.05 ± 1.1	14.75 ± 0.3	26.50 ± 2.1	12.53 ± 1.5
WP/Pec/TGase-coated	14.98 ± 0.8	16.03 ± 0.3	25.01 ± 2.8	12.84 ± 2.9
SP/WP-coated	17.90 ± 1.0	18.01 ± 2.1	30.48 ± 5.6	15.55 ± 2.2

Source: Reproduced with permission of Springer Nature (Rossi Marquez et al. 2014).
WP/Pec/TGase, whey protein/pectin/transglutaminase; SP/WP, soy protein/whey protein.

6.3.2 Chemical Parameters

Food is made of a complex mixture of biological substances, including lipids, proteins, carbohydrates, vitamins, minerals, phenolic compounds, and pigments. Plant foods are rich in carbohydrates, phenolic compounds, and pigments. However, significant variation exists both within the same types of food and between different foods. During food frying, food–food interactions cause the formation of several new organic compounds, which either impart characteristic properties to the food or cause negative acceptance for consumers. The stability or retention of these pre-existing or new compounds depends on several physical parameters, as already described. Thus, it is important to discuss them as they relate to the different food classes.

6.3.2.1 Vegetable Foods

Frying of vegetables induces changes in several classes of phytochemicals, such as polyphenols, carotenoids, chlorophylls, and vitamins. Palermo et al. (2014), Murador et al. (2014), and Ahmed and Eun (2017) reviewed the effects of different cooking methods on the phytochemicals contents of several vegetables. However, these studies are based on a number of different cooking procedures, including frying; therefore, updated knowledge on frying alone will be presented here. Frying of onion, green pepper, and cardoon in olive and sunflower oil was evaluated in terms of antioxidant capacity and (poly)phenolic compounds by Juániz et al. (2016). The main compounds were quercetin and isorhamnetin derivatives in onion, quercetin and luteolin derivatives in green pepper, and chlorogenic acids in cardoon. Frying was shown to enhance the phenolic compounds in these vegetables, suggesting a thermal destruction of the cell walls and subcellular compartments during the cooking process that favoured their release. The increase in chlorogenic acids during frying was significantly correlated ($r = 0.70$) with an increase in antioxidant capacity.

Ferracane et al. (2008) showed the major phenolic compounds present in raw artichoke samples were 5-*O*-caffeoylquinic, 1,5-di-*O*-caffeoylquinic acids, and apigenin-7-*O*-glucuronide. The total amount of all of the identified phenolic compounds (i.e. seven derivatives of caffeoylquinic acids and two of apigenin) significantly increased with frying, as shown in Figure 6.10.

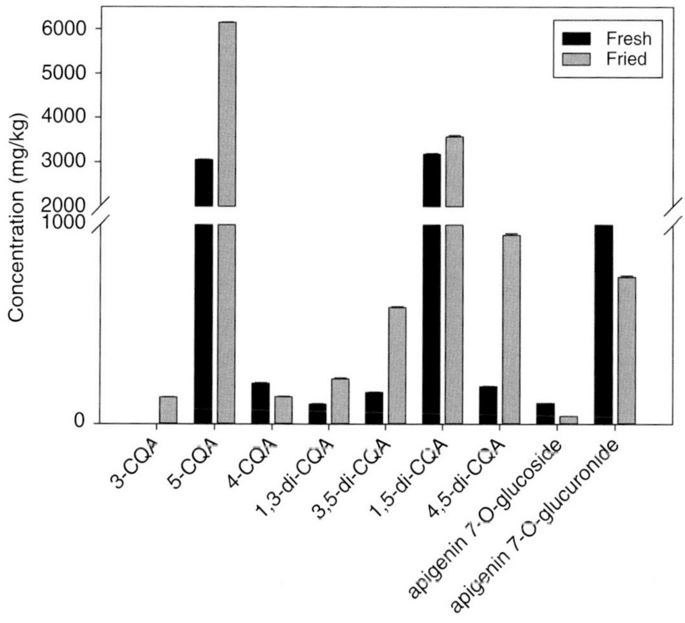

Figure 6.10 Effects on the polyphenolic compounds in artichoke of frying at 170 °C in peanut oil. Values are means of triplicate readings; error bar is the standard deviation. *Source:* Reproduced with kind permission of the American Chemical Society (Ferracane et al. 2008).

Zeb and Habib (2018) fried watercress leaves fried in sunflower oil and analysed them in terms of lipid oxidation and phenolic compounds using high-performance liquid chromatography with diode-array detection (HPLC-DAD). They observed that the TPC and radical scavenging activity (RSA) of the frying oil decreased significantly in the initial stages and then remained stable, indicating stability of the oil. A significant increase occurred in the chlorophyll content of the leaves and the oil. HPLC chromatograms showed a total of 12 phenolic compounds (Figure 6.11), containing caftaric acid (17.3–31.8 $\mu g\,g^{-1}$), sinapic acid derivative (11.2–14.4 $\mu g\,g^{-1}$), 4-*O*-caffeoylquinic acid (9.98–12.8 $\mu g\,g^{-1}$), 4-*O*-feruloyl quinic acid (8.47–13.4 $\mu g\,g^{-1}$), caffeoyl feruloyl quinic acid (7.31–11.7 $\mu g\,g^{-1}$), and 5-*O*-caffeoylquinic acid (3.73–6.51 $\mu g\,g^{-1}$) as major compounds. A significant increase in caffeic acid, caffeoyl hexose, chlorogenic acid, caftaric acid, and coumaric acid derivatives and 3,4-di-*O*-caffeoylquinic acid was observed during frying, while there was a decrease in sinapic acid derivative, 5-*O*-caffeoylquinic acid, 4-*O*-caffeoylquinic acid, 4-*O*-feruloyl quinic acid, and caffeoyl feruloyl quinic acid (Table 6.3) (Zeb and Habib 2018). Frying of watercress leaves stabilized the frying oil and increased the amount of medicinally important phenolic compounds.

Several other vegetables, such as sweet potato, green beans, Tommy Atkins mango, and blue potato, were fried both in vacuum at 120–130 °C and under atmospheric frying at 160–165 °C (Da Silva and Moreira 2008). The anthocyanin content was 60% higher in vacuum-fried blue potato chips than in those produced via atmospheric frying. The total carotenoid amount was higher in green beans (by 18%), mango chips (19%), and sweet potato chips (by 51%). This study supports the applicability of vacuum frying technology in providing high-quality fruit and vegetable snacks.

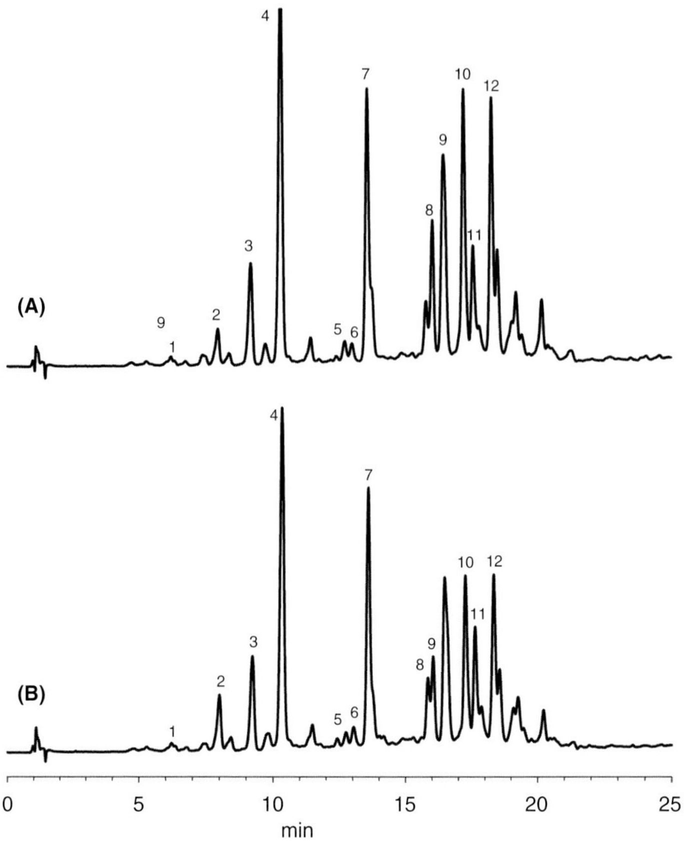

Figure 6.11 Representative HPLC-DAD chromatogram of phenolic compounds in watercress leaves at 360 nm. (A) Control unfried sample, and (B) fried leaves sample (25 min). Details concerning the method of sample preparation and analysis can be found in Zeb (2015). *Source*: Reproduced with kind permission of Springer Nature (Zeb and Habib 2018).

Skrede et al. (1997) reported the carotenoid retention in carrot (*Daucus carota* L.) varieties grown at six different latitudes in Norway. The authors showed that fried carrot chips are good sources of carotenes. Sulaeman et al. (2001) studied the effects of deep frying using different oils and temperatures on the carotenoid content of carrot chips. Sliced carrots were first blanched using steam, then cooled, and finally soaked in sodium metabisulfite (0.2%). The carrot slices were then deep fried at 165, 175, or 185 °C in canola, palm, or partially hydrogenated soybean oil (PHSO). It was found that frying temperature, but not oil type, significantly ($p < 0.05$) affected the α-carotene, β-carotene, and total carotenoid contents. Table 6.4 lists some recent studies on the effects of frying on the carotenoid composition of different vegetables. It has been observed that with the help of HPLC, the interactions during frying of several carotenoids can be determined.

Kidmose et al. (2006) studied the variation in β-carotene and vitamin A in commonly consumed vegetables in Asia. The retention of β-carotene and formation of *cis* isomers were investigated during stir frying. In sweet bell pepper, sweet potato, and tomato fried for 12–13 minutes, the retentions of *all-trans*-β-carotene varied between 73 and 98%. In sweet potato, the major *cis* isomer was 13-*cis*-β-carotene, while minor amounts

Table 6.3 Changes in the phenolic compounds in watercress leaves during frying at 180 °C for 10, 15, 20, and 25 minutes.

Peak	Compounds	Amount ($\mu g\, g^{-1}$)[a]					
		p < value	Control	10	15	20	25
1	Caffeic acid	0.01	0.64 ± 0.05a	0.96 ± 0.11b	1.62 ± 0.12c	2.22 ± 0.05d	3.18 ± 0.02e
2	Caffeoyl hexose	0.01	0.65 ± 0.01a	1.55 ± 0.03b	1.67 ± 0.02c	2.02 ± 0.08d	2.65 ± 0.05e
3	3-*O*-Caffeoylquinic acid	0.05	3.22 ± 0.02a	3.96 ± 0.03b	3.79 ± 0.09c	3.81 ± 0.11c	4.25 ± 0.05d
4	Caftaric acid	0.001	17.3 ± 0.3a	21.9 ± 0.75b	23.2 ± 0.11b	26.5 ± 0.2c	31.8 ± 0.6d
5	p-Coumaroyl hexose	0.01	0.72 ± 0.02a	0.78 ± 0.03a	0.83 ± 0.08a	0.67 ± 0.01b	0.86 ± 0.01a
6	p-Coumaric acid derivative	0.01	0.74 ± 0.01a	0.74 ± 0.01a	0.97 ± 0.02b	1.84 ± 0.03c	2.21 ± 0.02d
7	Sinapic acid derivative	0.05	14.4 ± 0.3a	11.5 ± 0.45b	11.2 ± 0.21b	11.3 ± 0.25b	13.3 ± 0.11c
8	5-*O*-Caffeoylquinic acid	0.01	6.51 ± 0.1a	4.86 ± 0.06b	4.25 ± 0.05b	4.0 ± 0.11b	3.73 ± 0.03c
9	4-*O*-Caffeoylquinic acid	0.01	12.8 ± 0.4a	10.9 ± 0.65b	11.2 ± 0.05b	11.0 ± 0.25b	9.98 ± 0.01b
10	4-*O*-Feruloylquinic acid	0.01	13.4 ± 0.3a	11.4 ± 0.41b	11.0 ± 0.15b	10.3 ± 0.15b	8.47 ± 0.03c
11	3,4-di-*O*-Caffeoylquinic acid	0.05	3.12 ± 0.11a	4.76 ± 0.16b	6.15 ± 0.85c	5.31 ± 0.05c	5.11 ± 0.01c
12	Caffeoyl feruloyl quinic acid	0.05	11.7 ± 0.11a	9.65 ± 0.15b	9.42 ± 0.21b	8.92 ± 0.21b	7.31 ± 0.11c

Source: Reproduced with kind permission of Springer Nature (Zeb and Habib 2018).
[a] Different letters (a–d) in the same row represent significant corresponding p values.

Table 6.4 Recent studies on the effects of frying on the carotenoids contents of different vegetables.

Vegetables	Frying Type	Frying Medium	Frying Temperature (°C)	Carotenoids	References
Sweet potato, Tomato, Sweet bell pepper, Amaranth leaves, Water convolvulus leaves, Sweet potato	Pan frying	Soybean oil	160	β-carotene and its Cis isomers	Kidmose et al. (2006)
Carrots, Courgettes, Broccoli	domestic deep fryer	Peanuts oil	170	Lutein, β-carotene, phytoene, phytofluene,	Miglio et al. (2008)
Artichokes	domestic deep fryer	Peanuts oil	170	Lutein, β-carotene, phytoene, phytofluene,	Ferracane et al. (2008)
Pumpkin	Stir frying	Hot oil (NM)	170	β-carotene	Azizah et al. (2009)
25 fresh vegetables (Taiwan)	Deep frying and stir frying	soybean oil	190	zeaxanthin, -carotene and their trans/cis (E/Z) isomers	Kao et al. (2012)
Spinach, Asia salads	Stir frying	Peanuts oil	NM	β-carotene, lutein	Eriksen et al. (2016)
Gim bugak	Shallow pan frying	Sesame oil, soybean oil, olive oil and palm oil	180	β-carotene, lutein	Jung and Choe (2017)
Spinach leaves	Pan frying	Sunflower oil	250	β-carotene –5,6-epoxide, All-*E*-neoxanthin, All-*E*-violaxanthin, All-*E*-luteoxanthin, All-*E*-mutatoxanthin, All-*E*-Lutein, its *Z*-isomers	Zeb and Nisar (2017)
Maize meal	Pan frying	Peanuts oil	150	Lutein, zeaxanthin, β-cryptoxanthin, all-(*E*)-β-carotene	Sowa et al. (2017)

NM: Not mentioned.

of 9-*cis*- and 15-*cis*-β-carotene were also formed. These authors showed that the net amount of *cis* isomers of β-carotene formed during frying depended on the frying time and the size, with the highest amounts found in cubes that were fried for 3 minutes (1070 μg 13-*cis*-β-carotene/100 g fresh weight (FW)). In leafy vegetables, only 13-*cis*-β-carotene was detected during frying. Extraction of β-carotene into the frying oil was observed only in low amounts after 3 minutes frying of sweet potato shreds.

Miglio et al. (2008) reported the effects of different cooking methods (including frying in peanut oil) on the carotenoid composition of carrot, courgette, and broccoli. Ferracane et al. (2008) likewise studied their effects on the carotenoid composition of artichoke. Azizah et al. (2009) showed that the β-carotene and lycopene contents of pumpkin increased with stir frying. Similar observations were reported during the frying of spinach leaves and different Asian salads in peanut oil (Eriksen et al. 2016, 2017). In contrast, Kao et al. (2012) showed that stir frying of 25 commonly consumed vegetables decreased their total carotenoid contents and all-*trans* forms, and increased their *cis* isomers. Individual vegetable such as gim bugak (Jung and Choe 2017) and vegetable meals such as maize meal (Sowa et al. 2017) showed changes in their specific carotenoid content during frying.

Zeb and Nisar (2017) revealed the effects of high-temperature frying on the carotenoid, chlorophylls, and tocopherol contents of spinach leaves. Zeb (2017) identified 13 pigments in spinach leaves using HPLC-DAD. Equally cut spinach leaves were fried in sunflower oil at 250 °C for 15, 30, 45, and 60 minutes. Lutein, neoxanthin, violaxanthin, and β-carotene-5,6-epoxide were the major carotenoids, while chlorophylls a and b′ were present in high amounts, as shown in Figure 6.12.

Frying of spinach leaves significantly increased the amounts of α-tocopherol, β-carotene-5,6-epoxide, luteoxanthin, lutein and its *cis* isomers, and chlorophyll *b′* isomer present in them. There was a significant reduction in neoxanthin, violaxanthin, chlorophyll *b*, *b′*, and chlorophyll *a* with increasing frying time (Figure 6.13). It was found that the spinach leaves increased the stability of the frying oil by prolonging the thermal oxidation. This study can be used to improve the quality of vegetable leaves and their products undergoing high-temperature frying in the food industry. It suggests that the frying of vegetables produces changes in their chemical constituents that increase their stability toward oxidation.

6.3.2.2 Animal Foods

Foods obtained from animals or their products can also be fried using a variety of techniques. The major components of animal foods are fats and proteins, with few vitamins, carbohydrates, and other biomolecules. Thus, most studies of animal foods have focused on the changes in proteins and fats during frying. Table 6.5 shows some of the important types of organic compounds formed during the frying of protein-rich animal foods. Jakszyn et al. (2004) have developed a database providing detailed information on some of these compounds.

The formation of HAAs has been found to be strongly correlated with an increase in frying temperature. Murkovic et al. (1997) showed that lower amounts of HAAs were produced in poultry meat at lower temperatures. Later, Murkovic (2004) reviewed the formation and occurrence of HAAs in fried foods. They found that HAAs are formed during the frying of red meat, poultry, fish, and meat products. Several factors contribute to the formation of HAAs during the frying of animal foods.

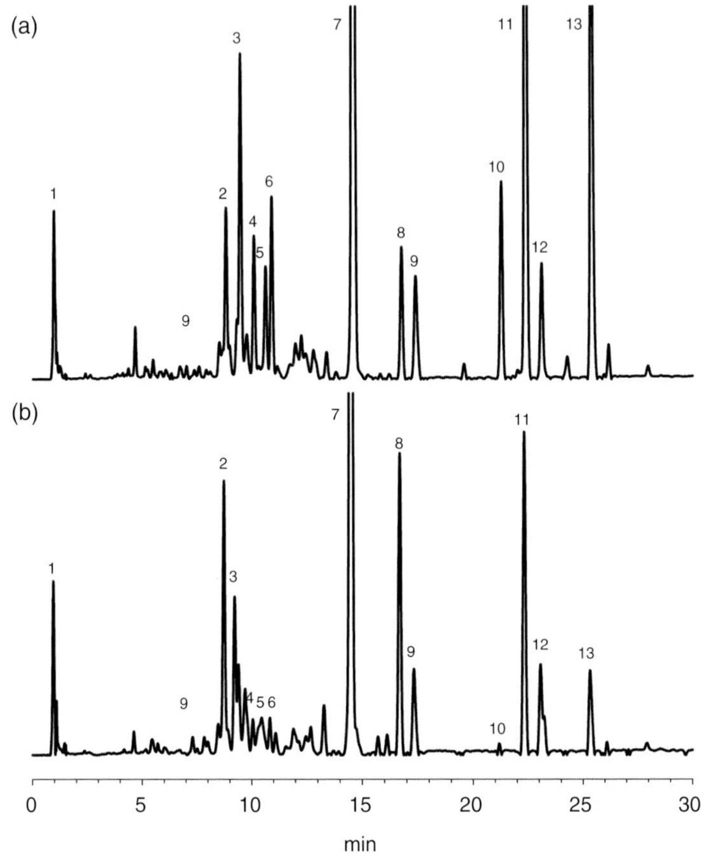

Figure 6.12 Representative HPLC-DAD chromatograms of spinach leaves at 450 nm (Zeb and Nisar 2017). (a) Unfried samples after 60 minutes. (b) Fried samples after 60 minutes. Details concerning the method of sample preparation and analysis can be found in Zeb (2017).

Gibis (2007) showed the effects of oil marinades containing garlic, onion, and lemon juice on the HAA content of fried beef patties, using two statistical design models. In a screening experiment, the addition of garlic caused significant ($p < 0.05$) inhibition of the formation of MeIQx in the patties. An increase in garlic from 2 to 20 g/100 g marinade produced a 70% reduction in the MeIQx content. In fried patties, the amount of MeIQx was in the range of 0.38–1.22 $ng\,g^{-1}$, that of 4,8-DiMeIQx was 0.0–0.45 $ng\,g^{-1}$, that of PhIP 0.0–0.09 $ng\,g^{-1}$, that of norharman 0.76–13.5 $ng\,g^{-1}$, and that of harman 2.9–21.5 $ng\,g^{-1}$. Statistical model analysis also showed a strong reduction in MeIQx with the addition of increasing amounts of garlic and onion, while a marginal reduction was caused by an increase in the amount of lemon juice. Optimal declines in HAAs were produced with 31.2% onion, 28.6% garlic, and 14.6% lemon juice in the marinade. Zhang et al. (2013) showed that more PhIP was pr0duced in patties than in meatballs and strips. A similar trend was reported for MeIQx and 4,8-DiMeIQx, as shown in Table 6.6.

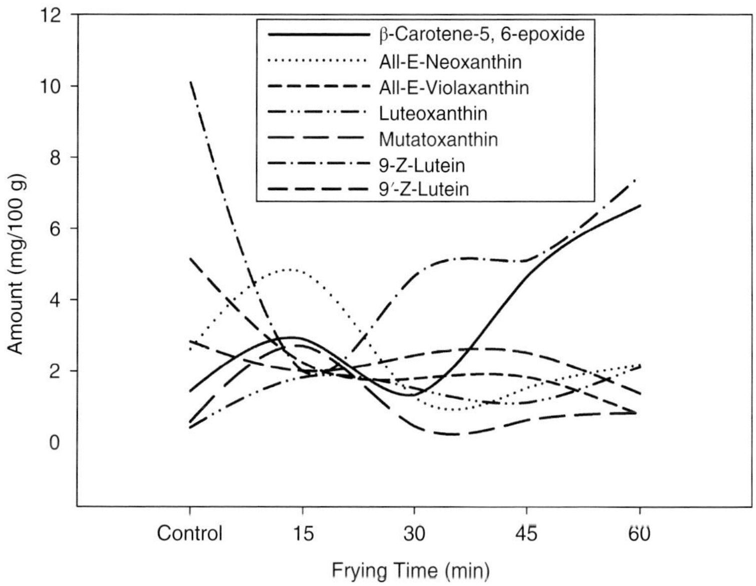

Figure 6.13 Interactions of carotenoids of spinach leaves during frying in sunflower.

Table 6.5 Different types of organic compounds formed from protein-rich animal foods.

Type of compound	Examples	Food
Heterocyclic aromatic hydrocarbons	Imidazoquinolines Imidazoxyquinolines Imidazopyridines Aminoimidazoquinolines	Meat and its products
Polycyclic aromatic hydrocarbons	Naphthalene Dibenzothiophene Phenanthrene	Beef, chicken, turkey, pork
Nitrosamines	*N*-Nitrosoproline, *N*-Nitrososarcosine *N*-Nitrosodimethylamine N-Nitrosopyrrolidine	Meat, bacon, sausages, mutton
Amides	Acrylamide	All animals foods
Polycyclic aromatic amines	2-Amino-3,8-dimethylimidazo[4,5-f]quinoxalines, 2-Amino-1-methyl-6-phenylimidazo [4,5-b] pyridine	Marine foods, pork

Table 6.6 Formation of HAAs (ng g^{-1}) in pork of different shapes fried for 5 minutes at 180 °C.

HAAs	Pork shape (%, RSD)		
	Patty	**Meatball**	**Strip**
MeIQx	3.22 (2.11)	1.02 (3.24)	0.960 (4.24)
4,8-DiMeIQx	0.706 (7.70)	0.310 (7.03)	0.277 (7.73)
PhIP	18.4 (11.5)	5.32 (2.74)	5.43 (7.54)

Source: Reproduced with permission of Taylor & Francis (Zhang et al. 2013).

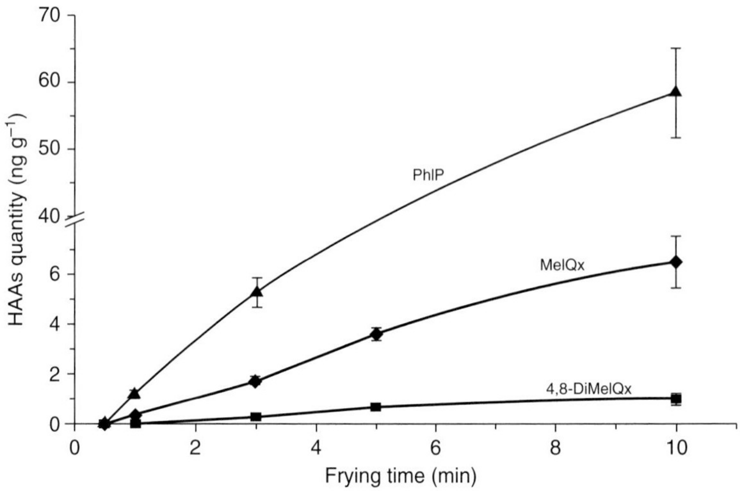

Figure 6.14 Effects of frying time on the formation of heterocyclic aromatic amines (HAAs) in pork fried at 180 °C. *Source:* Reproduced with kind permission of Taylor & Francis (Zhang et al. 2013).

The synthesis of MeIQx, 4,8-DiMeIQx, and PhIP was found to be highly dependent on the temperature and time of frying. Interestingly, none of the three HAAs was produced in noticeable amounts at below 160 °C in fried pork samples. When the temperature was increased to 220 °C, a significant increase in the formation of HAAs occurred. The amounts of selected HAAs also significantly increased with increasing frying time, as shown in Figure 6.14.

PAHs have been reported among the important mutagenic substances produced by the frying of meat. These compounds occur in different ratios in different parts of the food, and in different food shapes. Zahoor et al. (2015) analysed four tissues of Aseel chicken (Table 6.7). They found that the neck contained the highest amount (0.079 mg kg^{-1}) of anthracene, while the wings contained the lowest (0.006 mg kg^{-1}). The fluorene percentage increased in fried heart (225.04%), but fell in the wings 16.18%. Phenanthrene and pyrene, meanwhile, were maximal in the wings, at 0.249 mg kg^{-1} and 0.094 mg kg^{-1}, respectively. The results demonstrate that high-temperature treatment of chicken meat, such as microwave oven frying, may be responsible for the generation of these toxicants.

Table 6.7 Effects of microwave cooking on the formation of PAH (fluorene) in different unfried and fried organs of Aseel chicken.

	Concentration of PAH (mg kg^{-1})		
Organs	Unfried	Microwave oven fried	Increase (%)
Heart	0.02552	0.08295	225.0
Liver	0.06232	0.11677	87.3
Neck	0.15838	0.20991	32.5
Wings	0.37105	0.43107	16.1

Source: Reproduced with permission of Taylor & Francis (Zahoor et al. 2015).

6.4 Heat Transfer

Heat transfer is the principal factor responsible for changing the chemistry and other properties of frying. Heat is transferred from the frying apparatus to the frying oil either directly or indirectly, as shown in Figure 6.9. It is then transferred to the food, according to the food's structure and properties. It evaporates the water present in the food's body and pores as vapours, which enter the frying medium and produce turbulence. This turbulence is considered responsible for promoting heat transfer with the system. The vapours carry thermal energy from the system to the environment, decreasing the temperature. However, this decrease in temperature is not visible or easily detected, because of the complex nature of the oils and the frying process. The relatively constant temperature of the frying medium is the reason evaporation occurs, which does cause a minute decline in the temperature. Thus, heat transfer is responsible for several interactions during food frying.

As discussed in Chapter 2, heat transfer is dependent on several factors, among which thermal conductivity, thermal diffusivity, and the heat transfer coefficient have been extensively studied during frying. These properties have been found to increase with increasing temperature. Recent studies on heat transfer in gulab jamun (Franklin et al. 2014), Chhena jeili (Mondal and Dash 2017), loche and potato (Mauricio and Francisco 2017), Tulumba (Erim Kose and Dogan 2017), and batter-coated fried potato (Rahimi et al. 2017) indicate the importance of frying time and temperature. Thus, several types of oil–oil, food–food, and food–oil interaction may be considered important in the frying of foods.

6.5 Mass Transfer

Mass transfer during frying is responsible for several interactions, particularly between the food and the oil. It may include oil uptake by the food, and the transfer of organic and solid matter from the food to the oil. Oil uptake is a widely studied parameter as a function of mass transfer. It occurs based on several factors, such as oil temperature, frying time, food moisture content, food porosity, surface area, and atmospheric pressure. It has been experimentally proved that oil uptake is directly proportional to

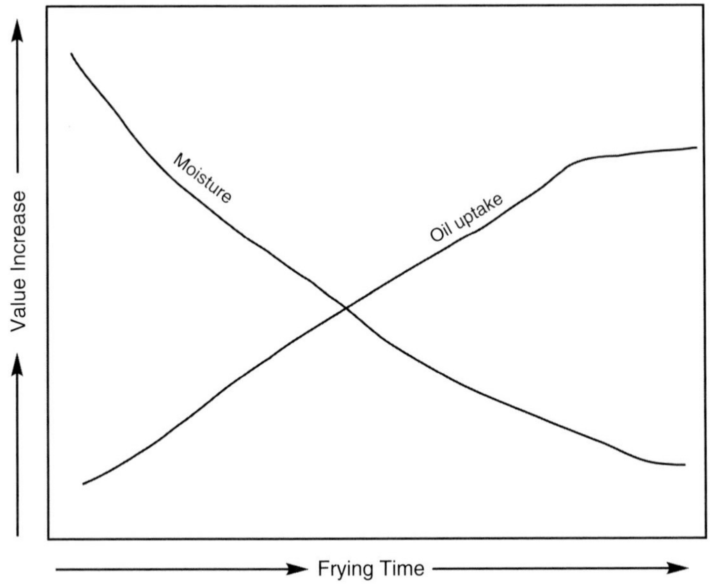

Figure 6.15 Correlation of moisture loss with oil uptake during food frying.

moisture loss. Moisture loss creates capillary pores, which are occupied by the oil via capillary action. The moisture loss decrease with increasing frying time until a steady state is reached, as shown in Figure 6.15.

In general, plant foods contain an abundant amount of moisture as compared to animal foods. A high moisture content leads to increased oil uptake and leaching of pigments during frying. For example, during the frying of spinach leaves, the amount of thiobarbituric acid reactive substances (TBARS) decreased in the oil and increased in the leaves (Figure 6.16) (Zeb and Nisar 2017). The strong correlation of TBARS value with frying time was a clear indicator that the TBARS content of the oil was assimilated in the leaf matrix. These results indicate that oil uptake may not be a simple matter of the transfer of all the oil contents, but involves several oxidized and toxic compounds.

6.6 Nutritional Value Retention

Food is made of complex biomolecules, with acceptability based on human perception. Not all foods are fried or cooked, with the majority of plant foods (e.g. fruits) eaten without any cooking procedures. In such case, the intestinal bioavailability of the nutrients they contain is the limiting factor in human nutrition. Among these nutrients, carbohydrates, lipids, proteins, vitamins, minerals, and natural antioxidants are the principal components. Animal foods are mostly cooked or fried, in order to increase digestibility and nutrient availability. The retention of nutrients is one of the main aims of modern food frying. This retention is calculated from the nutrient contents in the fried food per gram and the nutrient contents in the raw food per gram, expressed as a percentage. Fillion and Henry (1998) and Labuza and Tannenbaum (1972) reviewed the nutrient loss and gain of a variety of fried foods.

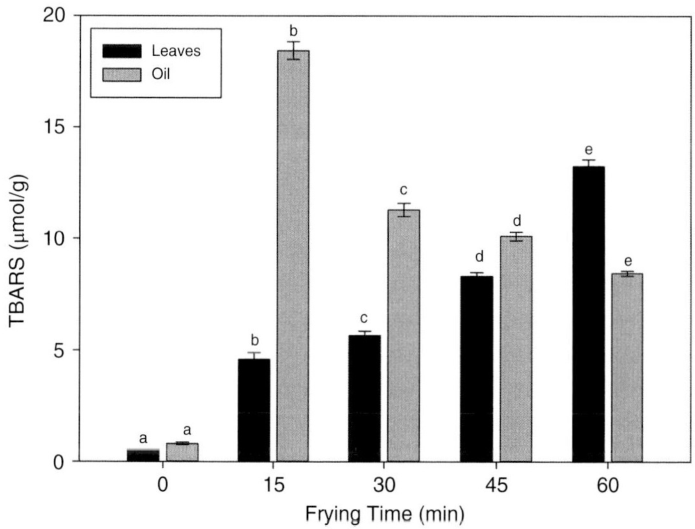

Figure 6.16 Changes in TBARS values during the frying of spinach leaves in sunflower oil at 250 °C (Zeb and Nisar 2017). Different letters (a–e) in the same type of bar represents significance at $p < 0.05$ of $n = 5$. Values are expressed as mean with SD of replicate readings. The error bars in each column represent its SD value.

6.6.1 Vitamins

Water-soluble vitamins such as B-complex and vitamin C are usually present in high amounts in plant foods. However, these vitamins are vulnerable to heat and thus are destroyed during frying, unlike fat-soluble vitamins (A, D, E, and K). A study of the retention of vitamin C during the frying of a number of vegetables from Thailand showed losses of 14–95% (Somsub et al. 2008). Rojas-Gonzalez et al. (2006) showed a vitamin C loss of 28% during the deep frying of plantains after 13 minutes at 140 °C, 7 minutes at 160 °C, and 4 minutes at 180 °C.

Carlson and Tabacchi (1986) showed that vitamin E is more stable than vitamin C in frying oil. Virgin olive oil (VOO), sunflower oil, and vegetable shortening were used for eight successive deep fryings and pan fryings of potato (Andrikopoulos et al. 2002). It was found that the retention of α- and (β + γ)-tocopherols decreased with each frying. During the first frying session, retention ranged from 85 to 90%; by the eighth, it declined to 15–40% during – except in the case of (β + γ)-tocopherols in sunflower oil, which completely vanished after the sixth session.

6.6.2 Carotenoids

Carotenoids such as carotene and xanthophylls are present in plant oils and foods. They are highly significant in terms of nutrition and have roles in several metabolic processes. Frying of vegetables is therefore of significant importance in the food industries. Deep frying of plantains showed carotenoid retention in the range of 9–13% at all frying temperatures (Rojas-Gonzalez et al. 2006). In frying oils such as sunflower, soybean, and corn oil, β-carotene was stable at lower temperatures and acted as a pro-oxidant at higher ones (Zeb 2012).

6.6.3 Phenolic Compounds

Phenolic compounds are present in all plant foods, as well as edible oils. In edible oils, the amount of phenolic compounds is relatively higher in crude form and lower following refinement. Andrikopoulos et al. (2002) found that the polyphenols in VOO showed different resistances during frying. Tannic acid (Tan) and hydroxy-tyrosol-elenolic acid dialdehydic acid (HTyr-EDA) exhibited remarkable resistance to oxidation. Tan showed a decline to 90% during the first four sessions, then 50% in the fifth session, and finally 20% at the eighth session. HTyr-EDA was highly stable even after the eighth session, with 80% retention. The other phenolic compounds, such as tyrosol (Tyr), coumaric acid (Coum), ferulic acid (Fer), homo-vanillic acid (Hom), oleuropein (Oleu), and cinnamic acid (Cinn), were retained up to 50% after four to five frying sessions. It was also reported that Htyr-EDA remains almost unaffected during deep frying as compared with pan frying, as shown in Figure 6.17.

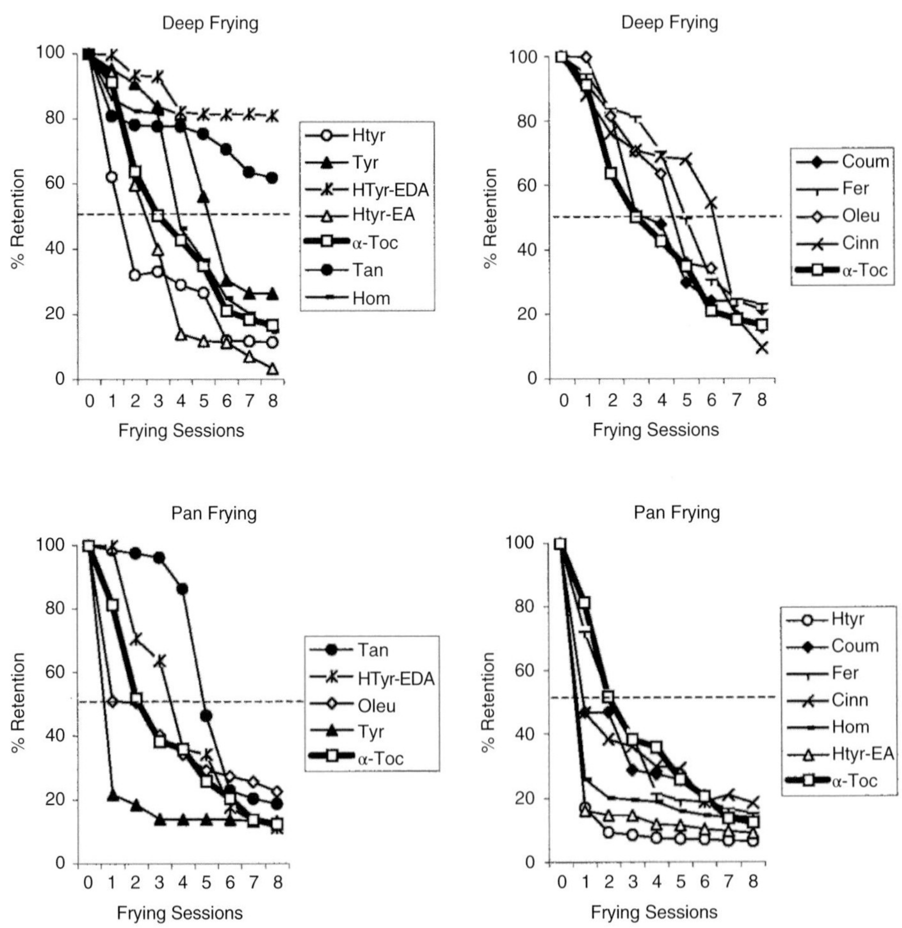

Figure 6.17 Retention of phenolic compounds in virgin olive oil (VOO) (plus a-Toc) during successive deep frying and pan frying of potatoes. The dotted line corresponds to 50% retention. *Source:* Reproduced with kind permission of Taylor & Francis (Andrikopoulos et al. 2002).

6.6.4 Fatty Acids and Triacylglycerols

TAGs constitute the major component of the frying medium and are thus highly affected by the high temperature of frying. During frying, saturated fatty acids (SFAs) are more stable than unsaturated FAs, resulting in a lower acceptability of polyunsaturated oils after prolonged frying (Sayyari and Farahmandfar 2017). Plant foods containing fewer polyunsaturated fatty acids (PUFAs) may take up high amounts during frying. Animal foods take up oil contents in different ratios depending upon the type of food fried. It is therefore warranted to check the stability of a frying oil, as it can affect the formation of polar compounds with toxic properties.

6.6.5 Other Nutrients

Several other nutrients, such as moisture, ash, proteins, carbohydrates, and total energy, are also affected by the frying of foods. Roncero-Ramos et al. (2017) showed the effects of different cooking methods on the proximate composition of a variety of mushrooms. Moisture, ash, proteins, and carbohydrates were reduced significantly in all samples during frying (Table 6.8). However, there was a significant increase in fat and energy contents, suggesting an increase in the nutritional value during frying. Tiwari et al. (2011) showed that these nutrients can be used to determine the quality characteristics and shelf lives of foods.

Table 6.8 Effects of deep frying on the proximate composition of different cultivated mushrooms.

Sample		Moisture	Ash	Protein*	Fat	Carbohydrates**	Energy**
Agaricus bisporus	Raw	89.7 ± 0.2^{a}	9.40 ± 0.06^{a}	24.64 ± 0.1^{a}	2.34 ± 0.03^{a}	63.6 ± 0.24^{a}	374.2 ± 0.5^{a}
	Fried	56.2 ± 1.0^{e}	5.01 ± 0.05^{e}	15.27 ± 0.1^{d}	45.08 ± 0.1^{d}	34.8 ± 0.10^{c}	605.3 ± 0.6^{d}
Lentinula edodes	Raw	87.83 ± 1.0^{a}	7.36 ± 0.01^{a}	16.82 ± 0.3^{a}	$2.06 \pm 0.01^{a,b}$	73.4 ± 0.21^{a}	380.9 ± 0.1^{a}
	Fried	39.4 ± 0.2^{d}	2.66 ± 0.02^{d}	5.90 ± 0.08^{b}	62.27 ± 0.7^{c}	29.2 ± 0.88^{c}	700.6 ± 3.7^{d}
Pleurotus ostreatus	Raw	89.4 ± 0.2^{a}	6.73 ± 0.05^{a}	12.55 ± 0.2^{a}	2.46 ± 0.01^{a}	78.3 ± 0.36^{a}	382.5 ± 0.1^{a}
	Fried	54.0 ± 2.2^{d}	3.14 ± 0.01^{d}	5.96 ± 0.1^{b}	50.38 ± 0.4^{c}	40.4 ± 0.33^{d}	639.4 ± 2.4^{c}
Pleurotus eryngii	Raw	88.1 ± 0.1^{a}	5.39 ± 0.04^{a}	12.30 ± 0.04^{a}	$1.60 \pm 0.02^{a,b}$	80.7 ± 0.02^{a}	386.5 ± 0.2^{a}
	Fried	45.9 ± 1.4^{d}	2.93 ± 0.04^{e}	7.38 ± 0.2^{d}	51.03 ± 0.02^{d}	38.5 ± 0.16^{d}	643.5 ± 0.4^{e}

Source: Part of the data is reproduced with kind permission of Taylor & Francis (Roncero-Ramos et al. 2017). Values are mean ± SD, $n = 3$. Different letters within a column indicate significant differences between raw and cooked samples of each mushroom ($p < 0.05$).

* $N \times 4.38$.

** Calculated by difference.

6.7 Key Concepts

- Interactions during frying are classified as either food–food, food–oil, or oil–oil.
- The interactions that occur within the different classes of biomolecules affect the chemistry, biochemistry, and safety of frying.
- Frying medium is affected by temperature, S/V ratio of food and oil, TAG or FA composition, and moisture content.
- Antioxidants (e.g. tocopherols, tocotrienols, carotenoids, and polyphenols) contribute greatly to the stability and interactions during frying.
- Physical parameters affecting food frying include the temperature, air, frying apparatus, S/V ratio, and type of food.
- Chemical parameters affecting food frying include the biomolecules (e.g. lipids, proteins, carbohydrates, vitamins, minerals, phenolic compounds, and pigments).
- Interactions among the chemical parameters define the actual food being fried.
- Chemical interactions may or may not be different in vegetable versus animal foods.
- Heat transfer is the principal process that causes all physical and chemical changes during the frying of foods.
- Mass transfer includes the transfer of biomolecules such as pigments and polyphenols from the frying medium to the food, and vice versa.
- Retention of nutritional parameters such as vitamins, carotenoids, phenolic compounds, FAs or TAGs, and proximate composition is key to the acceptability of foods by consumers.
- It is important to consider all possible interactions in the frying medium and food during frying so that a more acceptable, tasty, and healthy product can be prepared.

References

Ahmed, M. and Eun, J.-B. (2017). Flavonoids in fruits and vegetables after thermal and nonthermal processing: a review. *Critical Reviews in Food Science and Nutrition* 16: 1–30.

Aladedunye, F., Przybylski, R., and Matthaus, B. (2017). Performance of antioxidative compounds under frying conditions: a review. *Critical Reviews in Food Science and Nutrition* 57: 1539–1561.

Andreu-Sevilla, A.J., Hartmann, A., Burló, F. et al. (2009). Health benefits of using red palm oil in deep-frying potatoes: low acrolein emissions and high intake of carotenoids. *Food Science and Technology International* 15: 15–22.

Andrikopoulos, N.K., Dedoussis, G.V., Falirea, A. et al. (2002). Deterioration of natural antioxidant species of vegetable edible oils during the domestic deep-frying and pan-frying of potatoes. *International Journal of Food Sciences and Nutrition* 53: 351–363.

Aniołowska, M. and Kita, A. (2016). The effect of frying on glycidyl esters content in palm oil. *Food Chemistry* 203: 95–103.

Azizah, A., Wee, K., Azizah, O., and Azizah, M. (2009). Effect of boiling and stir frying on total phenolics, carotenoids and radical scavenging activity of pumpkin (*Cucurbita moschato*). *International Food Research Journal* 16: 45–51.

Ben Hammouda, I., Zribi, A., Ben Mansour, A. et al. (2017). Effect of deep-frying on 3-MCPD esters and glycidyl esters contents and quality control of refined olive pomace oil blended with refined palm oil. *European Food Research and Technology* 243: 1219–1227.

Bhardwaj, S., Passi, S.J., Misra, A. et al. (2016). Effect of heating/reheating of fats/oils, as used by Asian Indians, on trans fatty acid formation. *Food Chemistry* 212: 663–670.

Carlson, B.L. and Tabacchi, M.H. (1986). Frying oil deterioration and vitamin loss during foodservice operation. *Journal of Food Science* 51: 218–221.

Chiou, A., Salta, F.N., Kalogeropoulos, N. et al. (2007). Retention and distribution of polyphenols after pan-frying of French fries in oils enriched with olive leaf extract. *Journal of Food Science* 72: S574–S584.

Dana, D., Blumenthal, M.M., and Saguy, I.S. (2003). The protective role of water injection on oil quality in deep fat frying conditions. *European Food Research and Technology* 217: 104–109.

Da Silva, P.F. and Moreira, R.G. (2008). Vacuum frying of high-quality fruit and vegetable-based snacks. *LWT – Food Science and Technology* 41: 1758–1767.

Da Silva, T.O. and Pereira, P.A.D.P. (2008). Influence of time, surface-to-volume ratio, and heating process (continuous or intermittent) on the emission rates of selected carbonyl compounds during thermal oxidation of palm and soybean oils. *Journal of Agricultural and Food Chemistry* 56: 3129–3135.

Dobarganes, C., Márquez-Ruiz, G., and Velasco, J. (2000). Interactions between fat and food during deep-frying. *European Journal of Lipid Science and Technology* 102: 521–528.

Duda-Chodak, A., Wajda, L., Tarko, T. et al. (2016). A review of the interactions between acrylamide, microorganisms and food components. *Food & Function* 7: 1282–1295.

Dueik, V. and Bouchon, P. (2011). Vacuum frying as a route to produce novel snacks with desired quality attributes according to new health trends. *Journal of Food Science* 76: E188–E195.

Dueik, V., Robert, P., and Bouchon, P. (2010). Vacuum frying reduces oil uptake and improves the quality parameters of carrot crisps. *Food Chemistry* 119: 1143–1149.

Eriksen, J.N., Luu, A.Y., Dragsted, L.O., and Arrigoni, E. (2016). In vitro liberation of carotenoids from spinach and Asia salads after different domestic kitchen procedures. *Food Chemistry* 203: 23–27.

Eriksen, J.N., Luu, A.Y., Dragsted, L.O., and Arrigoni, E. (2017). Adaption of an in vitro digestion method to screen carotenoid liberation and in vitro accessibility from differently processed spinach preparations. *Food Chemistry* 224: 407–413.

Erim Kose, Y. and Dogan, I.S. (2017). Determination of simultaneous heat and mass transfer parameters of tulumba dessert during deep-fat frying. *Journal of Food Processing and Preservation* 41: e13082.

Ferracane, R., Pellegrini, N., Visconti, A. et al. (2008). Effects of different cooking methods on antioxidant profile, antioxidant capacity, and physical characteristics of Artichoke. *Journal of Agricultural and Food Chemistry* 56: 8601–8608.

Fillion, L. and Henry, C.J.K. (1998). Nutrient losses and gains during frying: a review. *International Journal of Food Sciences and Nutrition* 49: 157–168.

Franklin, M.E.E., Pushpadass, H.A., Ravindra Menon, R. et al. (2014). Modeling the heat and mass transfer during frying of gulab jamun. *Journal of Food Processing and Preservation* 38: 1939–1947.

Gere, A. (1983). Study of some factors affecting frying fat deterioration. *Fette, Seifen, Anstrichmittel* 85: 18–23.

Gibis, M. (2007). Effect of oil marinades with garlic, onion, and lemon juice on the formation of heterocyclic aromatic amines in fried beef patties. *Journal of Agricultural and Food Chemistry* 55: 10240–10247.

Gökmen, V. and Palazoğlu, T.K. (2009). Measurement of evaporated acrylamide during frying of potatoes: effect of frying conditions and surface area-to-volume ratio. *Journal of Food Engineering* 93: 172–176.

Gordon, M.H. and Kourkimskå, L. (1995). The effects of antioxidants on changes in oils during heating and deep frying. *Journal of the Science of Food and Agriculture* 68: 347–353.

Hudson, B.J. (2012). *Food Antioxidants*. Berlin: Springer Science+Business Media.

Jakszyn, P., Agudo, A., Ibáñez, R. et al. (2004). Development of a food database of nitrosamines, heterocyclic amines, and polycyclic aromatic hydrocarbons. *The Journal of Nutrition* 134: 2011–2014.

Juániz, I., Ludwig, I.A., Huarte, E. et al. (2016). Influence of heat treatment on antioxidant capacity and (poly) phenolic compounds of selected vegetables. *Food Chemistry* 197: 466–473.

Jung, L. and Choe, E. (2017). Lipid oxidation-related characteristics of gim bugak (Korean fried cuisine with Porphyra) affected by frying oil. *Food Science and Biotechnology* 26: 623–631.

Kao, F.J., Chiu, Y.S., Tsou, M.J., and Chiang, W.D. (2012). Effects of Chinese domestic cooking methods on the carotenoid composition of vegetables in Taiwan. *LWT – Food Science and Technology* 46: 485–492.

Kavousi, P., Mirhosseini, H., Ghazali, H., and Ariffin, A.A. (2015). Formation and reduction of 5-hydroxymethylfurfural at frying temperature in model system as a function of amino acid and sugar composition. *Food Chemistry* 182: 164–170.

Kidmose, U., Yang, R.-Y., Thilsted, S. et al. (2006). Content of carotenoids in commonly consumed Asian vegetables and stability and extractability during frying. *Journal of Food Composition and Analysis* 19: 562–571.

Krokida, M.K., Oreopoulou, V., and Maroulis, Z.B. (2000). Water loss and oil uptake as a function of frying time. *Journal of Food Engineering* 44: 39–46.

Labuza, T.P. and Tannenbaum, S.R. (1972). Nutrient losses during drying and storage of dehydrated foods. *C R C Critical Reviews in Food Technology* 3: 217–240.

Lampi, A.-M. and Kamal-Eldin, A. (1998). Effect of α- and γ-tocopherols on thermal polymerization of purified high-oleic sunflower triacylglycerols. *Journal of the American Oil Chemists' Society* 75: 1699–1703.

Lee, J., Kim, M., Park, K., and Choe, E. (2003). Lipid oxidation and carotenoids content in frying oil and fried dough containing carrot powder. *Journal of Food Science* 68: 1248–1253.

Ling, B., Tang, J., Kong, F. et al. (2015). Kinetics of food quality changes during thermal processing: a review. *Food and Bioprocess Technology* 8: 343–358.

Márquez-Ruiz, G., Ruiz-Méndez, M.V., and Velasco, J. (2014). Antioxidants in frying: analysis and evaluation of efficacy. *European Journal of Lipid Science and Technology* 116: 1441–1450.

Mauricio, V.-R.J. and Francisco, S.-V.W. (2017). Modeling heat transfer during blanching of cubic particles of Loche (*Cucurbita moschata* Duch.) and potato (*Solanum tuberosum* L.) using finite difference method. *Journal of Food Process Engineering* 40: e12451.

Miglio, C., Chiavaro, E., Visconti, A. et al. (2008). Effects of different cooking methods on nutritional and physicochemical characteristics of selected vegetables. *Journal of Agricultural and Food Chemistry* 56: 139–147.

Mondal, I.H. and Dash, K.K. (2017). Textural, color kinetics, and heat and mass transfer modeling during deep fat frying of Chhena Jhili. *Journal of Food Processing and Preservation* 41: e12828.

Murador, D.C., da Cunha, D.T., and de Rosso, V.V. (2014). Effects of cooking techniques on vegetable pigments: a meta-analytic approach to carotenoid and anthocyanin levels. *Food Research International* 65: 177–183.

Murkovic, M. (2004). Chemistry, formation and occurrence of genotoxic heterocyclic aromatic amines in fried products. *European Journal of Lipid Science and Technology* 106: 777–785.

Murkovic, M., Friedrich, M., and Pfannhauser, W. (1997). Heterocyclic aromatic amines in fried poultry meat. *Zeitschrift für Lebensmitteluntersuchung und -Forschung A* 205: 347–350.

Nayak, P.K., Dash, U., Rayaguru, K., and Krishnan, K.R. (2016). Physio-chemical changes during repeated frying of cooked oil: a review. *Journal of Food Biochemistry* 40: 371–390.

Nogala-Kalucka, M., Korczak, J., Elmadfa, I., and Wagner, K.-H. (2005). Effect of α- and δ-tocopherol on the oxidative stability of a mixed hydrogenated fat under frying conditions. *European Food Research and Technology* 221: 291–297.

Normand, L., Eskin, N., and Przybylski, R. (2001). Effect of tocopherols on the frying stability of regular and modified canola oils. *Journal of the American Oil Chemists' Society* 78: 369–373.

Palermo, M., Pellegrini, N., and Fogliano, V. (2014). The effect of cooking on the phytochemical content of vegetables. *Journal of the Science of Food and Agriculture* 94: 1057–1070.

Peri, I. and Saguy, I.S. (2015). Continuous injection of water and antioxidants possible roles on oil quality during frying. *LWT – Food Science and Technology* 64: 919–925.

Perkins, E.G. and Erickson, M.D. (1996). *Deep Frying: Chemistry, Nutrition, and Practical Applications*. Champaign, IL: AOCS Press.

Rahimi, J., Adewale, P., Ngadi, M. et al. (2017). Changes in the textural and thermal properties of batter coated fried potato strips during post frying holding. *Food and Bioproducts Processing* 102: 136–143.

Rojas-Gonzalez, J.A., Avallone, S., Brat, P. et al. (2006). Effect of deep-fat frying on ascorbic acid, carotenoids and potassium contents of plantain cylinders. *International Journal of Food Sciences and Nutrition* 57: 123–136.

Roncero-Ramos, I., Mendiola-Lanao, M., Pérez-Clavijo, M., and Delgado-Andrade, C. (2017). Effect of different cooking methods on nutritional value and antioxidant activity of cultivated mushrooms. *International Journal of Food Sciences and Nutrition* 68: 287–297.

Rossi, M., Alamprese, C., and Ratti, S. (2007). Tocopherols and tocotrienols as free radical-scavengers in refined vegetable oils and their stability during deep-fat frying. *Food Chemistry* 102: 812–817.

Rossi Marquez, G., Di Pierro, P., Esposito, M. et al. (2014). Application of transglutaminase-crosslinked whey protein/pectin films as water barrier coatings in fried and baked foods. *Food and Bioprocess Technology* 7: 447–455.

Saeleaw, M. and Schleining, G. (2011). Effect of frying parameters on crispiness and sound emission of cassava crackers. *Journal of Food Engineering* 103: 229–236.

Saguy, I.S. and Dana, D. (2003). Integrated approach to deep fat frying: engineering, nutrition, health and consumer aspects. *Journal of Food Engineering* 56: 143–152.

Sayyari, Z. and Farahmandfar, R. (2017). Stabilization of sunflower oil with pussy willow (*Salix aegyptiaca*) extract and essential oil. *Food Science & Nutrition* 5: 266–272.

Skrede, G., Nilsson, A., Baardseth, P. et al. (1997). Evaluation of carrot varieties for production of deep fried carrot chips – III. Carotenoids. *Food Research International* 30: 73–81.

Somsub, W., Kongkachuichai, R., Sungpuag, P., and Charoensiri, R. (2008). Effects of three conventional cooking methods on vitamin C, tannin, myo-inositol phosphates contents in selected Thai vegetables. *Journal of Food Composition and Analysis* 21: 187–197.

Sowa, M., Yu, J., Palacios-Rojas, N. et al. (2017). Retention of carotenoids in biofortified maize flour and β-cryptoxanthin-enhanced eggs after household cooking. *ACS Omega* 2: 7320–7328.

Stier, R.F. (2000). Chemistry of frying and optimization of deep-fat fried food flavour – an introductory review. *European Journal of Lipid Science and Technology* 102: 507–514.

Su, Y., Zhang, M., and Zhang, W. (2016). Effect of low temperature on the microwave-assisted vacuum frying of potato chips. *Drying Technology* 34: 227–234.

Sulaeman, A., Keeler, L., Giraud, D.W. et al. (2001). Carotenoid content and physicochemical and sensory characteristics of carrot chips deep-fried in different oils at several temperatures. *Journal of Food Science* 66: 1257–1264.

Tiwari, U., Gunasekaran, M., Jaganmohan, R. et al. (2011). Quality characteristic and shelf life studies of deep-fried snack prepared from rice brokens and legumes by-product. *Food and Bioprocess Technology* 4: 1172–1178.

Velisek, J. (2014). *The Chemistry of Food*. Chichester: John Wiley & Sons.

Vitrac, O., Trystram, G., and Raoult-Wack, A.L. (2000). Deep-fat frying of food: heat and mass transfer, transformations and reactions inside the frying material. *European Journal of Lipid Science and Technology* 102: 529–538.

Warner, K. and Moser, J. (2009). Frying stability of purified mid-oleic sunflower oil triacylglycerols with added pure tocopherols and tocopherol mixtures. *Journal of the American Oil Chemists' Society* 86: 1199.

Yagua, C.V. and Moreira, R.G. (2011). Physical and thermal properties of potato chips during vacuum frying. *Journal of Food Engineering* 104: 272–283.

Zahoor, A.F., Yousaf, M., Awais, M. et al. (2015). Synthesis, extraction, and identification of light polycyclic aromatic hydrocarbons (PAH) in microwave oven fried Aseel chicken tissues. *Toxicological & Environmental Chemistry* 97: 145–154.

Zeb, A. (2012). Thermal degradation of β-carotene in food oils. In: *Vitamin A and Carotenoids: Chemistry* (ed. V.R. Preedy). London: The Royal Society of Chemistry.

Zeb, A. (2015). Chemistry and liquid chromatography methods for the analyses of primary oxidation products of triacylglycerols. *Free Radical Research* 49: 549–564.

Zeb, A. (2017). A simple, sensitive HPLC-DAD method for simultaneous determination of carotenoids, chlorophylls and α-tocopherol in leafy vegetables. *Journal of Food Measurement and Characterization* 11: 1–8.

Zeb, A. and Habib, A. (2018). Lipid oxidation and changes in the phenolic profile of Watercress (*Nasturtium officinale* L.) leaves during frying. *Journal of Food Measurement and Characterization* https://doi.org/10.1007/s11694-018-9885-2.

Zeb, A. and Murkovic, M. (2010). Characterization of the effects of β-carotene on the thermal oxidation of triacylglycerols using HPLC-ESI-MS. *European Journal of Lipid Science and Technology* 112: 1218–1228.

Zeb, A. and Murkovic, M. (2011). Carotenoids and triacylglycerols interactions during thermal oxidation of refined olive oil. *Food Chemistry* 127: 1584–1593.

Zeb, A. and Nisar, P. (2017). Effects of high temperature frying of spinach leaves in sunflower oil on carotenoids, chlorophylls, and tocopherol composition. *Frontiers in Chemistry* 5: 19.

Zhang, Q., Saleh, A.S., Chen, J., and Shen, Q. (2012). Chemical alterations taken place during deep-fat frying based on certain reaction products: a review. *Chemistry and Physics of Lipids* 165: 662–681.

Zhang, Y., Yu, C., Mei, J., and Wang, S. (2013). Formation and mitigation of heterocyclic aromatic amines in fried pork. *Food Additives & Contaminants, Part A: Chemistry, Analysis, Control, Exposure & Risk Assessment* 30: 1501–1507.

Zhou, H., Jin, Q., Wang, X., and Xu, X. (2014). Effects of temperature and water content on the formation of 3-chloropropane-1,2-diol fatty acid esters in palm oil under conditions simulating deep fat frying. *European Food Research and Technology* 238: 495–501.

7

Analysis of Frying

7.1 Introduction

Frying produces complex changes in the chemistry of both the food and the frying medium. Thermal oxidation, polymerization, hydrolysis, degradation, and Maillard reactions all produce several classes of new compounds. These compounds are analysed via a number of analytical methods, either qualitatively or quantitatively. The aim of analysis is to obtain composition data concerning the new compounds formed, in order to elucidate any nutritional changes they may have caused. This chapter, therefore, looks only at analytical methods related to composition studies, mostly chromatographic studies; spectroscopic studies are beyond its scope. The limitations of each method may warrant the development of new or improved methodology. The following limitations are considered while aiming to develop a new analytical method:

1) The lower limit of detection or quantification.
2) The high cost of the chemical or equipment.
3) The reliability or reproducibility of the method.
4) Low accuracy or validity.
5) Low reproducibility.
6) The complex nature of the frying medium and of the food.
7) Different target analytes.
8) The nature of the separation mode.
9) The nature of the analytes.

Several review papers have aimed to collate the knowledge relevant to each topic or component of food frying. For example, Paul and Mittal (1997) reviewed the various regulations on the use of degraded oil or fats in the deep frying of foods; Hindra and Baik (2006) reviewed the kinetics of changes in the quality of the frying medium during food frying; Melton et al. (1994) reviewed the measurement of the stability of frying oils and of flavour; and Bansal et al. (2010) reviewed the rapid tests available to measure quality changes in frying oils and compared them with standard methods. Meanwhile, Zhang et al. (2015) reviewed the applications of different chromatographic techniques to the detection and identification of new constituents formed during food frying. Similarly, Zeb (2015a) reviewed the uses of different classes of chromatographic techniques, for the analysis of primary oxidation products of triacylglycerols (TAGs).

Food Frying: Chemistry, Biochemistry, and Safety, First Edition. Alam Zeb.

7.2 Analysis of Triacylglycerols

TAGs constitute the major part of the frying medium, and to a lesser extent of foods. Therefore, the majority of studies conducted to date have been based on the analysis of TAGs. Ruiz-Gutierrez and Barron (1995) reviewed the different methods of analysing TAGs in a variety of matrices. They also summarized the most interesting biomedical applications of TAGs analysis. However, at that time, other chromatographic (especially liquid chromatography) and spectrometric techniques were not yet widely used.

Table 7.1 shows the different methods of TAG analysis developed since 2010. These methods were conducted in a number of different frying media, including virgin olive, sunflower, soybean, corn, fish, hazelnut, peanut, and Menhaden oil. Reversed-phase liquid chromatography (RP-LC) was found to be widely used in stationary phase, including C-8 and C-18. The majority of separations were achieved by gradient elution, mostly with acetonitrile, methanol, and 2-propanol as the major solvents. Different detectors were used, including an evaporative light scattering detector (Adewuyi and Oderinde 2012; Solaesa et al. 2014; Beccaria et al. 2015; Abdallah et al. 2016; Tamba Sompila et al. 2017), ultraviolet (UV) detector (Gotoh et al. 2011; Lerma-García et al. 2011; Abdallah et al. 2016), differential refractometric detector (Kıralan et al. 2015), and refractive index detector (Endo et al. 2011). These detectors have a low sensitivity, but are cheap and easy to use.

Mass spectrometric detectors (MSDs) are widely used. They utilize different ionizations, such as electrospray ionization (ESI) (Zeb and Murkovic 2010a; Gao et al. 2017; La Nasa et al. 2018; Pagliuca et al. 2018) and atmospheric pressure chemical ionization (APCI) (Fasciotti and Pereira Netto 2010; Leskinen et al. 2010; Gotoh et al. 2011; Dugo et al. 2012; Bhuiyan et al. 2013; Lísa and Holčapek 2013; Řezanka et al. 2013; Zommick et al. 2013; Baiocchi et al. 2015; Beccaria et al. 2015; Ben Arfa et al. 2017; Syed Idrus et al. 2017). Most studies are based on APCI-MS detection, with very few using ESI-MS. In most cases, APCI-MS do not give protonated molecular ions of TAG, so only adducts of the TAGs are widely used. In ESI-MS, the molecular ion and its ammonium adducts – along with its diacylglycerol (DAG) or fatty acid (FA) moiety – are obtained as shown in Figure 7.1.

In liquid chromatography coupled to mass spectrometry (LC-MS), the oils or extracted lipids from foods are usually not processed or derivatized. Sample preparation mostly involves the dissolution of a few milligrams of oil in high-performance liquid chromatography (HPLC) solvent, mostly acetonitrile (ACN) or 2-propanol. The samples are then injected into the system to obtain results. The separation of TAGs, however, depends on several factors, including stationary and mobile phase. Due to its high reliability and sensitivity, the use of HPLC-MS is considered an absolute analytical tool, but the high cost of the MS device and the need for a skilled operator provide disadvantages to its use.

7.2.1 Analysis of Oxidized TAGS

During frying, TAGs are oxidized to form hydroperoxides, epoxide, and other oxidized species. These species are responsible for several properties of oils and fried foods. Therefore, analysis of oxidized TAGs is considered a prime target in food frying.

Table 7.1 Analytical methods of triacylglycerol (TAG) determination in different frying media and foods.

Sample	Mode of separation	Detection system	Solvent system	Column	References
Several oils and fats	RP-UHPLC	PDA-ESI-TOF-MS	Isocratic: 2-propanol : ACN : water (80 : 15 : 5)	Acquity BEHC-18 (50 × 2.1 mm, 1.7 μm)	Pagliuca et al. (2018)
Extra virgin oil, sansa olive oil	RP	ESI-Q-ToF	Gradient: methanol/water (85 : 15) and 2-propanol	Poroshell 120 EC-C-18 (3.0 × 50 mm, 2.7 μm)	La Nasa et al. (2018)
Sunflower, corn, and soybean oils	RP	ESI-Q-ToF	Gradient: methanol and CO_2	Two C-18 columns	Gao et al. (2017)
Sesame Seed	RP-UHPLC	APCI-MS	Isocratic: acetonitrile/butanol (74/26)	Acquity UPLC HSS C-18 (100 × 2.1 mm, 1.8 μm)	Ben Arfa et al. (2017)
Corn oil, lard, olive oil, palm oil, beef fat	RP	APCI-MS	Gradient: ACN and 2-propanol	Accucore C-18 (2.1 × 100 mm)	Idrus et al. (2017)
Oils and fats	RP-UHPLC	ELSD	Gradient: acetonitrile/2-propanol	Six C-18 columns	Tamba Sompila et al. (2017)
Olive oils	RP	ELSD, UV	Gradient: ACN and *n*-pentanol	Kinetex C-18 100 (150 × 4.6 mm, 2.6 μm)	Abdallah et al. (2016)
Marine foods	RP	APCI-MS	Gradient: ACN and DCM	Luna C-18 (2) (Phenomenex) columns (150, 2.1 mm, 3 μm)	Baiocchi et al. (2015)
Fish oil	Ag-LC and RP-LC	ELSD, APCI-MS	Gradient: modified hexane for Ag-LC and ACN, IPA	Nucleosil SA, 100 (150 × 1.0 mm, 5 μm)	Beccaria et al. (2015)
Hazelnut oil	RP	DFRD	Isocratic: ACN/2-propanol/hexane (500 : 118 : 100)	Nucleosil 100 C-18 (25 × 0.46 cm, 5-μm)	Kıralan et al. (2015)
Sardine oil	RP	ELSD	Gradient: acetone : ACN	Lichrospher 100 RP 18, (25 x 4 mm, 5 μm)	Solaesa et al. (2014)
Brewer's yeast	RP	APCI-MS	Gradient: ACN and 2-propanol	HIRPB-250AM (C-8/C-18 multi-alkyl phase)	Řezanka et al. (2013)

(*Continued*)

Table 7.1 (Continued)

Sample	Mode of separation	Detection system	Solvent system	Column	References
Hazelnut oil	NP	APCI-MS	Gradient: hexane and hexane-2-propanol (99 : 1)	Lux cellulose-1 (250 × 4.6 mm, 3 μm)	Lísa and Holčapek (2013)
Peanut oil	RP	APCI-MS	Gradient: ACN and 2-propanol	Zorbax eclipse plus C-18 (150 × 4.6 mm, 5 μm)	Hu et al. (2015)
Yeast	RP	APCI-MS	Gradient: DCM/ACN	Alltima HP C-18 hi-load columns (150 × 2.1 mm, 3 μm)	Bhuiyan et al. (2013)
*Luffa cylindrica*l and *Adenopus breviflorus* oils	RP	ELSD	Isocratic: acetone/ 2-propanol (95 : 5)	SGERP-C-18 (250 × 4.6, 5 μm)	Adewuyi and Oderinde (2012)
Menhaden oil	RP	APCI-IT-TOF-MS	Gradient: ACN and 2-propanol	Ascentis express fused-core C-18 (150 × 4.6 mm, 2.7 μm)	Dugo et al. (2012)
Soybean, rapeseed, and palm oils	RP	RID	Gradient: acetone : ACN	C-22 and C-30 columns	Endo et al. (2011)
Fishes and marine mammals	RP	APCI-MS, UV	Gradient: ACN/2-propanol (3 : 7)	Intersil ODS, 250 × 10.0 mm, 10 μm)	Gotoh et al. (2011)
36 Vegetable oils	RP	UV	Isocratic: ACN/n-pentanol (70 : 30)	Kinetex C-18 100 (150 × 4.6 mm, 2.6 μm)	Lerma-García et al. (2011)
Rapeseed, sunflower, and palm oils	UHPLC	APCI-MS/MS	Acetone: ACN	Acquity UPLC BEH C-18 (100 × 2.1 mm, 1.7 mm)	Leskinen et al. (2010)
Corn, soybean, and sunflower oils	RP	ESI-MS	Isocratic: methanol : 2-Propanol (82 : 18)	Phenomenex C-18 (150 x 3 mm)	Zeb and Murkovic (2010a)
Extra virgin olive oil, soybean oil, and blends	RP	APCI-MS	Gradient: ACN and 2-propanol	LiChrospher (ODS) (250 × 3 mm, 5 μm)	Fasciotti and Pereira Netto 2010

RP, reversed phase; NP, normal phase; HPLC, high-performance liquid chromatography; UPLC, ultra-pressure liquid chromatography; UHPLC, ultra high-pressure liquid chromatography; PDA, photo diode array; ESI, electrospray ionization; APCI, atmospheric pressure chemical ionization; MS, mass spectrometry; ELSD, evaporative light-scattering detector, Q-TOF-MS, quadrupole time-of-flight mass spectrometry; UV, ultraviolet; DFRD, differential refractometric detector; RID, refractive index detector; ACN, acetonitrile; DCM, dichloromethane.

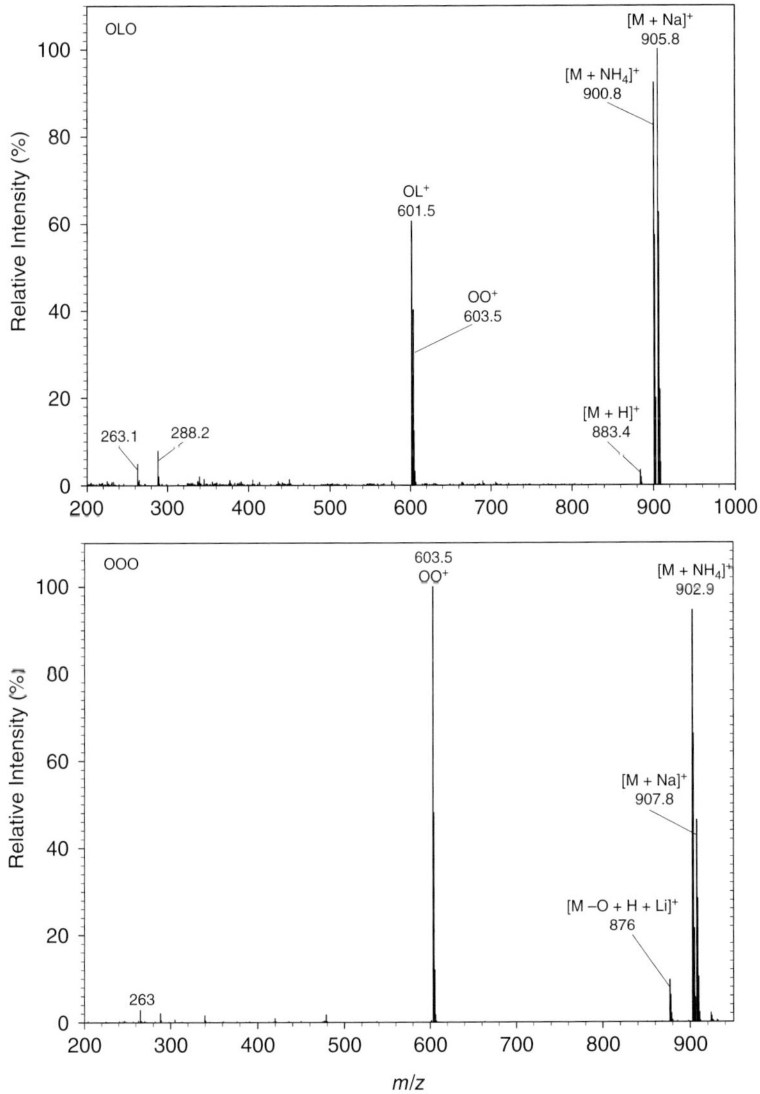

Figure 7.1 Representative electrospray ionization mass spectrometry (ESI-MS) spectra of triacylglycerols (TAGs) (OLO and OOO) showing protonated molecular ions, sodium and ammonium adducts, diacylglycerol (DAG) ions, and fatty acid (FA) ions.

This section summarizes the most widely used liquid chromatographic methods for the analysis of oxidized TAGs, as reproduced from Zeb (2015a) with the addition of recent literature.

7.2.1.1 Thin-Layer Chromatography

Thin-layer chromatography (TLC) is a widely used, fast, and relatively inexpensive chromatographic method for the separation of lipids, especially TAG or its oxidized products. TLC was used as starting method for the fractionation and isolation of TAGs and other lipids of biological significance (Zeb and Murkovic 2010c). Oette (1965) analysed several lipid peroxides using silica gel and various ratios of ethyl ether and petroleum ether as the solvent system. Perkins and Anfinsen (1971) reported a scheme for the extraction of nonvolatile oxidized compounds from the thermal oxidation of triolein, as described in Chapter 4. Different fractions of the oxidized compounds were separated using TLC. However, this method is very difficult to handle and requires extensive laboratory work.

Normal-phase TLC can also be used to purify the oxidation products of the synthetic TAGs (Sjövall et al. 1997, 2001b) from the oils (Sjövall et al. 2001a), as well as from animal tissues (Skipski et al. 1968). For this purpose, a mobile phase of heptane/di-isopropyl ether/acetic acid (60 : 40 : 4, v/v) solution was used. The band representing normal TAGs was eluted earlier, while the later band which contains the oxidized TAG molecules was scraped off from the plates. The selected lipids were recovered by extraction with chloroform/methanol (2 : 1, v/v) and washed with pure distilled water. The TAGs and their oxidation products were then detected in UV light after spraying with 2,7-dichlorofluorescein. The same method was applied to the analysis of oxidized TAGs in serum low-density lipoproteins (LDL). TAGs and their oxidized products, such as hydroxy, hydroperoxy, keto, and epoxy derivatives, were detected in UV light (Suomela et al. 2004a, 2004b, 2005). Table 7.2 shows the normal and reversed-phase TLC methods for the analysis of oxidized TAGs (Sánchez-Muniz et al. 1993; Chao et al. 2001).

Different solvent systems were used in normal-phase TLC, such as diethyl ether-hexane (40 : 60 or 80 : 20) by Neff et al. (1990) and Ravi Kiran et al. (2015); hexane-diethyl ether-acetic acid (80 : 20 : 1 or 85 : 15 : 1 or 30 : 70 : 1) by Sánchez-Muniz et al. (1993), Correia et al. (2015), and Schröter et al. (2016); hexane-isopropyl alcohol-acetic acid (60 : 40 : 4) by Sjövall et al. (2001b, 2002) and Suomela et al. (2005); and petroleum ether-diethyl ether-acetic acid (70 : 40 : 1) by Karimi et al. (2017). In reversed-phase TLC, methylene chloride-acetonitrile (30 : 70) (Neff et al. 1990), acetonitrile-methanol–water (6 : 3 : 1) (Chao et al. 2001), and petroleum ether-diethyl ether-acetic acid (70 : 40 : 1) (Karimi et al. 2017) were used. It was observed that normal-phase TLC was more widely than reversed-phase TLC. Diethyl ether-hexane was the typical solvent for elution. All TLC methods reported were based on isocratic elution, but there is a lack of information on the gradient system that is today widely used in TLC (Stolyhwo and Privett 1973).

The disadvantages of TLC in the analysis of TAG oxidation products include the sensitivity of lipid peroxidation products to silica gel and oxygen in the air. An inert environment such as nitrogen is therefore necessary to use this method. From a biological perspective, another disadvantage is the relatively impossibility of separating a complex mixture of oxidized TAGs. Due to the advent of the novel, easy-to-use technologies of separation, it is therefore important to use other liquid chromatographic techniques such as HPLC with reliable methods of detections.

7.2.1.2 High-Performance Liquid Chromatography

HPLC is one of the most reliable methods of separation for TAGs and their oxidation products. Kinter (1995) reviewed the different analytical techniques available for lipid oxidation products, showing liquid chromatography to be a reliable method of analysis of lipid oxidation chemistry. However, that review lacks information regarding the oxidation products of TAGs. On the basis of stationary phase, this chromatographic technique is classed into two main types: normal-phase and reversed-phase.

Table 7.2 Thin-layer chromatographic (TLC) methods for the analysis of oxidized triacylglycerols (TAGs).

	Stationary phase	Solvent system	Solvent ratio (v/v)	Detection	References
NP-TLC	Silica gel	Toluene-diethyl ether	4:1	Spraying of ferrous thiocynate	Peers and Coxon (1983)
	Silica gel	Hexane-diethyl ether	3:2	Spraying of ferrous thiocynate	Peers and Coxon (1983)
	Silica gel	Diethyl ether-hexane	40:60	UV light	Neff et al. (1990)
	Silica gel F 60	Hexane-diethyl ether-acetic acid	80:20:1	Iodine vapour	Sánchez-Muniz et al. (1993)
	Silica gel H	Hexane isopropyl alcohol-acetic acid	60:40:4	2,4-dinitropheny hydrazine	Sjövall et al. (2001a, 2001b)
	Silica gel G	Heptane-isopropyl ether-acetic acid	60:40:4	UV light	Suomela et al. (2005)
	Silica gel	Hexane-diethyl ether	80:20	Iodine vapour	Ravi Kiran et al. (2015)
	Silica gel 60 F254	Hexane-diethyl ether-acetic acid	85:15:1	Scanning densitometry	Correia et al. (2015)
	Silica gel 60 F254	Hexane-diethyl ether-acetic acid	30:70:1	Primuline (Direct Yellow 59) and scanning densitometry	Schröter et al. (2016)
	Silica gel	Petroleum ether-diethyl ether-acetic acid	70:40:1	UV light	Karimi et al. (2017)
RP-TLC	RP-18	Methylene chloride-acetonitrile	30:70	UV light	Neff et al. (1990)
	RP-18	Acetonitrile-methanol–water	6:3:1	Iodine vapour	Sánchez-Muniz et al. (1993)
	RP-18	Either hexane-diethyl ether-acetic acid	30:70:1	Primuline (Direct Yellow 59) and scanning densitometry	Schröter et al. (2016)

Source: Reproduced with kind permission of Taylor & Francis (Zeb 2015a).
NP-TLC, normal-phase thin-layer chromatopgraphy; UV, ultraviolet; RP-TLC, reversed-pahse thin-layer chromatopgraphy.

7.2.1.2.1 Normal-Phase Liquid Chromatography

In NP-LC, a polar stationary phase and a relatively nonpolar solvent system are used. NP-LC has been used by a limited number of scientists for the study of oxidation compounds of TAGs. Table 7.3 lists some studies utilizing silica columns and nonpolar solvents. Peers and Coxon (1983) used an isocratic method of elution for the separation of hydroperoxide isomers produced by the auto-oxidation of TAGs.

Table 7.3 Normal-phase high-performance liquid chromatographic (NP-HPLC) methods for the analysis of oxidized triacylglycerols (TAGs).

Sample	Type of oxidized TAG	Column, size	Solvent system	Detector	References
Purified TAG mixtures	Isomers of hydroperoxides	Partisil 5, 25 × 0.94 mm	Isocratic, ethanol (0.4%) in hexane	UV	Peers and Coxon (1983)
Soybean seeds	Hydroperoxides	Ultrasphere silica column	Isocratic, 0.75% isopropyl alcohol in hexane	UV	Clark and Snyder (1991)
Pure TAGs	Isomers of hydroperoxides	Supelcosil LC-Si, 250 × 2.1 mm	Isocratic, hexane-diethyl ether (500 : 34)	DAD/FL	Ohshima et al. (1996)
Triolein, peanuts	Primary hydroperoxides	Nucleosil 100 Si, 250 × 4.0 mm	Isocratic, 2% propanol in hexane	UV	Gladovič et al. (1997)
Vegetable oils	Hydroperoxides, epoxides, oxo products	Waters S3W (15 × 2.1 mm) and waters diol (20 × 3.0 mm), and waters S5W (250 × 2.1 mm)	Gradient, hexane and hexane-MTBE	DAD	Steenhorst-Slikkerveer et al. (2000)
Sunflower oils	Hydroperoxides, hydroxides, keto	Silica 60 column	Isocratic, n-heptane-diethyl ether (82 : 18)	UV	Morales et al. (2010, 2012b)
Edible oils	*Cis/trans* epoxides, hydroperoxides, hydroxides, ketodienes	Silica 60 column	Isocratic, n-heptane-diethyl ether (82 : 18)	UV/ELS	Morales et al. (2012b)

Source: Reproduced with kind permission of Taylor & Francis (Zeb 2015a).

The auto-peroxidation products were purified by TLC and quantified through the UV detection of preparative normal-phase HPLC. Similarly, another isocratic method was reported for the hydroperoxides produced in soybean seeds after storage (Clark and Snyder 1991). The hydroperoxides were eluted using 0.75% isopropyl alcohol in n-hexane. The authors suggested that the methods will be helpful for monitoring seed germination and any changes during storage.

An HPLC method utilizing chemiluminescence (CL) detection was used for the determination of lipid hydroperoxides in LDL present in human blood plasma (Miyazawa et al. 1990). This method involved the separation of LDL-total lipids with normal-phase silica gel HPLC. The elution was carried out using isocratic methanol-chloroform-1-propanol-water (9:1:2:0.1) with the post-column reaction of hydroperoxides with cytochrome c-haem monitored by CL. The authors suggested that the method can be used in patients with atherosclerosis and hyperlipidaemia. Similarly, Noguchi et al. (1998) studied the hydroperoxides in the human blood plasma LDL. Separation was carried out using gradient elution with hexane-2-propanol (90:10) and hexane-t-butyl alcohol (70:30) with the help of silica column. Akasaka et al. (1993) developed an HPLC post-column detection method for the determination of hydroperoxides in human plasma. The separation was carried out using gradient elution with hexane and hexane-1-butanol. The mechanism of the method is the post-column reaction of hydroperoxides with diphenyl-l-pyrenylphosphine (DPPP) (Akasaka et al. 1992; Akasaka and Ohrui 2000). The phenyl phosphine reacts with hydroperoxides and forms an oxide (Figure 7.2), which exhibits fluorescence and thus can be detected using a fluorescence detector. The hydroperoxides produced by the oxidation of a pure TAG or pure FA mixture were measured accurately by HPLC with post-column detection using DPPP (Ohshima et al. 1996). This method was used for the quantitative and qualitative determination of isomeric lipid hydroperoxides using isocratic hexane-diethyl ether (500:34) as solvent.

In another study, the hydroperoxides from the triolein in peanuts were determined using silica column and isocratic elution with 2% propanol in hexane (Gladovič et al. 1997). This paper describes the uses of UV detection, fast atom bombardment mass spectrometry (FAB-MS), and nuclear magnetic resonance (NMR) spectroscopy for the structural elucidation of novel hydroperoxides. The method is applicable for samples with a peroxide value (PV) more than 10 meq kg^{-1}; it is thus of the limited application with respect to samples containing a small amount of peroxides, such as biological samples. A much more sensitive method, such as a diode array detector (DAD), can also be used. Steenhorst-Slikkerveer et al. (2000) reported an HPLC-DAD method for the analysis of hydroperoxides in vegetable oils. The elution was carried out using normal-phase column. The authors also used MS detection for the characterization of

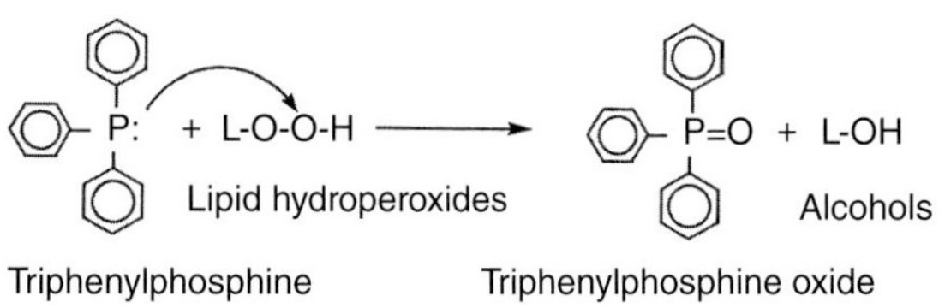

Figure 7.2 Reaction of triphenylphosphine with lipid hydroperoxide. This reaction is mostly used as a post-column derivatization.

oxidation compounds. They reported hydroperoxides, epoxy and oxo TAGs, mono- and dihydroxy TAGs, and other oxo products. They suggested that their method is very suitable for the characterization and quantitation of nonvolatile TAG oxidation products in reference samples or real oxidized samples of vegetable oils and other fat-based food products. *Cis*/*trans* epoxides, hydroperoxydienes, ketodienes, and hydroxydienes could be measured during autoxidation at 40 and 80 °C of FA methyl esters produced from edible oils with different degrees of unsaturation, such as high linoleic sunflower oil and high oleic sunflower oil (Morales et al. 2010, 2012a, 2012b).

In conclusion, the NP-LC methods reported here were mostly isocratic elution with hexane as major solvent, while the UV detector was the most used method of detection. Separation and structural elucidation, however, were not possible without the use of other spectroscopic techniques, such as NMR or MS. The major disadvantages of normal-phase LC for the determination of oxidation products of TAGs include the sensitivity of the oxidized compounds, the high retention in the column, and a possible oxidation. In order to avoid such cases, either pre-column derivatization or reversed-phase columns are preferred.

7.2.1.2.2 Reversed-Phase Liquid Chromatography

RP-LC employs nonpolar stationary phases and relatively polar solvents. Dorsey and Dill (1989) reviewed in detail the molecular mechanism underlying the separation on reversed-phase stationary phases. Later, Claessens and van Straten (2004) extensively reviewed the chemical and thermal stability of stationary phases for the preparation of reversed-phase chromatography. Rafferty et al. (2007) recently updated their review on the molecular mechanism of separation of reverse-phase chromatography. These reviews, however, lack information on the applications of RP-LC to the individual class of compounds. RP-LC has been used extensively in lipid analysis, but the corresponding review (Ruiz-Gutierrez and Barron 1995) focused on the reversed-phase methodologies used for the analysis of oxidized TAGs. A discussion of the molecular mechanism of the separation of oxidized TAGs on the reversed phase is beyond the scope of this section. Table 7.4 shows several RP-HPLC methods reported for the analysis of oxidized TAGs.

Pure TAGs were oxidized and their oxidation products were separated on isocratic methanol-water and detected using electrochemical (Yamada et al. 1987) or UV detectors (Terao et al. 1988). The reversed-phase separation and UV detections revealed hydroperoxides and hydroxides as major classes of oxidized TAGs. Similarly, pure synthetic trilinolein and trilinolenin were auto-oxidized and analysed using RP-HPLC (Miyashita et al. 1990). The separation was carried using an isocratic elution with ACN-MC-MeOH (85 : 15 : 1) with UV and flame ionization detector (FID) detection. The oxidation products were TAG hydroperoxides, hydroperoxides, and epidioxides. Mono-, bis-, and tris-hydroperoxides were reported to form in the pure trilinolein oxidation (Neff et al. 1990). Similar products were also reported from the oxidation of pure trilinolenin and other TAGs using the same method (Frankel et al. 1990, 1992). Similar studies reported the oxidation products of TAG pure standards such as trilinolenin, trilinolein, and triolein, as well as rapeseed oil TAG (Viinanen and Hopia 1994). Novapak C-18 column and isocratic elution with ACN-DCM-MeOH (80 : 10 : 10) was used with UV and evaporative light-scattering detectors (ELSDs). Neff and Byrdwell (1998), with the help of the previously mentioned elution system, reported mono-, bis-, and

Table 7.4 Reversed-phase high-performance liquid chromatographic (RP-HPLC) methods for the analysis of oxidized triacylglycerols (TAGs).

Column, size	Solvent system	Detection	Type of oxidized TAGs	Sample	References
Inertsil C-8, 150 × 4.6 mm	Isocratic, MeOH-water	UV	Hydroperoxides, hydroxyl products	Pure TAGs	Terao et al. (1988)
Zorbax C-18, 25 × 0.46 cm	Isocratic, ACN-MC-MeOH (85 : 15 : 1)	UV/FID	Hydroperoxides, hydroperoxides, epidioxides	Pure trilinolein and trilinolenin	Miyashita et al. (1990)
Dynamax C-18, 25 × 2.14 cm	Isocratic, MC-ACN (30 : 70)	RI/UV	Mono-, bis-, tris-hydroperoxides	Pure trilinolein	Neff et al. (1990)
Dynamax C18, 25 × 2.14 cm	Isocratic, MC-ACN (30 : 70)	RI/UV	Mono-, bis-, tris-hydroperoxides	Pure trilinolenin	Frankel et al. (1990, 1992)
Nova-Pak C-18, 150 × 3.9 mm	Isocratic, ACN-DCM-MeOH (80 : 10 : 10)	UV/ELSD	Total oxidized TAGs	Pure triolein, trilinolein, and trilinolenin	Viinanen and Hopia (1994)
Inertsil C-8, 150 × 4.6 mm	Isocratic, MeOH-water (98 : 2)	UV	Hydroperoxides, hydroxyl products	Human plasma	Araujo et al. (1995)
Ultrastyragel, 25 × 0.94 cm	Isocratic, THF	RI	Oxidized TAG momers	Frying oils	Márquez-Ruiz et al. (1993)
Finepak SIL C-18, 250 × 4.6 mm	Isocratic MeOH or MeOH-EtOH (5 : 1)	CL/UV	Mono-, bis-, tris-hydroperoxides	Vegetable oils	Miyazawa et al. (1995)
Nova-Pak C-18, 150 × 3.9 mm Spherisorb S5 ODS2, 250 × 4.6 mm	Isocratic, MeOH-2-propanol (90 : 10) and MeOH:2-propanol-DCM (80 : 10 : 10)	CL/UV/ELSD	Hydroperoxides	Purified TAGs	Mäkinen et al. (1996)
Supelcosil LC-18 DB,	Isocratic, T-Butanol-MeOH (1 : 1)	CL	Hydroperoxides	Butter and dairy spreads	Christensen and Holmer (1996)
Vydac ODS, 250 × 4.6 mm	Isocratic, ACN-MC-MeOH (90 : 5 : 5)	UV	Mono-, bis-, tris-hydroperoxides	Triolein, trilinolein, and trilinolenin	Neff and Byrdwell (1998)

(Continued)

Table 7.4 (Continued)

Column, size	Solvent system	Detection	Type of oxidized TAGs	Sample	References
AsahiPak C8P50, 150 × 4.6 mm	Gradient, water-ACN-MeOH (45 : 35 : 20) and ACN-MeOH (80 : 20)	UV	Hydroperoxides	Pure TAG	Sugino (1999)
Spherisorb S5W ODS2, 200 × 4.6 mm	Gradient, ACN, and MTBE	UV/ELSD	Mono-, bis-hydroperoxides, hydroxides	Vegetable oils	Bauer-Plank and Steenhorst-Slikkerveer (2000)
Prevail C-18, 250 × 4.6 mm	Isocratic, ACN-DCM (65 : 35)	DAD/ELSD	TAG oxidation	Pure TAGs	Christensen and Holmer (1996)
Phenogel 300, 300 × 7.5 mm	Isocratic, THF	ELSD/RI	Polar TAGs as oxidized TAGs	Frying oils	Caldwell et al. (2011)

Source: Reproduced with kind permission of Taylor & Francis (Zeb 2015a).

tris-hydroperoxides from the same TAG standard. However, these authors reported the structure of these oxidized compounds with the help of MS. Other studies also reported the oxidation of pure TAG mixtures producing hydroperoxides with the help of gradient elution with water-ACN-MeOH (45:35:20) and ACN-MeOH (80:20) (Sugino 1999), isocratic (ACN-DCM, 65:35) (Wijesundera et al. 2008), and two isocratic systems of MeOH-2-Propanol (90:10) and MeOH-2-propanol-DCM (80:10:10) (Mäkinen et al. 1996). The latter study suggested that CL and UV detectors are more suitable during early stages of oxidation, due to their good sensitivity and selectivity as compared with ELSD.

Araujo et al. (1995) reported an isocratic method with MeOH-water (98:2) as eluent, Inertsil C-8 column, and UV detection. Hydroperoxides and hydroxyl products were detected in the human plasma of a patient with hyperlipidaemia. The oxidative stress caused by these oxidized compounds was contributing to the hyperlipidaemia. The eating of a high-fat diet or thermally oxidized lipids is a contributing factor in this condition. Recent studies showed that frying oils or fats are major contributors to the development of fatty liver and hyperlipidaemia (Zeb and Mehmood 2012). They contain significant amounts of oxidized TAGs, which are digested through food and thus contribute to increases in reactive oxygen species (ROS) and oxidized LDL (Márquez-Ruiz et al. 1993; Dobarganes and Márquez-Ruiz 2003; Velasco et al. 2004, 2005). Frying oils have been found to contain oxidized compounds as polar compounds (Márquez-Ruiz et al. 1995). These polar compounds were separated using isocratic elution with THF on the reversed-phase column and detected using a refractive index (RI) detector. Auto- or photo-oxidation of oils produces mono-, bis-, and tris-hydroperoxides, which were detected using UV and CL detection (Miyazawa et al. 1995). These compounds were separated by isocratic elution with MeOH or MeOH-EtOH (5:1) on Finepak SIL C18 column. The CL detector was very helpful to the hydroperoxides produced in the butter and dairy spreads, as reported by Christensen and Holmer (1996). These authors used an isocratic, *tert*-Butanol-MeOH (1:1) as solvent system and CL as the detection method. The method was sensitive, fast, and reliable for the detection of primary oxidation compounds in dairy products. Bauer-Plank and Steenhorst-Slikkerveer (2000) also reported the use of RP-HPLC for the detection of TAG hydroperoxides and primary oxidation products. The gradient elution was carried out using ACN and MTBE, while oxidation compounds were detected using UV and ELSD. A linear correlation was obtained by plotting the PV and peak area of hydroperoxides, which shows that the method was in line with the standard iodometric method. However, as with other methods, it did not show the individual oxidation compounds. Recently, Caldwell et al. (2011) reported a rapid method for the determination of the total polar compounds (TPC) in several used frying oils. The authors concluded that the method was very similar to the official American Oil Chemists' Society (AOCS) method (Cd 22–91), and was capable of separating and quantifying polymerized TAG.

In reviewing all the reported RP-LC methods for the determination of oxidized TAGs, it was found that the majority used the isocratic method of elution with methanol and ACN as major solvents and a UV detector as detection methods. These reported methods were mostly focused on the standard TAGs or frying oils and fats, and limited papers showed oxidation of TAGs in other biological samples. The disadvantage of these methods, which is significant in terms of the real chemistry of oxidized TAGs, is the use of detection systems such as UV, RI, DAD, and ELSD-CL, which are not as easy to use in the study of the oxidation and formation of a single oxidized compound as novel MS detection systems.

7.2.1.3 Liquid Chromatography Mass Spectrometry

As evident from the literature discussed so far, only MS provides sufficient information about the structural elucidation of oxidized products produced by the oxidation of TAGs in various biological systems. MS is the most used method of detection for the analysis of lipids, especially TAGs (Dugo et al. 2005; Beermann et al. 2007; Zeb and Murkovic 2010a; Beccaria et al. 2014) and oxidized glycerolipids (Cheung et al. 1994; Harrison et al. 2000; Davis et al. 2006; Fang et al. 2010; Picariello et al. 2010; Choi et al. 2011; Hui 2014). The use of MS for the analysis of lipids or their products is thus classed as 'lipidomics' (Gross et al. 2005; Armstrong 2009; Blanksby and Mitchell 2010; Bou Khalil et al. 2010; Ekroos et al. 2010; Gonzalez-Covarrubias 2013).

Easy ambient sonic-spray ionization mass spectrometry (EASI-MS) was used to authenticate oxidation in edible oils (Simas et al. 2010). Similarly, transmission-mode direct analysis in real time ionization coupled with high-resolution mass spectrometry (TM-DART–HRMS) was used to monitor chemical changes in several vegetable oils during their thermal oxidation (Vaclavik et al. 2009, 2013). Table 7.5 shows the LC-MS methods reported for the determination of oxidation products of TAG in various biological samples. Two main types of ionization of MS – atmospheric pressure chemical ionization (APCI) and electrospray ionization (ESI) – are widely reported. These methods are discussed in this section, irrespective of the stationary phase classification.

7.2.1.3.1 HPLC APCI-MS

It is clear from Table 7.5 that the initial work reported was mostly focused on the oxidation of pure standard TAGs. For example, Neff and Byrdwell (1998) used a gradient method of elution (ACN and DCM) to auto-oxidize triolein, trilinolein, and trilinolenin in the dark at 50–60 °C until the oxidation products reached about 30% of the original samples. They then analysed the oxidation products using RP-HPLC-APCI-MS. Mono-, bis-, and tris-hydroperoxides, hydroperoxy epidioxides, and epoxides were found to be the major oxidation products. A similar gradient method of ACN and DCM was used to study the oxidation compounds produced by pure triolein and trilinolein (Warner et al. 2001). APCI-MS revealed the formation of hydroperoxides, epoxides, ketones, and dimeric products. The authors were also able to study the secondary oxidation compounds formed from hydroperoxides. This study was helpful in explaining the source of the deep-fried flavour that is characteristic of high linoleic frying oils. Similarly, the oxidation products produced by the pure triolein were reported by Zhang et al. (2002). The separation was carried out using Novapak C-18 column and isocratic elution with MeOH-Hexane (85 : 15). The major oxidation products were monohydroperoxides, bishydroperoxides, and epoxides. This paper shows the first comprehensive investigation into the interaction between Laccase enzymes and lipids containing unsaturated FAs. Similarly, Byrdwell and Neff (1999) determined mono- and bis-hydroperoxides, epoxides, and epidioxides from the oxidation of triolein at 190 °C. The oxidation products were separated using isocratic elution with ACN-DCM (40 : 60) and quantified using APCI-MS. Byrdwell and Neff (2001) also showed the uses of gradient elution with ACN and DCM for the determination of TAG oxidation products formed

Table 7.5 High-performance liquid chromatography coupled with mass spectrometry (HPLC-MS) for the analysis of oxidized triacylglycerols (TAGOXs).

Column, size	Solvent system	Detection	Type of oxidized TAGs	Sample	References
Inerstil ODS2, 250 × 4.6 mm	Gradient: ACN and DCM	ELSD/ APCI-MS	Mono-, bis-, tris-hydroperoxides, hydroperoxy epidioxides	Pure triolein, trilinolein, and trilinolenin	Neff and Byrdwell (1998)
Dynamax C-18, 30 × 2.25 cm	Isocratic: ACN-DCM (40 : 60)	RI/APCI-MS	Hydroperoxides, epoxy ketones	Triolein model TAGs	Byrdwell and Neff (1999)
Inerstil ODS2, 25 × 4.6 mm	Gradient: ACN and DCM	ELSD/ APCI-MS	Mono-, bis-hydroperoxides, epoxides, epidioxides	Canola oil	Byrdwell and Neff (2001)
Inerstil ODS-80-A, 25 × 4.6 cm	Gradient: ACN and DCM	APCI-MS	Hydroperoxides, epoxides, ketones, and dimeric products	Pure triolein and trilinolein	Warner et al. (2001)
Inerstil ODS3, 25 × 4.6 mm	Gradient: ACN and DCM	ELSD/APCI and ESI-MS	Mono-, bis-hydroperoxides, epoxides, epidioxides	Canola oil	Byrdwell and Neff (2002)
Nova-Pak C-18, 150 × 3.9 mm	Isocratic: MeOH-Hexane (85 : 15)	APCI-MS	Hydroperoxides, epoxides	Pure Triolein TAG	Zhang et al. (2002)
Supelcosil LC-18, 250 × 4.6 mm	Gradient: 2-propanol-MeOH, 20–80%	ELSD/ESI-MS	Hydroperoxides, diepoxides, hydroxides	Corn oil, sunflower oil	Sjövall et al. (1997, 2001a, 2001b, 2003)
Inerstil ODS3, 25 × 4.6 mm	Gradient: ACN and DCM	ESI-MS	Oxidized tristearin monomers, dimers, chain addition/shortening products	Triolein	Byrdwell and Neff (2004)
Ultra RP 18, 250 × 2.1 mm	Gradient: ACN and chloroform	ESI-MS	Hydroperoxides, epoxides	Pure TAGs	Giuffrida et al. (2004)
Discovery HS C18, 250 × 4.6 mm	Gradient: 2-propanol-MeOH, 20–80%	ELSD / ESI-MS	Hydroperoxides, epoxides, hydroxides, and keto products	Sunflower oil	Suomela et al. (2005)
Acquity UPLC BEH 150 × 1 mm	Gradient: ACN-MeOH-water (19 : 19: 2) and Isopropanol	ESI-MS	Hydroperoxides	Mouse liver	Ikeda et al. (2009)

(*Continued*)

Table 7.5 (Continued)

Column, size	Solvent system	Detection	Type of oxidized TAGs	Sample	References
Phenomenex C-18, 250 × 4.6 mm	Gradient: 2-propanol-MeOH	ESI-MS	Mon-, di-, tri-hydroxides	Castor oil	Lin and Chen (2010)
Nucleodur ISIS C-18, 250 × 4.6 mm	Gradient: ACN-water-formic acid (50 : 50 : 0.1) and isopropanol-acetic acid-formic acid (90 : 10 : 1)	ELSD/ESI-MS	Mon-, di-, tri-, tetrahydroperoxides	Pure TAGs, rapeseed oil, and fish	Tarvainen et al. (2010, 2015)
Kinetex C-18, 120 × 2.1 mm	Gradient: ACN and acetone-ACN	ESI-MS/MS	Hydroperoxides,hydroxides, diepoxides	Pure TAGs	Suomela et al. (2011)
Phenomenx C-18, 150 × 3.0 mm	Isocratic: 2-propanol-MeOH (18 : 82)	ESI-MS	Mono-, bis-hydroperxides, mono-, bis-epoxides, epoxy epidioxides, hydroxyl epidioxides	Model TAGs, edible oils	Zeb and Murkovic (2010b)
PL gel MIXED-E SEC, 7.5 × 300 mm	Isocratic: THF	TM-DART–HRMS	Oxidized TAG, TAG oligomers	Edible oils	Vaclavik et al. (2009, 2013)

Source: Reproduced with kind permission of Taylor & Francis (updated and modified from Zeb 2015a).

by the auto-oxidation of normal and genetically modified canola oils. Mono- and bis-hydroperoxides, epoxides, and epidioxides were detected and quantified using RP-LC-APCI-MS. The APCI spectra revealed the protonated molecular ions $[M+H]^+$, $[M-18]^+$, and $[M+90]^+$ adducts.

The major disadvantages of the use of APCI-MS in oxidation studies include aldehydes, which do not produce abundant protonated molecular ions in APCI-MS. The hydroxyl containing TAG is not detected as well as with hydroperoxides, while the protonated molecular ions of nearly all normal or oxidized TAGs are found in very low abundance.

7.2.1.3.2 HPLC ESI-MS

ESI-MS is more widely used in TAG oxidation studies than is APCI-MS, as shown in Table 7.5. Byrdwell and Neff (2002) used both APCI-MS and ESI-MS for the analysis of oxidation products of canola oil TAGs. The gradient elution was carried out with ACN and DCM using Inertsil ODS-3 column. The major oxidation products were mono- and bis-hydroperoxides, epoxides, and epidioxides. The authors showed that the product ion mass spectra produced by the ESI-MS/MS were very similar to the conventional APCI-MS spectra. Both systems allowed the correct identification of positional isomers based on the ratios of fragment ions of DAGs. Another study by the same authors showed that oxidized tristearin monomers, dimers, and chain addition or shortening products were oxidized products of triolein at the frying temperature (Byrdwell and Neff 2004). Another gradient method of elution (ACN and chloroform) was reported by Giuffrida et al. (2004). These authors identified epoxy- and hydroperoxy-TAGs, which were formed by the air and $^{18}O_2$ oxidation of 1,2-dipalmitoyl-3-oleoyl-glycerol (PPO) and 1,3-dipalmitoyl-2-oleoyl-glycerol (POP). The analyses were carried out using RP-LC-ESI-MS. The MS was a triple quadruple mass analyser in positive ion mode. In the MS spectra, the fragmentation of hydroperoxy-TAGs was distinct from their epoxy-TAGs homologues and consisted of simultaneous losses of hydrogen peroxide and water.

The frying of sunflower and corn oils was reported to produce hydroperoxides, diepoxides, and hydroxides (Sjövall et al. 1997, 2001a, 2001b, 2002, 2003). These studies showed a gradient elution with 2-propanol-MeOH (20–80%) on the Supelcosil LC-18 column. The oxidized products were hydroperoxides, diepoxides, and hydroxides. Sjövall et al. (2003) showed that the rapid peroxidation with tert-butyl hydroperoxide was an effective LC-ESI-MS method of enriching natural oils and fats in TAG core aldehydes for biochemical and metabolic testing. Suomela et al. (2005) reported a method based on RP-HPLC and ESI-MS for the analysis of oxidized TAGs in chylomicrons and very low-density lipoproteins (VLDLs) of growing pigs. The gradient elution with 2-propanol-MeOH (20–80%) was used on Discovery HS C18 column. The reported oxidation compounds were hydroperoxides, epoxides, hydroxides, and keto products. Recently, Ikeda et al. (2009) reported an effective method for the global analysis of TAG molecular species from complex lipid mixtures of mouse liver and white adipose tissue using RP-LC coupled with ESI-quadruple/time-of-flight hybrid mass spectrometer (QTOF-MS). The gradient elution was based on ACN-MeOH-water (19:19:2) and isopropanol. The authors showed that several new oxidized TAGs were detected in mouse adipose tissue, along with non-oxidized TAGs in mouse liver.

A Phenomenex C-18 column and gradient elution with 2-propanol-MeOH was used along with ESI-MS detection for the determination of hydroxides in castor oil (Lin and

Chen 2010). A model TAG mixture was thermally oxidized at 110 °C, and the products were studied with the help of an isocratic method consisting of 2-propanol-MeOH (18 : 82) (Zeb and Murkovic 2010b). Identification and quantification of the oxidized species were carried out using ESI-MS. The same method was applied to different edible oils with similar (Zeb and Murkovic 2011) and different (Zeb and Murkovic 2013a, 2013b) TAG compositions as compared to the model TAGs in the presence of synthetic antioxidants and thermal stress. It was found that the major oxidation compounds of the TAG mixtures were mono- and bis-hydroperoxides, mono- and bis-epoxides, epoxy epidioxides, and hydroxyl epidioxides. TAGs with similar FA compositions showed similar oxidation compounds, irrespective of the oxidation treatment applied (Zeb 2012). The advantage of this method was that it required only a single run to provide a complete profile. However, it was unable to separate the isomers of individual oxidized TAG.

A gradient elution with ACN-water-formic acid (50 : 50 : 0.1) and isopropanol-acetic acid-formic acid (90 : 10 : 1) was used to separate and identify oxidized molecular species of pure TAGs, rapeseed oil, and fish samples using UHPLC (Tarvainen et al. 2010, 2015). ESI-MS spectra revealed the presence of mono-, di-, tri-, and tetra-hydroperoxides and vast amounts of several oxidation products. This UHPLC-ESI-MS method proved to be fast, highly selective, and sensitive, consuming one-tenth as many solvents as a previously used high-performance LC-ESI-MS method (Kuksis et al. 2009a, 2009b). Recently, Suomela et al. (2011) showed the separation of region-isomers of oxidized compounds of TAGs using gradient elution with ACN and acetone-ACN. The analysis was carried out using UHPLC coupled with tandem MS. It was found that the method was helpful in the separation of region-isomers of oxidized TAGs.

Ito et al. (2017a) developed a chiral stationary phase for separation of oxidized TAGs using LC-MS. The authors proved that the combination of chiral stationary phase coupled to an LC-MS/MS method with lipase enzyme was a powerful tool for evaluating the mechanisms of lipid oxidation during food deterioration. The downsides included the high cost of MS, the need for a highly skilled worker to conduct the analysis of oxidized lipids (especially oxidized TAGs), and the unavailability of a spectral database.

7.2.2 Quantification of Oxidized TAGs

The oxidized products of TAGs can be quantified using different liquid chromatographic methods. Sjövall et al. (2001a) quantitatively estimated the residual TAGs and mono-, di-, and tri-tertiary butyl hydroperoxide (TBHP) adducts from the TLC band using mol% of the total ion current. The mono-TBHP was 11.7 mol%, di-TBHP was 8.7 mol%, while tri-, tetra-, and penta-TBHP adducts were present in trace amounts. This study showed that TBHP adducts of the TAG oxidation products can be quantified easily using TLC with MS. Our previous study showed that hydroxy hydroperoxides, hydroxy epidioxides, epoxy epidioxides, hydroperoxides, and epoxides were formed in amounts less than 5.0% in the Rancimat at 110 °C (Zeb and Murkovic 2011), with epoxy epidioxides produced (about 8%) in greater amounts than the previously mentioned oxidation products. The addition of synthetic antioxidant was found to produce significant changes in the formation of these products. Another study (Zeb 2012) showed the formation and quantification of auto-oxidation products of TAG in Camellia oil. The amount of epoxy hydroperoxides was 1.12%, that of epoxy epidioxides 5.5%, and that of epoxides 6.89%. MS was highly useful for the quantification of individual oxidation compounds in the simple or complex system. However, the high cost of MS and the difficult interpretation of individual MS spectra of oxidized products provide disadvantages to such studies. Therefore, UV and other detection systems are used for quantification and detection purposes. Morales et al. (2014) showed that oxidized products of fatty acid methyl esters (FAME) can be analysed using HPLC-UV. These authors showed that when high linoleic sunflower oil was oxidized for 35 days at 40 °C, 24.12 mg g^{-1} of hydroperoxy dienes were produced. Similar amounts of hydroperoxides (25.25 mg g^{-1}) were also produced along with relatively small amounts of keto- and hydroxy-dienes. The autoxidation of high oleic sunflower oil produced 12.21 mg g^{-1} of hydroperoxy dienes, 13.27 mg g^{-1} of hydroperoxides, and a small amount of keto- and hydroxy-dienes. These results clearly show that the previously mentioned oxidized products were produced in double amounts when another double bond was present in the FA. It is clear that the process of oxidation and the compositions of TAGs play a key role in the formation of oxidation compounds.

7.3 Analysis of FA Oxidation Products

FAs produced by the hydrolysis of TAGs form several minor oxidation products. The details given here are reproduced from Xia and Budge (2017), who reviewed the analysis of epoxides, alcohols, and ketones. For example, Velasco et al. (2004) determined monoepoxy FAs in oils heated at the same temperature and for the same time as reported by Berdeaux et al. (1999). The total amounts of the six common monoepoxy FAs were 14.24 and 9.44 mg g^{-1} in olive and sunflower oils, respectively (Velasco et al. 2004). The authors determined methyl epoxystearates and methyl epoxyoleates in 10 used frying oils and concluded that: (i) more *trans* epoxides than *cis* epoxides were produced in monounsaturated oils; and (ii) a greater concentration of epoxides was found in monounsaturated oils than in polyunsaturated oils, consistent with Berdeaux et al. (1999); this may be attributed to a lower tendency of epoxystearates to form polymers and to participate in further oxidative reactions (Velasco et al. 2004).

In addition to frying oils, epoxy FAs have been determined at much lower concentrations (approximately 1000-fold lower) in fresh oil, with *cis*-9,10-epoxystearate, *cis*-12,13-epoxyoleate, and *cis*-9,10-epoxyoleate being the most prominent (Mubiru et al. 2013). *Trans* epoxy FAs were generally present at lower levels than *cis* epoxy FAs in fresh oils. For example, in sunflower oils, individual *cis* epoxy FAs ranged from 55 to 1430 μg g^{-1}, while *trans* epoxy FAs ranged from 4 to 33 μg g^{-1}. Moreover, a strong correlation was reported between the ratio of epoxystearate to epoxyoleate and the ratio of 18 : 1–18 : 2 FAs.

Mubiru et al. (2014) measured the six common monoepoxy FAs in food matrices including biscuits, meat, butter, and nuts. A range of 3–171 μg g^{-1} total epoxy FAs was determined in the foods, and no relationship was found between the sample type and the total epoxy FA concentrations. A large amount of *trans*-9,10 epoxystearate (53.66 μg g^{-1}) was reported for the butter sample; the other food matrices had lower concentrations, ranging from 0.75 to 17.33 μg g^{-1}.

Considering the locations of the epoxy groups in the six most measured monoepoxy FAs, in which the epoxy groups take the place of the original double bond in position 9 or 12 in methyl oleate and linoleate, these epoxy FAs seem to be formed by peroxide addition to double bonds. Positional isomers, such as 8, 9-epoxystearate and 10, 11-epoxystearate (Frankel 1984), may be formed by the cyclization of alkoxy radicals.

The preceding results reported for epoxides were determined by gas chromatography (GC); therefore, individual epoxy FAs were identified and quantified. In contrast, NMR methods determine total epoxide content. For instance, Goicoechea and Guillen (2010) determined monoepoxides and diepoxides in sunflower oils heated at 70 and 100 °C with aeration. At 100 °C, both monoepoxides and diepoxides reached a maximum value at 161 and 107 mmol L^{-1} oil (approximately 56 and 39 mg g^{-1}) and showed a declining trend afterwards; at 70 °C, they reached 41 and 123 mmol L^{-1} oil (approximately 14 and 45 mg g^{-1}).

Available data on the occurrence of alcohols and ketones are scarcer than for epoxides. Hydroxy FAs were first quantified in hydrogenated groundnut oil, and individual C18 hydroxy FAs were found to be present at between 2 and 494 µg g^{-1}, with a total amount of 1422 µg g^{-1} (Wilson et al. 1997). In fried sunflower oil (180 °C for 10 hours), total C18 hydroxy FAME ranged between 1.9 and 5.5 mg g^{-1} oil; C18 keto FAME and C18 epoxy FAME were also measured, determined at 0.5–2.5 and 1.3–4.4 mg g^{-1} oil, respectively (Marmesat et al. 2008). In extra virgin olive oil heated at 190 °C for 7.5–32 hours, primary alcohols were determined by NMR at 2–5 mmol mol^{-1} TAG (approximately 0.7–1.7 mg g^{-1} TAG), while secondary alcohols were between 4 and 20 mmol mol^{-1} TAG (approximately 1.4–7.0 mg g^{-1} TAG) (Martínez-Yusta and Guillen 2014a). In this experiment, (*E*)- and (*Z*)-epoxides reached concentrations of 20 and 12 mmol mol^{-1} TAG (approximately 7.0 and 4.2 mg g^{-1}), respectively. Three other food types, Spanish doughnut, pork adipose tissue, and farmed salmon fillets, were exposed to the same experimental conditions and did not show great differences in alcohol or epoxide content. However, in a later study, the same authors (Martínez-Yusta and Guillén 2014b) compared the formation of primary alcohols in fried soybean oils and the same three food types (doughnuts, adipose tissue, and salmon fillets) and found that primary alcohols had much higher concentrations in the salmon fillets than the other three media. The authors also reported that the primary alcohols began to be detectable after 7.5 hours of frying in soybean oil and Spanish doughnut and after 10 hours of frying in pork adipose tissue and salmon fillet.

Hydroxy dienes and keto dienes were monitored during oxidation of sunflower oil-derived FAME at 40 °C (Morales et al. 2010). Due to the limits of quantification, which were 0.05 mg g^{-1} for hydroxydienes and 0.02 mg g^{-1} for ketodienes, hydroxydienes and ketodienes were detectable but not quantifiable at the beginning of the experiments and increased in concentration over time. After 91 hours of heating, 3.1 mg g^{-1} of ketodienes and 1.3 mg g^{-1} of hydroxy dienes were found in the FAME derived from high linoleic sunflower oil. Under the conditions of the study, more ketodienes than hydroxydienes were found in FAME derived from both high linoleic and high oleic sunflower oils throughout the oxidation period. For high linoleic sunflower oil oxidized under similar conditions, the same group found concentrations of ketodienes and hydroxydienes up to 1.6 and 0.7 mg g^{-1}, respectively (Morales et al. 2014).

Steenhorst-Slikkerveer et al. (2000) determined the concentrations of epoxy-, oxo-, and hydroxy-TAGs in oxidized rapeseed oil and linseed/safflower oil. The oils were analysed when received, without any artificial oxidation. Epoxy- and oxo-TAGs co-eluted, so they were determined together, at 2734 ppm (approximately 2.7 mg g^{-1}) in rapeseed oil and 28 701 ppm (approximately 28.7 mg g^{-1}) in linseed/safflower oil.

The concentrations of hydroxy TAGs were 8759 and 3549 ppm (approximately 8.8 and 3.5 $mg\,g^{-1}$) in rapeseed oil and linseed/safflower oil, respectively.

Quantitative results on the occurrence of epoxides, alcohols, and ketones are summarized in Table 7.6. It should be noted that possible discrepancies exist among the results, due to the different techniques and oxidation conditions employed. The quantitative data show clear trends, where most oxidized oils have epoxy, hydroxy, and keto compounds at concentrations in the range of $mg\,g^{-1}$. In un-oxidized oils, these oxygenated compounds were quantifiable using chromatographic techniques but were present at much lower levels, in the range of $\mu g\,g^{-1}$.

7.3.1 Quantification of Oxygenated FAs

While GC-MS may help to identify the structures of oxygenated FAs, standards of commercial FAs are used as references to determine retention times and response factors in GC-FID. The major challenge in measuring oxygenated FAs in oils is that they are present at low concentrations. In order to make them detectable, it is necessary to extract and concentrate them from large amounts of the unaltered FAs. One useful approach to such separation employs solid-phase extraction (SPE) with a silica column as the stationary phase. When this technique was applied to oxidized fats and oils directly, the relatively polar compounds, including DAGs, monoacylglycerols (MAGs), and free fatty acids (FFAs), were separated from TAGs (Dobarganes et al. 2000). For SPE separation between non-oxygenated and oxygenated FAME, the first elution solvent, comprising a less polar solvent mixture such as hexane:ethyl acetate 98:2 (v/v) (Jenske and Vetter 2008), hexane:diethyl ether 95:5 (v/v) (Velasco et al. 2002), and hexane:diethyl ether 98:2 (v/v) (Marmesat et al. 2008), removes unaltered FAME, so oxygenated FAME are retained. The second elution solvent, normally ethyl acetate or diethyl ether, is sufficiently polar to elute the oxygenated FAME from the silica column. By removing the unaltered FAME, SPE acts as a concentration step and has been useful in quantifying trace levels of oxygenated FAs (Velasco et al. 2002; Jenske and Vetter 2008; Marmesat et al. 2008). The SPE step can also be modified for specific purposes. For example, Wilson et al. (1997) used three elutions on a silica column to separate monohydroxy FAs from dihydroxy FAs and other more polar compounds. Such SPE separations can be particularly useful in improving resolution with GC. For instance, Mubiru et al. (2013) also used three elutions on a silica column to elute epoxy FAs with n-hexane:diethyl ether (90:10, v/v), and successfully removed co-eluting hydroxy FA interferences from monoepoxy FAs. The pretreatment methods and detection conditions are summarized in Table 7.7.

With appropriate pretreatment, including transmethylation and SPE, GC-FID has been reported to be a powerful tool in simultaneously determining epoxy, hydroxy, and keto FAs (Marmesat et al. 2008), as well as monoepoxy FAs alone (Berdeaux et al. 1999; Velasco et al. 2002). In these studies, hydrogenation led to simpler chromatograms and better peak shapes for epoxy FAs. This step was particularly important for the resolution of keto and hydroxy FAs by GC, as their peaks severely co-eluted in their unsaturated forms. An obvious disadvantage is that hydrogenation resulted in a loss of information about the number and position of double bonds in the hydroxy and keto FAs. Unsaturated oxygenated FAs have been discussed, but the limited information concerning their structures in nonmodel systems has prevented easy quantification by GC. Measurements of unsaturated monoepoxy FAs in fresh oils have been conducted

Table 7.6 Occurrence of epoxides, alcohols, and ketones in FAME, TAGs, oils, and foods with or without artificial oxidation.

Sample	Oxidation condition	Quantitative technique	Analyte	Range of quantity	References
FAME					
Methyl oleate	Heated at 180 °C for 5–15 hours	GC-FID	Monoepoxy FAs	14.2–35.2 mg g^{-1}	Berdeaux et al. (1999)
Methyl linoleate	Heated at 180 °C for 5–15 hours	GC-FID	Monoepoxy FAs	9.2–19.1 mg g^{-1}	Berdeaux et al. (1999)
FAME derived from high linoleic sunflower oil	Heated at 40 °C for 0–91 hours	HPLC-UV	Ketodienes	Up to 3.094 mg g^{-1}	Morales et al. (2010)
		HPLC-UV	Hydroxydienes	Up to 1.338 mg g^{-1}	Morales et al. (2010)
FAME derived from high oleic sunflower oil	Heated at 40 °C for 0–192 hours	HPLC-UV	Ketodienes	Up to 1.726 mg g^{-1}	Morales et al. (2010)
		HPLC-UV	Hydroxydienes	Up to 0.590 mg g^{-1}	Morales et al. (2010)
FAME derived from high linoleic sunflower oil	Heated at 80 °C for 0–6 hours	HPLC-UV	Ketodienes	Up to 0.38 mg g^{-1}	Morales et al. (2012a)
		HPLC-UV	Hydroxydienes	Up to 0.46 mg g^{-1}	Morales et al. (2012a)
FAME derived from high oleic sunflower oil	Heated at 80 °C for 0–17 hours	HPLC-UV	Ketodienes	Up to 0.78 mg g^{-1}	Morales et al. (2012a)
		HPLC-UV	Hydroxydienes	Up to 0.31 mg g^{-1}	Morales et al. (2012a)
		HPLC-ELS	Methyl *trans*-epoxystearate	Up to 1.33 mg g^{-1}	Morales et al. (2012a)
		HPLC-ELS	Methyl *cis*-epoxystearate	Up to 1.52 mg g^{-1}	Morales et al. (2012a)
TAGs					
Triolein	Heated at 180 °C for 5–15 hours	GC-FID	Monoepoxy FAs	13.3–35.7 mg g^{-1} after CH_3ONa-catalysed transmethylation	Berdeaux et al. (1999)

Trilinolein	Heated at 180 °C for 5–15 hours	GC-FID	Monoepoxy FAs	6.9–18.3 mg g^{-1} after CH_3ONa-catalysed transmethylation	Berdeaux et al. (1999)
TAG mixture	Heated at 110 °C for 1–14 hours	HPLC-ESI-MS	Oxidized TAGs		Zeb and Murkovic (2010c)
Oils					
Hydrogenated ground nut oil	—	GC-MS-SIM	C-18 hydroxy FA isomers (C-6–C-17 substitution)	2–494 μg g^{-1}	Wilson et al. (1997)
Olive oil	Heated at 180 °C for 15 hours	GC-FID	Monoepoxy FAs	13.52 mg g^{-1} with on-column injection, 13.72 mg g^{-1} with split injection	Velasco et al. (2002)
Sunflower oil	Heated at 180 °C for 15 hours	GC-FID	Monoepoxy FAs	10.88 mg g^{-1} with on-column injection, 10.87 mg g^{-1} with split injection	Velasco et al. (2002)
Olive oil	Heated at 180 °C for 5–15 hours	GC-FID	Monoepoxy FAs	4.29–14.24 mg g^{-1}	Velasco et al. (2004)
Sunflower oil	Heated at 180 °C for 5–15 h	GC-FID	Monoepoxy FAs	5.10–9.44 mg g^{-1}	Velasco et al. (2004)
Used frying oils	—	GC-FID	Monoepoxy FAs	3.37–14 42 mg g^{-1}	Velasco et al. (2004)
Edible oils	—	GC-FID	*cis*-epoxy-oleic acid	110–2300 μg g^{-1}	Fankhauser-Noti et al. (2006)
		GC-FID	*cis*-epoxy-linoleic acid	150–1550 μg g^{-1}	Fankhauser-Noti et al. (2006)
		GC-FID	Diepoxy linoleic acid	0.2–1.5 μg g^{-1}	Fankhauser-Noti et al. (2006)
Frying oils	—	GC-FID	*cis*-epoxy-oleic acid	1900–7300 μg g^{-1}	Fankhauser-Noti et al. (2006)

(*Continued*)

Table 7.6 (Continued)

Sample	Oxidation condition	Quantitative technique	Analyte	Range of quantity	References
		GC-FID	*trans*-epoxy-oleic acid	2800–16 400 $\mu g\,g^{-1}$	Fankhauser-Noti et al. (2006)
		GC-FID	*cis*-epoxy-linoleic acid	1400–2000 $\mu g\,g^{-1}$	Fankhauser-Noti et al. (2006)
		GC-FID	*trans*-epoxy-linoleic acid	1800–3300 $\mu g\,g^{-1}$	Fankhauser-Noti et al. (2006)
		GC-FID	Diepoxy linoleic acid	4.9–16.0 $\mu g\,g^{-1}$	Fankhauser-Noti et al. (2006)
Sunflower oils	Heated at 180 °C for 10 hours	GC-FID	Epoxy FAs	1.3–4.4 $mg\,g^{-1}$	Marmesat et al. (2008)
		GC-FID	Keto FAs	0.5–2.5 $mg\,g^{-1}$	Marmesat et al. (2008)
		GC-FID	Hydroxy FAs	1.9–5.5 $mg\,g^{-1}$	Marmesat et al. (2008)
Fresh oils	—	GC-FID	Epoxy FAs	0.03–2 $mg\,g^{-1}$	Mubiru et al. (2013)
Used frying fat/oil	—	GC-FID	Epoxy FAs	0.05–16.57 $mg\,g^{-1}$	Brühl et al. (2016)
Pumpkin seed oil	—	GC-FID	Epoxy FAs	1.12–5.10 $mg\,g^{-1}$	Brühl et al. (2016)
Sweet almond oil	—	GC-FID	Epoxy FAs	0.72–2.65 $mg\,g^{-1}$	Brühl et al. (2016)
Sunflower oil	—	GC-FID	Epoxy FAs	0.06–1.90 $mg\,g^{-1}$	Brühl et al. (2016)
Groundnut oil	—	GC-FID	Epoxy FAs	0.10–7.25 $mg\,g^{-1}$	Brühl et al. (2016)
Olive oil	—	GC-FID	Epoxy FAs	0.10–0.32 $mg\,g^{-1}$	Brühl et al. (2016)
Rapeseed oil	—	HPLC-MS	Epoxy- and oxo-TAG	2734 ppm (2734 $\mu g\,g^{-1}$)[a]	Steenhorst-Slikkerveer et al. (2000)
		HPLC-MS	Oxo-$2^1/_2$-glycerides	506 ppm (506 $\mu g\,g^{-1}$)[a]	Steenhorst-Slikkerveer et al. (2000)
		HPLC-MS	Hydroxy-TAG	8759 ppm (8759 $\mu g\,g^{-1}$)[a]	Steenhorst-Slikkerveer et al. (2000)

		HPLC-MS	Dihydroxy-TAG	1049 ppm (1049 $\mu g\,g^{-1}$)[a]	Steenhorst-Slikkerveer et al. (2000)
Linseed/safflower oil	—	HPLC-MS	Epoxy- and oxo-TAG	28 701 ppm (28 701 $\mu g\,g^{-1}$)[a]	Steenhorst-Slikkerveer et al. (2000)
		HPLC-MS	Oxo-$2^1/_2$-glycerides	2146 ppm (2146 $\mu g\,g^{-1}$)[a]	Steenhorst-Slikkerveer et al. (2000)
		HPLC-MS	Hydroxy-TAG	3549 ppm (3549 $\mu g\,g^{-1}$)[a]	Steenhorst-Slikkerveer et al. (2000)
		HPLC-MS	Dihydroxy-TAG	1823 ppm (1823 $\mu g\,g^{-1}$)[a]	Steenhorst-Slikkerveer et al. (2000)
Sunflower oil	Heated at 100 °C with aeration	^{1}H NMR	Monoepoxides	Maximum level at 161 mmol L^{-1} (56 mg g^{-1})[b]	Goicoechea and Guillen (2010)
			Diepoxides	Maximum level at 107 mmol L^{-1} (39 mg g^{-1})[b]	Goicoechea and Guillen (2010)
Sunflower oil	Heated at 70 °C with aeration	^{1}H NMR	Monoepoxides	Maximum level at 41 mmol L^{-1} (14 mg g^{-1})[b]	Goicoechea and Guillen (2010)
			Diepoxides	Maximum level at 123 mmol L^{-1} (44 mg g^{-1})[b]	Goicoechea and Guillen (2010)
Extra virgin olive oil	Heated at 190 °C for periods of 8 hours day^{-1} for 5 days	^{1}H NMR	(*E*)-9,10-epoxystearic acyl groups	22.8 mmol L^{-1} (7.9 mg g^{-1})[b]	Guillén and Uriarte (2012)
		^{1}H NMR	(*Z*)-9,10-epoxystearic acyl groups	14.5 mmol L^{-1} (5.0 mg g^{-1})[b]	Guillén and Uriarte (2012)
Soybean oil	Heated at 100 °C for 4–20 days	^{1}H NMR	Epoxides	8.4–90.7 mmol kg^{-1} (2.6–28.3 mg g^{-1})[b]	Xia et al. (2016)
High oleic sunflower oil	Heated at 40 °C for 0–86 days	HPLC-UV	Ketodienes	0.13–1.34 mg g^{-1}	Morales et al. (2012b)
		HPLC-UV	Hydroxydienes	0.13–0.72 mg g^{-1}	Morales et al. (2012b)

(*Continued*)

Table 7.6 (Continued)

Sample	Oxidation condition	Quantitative technique	Analyte	Range of quantity	References
High linoleic sunflower oil	Heated at 40 °C for 0–41 days	HPLC-UV	Ketodienes	0.12–0.79 $mg\,g^{-1}$	Morales et al. (2012b)
		HPLC-UV	Hydroxydienes	0.36–1.12 $mg\,g^{-1}$	Morales et al. (2012b)
Soybean oil	Heated at 40 °C for 0–53 days	HPLC-UV	Ketodienes	0.07–2.97 $mg\,g^{-1}$	Morales et al. (2014)
		HPLC-UV	Hydroxydienes	1.02–3.53 $mg\,g^{-1}$	Morales et al. (2014)
Rapeseed oil	Heated at 40 °C for 0–46 days	HPLC-UV	Ketodienes	Up–2.1 $mg\,g^{-1}$	Morales et al. (2014)
		HPLC-UV	Hydroxydienes	Up–3.71 $mg\,g^{-1}$	Morales et al. (2014)
Foods					
Infant foods	—	GC-FID	Epoxidized soybean oil	9–86 $\mu g\,g^{-1}$	Fankhauser-Noti et al. (2005)
Oily sauces	—	GC-FID	Epoxidized soybean oil	47–580 $\mu g\,g^{-1}$	Fankhauser-Noti et al. (2005)
Products in oil	—	GC-FID	Epoxidized soybean oil	85–350 $\mu g\,g^{-1}$	Fankhauser-Noti et al. (2005)
Bakery foods	—	GC-FID	*cis*-epoxy oleic acid	150–4240 $\mu g\,g^{-1}$	Fankhauser-Noti et al. (2006)
		GC-FID	*cis*-epoxy linoleic acid	60–3460 $\mu g\,g^{-1}$	Fankhauser-Noti et al. (2006)
		GC-FID	Diepoxy linoleic acids	0.5–5.3 $\mu g\,g^{-1}$	Fankhauser-Noti et al. (2006)
Products in oil	—	GC-FID/ GC–MS	Epoxidized soybean oil	36–374 $\mu g\,g^{-1}$ (GC-FID) and 42–363 $\mu g\,g^{-1}$ (GC-MS)	Biedermann-Brem et al. (2007)

Biscuits	—	GC-FID	Epoxy FAs	46.95–74.49 μg g^{-1}	Mubiru et al. (2014)
Mayonnaise	—	GC-FID	Epoxy FAs	32.6–49.16 μg g^{-1}	Mubiru et al. (2014)
Nuts	—	GC-FID	Epoxy FAs	48.88–170.78 μg g^{-1}	Mubiru et al. (2014)
Butter	—	GC-FID	Epoxy FAs	115.34 μg g^{-1}	Mubiru et al. (2014)
Meat	—	GC-FID	Epoxy FAs	3.24–5.58 μg g^{-1}	Mubiru et al. (2014)
Chocolate, cocoa butter	—	GC-FID	Epoxy FAs	0.57–3.38 mg g^{-1}	Brühl et al. (2016)
Extra virgin olive oil fried with doughnut/pork/salmon fillets	Heated at 190 °C for 7.5–3.02 hours	^{1}H NMR	(*E*)-9,10-epoxystearic acyl groups	4–20 mmol mol^{-1} TAG (1.4–7.0 mg g^{-1})[c]	Martínez-Yusta and Guillén (2014b)
			(*Z*)-9,10-epoxystearic acyl groups	4–12 mmol mol^{-1} TAG (1.4–4.2 mg g^{-1})[c]	Martínez-Yusta and Guillén (2014b)
			Primary alcohols	2–5 mmol mol^{-1} TAG (0.7–1.7 mg g^{-1})[c]	Martínez-Yusta and Guillén (2014b)
			Secondary alcohols	4–20 mmol mol^{-1} TAG (1.4–7.0 mg g^{-1})[c]	Martínez-Yusta and Guillén (2014b)

Source: Data from Xia and Budge (2017).

[a] Values in units of ppm were converted into μg g^{-1} using 1 : 1 ratio.

[b] Values in units of mmol kg^{-1} or mmol L^{-1} were converted into mg g^{-1} for comparison, using the molecular weights of methyl epoxystearate and methyl diepoxystearate and assuming that all monoepoxides were methyl epoxystearate and all diepoxides were methyl diepoxystearate. The density of oil was assumed at 0.9 kg L^{-1} for calculation.

[c] Values in units of mmol mol^{-1} TAG were converted into mg g^{-1} for comparison, using the molecular weights of methyl epoxystearate for epoxides and methyl hydroxystearate for alcohols. The molecular weight of TAG was assumed at 900 g mol^{-1} for calculation.

—, information not available; GC, gas chromatography; FID, flame ionization detector; FA, fatty acid; FAME, fatty acid methyl esters; HPLC, high-performance liquid chromatography; UV, ultraviolet; ELS, evaporative light scattering; TAG, triacylglycerol; ESI, electrospray ionization; MS, mass spectrometry; SIM; NMR, nuclear magnetic resonance.

Table 7.7 Gas chromatographic (GC) quantitative methods and corresponding pretreatment and detection conditions for analysis of oxygenated fatty acids (FAs) in foods.

Analytes	Samples	Pretreatment	Extraction condition	Detection	Column	LOD/LOQ	References
Monohydroxy (plus hydroperoxy) FAs	Nut oil	Hydrogenation, transmethylation, SPE, and methylation of OH group	SPE: 1st elution: 95 : 5 hexane : ethyl acetate; 2nd elution: 80 : 20 hexane : ethyl acetate	EI-MS-SIM	CP-Sil 19 fused silica (25 m × 0.25 mm i.d.)	LOD = 0.2 ng (injection volume = 1 μL)	Wilson et al. (1997)
Mono epoxy FAs	Thermoxidized olive and sunflower oils	Transmethylation and SPE	SPE: 1st elution: 95 : 5 hexane : diethyl ether; 2nd elution: diethyl ether	FID	DB-Wax fused silica (30 m × 0.25 mm i.d.)	Not reported	Velasco et al. (2002)
Mono epoxy FAs	Thermoxidized olive and sunflower oils/ used frying oils	Transmethylation and SPE	SPE: 1st elution: 95 : 5 hexane : diethyl ether; 2nd elution: diethyl ether	FID	HP-Innowax fused silica (30 m × 0.25 mm i.d.)	Not reported	Velasco et al. (2004)
Methyl diepoxy linoleate	Oil/sauce/infant food	Transmethylation, NPLC extraction	NPLC column: packed with a cyano phase (25 cm × 0.2 mm i.d.) Mobile phase: 20% MTBE/ pentane	FID	A column coated with a 0.2 mm film of PS-255, a methyl polysiloxan (30 m × 0.25 mm i.d.)	LOD = 2 mg kg^{-1} extract and LOQ = 6 mg kg^{-1}	Fankhauser-Noti et al. (2005, 2006)
Monoepoxy FAs	Oils/snacks	Transmethylation	N/A	FID	SP-2560 (100 m × 0.25 mm i.d.)	Not reported	Fankhauser-Noti et al. (2006), Biedermann-Brem et al. (2007)

Methyl mono–/di-epoxy linoleate	Oily foods	Transmethylation	N/A	EI/CI-MS	Rtx 2330 (20 m × 0.25 mm i.d.)	LOD < 1 mg kg^{-1} (CI) LOQ = 3 mg kg^{-1} (EI)	Biedermann-Brem et al. (2007)
Epoxy, keto, and hydroxy FAs	Thermoxidized sunflower oils	Transmethylation, SPE, and hydrogenation	SPE: 1st elution: 98 : 2 hexane : diethyl ether; 2nd elution: diethyl ether	FID	DB-Wax fused silica (60 m × 0.25 mm i.d.)	LOQ = 1.6–2.1 μg mL^{-1}	Marmesat et al. (2008)
Keto and hydroxy FAs	Milk fat	Transmethylation and SPE	SPE: 1st elution: 98 : 2 hexane : diethyl ether; 2nd elution: diethyl ether	FID	DB-Wax fused silica (30 m × 0.32 mm i.d.)	Not reported	Márquez-Ruiz et al. (2011)
Mono epoxy FAs	Arachid/colza/ corn/ frying/ olive/soya/ sunflower/ salad/ mixed/ soya oils	Transmethylation and SPE	SPE: 1st elution: 98 : 2 hexane : diethyl ether; 2nd elution: 90 : 10 hexane : diethyl ether; 3rd elution: 70 : 30 hexane : diethyl ether	FID	CP-Sil 88 (60 m × 0.25 mm i.d.)	LOD = 1.45 μg g^{-1} and LOQ = 2.9 μg g^{-1}	Mubiru et al. (2013)
Mono epoxy FAs	Oil/crisps/pork/ milk powder	Modified Bligh and Dyer method, transmethylation, and SPE	SPE: 1st elution: 98 : 2 hexane : diethyl ether; 2nd elution: 90 : 10 hexane : diethyl ether	FID	CP-Sil 88 (50 m × 0.25 mm i.d.)	LOD = 1.7–10.2 μg g^{-1} and LOQ = 3.3–20.5 μg g^{-1}	Mubiru et al. (2014)

Source: Reproduced with kind permission of the Institute for Food Technologists (Xia and Budge 2017).
RSD, relative standard deviation; LOD, limit of detection; LOQ, limit of quantification.

by Mubiru et al. (2013). The limit of detection (LOD) and limit of quantification (LOQ) were 1.45 and 2.9 μg/g of oil, respectively. This method has also been applied to monitor epoxy FA formation in fats, oils, and chocolates (Brühl et al. 2016).

Methoxy derivatives of C-18 hydroxy FAME isomers were employed as a marker of lipid peroxidation (Wilson et al. 1997), where hydrogenation was conducted to reduce hydroperoxy groups to hydroxy groups and to eliminate double bonds; thus, this approach did not differentiate between hydroxy and hydroperoxy FAs. Two ion fragments resulted from cleavage at each side of the OCH_3 group, giving the location of the original hydroxy/hydroperoxy group and enabling quantification in selective ion monitoring (SIM) mode. This approach has been applied to food and oil samples, with a detection limit of 0.2 ng when monitoring a single isomer. It was further developed for simultaneous determination of hydroxy and epoxy FAs in human plasma, where such FAs were converted into methoxy and dimethoxy derivatives, respectively (Wilson and Lyall 2002); however, no applications have been reported for food.

Fluorinated reagents, such as fluorinated anhydride and pentafluorobenzoyl chloride, have been used to analyse hydroxy FAME in biological samples where hydroxy groups were derivatized into fluorinated groups. The utilization of fluorinated derivatives with NCI increased sensitivity and selectivity for hydroxy FAs (Stan and Scheutwinkel-Reich 1980; Jenske and Vetter 2008). A recent study reported the detection limits to be 300 fg–2 pg for 2-hydroxy FAME and 50–500 fg for 3-hydroxy FAME in bovine milk fat (Jenske and Vetter 2008). However, since fluorinated derivatives of hydroxy FAME have seldom been used for the study of lipid oxidation, their applicability to unsaturated FAs needs verification. Like hydroxy FAME, the use of fluorinated reagents facilitated the quantification of keto FAME. With improved sensitivity in negative CI (NCI) mode, PFB derivatives have been mainly applied for quantification, with a detection limit of 0.1 pg (Hachey et al. 1991). No applications of these fluorinated derivatives have been reported for the study of lipid oxidation in food and oil.

In summary, GC has shown excellent sensitivity for the determination of oxygenated FAs as minor products of FAs oxidation, but there remain unresolved problems in co-elution and identification of unknown compounds. Co-elution results in even fewer characteristic spectra and further difficulties in the identification of the peaks. Instead of traditional one-dimensional GC, two-dimensional GC, employing two columns of distinct polarity, has also been used to analyse food FAs, which significantly improved the separations of both the major and the minor components (Hyötyläinen et al. 2004). This technique has potential in obtaining improved separations of complex oxygenated oxidation products in GC.

7.4 Analysis of Sterol Oxidation

Guardiola (2002) published a very exciting book on the analysis, occurrence, and biological effects of cholesterol and phytosterol oxidation products. The advent of modern analytical instrumentation and techniques, however, has given rise to the most precise, correct, and reproducible results on sterols and their oxidation products during frying. It is worth mentioning that not all of the scientific reporting on this subject is based on frying alone, and studies on similar food compositions and frying conditions can be used to evaluate data or extrapolate them to the real food matrix during frying.

7.4.1 Analysis of Cholesterol and COPs

Foods of animal origin are largely formed of cholesterol, which is oxidized to form cholesterol oxidation products (COPs). Dinh et al. (2011) reviewed the cholesterol contents of different meat and poultry products. They found that much of the early data (through the early 1980s) was produced using either spectrophotometric, gravimetric, or enzymatic methods, many of which were originally developed to measure cholesterol in blood serum (Caldironi and Manes 2006). Cholesterol in foods has been determined gravimetrically by precipitation using digitonin or tomatin, with digitonin the more common (Bragagnolo 2009). Digitonin is washed and the precipitated cholesterol is weighed and calculated using several common factors to account for the proportion of cholesterol in the precipitate. A series of sample preparation steps can be followed from lipid extraction to purification, or direct precipitation of cholesterol from the extract can be performed.

Enzymatic determination of cholesterol has been used for food samples, but it is much more common in clinical laboratories (Bragagnolo 2009). Enzymatic methods employ cholesterol esterase to cleave any ester bond and cholesterol oxidase to oxidize cholesterol to peroxide. The peroxide then reacts with peroxidase and 4-aminophenazone (or other reagents) to produce a pigment that can be measured spectrophotometrically (Rambaldi et al. 2009). The enzymatic method has been used commonly for dairy products and occasionally for other foods, including meats (Caldironi and Manes 2006). It was found to yield comparable results to the GC technique when used to determine the cholesterol content of milk (not statistically tested) and processed foods containing primarily animal fat or fat of animal origin (Ulberth and Reich 1992). Nonetheless, this method lacks specificity, because other sterols with a 3β-OH group, including phytosterols, can also be oxidized to form similar pigments. Therefore, in processed foods containing primarily lipids of vegetable origin, the enzymatic method overestimates cholesterol content compared with GC. Jiang et al. (1991) evaluated four methods of cholesterol determination in egg yolk: colourimetric, enzymatic, GC, and HPLC. Results with the enzymatic method were found to be similar to those with GC and HPLC. All three achieved excellent precision; colourimetry alone did not.

In colour-based methods, the application of the Liebermann–Burchard reaction or other colour development reactions is usually the key step after extraction and de-esterification. Colouring reagents include acetic acid and concentrated sulfuric acid, which react in solvents such as chloroform and ether (Liebermann–Burchard reaction); para-toluene sulfonic acid or similar reagents containing glacial acetic acid and concentrated sulfuric acid; and iron salt (iron chloride or iron sulfate) in sulfuric acid and glacial acetic acid (Tonks 2006). Colour-producing reagents are added after extraction, saponification, and purification, or sometimes directly to the extract. Colourimetric methods usually suffer various interferences from unsaturated FAs, proteins, and other steroids, or even from vitamin A. Cholesterol esters, if not hydrolysed, can react with colour reagents in different ways. The colour reactions depend on a double bond system, other functional groups, and the presence of a nonpolar side chain. Colourimetric methods have commonly been used to measure cholesterol in meats; however, the lack of specificity and colour stability, the issue of temperature dependency, and the turbidity of the final colour mean there are grave concerns about their accuracy (Tonks 2006). Without saponification, Bohac et al. (1988) showed that the colourimetric method overestimated the cholesterol content of meats. Even with saponification, the presence of more unsaturated FAs (more double bonds) yielded greater cholesterol concentrations without antioxidant protection. The cholesterol contents of saponified samples with an antioxidant were similar to those analysed by GC. Gravimetric, colourimetric, and enzymatic methods require strict control of analytical conditions in order to provide precise and accurate results (Bragagnolo 2009).

7.4.1.1 Gas Chromatography

GC and HPLC have been studied extensively in recent years. Chromatography is preferred for cholesterol analysis in foods because of its specificity in separating cholesterol from other unsaponifiable compounds based on their differences in terms of physical and chemical properties and their interactions with stationary and mobile phases. Processed meats with non-meat ingredients contain not only cholesterol but also plant sterols, tocopherol, tocotrienol, saturated hydrocarbons, squalene, aliphatic alcohols, terpene alcohols, triterpene alcohols, and stearoyl esters, all of which are unsaponifiable (Fenton 1992). Recently, HPLC has been used more than GC, as it is thought to decrease cholesterol oxidation through its lower operational temperature and it allows for a nondestructive separation; however, GC is still preferred for its higher sensitivity (Bragagnolo 2009).

Because of the interference of other unsaponifiable materials and the cost and availability of highly specific detectors, the effectiveness of the chromatographic separation and quantification of cholesterol greatly depends on the heart of the chromatographic system: the column. GC is still the most commonly used technique for the determination of sterols, including cholesterol – especially when cholesterol is the sole compound of interest (Fenton 1992; Abidi 2001). GC columns have evolved dramatically, from simple packed columns to capillary columns with wide variation in polarity. Ubhayasekera et al. (2004) evaluated three methods of saponification and *trans*-esterification of tallow using GC and GC-MS, and showed that GC-MS is more sensitive than GC. Yen et al. (2010) developed a GC-MS method for the separation of cholesterol and eight COPs. Cardenia et al. (2012) developed a fast GC-MS method for the analysis of COPs in fried fish. They separated all COPs studied in 3.6 minutes with very good separation, as shown in Figure 7.3. The separated compounds were 7α-HC, 19-HC, 7β-HC, β-CE, α-CE, triol, 25-HC, and 7-KC.

The authors evaluated the sensitivity of the method in a standard COP mixture by defining the LOD and LOQ of each oxysterol. The LOD and LOQ of fast GC-MS method varied from 2.18 to 5.07 and from 6.93 to 16.90 $ng\,mL^{-1}$, respectively (Table 7.8). These results were in agreement with Guardiola (2002), who reported that a single COP detected by GC-MS in single ion-monitoring mode displayed different LODs and LOQs, probably due to different molecular structures and competitive interactions (Cardenia et al. 2012).

Souza et al. (2017) developed a novel microwave-assisted direct saponification method for the simultaneous determination of cholesterol and cholesterol oxides in shrimp. They showed very low detection ($\leq 0.57\,\mu g\,mL^{-1}$) and quantification ($\leq 1.73\,\mu g\,mL^{-1}$) limits, good repeatability (≤10.50% intraday and ≤8.56% interday), and low artifact formation. The LOD and LOQ were, however, lower than reported by Cardenia et al. (2012). The uses of solid phase microextraction and subsequent analysis by GC-FID can also be very effective for the analysis of cholesterol and its oxidation products (Narvaez-Rivas et al. 2014).

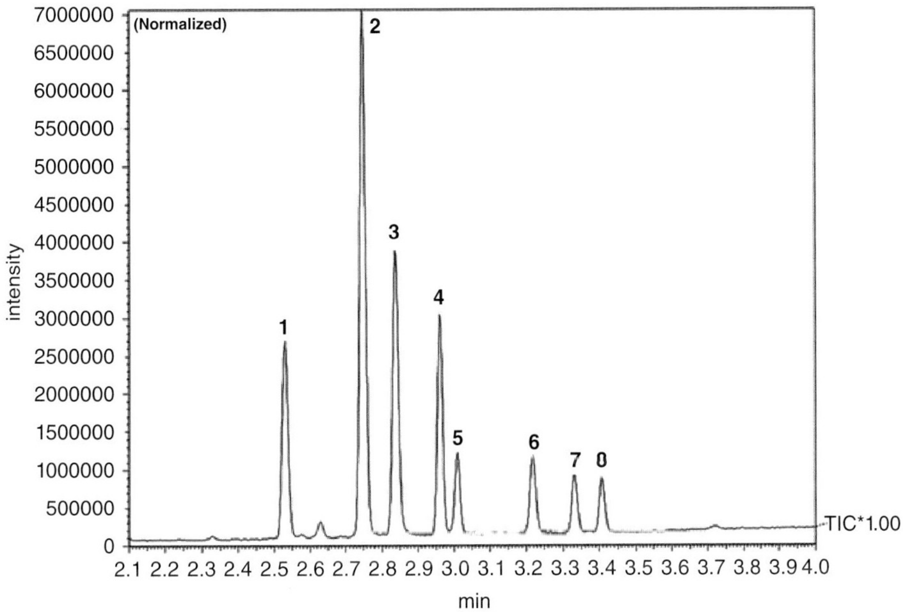

Figure 7.3 Fast GC-MS trace (TIC) of TMS derivatives of a standard COP mixture, obtained with a 10-m RTX-5 column. Peak identification: 1, 7α-HC; 2, 19-HC; 3, 7β-HC; 4, β-CE; 5,α-CE; 6, triol; 7, 25-HC; 8, 7-KC. *Source:* Reproduced with kind permission from Wiley-VCH Verlag GmbH & Co. KGaA. (Cardenia et al. 2012).

Table 7.8 Analytical parameters of fast and conventional GC-MS methods.

	Intraday (RSD)		Interday (RSD)		LOD (ng mL^{-1})		LOQ (ng mL^{-1})	
Analyte	Fast	Conventional	Fast	Conventional	Fast	Conventional	Fast	Conventional
7α-HC	2.36	2.37	2.36	4.38	2.18	2.82	6.93	9.40
7β-HC	2.85	3.72	4.02	5.01	3.68	3.11	12.27	10.37
β-CE	1.35	4.01	5.79	4.34	4.97	4.19	16.57	13.97
α-CE	3.09	0.49	4.12	2.34	5.07	5.14	16.90	17.13
Triol	3.35	4.28	5.40	4.30	2.65	2.03	8.83	6.77
25-HC	4.29	2.82	3.66	3.66	2.85	2.73	9.50	9.10
7-KC	1.81	4.02	4.61	4.61	3.92	4.84	13.07	16.13

Source: Reproduced with kind permission WILEY-VCH Verlag GmbH & Co. KGaA (Cardenia et al. 2012).

RSD, relative standard deviation; LOD, limit of detection; LOQ, limit of quantification.

7.4.1.2 Liquid Chromatography

There are many published HPLC methods for cholesterol determination; however, most were not developed for routine analysis, but rather to separate cholesterol from other unsaponifiable compounds for specific research purposes (Mestre Prates et al. 2006;

Daneshfar et al. 2009). Because of the slight polarity caused by the hydroxyl group, either normal-phase (NP) or reversed-phase (RP) HPLC can be used for analysis of cholesterol. NP-HPLC has been used to separate triglycerides, diglycerides, and cholesterol (Fenton 1992). As mentioned previously, the GC system is ineffective for separating cholesterol from tocopherols; however, Katsanidis and Addis (1999) successfully separated and quantified cholesterol, vitamin E, and vitamin E homologues using an NP-HPLC technique with a 25 cm Zorbax RX-Sil column (particle size = 5 μm). The compounds were detected with a UV detector at 295 nm for vitamin E and 202 nm for cholesterol. The column was made of ultra-clean porous silica microparticles. The mobile phase was 99% hexane and 1% isopropanol. Most NP-HPLC methods use an NP (polar stationary phase) column made of highly pure, porous silica microparticles, such as μPorasil (Kermasha et al. 1994), Zorbax RX-Sil (Katsanidis and Addis 1999), InertSil ODS-2 (Sion et al. 2001), Zorbax RX-Sil (Ponte et al. 2008), or Spherisorb S5W (Costa et al. 2006). Generally, the particle size is 5 μm; however, the column used by Kermasha et al. (1994) was 10 μm. Smaller particle size leads to significantly increased column pressures, but also increases the resolution (theoretical plate number). More polar stationary phases, such as cyanopropylsilica and alcohol-bonded silica, have been used (Abidi 2001). The mobile phase for NP-HPLC has primarily been an isocratic phase of 1–3% isopropanol in hexane (Kermasha et al. 1994; Katsanidis and Addis 1999; Abidi 2001; Ponte et al. 2008). In order to use light-scattering detection and improve peak shape, Sion et al. (2001) used a gradient mobile phase made of 98:2 methanol/water (v/v), 30:6:10 chloroform/methanol/water (v/v/v), and 50:50 methanol/chloroform (v/v). A constant flow of 1 mL min^{-1} or lower has been used in most studies.

Grun and Besseau (2016) developed NP- LC coupled to an atmospheric pressure photoionization MS method for the determination of cholesterol and phytosterol oxidation products. They reported the separation of 15 sterol oxidation products, as shown in Figure 7.4. The best chromatographic separation with the highest sensitivity was

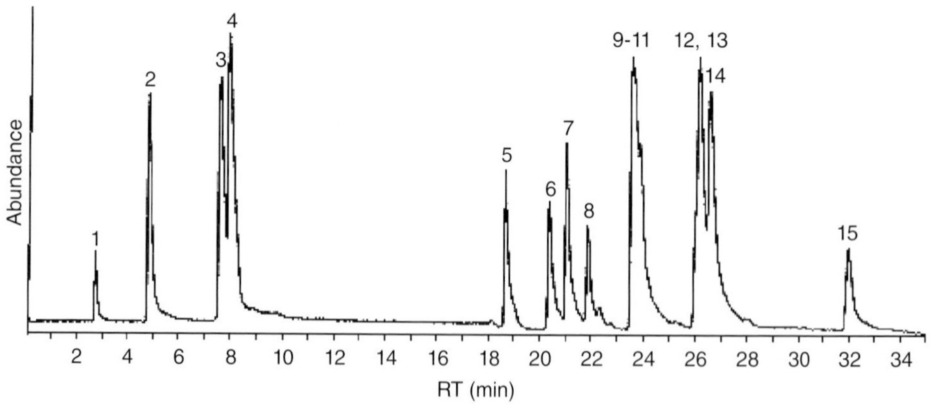

Figure 7.4 NP-LC-APPI-MS analysis of cholesterol oxidation products. (1) cholesta-3,5-diene, (2) 5-cholesten-3-one, (3) 4-cholesten-3-one, (4) cholesta-4,6-dien-3-one, (5) cholesterol, (6) 22-ketocholesterol, (7) 5 α,6α-epoxycholesterol, (8) 20α-hydroxycholesterol, (9) 7-ketocholesterol, (10) 25-hydroxycholesterol, (11) 22(S)-hydroxycholesterol, (12) 19-hydroxycholesterol, (13) 7 α-hydroxycholesterol, (14) 7 β-hydroxycholesterol, and (15) 5 α,6β-dihydroxycholesterol. *Source:* Reproduced with kind permission of Elsevier (Grun and Besseau 2016).

accomplished by employing a solvent system consisting of hexane and isopropanol on a diol-bonded silica column. The dopant of choice was chlorobenzene, added post-column to the solvent stream at 10% of the flow rate. The developed NP-LC-APPI-MS method proved to be a valuable tool for the separation and detection of sterol oxidation products.

Although NP-HPLC was the first technique used successfully to separate cholesterol in meat such as beef (Abreu et al. 2016), bovine tissues (Katsanidis and Addis 1999), and chicken (Ponte et al. 2008), RP-HPLC has been preferred because it offers a wider range of column selectivity and separable compounds based on polarity. The RP-HPLC method uses a nonpolar column or a column with low polarity and a polar mobile phase. The simple addition of some polar solvents, such as methanol, ethanol, or water, can dramatically change mobile phase polarity and alter elution (Abidi 2001). Most RP-HPLC methods of cholesterol determination use acetonitrile as the polar organic solvent, with the addition of either ethanol (Daneshfar et al. 2009), methanol (Simsek et al. 2009), or isopropanol (Komprda et al. 2003). They have typically employed silica (Abidi 2001) as a 5 μm particle support material covalently bonded with either octadecyl (Salvatori et al. 2008) or octyl (Nakajima et al. 1995; Daneshfar et al. 2009). The RP-HPLC C-18 column was found to be superior to the C-8 column because it has a much stronger retention of cholesterol and other sterols (Fenton 1992). Therefore, a mobile phase that is less polar should be employed for cholesterol separation on a C-18 column. The HPLC technique, although superior to GC in terms of resolution of multiple unsaponifiable compounds, can be further improved by programmed temperature elution.

Constantinou et al. (2015) reported a baseline separation of seven COPs in several food samples. The separation was achieved with the help of a cyano-bonded HPLC column and isocratic mobile phase consisting of *n*-hexane/2-propanol/acetone (97 : 1.5 : 1.5, v/v), as shown in Figure 7.5. The LOD and LOQ were in the range 0.15–0.63 and 0.45–1.91 μg mL^{-1}, respectively, for the seven COPs.

Georgiou et al. (2016) used a novel approach for the fast determination of the most important COPs in several Cypriot foods (halloumi cheese, feta cheese, hiromeri, snail, etc.) using UPLC-ESI-MS (Table 7.9). Among them, 7-keto was detected in all of the samples analysed. This finding confirms that 7-keto can be used as an appropriate marker of cholesterol oxidation in food systems, especially fried products.

Another rapid and sensitive HPLC-APCI-MS/MS method for the determination of COPs in milk powder-based foods was reported by Gorassini et al. (2017). The method consists in the direct saponification of the sample and purification of oxysterols by reversed-phase C-18-SPE followed by HPLC-MS/MS analysis. By this procedure, the extraction and enrichment of oxysterols are combined in a unique step, reducing sample manipulation and the possible formation of artifacts. The LOD and LOQ were in the concentration ranges of 2–8 and 8–30 ng g^{-1}, respectively.

These studies suggest that HPLC-MS provides a very sensitive analytical tool to determine the cholesterol or COP contents in fried foods. However, it should be kept in mind that sample preparation is highly important: Georgiou et al. (2014) showed that it contributes to artifact formation, perhaps altering the chemical composition of foods under study. Further, no method used to date can be considered precise and accurate for routine analysis of COPs in foods.

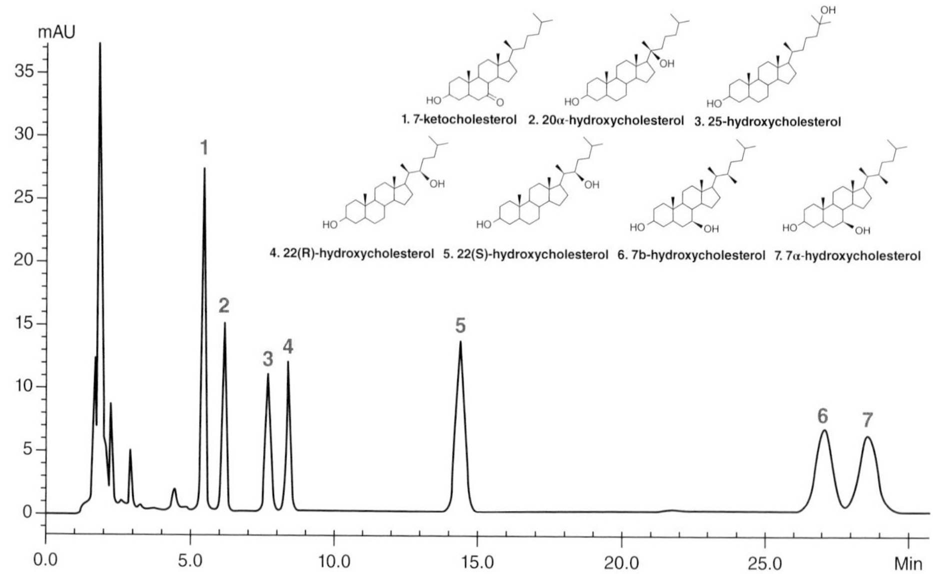

Figure 7.5 Optimum HPLC chromatographic separation of seven cholesterol oxidation products (COPs). Baseline separation was achieved by the use of *n*-hexane:2-propanol:acetone (97:1.5:1.5, v/v) at a flow rate of 1 ml min^{-1} and a temperature of 25 °C. (1) 20α-OH, (2) 22R-OH, (3) 22S-OH, (4) 25-OH, (5) 7-keto, (6) 7β-OH, (7) 7α-OH. *Source:* Reproduced with kind permission of Springer Nature (Constantinou et al. 2015).

Table 7.9 Levels of cholesterol oxidation products (COPs) detected in Cypriot food products.

	COPs (ng per 250 mg fat)			
Samples	**7-OH**	**7-Keto**	**5,6-EP**	**Total COPs**
Yogurt	41 ± 1.7	57 ± 5.2	1314 ± 1.7	1412
Feta cheese	1430 ± 1.8	73 ± 2.8	n.d	1503
Halloumi cheese	616 ± 1.1	82 ± 3.0	n.d	698
Anari cheese	105 ± 2.4	1099 ± 2.7	n.d	1204
Snail	1239 ± 4.7	1424 ± 3.0	n.d	2663
Hiromeri	395 ± 2.2	155 ± 2.8	105 ± 2.7	655
Sausage	444 ± 1.7	490 ± 4.7	n.d	934
Salami	972 ± 2.5	141 ± 3.7	442 ± 3.0	1555
Caul fat	122 ± 1.7	331 ± 3.6	1406 ± 2.3	1859
Bacon (raw)	1104 ± 1.5	158 ± 3.0	n.d.	1262
Bacon (fried)	2166 ± 1.5	237 ± 1.6	232 ± 2.8	2635
Rabbit (fried)	843 ± 1.8	681 ± 1.4	202 ± 2.6	1726
Rabbit (oven-cooked)	2530 ± 0.3	938 ± 1.8	1594 ± 2.7	5062

Source: Reproduced with kind permission of WILEY-VCH Verlag GmbH & Co. KGaA (Georgiou et al. 2016). Values are mean with standard deviations (n = 4).

7.4.2 Analysis of Phytosterol Oxidation

Plant foods also contain sterols called phytosterols, which can be oxidized to form phytosterol oxidation products (POPs). Thermal oxidation of phytosterols occurring in foods encompasses a sequence of reactions resulting in primary (hydroperoxides), secondary (polar: ketones, alcohols, epoxides; unpolar: steradienes, steratrienes), and tertiary (dimers, oligomers, polymers) oxidation products. Due to the analytical capabilities of each lab, the focus has almost exclusively been put on the secondary polar POPs. Accordingly, the term 'POPs' here refers to the keto-, epoxy-, and hydroxy-compounds derived from the respective sterols/stanols.

There are a number of analytical methodologies available, based on (i) lipid extraction and saponification or transesterification, (ii) isolation and purification via TLC or solid phase extraction, (iii) derivatization to trimethylsilyl ethers, and (iv) detection via HPLC- and GC-based techniques (Guardiola et al. 2004). GC-based analysis of POPs is usually not preferred at high temperatures. GC-FID has been used for phytosterol analysis (Duong et al. 2016). The use of MS detectors has improved the sensitivity of phytosterols in foods (Inchingolo et al. 2014).

These methods have been employed to generate a broad spectrum of analytical data on model systems involving the thermal treatment of phytosterol standards (Oehrl et al. 2001; Soupas et al. 2004; Barriuso et al. 2012). Thermal oxidations under different time and temperature conditions revealed that some of the secondary POPs constitute intermediates, which are further transformed or degraded in the course of the reaction (Rudzińska et al. 2009). Initial attempts to isolate fractions containing dimers, trimers, and tetramers via size-exclusion chromatography (SEC) have been described, and structures for sterol dimers have been proposed (Lampi et al. 2009; Rudzińska et al. 2010; Sosińska et al. 2013, 2014).

Heating of stigmasterol at 180 °C for 3 hours resulted in a loss of the intact sterol of 61%. Polar, midpolar, and nonpolar oxidation products accounted for 39% of the loss; the formation of dimers and polymers accounted for 30%. This means that there is a gap in the mass balance, leaving 31% of the stigmasterol loss unexplained (Menéndez-Carreño et al. 2010).

There are data available indicating qualitative and quantitative differences in oxidation profiles between free and esterified phytosterols (Soupas et al. 2004; Lehtonen et al. 2011). Complex mixtures are to be expected, owing to potential oxidations in the sterol as well as in the FA moiety of phytosteryl esters of unsaturated FAs (Lehtonen et al. 2012). Very recent studies describe approaches to the analysis of intact oxidized phytosteryl FA esters via HPLC-ESI-MS (Julien-David et al. 2014).

Sun et al. (2017) used 4'-carboxy-substituted rosamine (CSR) as a derivative reagent for the determination of phytosterols. The CSR reacted with the hydroxyl group of phytosterol to form a CSR-phytosterol complex (Figure 7.6).

Agilent SB C-18 column with the specification 2.1 × 50 mm, 1.8 μm was used for separation via UHPLC-MS/MS. With the help of 0.1% formic acid in the mobile phase, better separations were obtained for five phytosterols and diosgenin derivatives using acetonitrile versus methanol (Figure 7.7). The major product ions for all CSR derivatives of phytosterols and diosgenin are m/z 443.1 and 399.0. In this work, the product ion at m/z 443.1 was used for quantitation, and the product ion at m/z 399.0 was employed to confirm the identification.

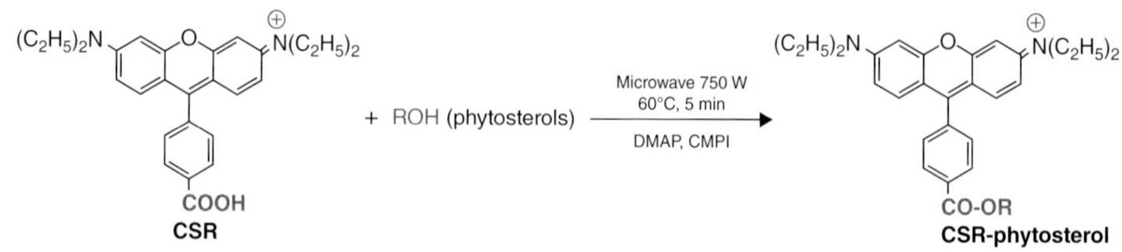

Figure 7.6 Derivatization reaction of phytosterols with 4′-carboxy-substituted rosamine (CSR). *Source:* Reproduced with kind permission of Wiley-VCH Verlag GmbH & Co. KGaA (Sun et al. 2017).

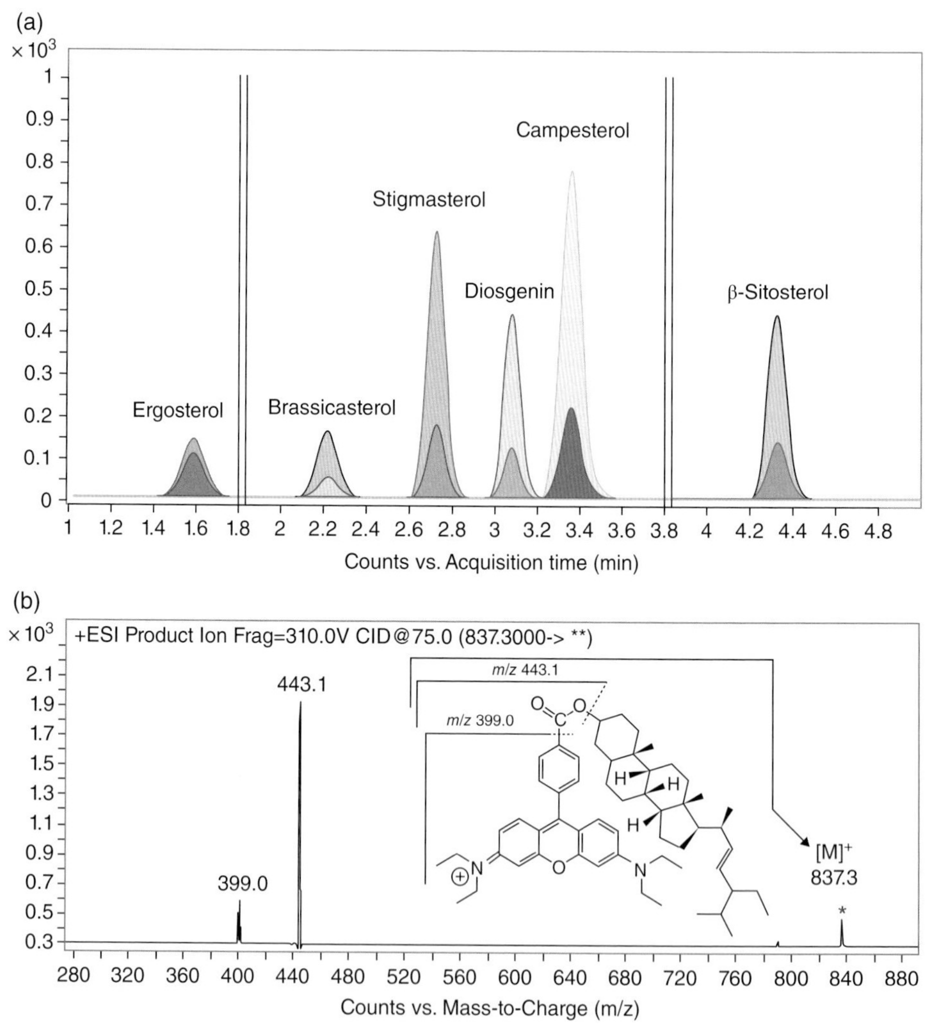

Figure 7.7 (a) Representative multiple reaction monitoring (MRM) chromatograms of 4′-carboxy-substituted rosamine (CSR) derivatives of five phytosterols and diosgenin in corn oil. (b) Product ion spectrum and proposed fragmentation pathways of a CSR-stigmasterol derivative. *Source:* Reproduced with kind permission of Wiley-VCH Verlag GmbH & Co. KGaA (Sun et al. 2017).

Delgado-Zamarreño et al. (2016) proposed an LC-MS method, which was successfully applied to the analysis of free phytosterols in almonds, cashew nuts, hazelnuts, peanuts, tiger nuts, sunflower seeds, and pistachios. Another method was reported by Yuan et al. (2017) for the simultaneous determination of tocopherols and phytosterols in edible oil. The authors determined the content of plant sterols, which ranged from 94.9 mg/100 g in camellia oil to 314.1 mg/100 g in corn oil, 164.1 mg/100 g in soybean oil, and 127.8 mg/100 g in olive oil. Ito et al. (2017b) also reported an HPLC-FLD method for determining the phytosterol contents of several plant foods. There is a large amount of literature available regarding phytosterol and its oxidation products in foods, suggesting that HPLC with different detection systems offers superior analytical tools for studying the frying of foods.

7.5 Analysis of Sensory Metabolites

Zhang et al. (2015) showed that a palatable aroma, which is mainly attributed to the volatile compounds formed during the frying process, is one of the reasons that fried food is so popular. The two major sources of volatile compounds are the low-molecular-weight TAG oxidation decomposition compounds, such as aldehydes, alcohols, ketones, hydrocarbons, and acids, and the Maillard reaction products, such as sulfur- and nitrogen-containing compounds (Martin and Ames 2001). Due to their direct influence on the quality of fried food (both positive and negative), the identification of these volatile compounds in the frying process is useful and vital (Romano et al. 2013; Aladedunye and Matthäus 2014).

Because of their volatility, the determination of volatile compounds during or at the end of the frying process is not easy. However, precisely designed devices for identifying volatile products have been reported to overcome this difficulty. Early isolation of volatiles was mainly achieved using a Likens–Nickerson apparatus modified to perform simultaneous steam distillation and extraction (SDE) with an organic solvent such as hexane or diethyl ether. As a result, the volatiles from deep-fat-fried, microwave-heated, and oven-baked garlic slices were successfully analysed (Yu et al. 1993). Briefly, samples were mixed with distilled water to be steam-distilled and were then extracted into 60 mL pentane/ether (1 : 1, v/v) solvent in a modified Likens–Nickerson apparatus for 3 hours. Before the drying treatment of the isolate with anhydrous sodium sulfate, a certain amount of 1-propyl butyrate stock solution was added as the internal standard. Then, the distillate was concentrated to a minimum volume using a spinning-band distillation apparatus. Finally, the distillate was concentrated by blowing nitrogen gas to a volume of 0.2 mL for GC-FID and GC-MS analysis. A similar sample pretreatment was also performed to monitor the volatile compounds released from used frying oil (Takeoka et al. 1996). Later, the formation of volatiles from the Maillard reaction that occurs between lysine and xylose or glucose and that is influenced by various types of vegetable oils was investigated using the same volatile extraction method (Negroni et al. 2001). This method was also utilized to effectively investigate the flavour of deep-fried shallots as influenced by various types of frying oil (Chyau and Mau 2001).

Another simultaneous gas-purging and SE apparatus for the isolation of volatiles was designed to use nitrogen in the transfer of volatiles. With the help of this tool, volatile compounds produced in heated peanut oil (Chung et al. 1994) and menhaden fish oil

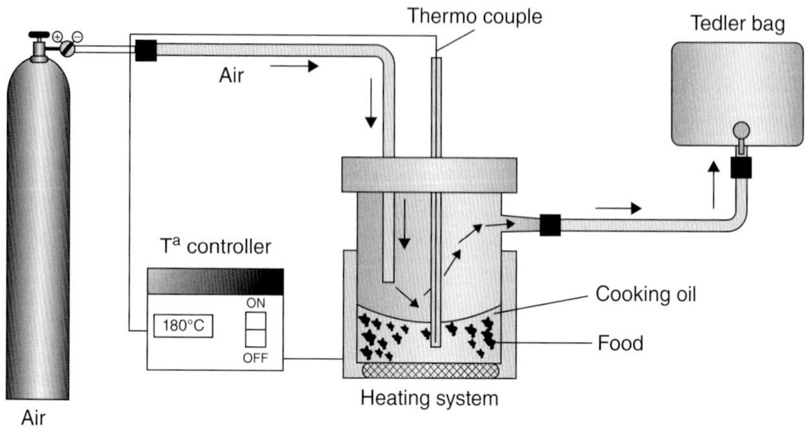

Figure 7.8 Schematic representation of an experimental system used in a frying simulation. *Source:* Reproduced with kind permission of Elsevier Ltd (Katragadda et al. 2010).

(Horiuchi et al. 1998) with different amounts of cysteine were precisely measured by GC-MS. Volatile compounds that were trapped in a Tenax trap prior to GC-MS were measured in order to study the formation of Strecker reaction aldehydes and pyrazines in a fried potato model system. The results showed that both sugars and amino acids present in the potato were crucial to the development of flavour compounds in the final product (Martin and Ames 2001).

Recently, an extraction apparatus was designed to determine the volatile aldehydes formed when coconut oil, safflower oil, canola oil, and extra virgin olive oil were heated for 6 hours at 180, 210, 240, and 270 °C, as shown in Figure 7.8 (Katragadda et al. 2010). By collecting the fumes using Tedlar bags and later analysing them by GC-MS, saturated aldehydes such as acetaldehyde, propanal, butanal, pentanal, heptanal, octanal, and nonanal and unsaturated aldehydes such as acrolein, 2-hexenal, 2-heptenal, 2-octenal, 2-decenal, 2,4-heptadienal, and 2,4-decadienal were identified in all frying oils. Some of these volatiles, such as 2-hexenal and 2,4-decadienal, endow the oily flavour of fried food; however, some of them, such as acrolein and 2,4-decadienal, are harmful to consumers. Previously, the same fume-trapping method and similar aldehyde emissions were reported when canola oil, olive oil, and extra virgin olive oil were heated at 180 and 240 °C in a closed Pyrex Instatherm reaction flask (Fullana et al. 2004).

Different volatile trapping methods, solvent extraction (SE), SDE, and nitrogen purge and steam distillation (NPSD) were tested for the collection of flavour substances from a traditional Dalmatian smoked ham that had previously undergone a dry-curing and frying treatment (Jerković et al. 2007). After GC and GC-MS analysis, it was shown that SE extracted the most isolated volatiles, due to its thorough extraction of medium-volatile compounds, which were only partially isolated by SDE. NPSD extracted the fewest, because it could only extract headspace volatiles. Volatile compounds formed in a simulated frying process can be thoroughly identified by means of these methods. However, practical food frying occurs in an open environment, in which the volatile compounds are released throughout the high-temperature treatment process. Therefore, determination of the volatile compounds produced at the end of the frying process captures only a portion of the total volatiles produced.

In addition to the foregoing static and dynamic SE, SDE, and NPSD methods, volatile compounds can also be absorbed by absorption fibres coated with special materials, such as carbowax (CW), polydimethylsiloxane (PDMS), divinyl benzene (DVB), carboxen (CAR), and polyethylene glycol. This isolation method is usually known as headspace solid-phase micro-extraction (SPME). SPME is recognized as a rapid and convenient sample preparation tool when compared with the traditional methods (Doleschall et al. 2003), and it has been extensively applied to the study of the volatiles of oils and fats and the evaluation of their stability (Gromadzka and Wardencki 2010; Petersen et al. 2012).

The odour released from palm olein deep fried with frozen, breaded chicken at 182 °C for a total of 136 hours (8 hours per day for 17 consecutive days) was investigated by the SPME-GC-MS technique (Osawa et al. 2013). Briefly, 5 g of oil was placed in a glass vial with a Teflon septum and an aluminium cap containing a magnetic stir rod. A 50/30 μm DVB/CAR/PDMS fibre with a length of 2 cm coupled to a holder was inserted into the headspace of the vial and positioned approximately 1 cm above the oil level. A total of 45 minutes was timed for the capture of the volatiles by agitation at 500 rpm. A water bath was used to maintain the samples at 80 °C. Then, the fibre was removed from the vial and immediately desorbed into the injector of the GC equipment. A similar isolation procedure was effectively used to collect the volatile compounds released during the storage of potato chips that had previously been fried in mid oleic sunflower oil (Lee and Pangloli 2013). Recently, a new sampling system based on the use of SPE-GC-MS to determine the volatile compounds generated during the heating of sunflower and extra virgin olive oils above their smoking point was developed by Ontanón et al. (2013). The relatively sealed environment and the SPE device guaranteed the comprehensiveness of the volatiles formed during the heating process. The main volatiles formed during food frying can be successfully measured using this method.

It is very important to achieve maximum extraction of volatiles in order to analyse them via GC or GC-MS. Many factors, such as the variety of fibre coating, the extraction temperature and time, the addition of salt to promote volatilization, and whether the sample is stirred can influence the final extraction result. Furthermore, the time gap between food frying and sample extraction should be considered, because it is easy to lose some volatile compounds during this interval. Therefore, conducting SPME of the oil sample as soon as possible after frying and selecting appropriate SPME conditions are crucial for the analysis of volatiles by SPME-GC-MS during or at the end of the food frying process.

The analysis of a single volatile compound or of a single class of volatiles has also been studied (Mubiru et al. 2014). Acrolein is a typical low-molecular-weight volatile aldehyde formed when oils and fats undergo frying. Derivatization with 2,4-dinitrophenylhydrazine (2,4-DNPH) before SPME treatment was conducted to study the acrolein profile in potato chips that had been fried in soybean, corn, canola, sunflower, or palm oils (Osório and De Lourdes Cardeal 2011). This sample pretreatment promoted the extraction of acrolein and improved the accuracy, sensitivity, and linearity of the SPME-GC-MS method in the range of interest. The evolution of oxygenated α,β-unsaturated aldehydes in extra virgin olive oil, sunflower oil, and virgin linseed oil heated at 190 °C for periods of 8 h day^{-1} was monitored using the SPME-GC-MS method. As a result, 4-hydroxy-(*E*)-2-nonenal, (*E*)-4,5-epoxy-(*E*)-2-decenal, 4-hydroxy-(*E*)-2-hexenal, and 4-oxo-(*E*)-2-nonenal were first detected using standards or taking into account their retention times, mass spectra, and matching rates with a commercial mass spectrum library or with mass spectra published in previous references (Guillén and Uriarte 2012).

Nitrogen-containing heterocyclic compounds are typical volatile products that result from the interactions that occur between the constituents of food materials and the components of frying oil during the food frying process. Three methods – GC-ion trap (IT)/MS, GC-time-of-flight (ToF)/MS, and GC × GC-ToF/MS – combined with SPME were compared for the critical measurement of substituted pyrazines and related substances in potato chips prepared using different frying techniques (Lojzova et al. 2009). Among these compounds, pyrazines, pyrroles, pyridines, thiazoles, thiazolines, oxazoles, oxazolines, thiophenes, and pyrrolidinones were reported. These volatiles are the typical components of the meat flavour generated through the Maillard reaction. Although some pairs of pyrazine isomers, such as 2-ethyl-5(6)-methylpyrazine (4(5)) + 2,3,5-trimethylpyrazine (6) and 2-ethyl-3,5(6)-dimethylpyrazine (8(9)) + 2,3-diethylpyrazine (10) were not fully separated from each other, the use of GC × GC-ToF/MS offered the best solution because it had the lowest quantification limits for all of the targeted alkyl pyrazines among the three methods. Accordingly, the results shown using a two-dimensional diagram were more visual and effective than those shown in a one-dimensional diagram.

The formation and emission of volatile compounds, including the aldehydes and some toxic compounds of oil samples, pure refined olive-pomace oil (ROPO), and the blended ROPO/ refined coconut oil (RCO) (80–20%), were carried out during deep frying at 180 °C (Ben Hammouda et al. 2017). The volatile profile of both oil samples was evaluated using an optimized HS-SPME-GC/MS method, before and after 20, 40, and 60 successive sessions of deep frying. The blended ROPO/RCO revealed fewer formations of unsaturated aldehydes, including toxic ones, such as acrolein, and showed a greater stability against oxidative thermal degradation compared to ROPO pure.

In addition to the GC-MS analysis, volatile aldehydes were also identified using the HPLC-ultraviolet visible (UV/VIS) technique. Non-prefried frozen sliced potatoes were fried in DAG-rich oil and TAG-rich oil with different unsaturated degrees of FA moieties in a round-bottom flask at an average temperature of 170 °C (Katsuta et al. 2008). Quantitative analysis of aldehydes was achieved through an external standard method. The same derivatization pretreatment and HPLC-DAD-MS method was successfully used to monitor the carbonyl compounds formed during the thermal oxidation of canola oil treated for different heating times and using different metal ions (Bastos and Pereira 2010). As a result, formaldehyde, acetaldehyde, acrolein, propanal, butanal, hexanal, (*E*)-2-heptenal, and octanal were identified in the emissions of the heating system. 2,4-Decadienal is a dominant volatile compound mainly formed during the peroxidation of linoleic acid and arachidonic acid. 2,4-Decadienal was extracted from the used frying oil by a methanol/water solution and redissolved in acetonitrile. The acetonitrile solution was further purified by RP-HPLC separation. Subsequently, the 2,4-decadienal was identified by GC-MS analysis (Andrikopoulos et al. 2004). The DNPH derivatization of aldehydes for HPLC analysis was shown to sensitively and precisely identify the volatile aldehydes formed during the frying process. Wang et al. (2016) reported an LC-MS-based method involving derivatization of volatiles formed in heated frying oils and French fries with 2-hydrazinoquinoline. The authors showed that aldehyde formed in French fries is comparable with that formed in frying oil.

7.6 Analysis of Heterocyclic Amines

Heterocyclic amines (HCAs) are Maillard reaction products with potential mutagenicity and carcinogenicity that are found in fried food. Generally, the analysis of HCAs includes sample pretreatments, such as extraction, enrichment, and purification; and qualitative and quantitative analyses using HPLC-based methods. SPE and other clean-up procedures have typically been adapted to improve the sensitivity and chromatographic resolution of the HPLC analysis of HCAs formed in fried meat products (Gibis and Weiss 2012; Keşkekoğlu and Üren 2014). The quantitative analysis of HCAs by SPE and HPLC coupled with photodiode-array UV and FLD was investigated to evaluate the influence of oil type on the evolution of HCAs in fried beef burgers (Johansson et al. 1995). An ODS 80TM column (250 × 4.6 mm; 5 μm) and an LC-18-DB precolumn (20 × 4.6 mm) were combined to perform the separation. Three solvents, 10 mM triethylamine with pH 3.2, 10 mM triethylamine with pH 3.6, and acetonitrile, were selected to constitute the mobile phase in linear gradient mode. Identification and quantitative analysis were achieved by comparing the spectra and a peak area of the spiked HCAs. As a result, the amounts of HCAs in the meat and the pan residue were significantly lower when sunflower oil and margarine were used as the frying media than when butter, liquid margarine, margarine fat phase, or rapeseed oil was used.

The HCAs formed in beef steak roasted with extra virgin olive oil for 5 minutes at 200 °C were quantitatively and qualitatively investigated by HPLC-ESI-MS (Lee et al. 2011). HCAs were extracted by a series of sample pretreatments, which included sample extraction with sodium hydroxide by sonicating and shaking, elution with dichloromethane/toluene (95 : 5, v/v) on a ChemElut-20 ML column, SPE with ammonium acetate on a polysulfonic acid cartridge, and SPE with methanol on a C-18 cartridge. A Zorbox RX-C-8 column and a mobile phase consisting of ammonium formate buffer (pH 3.7) and acetonitrile in gradient mode were selected to perform the chromatographic separation. By comparing the retention times of the spiked HCAs, 3-amino-1,4-dimethyl-5*H*-pyrido-[4,3-*b*]indole,3-amino-1-methyl-5*H*-pyrido[4,3-*b*]indole,9*H*-pyrido[3,4-*b*]indole,1-methyl-9*H*-pyrido[3,4-*b*]indole, 2-amino-9*H*-pyrido[2,3-*b*]indole, 2-amino-3-methyl-9*H*-pyrido[2,3-*b*]indole, 2-amino-3,8-dimethylimidazo[4,5-*f*]-quinoxaline, and 2-amino-1-methyl-6-phenylimidazo [4,5-*b*]-pyridine were detected in all the cooked beef steaks.

An alternative HPTLC technique coupled with UV/FLD was selected to quantify the HCAs formed in beef patties fried at 230 ± 1 °C after SPE treatment (Jautz et al. 2008). The chamber was initially automatically saturated with a solution of ammonia aqueous solution/ultrapure water (1 : 4, v/v) for 20 minutes. The plate activity was adjusted to 34% relative humidity with a saturated magnesium chloride aqueous solution. The chromatographic separation was performed up to a migration distance of 60 mm with a solution of methanol/chloroform (1 : 9, v/v) at ambient temperature for 30 minutes. Immersion of the layer in the solution of paraffin/n-hexane (1 : 1, v/v) was adopted to enhance the fluorescence of the analytes. Multi-wavelength scanning at 262 nm for absorbance measurements and at 313/>340 and 366/>400 nm for FLD was selected to perform the densitometry. The concentration ranges of HCAs obtained by the HPTLC- and the HPLC-based methods were similar, based on highly satisfactory correlations between the two (R^2 = 0.8875–0.9751). However, the HPTLC-based method was faster and less expensive than the HPLC-based method.

Ultra-performance LC-MS/MS was used to determine the HCA contents of roasted beef patties (Zeng et al. 2016). Another method was reported for HCA analysis in hamburger patties, involving microwave-assisted extraction of HCA and analysis using HPLC (Aeenehvand et al. 2016). It has been observed that HPLC with specific extraction procedures is the best choice for measurement of HCA in fried foods.

7.7 Analysis of Acrylamide

The quantitative analysis of acrylamide (AA) in foodstuffs or fried foods has been extensively studied and reviewed (Oracz et al. 2011; Tekkeli et al. 2012; Matthäus and Haase 2014; Hu et al. 2015). According to previous studies, GC and HPLC coupled with appropriate detectors, such as an electron-capture detector (ECD) or a DAD and MS or MS/MS, are conventionally used to monitor AA.

The characterization of AA formed in fried foods has been reported mainly in starch-based foods, such as potato chips, due to the extensive market for these products. Initial AA analysis was unsuccessful when GC-based methods in starch-based samples were used, due to the strong gelatinization effect. To overcome this obstacle, the use of GC-MS in SIM mode after the bromination of AA to form a derivative with high-mass ions after fragmentation was investigated. After a series of sample preparations, AA in chips was successfully identified and quantified by the GC-MS method with a DB-5MS capillary column and monitored ions of m/z 106 and 109 (Gertz and Klostermann 2002). A DB-23 capillary column with a programmed temperature and selected ions of m/z 74, 71, 58, and 55 was also adopted to investigate the influences on the AA concentration during the deep fat frying of potatoes. In addition, a DB-17 capillary column and selected ions of m/z 106, 108, 150, and 152 for the brominated products and 120 and 122 for the brominated methacrylamide were investigated and used to monitor the formation and amount of AA in French fries (Romani et al. 2009).

A modified sample pretreatment method was studied, consisting in the addition of ^{13}C3-labelled AA as the internal standard, defatting with hexane, extraction with a sodium chloride aqueous solution, derivatization with potassium bromate and potassium bromide, and liquid–liquid extraction with ethyl acetate. The final extract was quantitatively measured by the GC-ECD method with an HP-INNOWax capillary column (30 m × 0.32 mm × 25 μm, 100% polyethyleneglycol). After further purification of the derivatized extracts with OASIS HLB cartridges, the analytes were confirmed by GC-MS with an HP5-MS capillary column and selected ions of m/z 70, 149, and 151 for 2-bromopropenamide and m/z 110 and 154 for 2-bromo(^{13}C3)-propenamide (Zhang et al. 2006). With this method, a low limit of detection of 0.1 $\mu g\,kg^{-1}$ was estimated. According to the quantification and identification results, this method represents a robust and low-cost alternative for the conventional characterization of AA in fried food.

In addition to GC analysis, HPLC-UV and HPLC-MS/MS methods have been conventionally used to monitor AA formed in fried food (Wang et al. 2013; Lim et al. 2014). As with the preparation procedures already described, the general pretreatment process for AA analysis in fried food via HPLC includes defatting, centrifugation, ultrafiltration, and further purification by SPE columns or cartridges. To investigate the influence of oil type on the concentration of AA formed in the model system, as well as in French fries, two SPE columns – an Oasis HLB column and a Bond Elut-Accucat column – were successively used to perform a complete extraction (Mestdagh et al. 2005). As a result, no significant difference in the formation of AA was found among the nine lipids investigated, although the homemade potato powder mixture containing vegetable oil may significantly have influenced the AA concentration. By using this method, degradation products (such as glycerol, MAGs, and DAGs) of soybean oil, corn oil, soybean fat, and commercial palm fat mixtures (80% palm oil and 20% palm stearin) were confirmed to have no significant influence on the formation of AA during either the frying of French fries or the heating of an artificial mixture (Mestdagh et al. 2007). In addition, the colour of the fried potato strips, represented by L^* and a^*, was also closely correlated with the concentration of AA (Pedreschi et al. 2006). The effect of frying time and the shape of the potato sample on the formation of AA in deep-fat-fried potato cubes was also investigated using this joint HPLC-MS/MS method (Carrieri et al. 2010). The detection of AA by MS/MS can alternatively be achieved by UV detection at 210 and 225 nm (Wang et al. 2008) and DAD at 226 nm with continuous monitoring of the peak spectra within the range 190–350 nm (Gokmen et al. 2005).

Both GC- and HPLC-based methods can be utilized, as they gave satisfactory qualitative and quantitative analyses of AA with high sensitivity in a food matrix. However, no derivatization treatment is needed in the HPLC-based method, which makes the sample preparation for AA analysis easier (Zhang et al. 2015).

7.8 Analysis of Tocopherols

Tocopherols and tocotrienols are important antioxidants that are found both in frying medium and in foods. Saini and Keum (2016) reviewed some methods for the extraction, chromatographic separation, and detection of tocopherols in plants or their products. Supercritical fluid extraction is an emerging technique for the extraction of tocols, while RP-HPLC and MS in negative mode are sensitive tools. Extraction may be required in fried samples, but not in frying oils. Several GC methods have been reported for the determination of tocopherols in edible oils or foods (Schwartz et al. 2008; Mitei et al. 2009). However, due to their stable and nonpolar nature, HPLC is a better choice (Table 7.10). Only two NP-HPLC methods have been reported since 2011 (Cruz and Casal 2013; Górnaś et al. 2014). The solvent in this method was 1,4-dioxane and n-hexane, while FLD and DAD were used as detectors for the analysis of tocopherols in green vegetables and butter.

RP-HPLC is a commonly applied technique for the separation of different tocopherols and tocotrienols in food samples. The most widely used detector in HPLC was found to be FLD for marine and freshwater fishes (Özogul et al. 2011), grape seeds (Tangolar et al. 2011), cereals (Irakli et al. 2012), butter samples (Górnaś et al. 2014, 2015), vegetables (Knecht et al. 2015), rice (Shammugasamy et al. 2015), vegetable oils (Bakre et al. 2015), animal feed and food (Claeys et al. 2016), and other foods and fruits (Figueira et al. 2017; Sunarić et al. 2017). A DAD or UV detector (Cruz and Casal 2013; Gong et al. 2014; Knecht et al. 2015; Shammugasamy et al. 2015; Claeys et al. 2016; Sunarić et al. 2017) and an ESI-MS detector (Ansolin et al. 2017) have also been successfully used. These showed that acetonitrile and methanol were the major solvents used during separation on C-18 columns. Irakli et al. (2012) demonstrated that column temperatures affect the elution and separation of tocopherols, as shown in Figure 7.9.

Table 7.10 Analytical chromatographic methods for the determination of vitamin E in frying medium and foods.

Sample	Mode of separation	Detection system	Solvent system	Column	References
Marine and freshwater fish Species	RP	FLD	Isocratic: ACN-methanol (50 : 50)	Develosil RPAQUEOUS (150 × 4.6 mm, 5 μm)	Özogul et al. (2011)
Grape	RP	FLD	Gradient: ACN and methanol	Develosil RPAQUEOUS (150 × 4.6 mm, 5 μm)	Tangolar et al. (2011)
Cereals	RP	FLD	Gradient: 2-propanol and water	PerfectSil Target ODS-3 (250 × 4.6 mm, 3 μm)	Irakli et al. (2012)
Green vegetables	NP	FLD and DAD	Isocratic: 1,4-dioxane in n-hexane (2.5%, v/v)	Supelcosil LC-SI (7.5 cm × 3 mm, 3 μm)	Cruz and Casal (2013)
Tropical fruits	UPC	DAD	Gradient: CO_2 and methanol	ACQUITY UPC2 BEH (100 × 3.0 mm, 1.7 μm)	Gong et al. (2014)
Butter and seed samples	RP	FLD	Isocratic: methanol-water (93 : 7, v/v)	Luna PFP (150 × 4.6 mm, 3 μm)	Górnaś et al. (2014, 2015)
Butter samples	NP	FLD	Isocratic: *n*-hexane with 1,4-dioxane (96 : 4, v/v)	LiChrosorb Si60 (4.6 × 250 mm, 5 μm)	Górnaś et al. (2014)
Vegetables	RP	FLD and DAD	Gradient: methanol-water (85 : 15) and *tert*-methylbutyl- ether-methanol-water (80 : 18 : 2)	Develosil RP aqueous C-30 (150 × 3 mm, 3 μm) and Kinetex PFP (150 × 3 mm, 2.6 μm)	Knecht et al. (2015)
Rice	RP	DAD and FLD	Gradient: water and methanol	Kinetex PFP (250 × 4.6 mm, 5 μm)	Shammugasamy et al. (2015)
Sunflower and olive oils	RP	FLD	Gradient: ACN and methanol	C-18 (250 × 4.6 mm, 5 μm)	Bakre et al. (2015)

Animal feed and food	RP	UV	Gradient: methanol-water solution (94 : 6, v/v) and methanol	Supelcosil LC-18 (25 cm × 4.6 mm, 5 μm)	Claeys et al. (2016)
Foods	RP	FLD and UV	Isocratic: ACN 100%	Restek Ultra IBD (150 × 3 mm, 3 μm)	Sunarić et al. (2017)
Vegetable oils	RP	ESI-MS	Gradient: methanol-water-ammonium hydroxide (99 : 1 : 0.1, v/v/v) and 2-propanol	Acquity UPLC BEH C-18 (100 × 2.1 mm, 1.7 μm)	Ansolin et al. (2017)
Tomatoes	RP	FLD	Gradient: ACN and methanol	Acquity UPLC BEH C-18 (50 × 2.1 mm, 1.7 μm)	Figueira et al. (2017)

RP, reversed phase; NP, normal phase; HPLC, high-performance liquid chromatography; UPLC, ultra-pressure liquid chromatography; DAD, diode array detector; ESI, electrospray ionization; MS, mass spectrometry; UV, ultraviolet; FLD, fluorescence detector; ACN, acetonitrile.

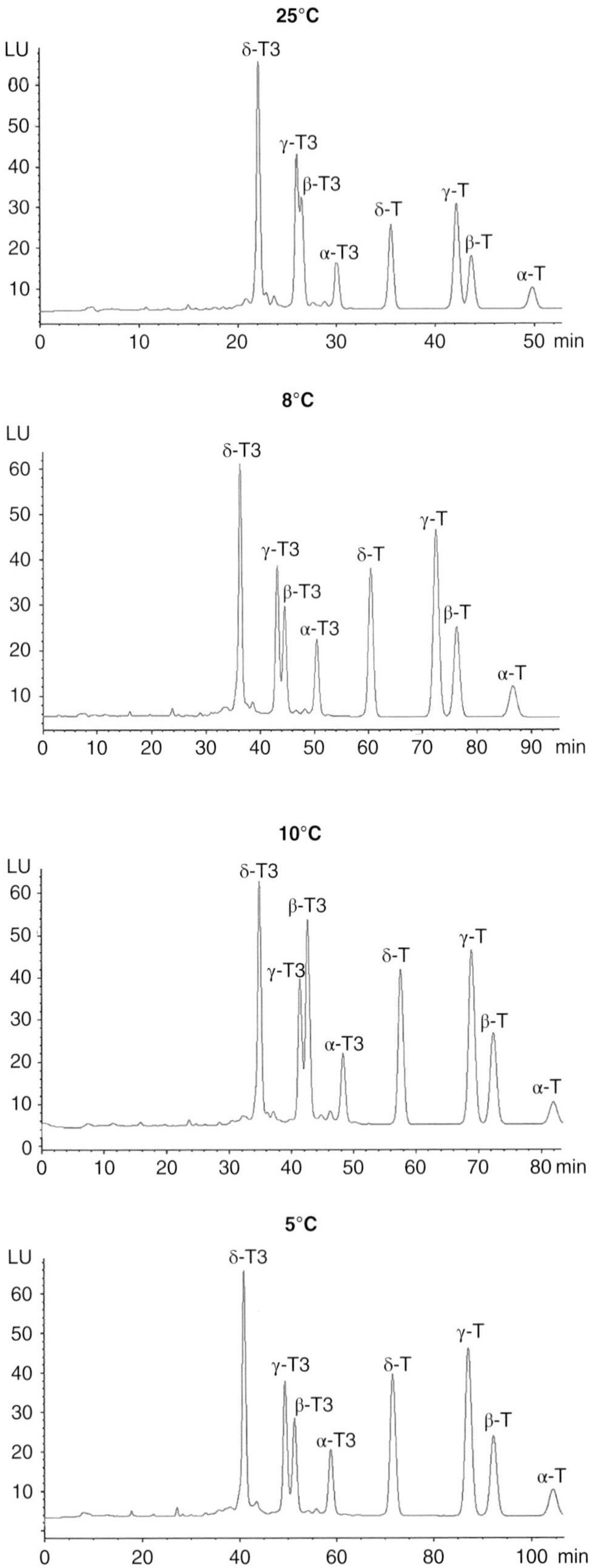

Figure 7.9 HPLC separation of T and T3 isomers at different temperatures with PerfectSil Target stationary phase. Mobile phase, isopropanol-water (75 : 25, v/v); flow rate, 0.3 mL min^{-1}. *Source:* Reproduced with kind permission of American Chemical Society (Irakli et al. 2012).

Tocopherol has been identified simultaneously with carotenoids or with phenolic compounds. In green vegetables, tocopherols were extracted with carotenoids and chlorophylls and separated using a C-18 column and gradient elution of methanol, water, and tertiary methyl butyl ether (MTBE) (Zeb 2017). Similarly, tocopherols have been identified and quantified simultaneously with carotenoids and free and esterified sterols in canola oils (Flakelar et al. 2017) using NP-HPLC. These results show that HPLC with FLD is the best choice for the analysis of tocopherols during frying.

7.9 Analysis of Polyphenolic Compounds

Phenolic compounds are found in a variety of plant foods. The most widely used means of determining total polyphenol contents is a spectrophotometric method using Folin Ciocaltaeu reagent. The modes of analysis, extraction, and detection have been reviewed by Ignat et al. (2011). Details regarding polyphenols are usually obtained via separation by HPLC and detection with a DAD detector. The RP-HPLC-DAD method was developed for the determination of polyphenolic compounds in plant leaves (Zeb 2015b) and was successfully used for the determination of changes in polyphenolic compounds during frying of chapli kebab (Zeb and Haq 2017). For phenolic compounds in the presence or absence of an authentic standard, the use of MS with chromatography is warranted. Oueslati et al. (2017) reported a method using HPLC-DAD-ESI-TOF-MS for the determination of changes in phenolic compounds during heating or microwave heating. These studies show that phenolic compounds and their variations can be accurately assessed with the help of LC and MS detectors.

7.10 Analysis of Other Minor Compounds

In addition to the previously described GC and HPLC analyses of the common products formed during oil heating and food frying, products present in low concentrations have also been reported. These products, such as α-tocopherol oxides and polycyclic aromatic hydrocarbons (PAHs), directly or indirectly influence the final quality of the frying oil and fried food. Those compounds, such as *trans* fatty acids (TFAs), that do not have a significant effect on the final food will not be discussed here. By using specialized sample pretreatment and an appropriate detector, the HPLC-based method is commonly used to conduct qualitative and quantitative analyses of minor compounds of interest.

The typical variation of lipid microcomponents involves changes in antioxidants during the frying process (Hwang et al. 2013; Aladedunye et al. 2017). The frying process causes the number of antioxidants, and thus the corresponding antioxidative activity, to decrease, which reduces the ability of the system to prevent the deterioration of oil and to inhibit the formation of products with potential health risks. These changes can be attributed to reactions involving the antioxidants and to the loss of the antioxidative configurations during the frying process. Oxidation can also take place in natural antioxidants. The oxidation products of α-tocopherol formed in a model triolein system with 0.35% α-tocopherol heated at 240 °C for 90 minutes were analysed by

NP-HPLC-FLD with a silica column and an isocratic mobile phase consisting of 99.5% hexane and 0.5% isopropyl alcohol at 1.5 mL min^{-1}. The excitation and emission wavelengths of FLD were set at 230 and 292 nm, respectively. The quantification and identification of the oxidative products were achieved with the help of pure standards and the aforementioned synthetic α-tocopherol oxidation products. As a result, α-tocopherol quinone, 4a,5-epoxy-α-tocopherol quinone, and 7,8-epoxy-α-tocopherol quinone were confirmed to be present during the heating process (Verleyen et al. 2001). In addition, α-tocopherol quinone (1.4–7.7%) and epoxy-α-tocopherol quinone (4.3–34.8%) were quantitatively monitored in the model system containing a triolein and tripalmitin mixture with 0.1% α-tocopherol heated at 150–250 °C over a certain heating period using this method. Moreover, the amounts of the oxidation products of α-tocopherol were changeable during the heating process, and the fate of the α-tocopherol itself depended on the heating time and temperature.

As a class of toxicants, PAHs are ubiquitous in the environment and in human bodies (Ramesh et al. 2004). A number of analyses of PAHs during the heating of vegetable oils or the frying process have been reported in recent years (Sjaastad et al. 2010). As with the analysis method used for HCAs, alkaline hydrolysis, clean-up or purification by SPE or column chromatography, and HPLC-UV/FLD are usually involved in the preparation and analysis of PAHs formed in fried meat (Rivera et al. 1996). A purified extract was separated by a Hypersil Green PAH column (250 × 4.6 mm; 5 μm) connected with a guard column (10 × 4 mm; 5 μm) and an isocratic mobile phase (84% acetonitrile and 16% water). Through the addition of PAH standards, and using the external calibration curve method, benzo[b]fluoranthene, benzo[k]fluoranthene, benzo[a]anthracene, benzo[a]pyrene, dibenzo[a,h]anthracene, and benzo[g,h,i]perylene were quantitatively characterized in fried pork meat and gravy (Janoszka 2011). Naphthalene, acenaphthylene, fluorine + acenapthene, phenanthrene, anthracene, fluoranthene, and pyrene, benzo(*a*)anthracene + chrysene, benzo(*k*)fluoranthene, benzo-(*b*)fluoranthene, B(*a*)P, dibenzo(*a,h*)anthracene, indino(1,2,3-*cd*)pyrene, and benzo(*g,h,i*)perylene were also quantitatively analysed in repeatedly boiled sunflower oil using this method (Srivastava et al. 2010). In addition, the concentrations and particle size distributions of 19 major PAHs in the oil mist emitted by grilled pork, trout, beef, prawns, and corn were also studied by GC-MS with a special HAC collection apparatus (Saito et al. 2014).

Other compounds, such as 3-monochloropropane-1,2-diol (3-MCPD) and 2-monochloropropane-1,3-diol (2-MCPD), are also present as contaminants in foods. Xu et al. (2013) reported a GC-MS method for the determination of 3-MCPD. Similarly, Karl et al. (2016) developed GC-MS method for the determination of 3-MCPD in fishes. The 3-MCPD was found to be affected by salt concentration during frying (Wong et al. 2017).

7.11 Key Concepts

- Several analytical methods for TAG determination have been reported recently, which show that RP-HPLC provides high sensitivity.
- Oxidized TAGs have been studied using normal- and reversed-phase TLC and HPLC, while limited studies are available using GC-MS.
- The best technique for the analysis of oxidized TAGs is RP-LC coupled with MS.

- FA oxidation products can be determined with GC-MS, which has superior sensitivity.
- Better extraction methods such as SPE and derivatization can give more sensitive results for oxygenated FAs.
- Cholesterol and phytosterols have largely been analysed using GC-based techniques; HPLC coupled to MS is also a good option.
- Secondary metabolites such as volatile products are usually collected and isolated using specific extraction techniques and analysed using GC.
- GC coupled to a variety of detectors, including MS, provides highly sensitive and accurate results.
- Heterocyclic aromatic amines (HAAs) have been analysed using RP-HPLC with DAD or MS detection.
- GC- and HPLC-based methods can be utilized, as they give satisfactory qualitative and quantitative analyses of AA with high sensitivity in a food matrix.
- Tocopherols have been determined using HPLC with FLD detection, which provides superior sensitivity.
- Polyphenols can be analysed using RP-HPLC, which gives better sensitivity on MS detection.
- Several minor compounds can be identified and quantified using GC- or HPLC-based techniques.

References

Abdallah, M., Vergara-Barberán, M., Lerma-García, M.J. et al. (2016). Use of triacylglycerol profiles established by HPLC-UV and ELSD to predict cultivar and maturity of Tunisian olive oils. *European Food Research and Technology* 242: 1607–1619.

Abidi, S.L. (2001). Chromatographic analysis of plant sterols in foods and vegetable oils. *Journal of Chromatography A* 935: 173–201.

Abreu, F.F., Souza, A.C., Teixeira, S.A. et al. (2016). Elucidating the role of oxidative stress in the therapeutic effect of rutin on experimental acute pancreatitis. *Free Radical Research* 50: 1350–1360.

Adewuyi, A. and Oderinde, R.A. (2012). Analysis of the lipids and molecular speciation of the triacylglycerol of the oils of Luffa cylindrica and Adenopus breviflorus. *CyTA Journal of Food* 10: 313–320.

Aeenehvand, S., Toudehrousta, Z., Kamankesh, M. et al. (2016). Evaluation and application of microwave-assisted extraction and dispersive liquid–liquid microextraction followed by high-performance liquid chromatography for the determination of polar heterocyclic aromatic amines in hamburger patties. *Food Chemistry* 190: 429–435.

Akasaka, K. and Ohrui, H. (2000). Development of phosphine reagents for the high-performance liquid chromatographic-fluorometric determination of lipid hydroperoxides. *Journal of Chromatography A* 881: 159–170.

Akasaka, K., Ijichi, S., Watanabe, K. et al. (1992). High-performance liquid chromatography and post-column derivatization with diphenyl-1-pyrenylphosphine for fluorimetric determination of triacylglycerol hydroperoxides. *Journal of Chromatography* 596: 197–202.

Akasaka, K., Ohrui, H., Meguro, H., and Tamura, M. (1993). Determination of triacylglycerol and cholesterol ester hydroperoxides in human plasma by high-performance liquid chromatography with fluorometric postcolumn detection. *Journal of Chromatography* 617: 205–211.

Aladedunye, F. and Matthäus, B. (2014). Phenolic extracts from *Sorbus aucuparia* (L.) and *Malus baccata* (L.) berries: antioxidant activity and performance in rapeseed oil during frying and storage. *Food Chemistry* 159: 273–281.

Aladedunye, F., Przybylski, R., and Matthaus, B. (2017). Performance of antioxidative compounds under frying conditions: a review. *Critical Reviews in Food Science and Nutrition* 57: 1539–1561.

Andrikopoulos, N.K., Chiou, A., Mylona, A. et al. (2004). Monitoring of 2,4-decadienal in oils and fats used for frying in restaurants in Athens, Greece. *European Journal of Lipid Science and Technology* 106: 671–679.

Ansolin, M., de Souza, P.T., de Almeida Meirelles, A.J., and Batista, E.A.C. (2017). Tocopherols and tocotrienols: an adapted methodology by UHPLC/MS without sample pretreatment steps. *Food Analytical Methods* 10: 2165–2174.

Araujo, F.B., Barbosa, D.S., Hsin, C.Y. et al. (1995). Evaluation of oxidative stress in patients with hyperlipidemia. *Atherosclerosis* 117: 61–71.

Armstrong, D. (2009). *Lipidomics: Methods and Protocols*. New York, NY: Humana Press.

Baiocchi, C., Medana, C., Dal Bello, F. et al. (2015). Analysis of regioisomers of polyunsaturated triacylglycerols in marine matrices by HPLC/HRMS. *Food Chemistry* 166: 551–560.

Bakre, S.M., Gadmale, D.K., Toche, R.B., and Gaikwad, V.B. (2015). Rapid determination of alpha tocopherol in olive oil adulterated with sunflower oil by reversed phase high-performance liquid chromatography. *Journal of Food Science and Technology* 52: 3093–3098.

Bansal, G., Zhou, W., Barlow, P.J. et al. (2010). Review of rapid tests available for measuring the quality changes in frying oils and comparison with standard methods. *Critical Reviews in Food Science and Nutrition* 50: 503–514.

Barriuso, B., Otaegui-Arrazola, A., Menéndez-Carreño, M. et al. (2012). Sterols heating: degradation and formation of their ring-structure polar oxidation products. *Food Chemistry* 135: 706–712.

Bastos, L.C.S. and Pereira, P.A.D.P. (2010). Influence of heating time and metal ions on the amount of free fatty acids and formation rates of selected carbonyl compounds during the thermal oxidation of canola oil. *Journal of Agricultural and Food Chemistry* 58: 12777–12783.

Bauer-Plank, C. and Steenhorst-Slikkerveer, L. (2000). Analysis of triacylglyceride hydroperoxides in vegetable oils by nonaqueous reversed-phase high-performance liquid chromatography with ultraviolet detection. *Journal of the American Oil Chemists' Society* 77: 477–482.

Beccaria, M., Sullini, G., Cacciola, F. et al. (2014). High performance characterization of triacylglycerols in milk and milk-related samples by liquid chromatography and mass spectrometry. *Journal of Chromatography A* 1360: 172–187.

Beccaria, M., Costa, R., Sullini, G. et al. (2015). Determination of the triacylglycerol fraction in fish oil by comprehensive liquid chromatography techniques with the support of gas chromatography and mass spectrometry data. *Analytical and Bioanalytical Chemistry* 407: 5211–5225.

Beermann, C., Winterling, N., Green, A. et al. (2007). Comparison of the structures of triacylglycerols from native and transgenic medium-chain fatty acid-enriched rape seed oil by liquid chromatography--atmospheric pressure chemical ionization ion-trap mass spectrometry (LC-APCI-ITMS). *Lipids* 42: 383–394.

Ben Arfa, K., de Person, M., Hmida, D. et al. (2017). UHPLC-APCI-MS profiling of triacylglycerols in vegetable oils – application to the analysis of four North African sesame seed varieties. *Food Analytical Methods* 10: 2827–2838.

Ben Hammouda, I., Freitas, F., Ammar, S. et al. (2017). Comparison and characterization of volatile compounds as markers of oils stability during frying by HS–SPME-GC/MS and chemometric analysis. *Journal of Chromatography B: Analytical Technologies in the Biomedical and Life Sciences* 1068–1069: 322–334.

Berdeaux, O., Márquez-Ruiz, G., and Dobarganes, M. (1999). Characterization, quantitation and evolution of monoepoxy compounds formed in model systems of fatty acid methyl esters and monoacid triglycerides heated at high temperature. *Grasas y Aceites* 50: 53–59.

Bhuiyan, M., Tucker, D., and Watson, K. (2013). Determination and differentiation of triacylglycerol molecular species in Antarctic and non-Antarctic yeasts by atmospheric pressure-chemical ionization-mass spectrometry. *Journal of Microbiological Methods* 94: 249–256.

Biedermann-Brem, S., Biedermann, M., Fankhauser-Noti, A. et al. (2007). Determination of epoxidized soybean oil (ESBO) in oily foods by GC-FID or GC-MS analysis of the methyl diepoxy linolate. *European Food Research and Technology* 224: 309–314.

Blanksby, S.J. and Mitchell, T.W. (2010). Advances in mass spectrometry for lipidomics. *Annual Review of Analytical Chemistry* 3: 433–465.

Bohac, C.E., Rhee, K.S., Cross, H.R., and Ono, K. (1988). Assessment of methodologies for calorimetric cholesterol assay of meats. *Journal of Food Science* 53: 1642–1644.

Bou Khalil, M., Hou, W., Zhou, H. et al. (2010). Lipidomics era: accomplishments and challenges. *Mass Spectrometry Reviews* 29: 877–929.

Bragagnolo, N. (2009). Cholesterol and cholesterol oxides in meat and meat products. In: *Handbook of Muscle Food Analysis* (ed. L.M.L. Nollet and F. Toldra). Boca Raton, FL: CRC Press.

Brühl, L., Weisshaar, R., and Matthäus, B. (2016). Epoxy fatty acids in used frying fats and oils, edible oils and chocolate and their formation in oils during heating. *European Journal of Lipid Science and Technology* 118: 425–434.

Byrdwell, W.C. and Neff, W.E. (1999). Non-volatile products of triolein produced at frying temperatures characterized using liquid chromatography with online mass spectrometric detection. *Journal of Chromatography A* 852: 417–432.

Byrdwell, W.C. and Neff, W.E. (2001). Autoxidation products of normal and genetically modified canola oil varieties determined using liquid chromatography with mass spectrometric detection. *Journal of Chromatography A* 905: 85–102.

Byrdwell, W.C. and Neff, W.E. (2002). Dual parallel electrospray ionization and atmospheric pressure chemical ionization mass spectrometry (MS), MS/MS and MS/MS/MS for the analysis of triacylglycerols and triacylglycerol oxidation products. *Rapid Communications in Mass Spectrometry* 16: 300–319.

Byrdwell, W.C. and Neff, W.E. (2004). Electrospray ionization MS of high M.W. TAG oligomers. *Journal of the American Oil Chemists' Society* 81: 13–26.

Caldironi, H.A. and Manes, M.E. (2006). Proximate composition, fatty acids and cholesterol content of meat cuts from tegu lizard *Tupinambis merianae*. *Journal of Food Composition and Analysis* 19: 711–714.

Caldwell, J.D., Cooke, B.S., and Greer, M.K. (2011). High performance liquid chromatography–size exclusion chromatography for rapid analysis of total polar compounds in used frying oils. *Journal of the American Oil Chemists' Society* 88: 1669–1674.

Cardenia, V., Rodriguez-Estrada, M.T., Baldacci, E. et al. (2012). Analysis of cholesterol oxidation products by fast gas chromatography/mass spectrometry. *Journal of Separation Science* 35: 424–430.

Carrieri, G., Anese, M., Quarta, B. et al. (2010). Evaluation of acrylamide formation in potatoes during deep-frying: the effect of operation and configuration. *Journal of Food Engineering* 98: 141–149.

Chao, P., Chao, C., Lin, F., and Huang, C. (2001). Oxidized frying oil up-regulates hepatic acyl-CoA oxidase and cytochrome P450 4 A1 genes in rats and activates PPARα. *Journal of Nutrition* 131: 3166–3174.

Cheung, M., Young, A.B., and Harrison, A.G. (1994). O(−) and OH(−) chemical ionization of some fatty acid methyl esters and triacylglycerols. *Journal of the American Society for Mass Spectrometry* 5: 553–557.

Choi, J., Zhang, W., Gu, X. et al. (2011). Lysophosphatidylcholine is generated by spontaneous deacylation of oxidized phospholipids. *Chemical Research in Toxicology* 24: 111–118.

Christensen, T.C. and Holmer, G. (1996). Lipid oxidation determination in butter and dairy spreads by HPLC. *Journal of Food Science* 61: 486–489.

Chung, T.Y., Eiserich, J.P., and Shibamoto, T. (1994). Volatile compounds produced from peanut oil heated with different amounts of cysteine. *Journal of Agricultural and Food Chemistry* 42: 1743–1746.

Chyau, C.-C. and Mau, J.-L. (2001). Effects of various oils on volatile compounds of deep-fried shallot flavouring. *Food Chemistry* 74: 41–46.

Claessens, H.A. and van Straten, M.A. (2004). Review on the chemical and thermal stability of stationary phases for reversed-phase liquid chromatography. *Journal of Chromatography A* 1060: 23–41.

Claeys, E., Vossen, E., and De Smet, S. (2016). Determination of α-tocopherol by reversed-phase HPLC in feed and animal-derived foods without saponification. *Journal of the Science of Food and Agriculture* 96: 522–529.

Clark, P.K. and Snyder, H.E. (1991). Hydroperoxide formation in soybean seeds during storage. *Journal of the American Oil Chemists Society* 68: 346–347.

Constantinou, M.S., Georgiou, C.A., and Kapnissi-Christodoulou, C.P. (2015). Development of a reliable analytical protocol for the isolation of cholesterol oxidation products – a comparison of different lipid extraction and saponification methods. *Food Analytical Methods* 8: 1499–1507.

Correia, A.C., Dubreucq, E., Ferreira-Dias, S., and Lecomte, J. (2015). Rapid quantification of polar compounds in thermo-oxidized oils by HPTLC-densitometry. *European Journal of Lipid Science and Technology* 117: 311–319.

Costa, P., Roseiro, L.C., Partidário, A. et al. (2006). Influence of slaughter season and sex on fatty acid composition, cholesterol and α-tocopherol contents on different muscles of Barrosã-PDO veal. *Meat Science* 72: 130–139.

Cruz, R. and Casal, S. (2013). Validation of a fast and accurate chromatographic method for detailed quantification of vitamin E in green leafy vegetables. *Food Chemistry* 141: 1175–1180.

Daneshfar, A., Khezeli, T., and Lotfi, H.J. (2009). Determination of cholesterol in food samples using dispersive liquid–liquid microextraction followed by HPLC–UV. *Journal of Chromatography, B: Analytical Technologies in the Biomedical and Life Sciences* 877: 456–460.

Davis, T.A., Gao, L., Yin, H. et al. (2006). In vivo and in vitro lipid peroxidation of arachidonate esters: the effect of fish oil omega-3 lipids on product distribution. *Journal of the American Chemical Society* 128: 14897–14904.

Delgado-Zamarreño, M.M., Fernández-Prieto, C., Bustamante-Rangel, M., and Pérez-Martín, L. (2016). Determination of tocopherols and sitosterols in seeds and nuts by QuEChERS-liquid chromatography. *Food Chemistry* 192: 825–830.

Dinh, T.T.N., Thompson, L.D., Galyean, M.L. et al. (2011). Cholesterol content and methods for cholesterol determination in meat and poultry. *Comprehensive Reviews in Food Science and Food Safety* 10: 269–289.

Dobarganes, C. and Márquez-Ruiz, G. (2003). Oxidized fats in foods. *Current Opinion in Clinical Nutrition and Metabolic Care* 6: 157–163.

Dobarganes, M., Velasco, J., and Dieffenbacher, A. (2000). Determination of polar compounds, polymerized and oxidized triacylglycerols, and diacylglycerols in oils and fats: results of collaborative studies and the standardized method (technical report). *Pure and Applied Chemistry* 72: 1563–1575.

Doleschall, F., Recseg, K., Kemény, Z., and Kővári, K. (2003). Comparison of differently coated SPME fibres applied for monitoring volatile substances in vegetable oils. *European Journal of Lipid Science and Technology* 105: 333–338.

Dorsey, J.G. and Dill, K.A. (1989). The molecular mechanism of retention in reversed-phase liquid chromatography. *Chemical Reviews* 89: 331–346.

Dugo, P., Kumm, T., Lo Presti, M. et al. (2005). Determination of triacylglycerols in donkey milk by using high performance liquid chromatography coupled with atmospheric pressure chemical ionization mass spectrometry. *Journal of Separation Science* 28: 1023–1030.

Dugo, P., Beccaria, M., Fawzy, N. et al. (2012). Mass spectrometric elucidation of triacylglycerol content of *Brevoortia tyrannus* (menhaden) oil using non-aqueous reversed-phase liquid chromatography under ultra high pressure conditions. *Journal of Chromatography A* 1259: 227–236.

Duong, S., Strobel, N., Buddhadasa, S. et al. (2016). Rapid measurement of phytosterols in fortified food using gas chromatography with flame ionization detection. *Food Chemistry* 211: 570–576.

Ekroos, K., Janis, M., Tarasov, K. et al. (2010). Lipidomics: a tool for studies of atherosclerosis. *Current Atherosclerosis Reports* 12: 273–281.

Endo, Y., Ohta, A., Kido, H. et al. (2011). Determination of triacylglycerol composition in vegetable oils using high-performance liquid chromatography: a collaborative study. *Journal of Oleo Science* 60: 451–456.

Fang, L., Harkewicz, R., Hartvigsen, K. et al. (2010). Oxidized cholesteryl esters and phospholipids in zebrafish larvae fed a high cholesterol diet: macrophage binding and activation. *Journal of Biological Chemistry* 285: 32343–32351.

Fankhauser-Noti, A., Fiselier, K., Biedermann-Brem, S., and Grob, K. (2005). Epoxidized soy bean oil migrating from the gaskets of lids into food packed in glass jars - Analysis by on-line liquid chromatography-gas chromatography. *Journal of Chromatography A* 1082: 214–219.

Fankhauser-Noti, A., Fiselier, K., Biedermann-Brem, S., and Grob, K. (2006). Assessment of epoxidized soy bean oil (ESBO) migrating into foods: comparison with ESBO-like epoxy fatty acids in our normal diet. *Food and Chemical Toxicology* 44: 1279–1286.

Fasciotti, M. and Pereira Netto, A.D. (2010). Optimization and application of methods of triacylglycerol evaluation for characterization of olive oil adulteration by soybean oil with HPLC-APCI-MS-MS. *Talanta* 81: 1116–1125.

Fenton, M. (1992). Chromatographic separation of cholesterol in foods. *Journal of Chromatography A* 624: 369–388.

Figueira, J.A., Pereira, J.A.M., and Câmara, J.S. (2017). Quantification of δ-, γ- and α-tocopherol in tomatoes using an improved liquid-dispersive solid-phase extraction combined with ultrahigh pressure liquid chromatography. *Food Analytical Methods* 10: 2507–2517.

Flakelar, C.L., Prenzler, P.D., Luckett, D.J. et al. (2017). A rapid method for the simultaneous quantification of the major tocopherols, carotenoids, free and esterified sterols in canola (Brassica napus) oil using normal phase liquid chromatography. *Food Chemistry* 214: 147–155.

Frankel, E.N. (1984). Chemistry of free radical and singlet oxidation of lipids. *Progress in Lipid Research* 23: 197–221.

Frankel, E.N., Neff, W.E., and Miyashita, K. (1990). Autoxidation of polyunsaturated triacylglycerols. II. Trilinolenoylglycerol. *Lipids* 25: 40–47.

Frankel, E.N., Selke, E., Neff, W.E., and Miyashita, K. (1992). Autoxidation of polyunsaturated triacylglycerols. IV. Volatile decomposition products from triacylglycerols containing linoleate and linolenate. *Lipids* 27: 442–446.

Fullana, A., Carbonell-Barrachina, Á.A., and Sidhu, S. (2004). Volatile aldehyde emissions from heated cooking oils. *Journal of the Science of Food and Agriculture* 84: 2015–2021.

Gao, B., Luo, Y., Lu, W. et al. (2017). Triacylglycerol compositions of sunflower, corn and soybean oils examined with supercritical CO2 ultra-performance convergence chromatography combined with quadrupole time-of-flight mass spectrometry. *Food Chemistry* 218: 569–574.

Georgiou, C.A., Constantinou, M.S., and Kapnissi-Christodoulou, C.P. (2014). Sample preparation: a critical step in the analysis of cholesterol oxidation products. *Food Chemistry* 145: 918–926.

Georgiou, C.A., Constantinou, M.S., Andreou, R. et al. (2016). Novel approach to fast determination of cholesterol oxidation products in Cypriot foodstuffs using ultra-performance liquid chromatography-tandem mass spectrometry. *Electrophoresis* 37: 1101–1108.

Gertz, C. and Klostermann, S. (2002). Analysis of acrylamide and mechanisms of its formation in deep-fried products. *European Journal of Lipid Science and Technology* 104: 762–771.

Gibis, M. and Weiss, J. (2012). Antioxidant capacity and inhibitory effect of grape seed and rosemary extract in marinades on the formation of heterocyclic amines in fried beef patties. *Food Chemistry* 134: 766–774.

Giuffrida, F., Destaillats, F., Skibsted, L.H., and Dionisi, F. (2004). Structural analysis of hydroperoxy- and epoxy-triacylglycerols by liquid chromatography mass spectrometry. *Chemistry and Physics of Lipids* 131: 41–49.

Gladovič, N., Zupančič-Kralj, L., and Plavec, J. (1997). Determination of primary oxidation products of linoleic acid and triacylglycerols. *Journal of Chromatography A* 767: 63–68.

Goicoechea, E. and Guillen, M.D. (2010). Analysis of hydroperoxides, aldehydes and epoxides by 1H nuclear magnetic resonance in sunflower oil oxidized at 70 and 100 degrees C. *Journal of Agricultural and Food Chemistry* 58: 6234–6245.

Gokmen, V., Senyuva, H.Z., Acar, J., and Sarioglu, K. (2005). Determination of acrylamide in potato chips and crisps by high-performance liquid chromatography. *Journal of Chromatography A* 1088: 193–199.

Gong, X., Qi, N., Wang, X. et al. (2014). A new method for determination of α-tocopherol in tropical fruits by ultra performance convergence chromatography with diode array detector. *Food Analytical Methods* 7: 1572–1576.

Gonzalez-Covarrubias, V. (2013). Lipidomics in longevity and healthy aging. *Biogerontology* 14: 663–672.

Gorassini, A., Verardo, G., Fregolent, S.C., and Bortolomeazzi, R. (2017). Rapid determination of cholesterol oxidation products in milk powder based products by reversed phase SPE and HPLC-APCI-MS/MS. *Food Chemistry* 230: 604–610.

Górnaś, P., Siger, A., Czubinski, J. et al. (2014). An alternative RP-HPLC method for the separation and determination of tocopherol and tocotrienol homologues as butter authenticity markers: a comparative study between two European countries. *European Journal of Lipid Science and Technology* 116: 895–903.

Górnaś, P., Soliven, A., and Segliņa, D. (2015). Seed oils recovered from industrial fruit by-products are a rich source of tocopherols and tocotrienols: rapid separation of α/β/γ/δ homologues by RP-HPLC/FLD. *European Journal of Lipid Science and Technology* 117: 773–777.

Gotoh, N., Matsumoto, Y., Nagai, T. et al. (2011). Actual ratios of triacylglycerol positional isomers consisting of saturated and highly unsaturated fatty acids in fishes and marine mammals. *Food Chemistry* 127: 467–472.

Gromadzka, J. and Wardencki, W. (2010). Static headspace sampling and solid-phase microextraction for assessment of edible oils stability. *Chromatographia* 71: 81–86.

Gross, R.W., Jenkins, C.M., Yang, J. et al. (2005). Functional lipidomics: the roles of specialized lipids and lipid-protein interactions in modulating neuronal function. *Prostaglandins & Other Lipid Mediators* 77: 52–64.

Grun, C.H. and Besseau, S. (2016). Normal-phase liquid chromatography-atmospheric-pressure photoionization-mass spectrometry analysis of cholesterol and phytosterol oxidation products. *Journal of Chromatography A* 1439: 74–81.

Guardiola, F. (2002). *Cholesterol and Phytosterol Oxidation Products: Analysis, Occurrence, and Biological Effects*. Urbana, IL: The American Oil Chemists Society, AOCS Press.

Guardiola, F., Bou, R., Boatella, J., and Codony, R. (2004). Analysis of sterol oxidation products in foods. *Journal of AOAC International* 87: 441–466.

Guillén, M.D. and Uriarte, P.S. (2012). Aldehydes contained in edible oils of a very different nature after prolonged heating at frying temperature: presence of toxic oxygenated α,β unsaturated aldehydes. *Food Chemistry* 131: 915–926.

Hachey, D.L., Patterson, B.W., Reeds, P.J., and Elsas, L.J. (1991). Isotopic determination of organic keto acid pentafluorobenzyl esters in biological fluids by negative chemical ionization gas chromatography/mass spectrometry. *Analytical Chemistry* 63: 919–923.

Harrison, K.A., Davies, S.S., Marathe, G.K. et al. (2000). Analysis of oxidized glycerophosphocholine lipids using electrospray ionization mass spectrometry and microderivatization techniques. *Journal of Mass Spectrometry* 35: 224–236.

Hindra, F. and Baik, O.D. (2006). Kinetics of quality changes during food frying. *Critical Reviews in Food Science and Nutrition* 46: 239–258.

Horiuchi, M., Umano, K., and Shibamoto, T. (1998). Analysis of volatile compounds formed from fish oil heated with cysteine and trimethylamine oxide. *Journal of Agricultural and Food Chemistry* 46: 5232–5237.

Hu, Q., Xu, X., Fu, Y., and Li, Y. (2015). Rapid methods for detecting acrylamide in thermally processed foods: a review. *Food Control* 56: 135–146.

Hui, S.P. (2014). Analyses of lipid hydroperoxides and physiologically active lipids by mass spectrometry. *Rinsho Byori* 62: 283–290.

Hwang, H.-S., Winkler-Moser, J.K., Bakota, E.L. et al. (2013). Antioxidant activity of sesamol in soybean oil under frying conditions. *Journal of the American Oil Chemists' Society* 90: 659–666.

Hyötyläinen, T., Kallio, M., Lehtonen, M. et al. (2004). Comprehensive two-dimensional gas chromatography in the analysis of dietary fatty acids. *Journal of Separation Science* 27: 459–467.

Idrus, S.I.S., Latiff, A.A., and Ismail, M.N. (2017). Determination of triacylglycerols in food by high-performance liquid chromatography. *Instrumentation Science & Technology* 45: 577–591.

Ignat, I., Volf, I., and Popa, V.I. (2011). A critical review of methods for characterisation of polyphenolic compounds in fruits and vegetables. *Food Chemistry* 126: 1821–1835.

Ikeda, K., Oike, Y., Shimizu, T., and Taguchi, R. (2009). Global analysis of triacylglycerols including oxidized molecular species by reverse-phase high resolution LC/ESI-QTOF MS/MS. *Journal of Chromatography B: Analytical Technologies in the Biomedical and Life Sciences* 877: 2639–2647.

Inchingolo, R., Cardenia, V., and Rodriguez-Estrada, M.T. (2014). Analysis of phytosterols and phytostanols in enriched dairy products by fast gas chromatography with mass spectrometry. *Journal of Separation Science* 37: 2911–2919.

Irakli, M.N., Samanidou, V.F., and Papadoyannis, I.N. (2012). Optimization and validation of the reversed-phase high-performance liquid chromatography with fluorescence detection method for the separation of tocopherol and tocotrienol isomers in cereals, employing a novel sorbent material. *Journal of Agricultural and Food Chemistry* 60: 2076–2082.

Ito, M., Ishimaru, M., Shibata, T. et al. (2017b). High-performance liquid chromatography with fluorescence detection for simultaneous analysis of phytosterols (stigmasterol, β-sitosterol, campesterol, ergosterol, and fucosterol) and cholesterol in plant foods. *Food Analytical Methods* 10: 2692–2699.

Ito, J., Shimizu, N., Kobayashi, E. et al. (2017a). A novel chiral stationary phase LC-MS/MS method to evaluate oxidation mechanisms of edible oils. *Scientific Reports* 7: 10026.

Janoszka, B. (2011). HPLC-fluorescence analysis of polycyclic aromatic hydrocarbons (PAHs) in pork meat and its gravy fried without additives and in the presence of onion and garlic. *Food Chemistry* 126: 1344–1353.

Jautz, U., Gibis, M., and Morlock, G.E. (2008). Quantification of heterocyclic aromatic amines in fried meat by HPTLC/UV-FLD and HPLC/UV-FLD: a comparison of two methods. *Journal of Agricultural and Food Chemistry* 56: 4311–4319.

Jenske, R. and Vetter, W. (2008). Gas chromatography/electron-capture negative ion mass spectrometry for the quantitative determination of 2-and 3-hydroxy fatty acids in bovine milk fat. *Journal of Agricultural and Food Chemistry* 56: 5500–5505.

Jerković, I., Mastelić, J., and Tartaglia, S. (2007). A study of volatile flavour substances in Dalmatian traditional smoked ham: impact of dry-curing and frying. *Food Chemistry* 104: 1030–1039.

Jiang, Z., Fenton, M., and Sim, J.S. (1991). Comparison of four different methods for egg cholesterol determination. *Poultry Science* 70: 1015–1019.

Johansson, M.A., Fredholm, L., Bjerne, I., and Jagerstad, M. (1995). Influence of frying fat on the formation of heterocyclic amines in fried beefburgers and pan residues. *Food and Chemical Toxicology* 33: 993–1004.

Julien-David, D., Zhao, M., Geoffroy, P. et al. (2014). Analysis of sitosteryl oleate esters in phytosterols esters enriched foods by HPLC–ESI-MS2. *Steroids* 84: 84–91.

Karimi, S., Wawire, M., and Mathooko, F.M. (2017). Impact of frying practices and frying conditions on the quality and safety of frying oils used by street vendors and restaurants in Nairobi, Kenya. *Journal of Food Composition and Analysis* 62: 239–244.

Karl, H., Merkle, S., Kuhlmann, J., and Fritsche, J. (2016). Development of analytical methods for the determination of free and ester bound 2-, 3-MCPD, and esterified glycidol in fishery products. *European Journal of Lipid Science and Technology* 118: 406–417.

Katragadda, H.R., Fullana, A., Sidhu, S., and Carbonell-Barrachina, Á.A. (2010). Emissions of volatile aldehydes from heated cooking oils. *Food Chemistry* 120: 59–65.

Katsanidis, E. and Addis, P.B. (1999). Novel HPLC analysis of tocopherols, tocotrienols, and cholesterol in tissue. *Free Radical Biology and Medicine* 27: 1137–1140.

Katsuta, I., Shimizu, M., Yamaguchi, T., and Nakajima, Y. (2008). Emission of volatile aldehydes from DAG-rich and TAG-rich oils with different degrees of unsaturation during deep-frying. *Journal of the American Oil Chemists' Society* 85: 513–519.

Kermasha, S., Kubow, S., and Goetghebeur, M. (1994). Comparative high-performance liquid chromatographie analyses of cholesterol and its oxidation products using diode-array ultraviolet and laser light-scattering detection. *Journal of Chromatography A* 685: 229–235.

Keşkekoğlu, H. and Üren, A. (2014). Inhibitory effects of pomegranate seed extract on the formation of heterocyclic aromatic amines in beef and chicken meatballs after cooking by four different methods. *Meat Science* 96: 1446–1451.

Kinter, M. (1995). Analytical technologies for lipid oxidation products analysis. *Journal of Chromatography B: Biomedical Applications* 671: 223–236.

Kıralan, S., Yorulmaz, A., Şimşek, A., and Tekin, A. (2015). Classification of Turkish hazelnut oils based on their triacylglycerol structures by chemometric analysis. *European Food Research and Technology* 240: 679–688.

Knecht, K., Sandfuchs, K., Kulling, S.E., and Bunzel, D. (2015). Tocopherol and tocotrienol analysis in raw and cooked vegetables: a validated method with emphasis on sample preparation. *Food Chemistry* 169: 20–27.

Komprda, T., Zelenka, J., Fajmonová, E. et al. (2003). Cholesterol content in meat of some poultry and fish species as influenced by live weight and total lipid content. *Journal of Agricultural and Food Chemistry* 51: 7692–7697.

Kuksis, A., Suomela, J.P., Tarvainen, M., and Kallio, H. (2009a). Lipidomic analysis of glycerolipid and cholesteryl ester autooxidation products. *Molecular Biotechnology* 42: 224–268.

Kuksis, A., Suomela, J.P., Tarvainen, M., and Kallio, H. (2009b). Use of lipidomics for analyzing glycerolipid and cholesteryl ester oxidation by gas chromatography, HPLC, and on-line MS. *Methods in Molecular Biology* 580: 39–91.

Lampi, A.-M., Kemmo, S., Mäkelä, A. et al. (2009). Distribution of monomeric, dimeric and polymeric products of stigmasterol during thermo-oxidation. *European Journal of Lipid Science and Technology* 111: 1027–1034.

La Nasa, J., Degano, I., Brandolini, L. et al. (2018). A novel HPLC-ESI-Q-ToF approach for the determination of fatty acids and acylglycerols in food samples. *Analytica Chimica Acta* 1013: 98–109.

Lee, J.H. and Pangloli, P. (2013). Volatile compounds and storage stability of potato chips fried in mid-oleic sunflower oil. *International Journal of Food Properties* 16: 563–573.

Lee, J., Dong, A., Jung, K., and Shin, H.-S. (2011). Influence of extra virgin olive oil on the formation of heterocyclic amines in roasted beef steak. *Food Science and Biotechnology* 20: 159–165.

Lehtonen, M., Kemmo, S., Lampi, A.-M., and Piironen, V. (2011). Effects of esterification on the formation and decomposition of steryl hydroperoxides. *European Food Research and Technology* 232: 255–264.

Lehtonen, M., Lampi, A.-M., Riuttamäki, M.-A., and Piironen, V. (2012). Oxidation reactions of steryl esters in a saturated lipid matrix. *Food Chemistry* 134: 2030–2039.

Lerma-García, M.J., Lusardi, R., Chiavaro, E. et al. (2011). Use of triacylglycerol profiles established by high performance liquid chromatography with ultraviolet–visible detection to predict the botanical origin of vegetable oils. *Journal of Chromatography A* 1218: 7521–7527.

Leskinen, H.M., Suomela, J.P., and Kallio, H.P. (2010). Quantification of triacylglycerol regioisomers by ultra-high-performance liquid chromatography and ammonia negative ion atmospheric pressure chemical ionization tandem mass spectrometry. *Rapid Communications in Mass Spectrometry* 24: 1–5.

Lim, P.K., Jinap, S., Sanny, M. et al. (2014). The influence of deep frying using various vegetable oils on acrylamide formation in sweet potato (*Ipomoea batatas* L. Lam) chips. *Journal of Food Science* 79: T115–T121.

Lin, J.T. and Chen, G.Q. (2010). Acylglycerols containing trihydroxy fatty acids in castor oil and the regiospecific quantification of triacylglycerols. *Journal of the American Oil Chemists' Society* 87: 1371–1379.

Lísa, M. and Holčapek, M. (2013). Characterization of triacylglycerol enantiomers using chiral HPLC/APCI-MS and synthesis of enantiomeric triacylglycerols. *Analytical Chemistry* 85: 1852–1859.

Lojzova, L., Riddellova, K., Hajslova, J. et al. (2009). Alternative GC-MS approaches in the analysis of substituted pyrazines and other volatile aromatic compounds formed during Maillard reaction in potato chips. *Analytica Chimica Acta* 641: 101–109.

Mäkinen, M., Piironen, V., and Hopia, A. (1996). Postcolumn chemiluminescence, ultraviolet and evaporative light-scattering detectors in high-performance liquid chromatographic determination of triacylglycerol oxidation products. *Journal of Chromatography A* 734: 221–229.

Marmesat, S., Velasco, J., and Dobarganes, M.C. (2008). Quantitative determination of epoxy acids, keto acids and hydroxy acids formed in fats and oils at frying temperatures. *Journal of Chromatography A* 1211: 129–134.

Márquez-Ruiz, G., Perez-Camino, M.C., and Dobarganes, M.C. (1993). Evaluation of hydrolysis and absorption of thermally oxidized olive oil in non-absorbed lipids in the rat. *Annals of Nutrition & Metabolism* 37: 121–128.

Márquez-Ruiz, G., Tasioula-Margari, M., and Dobarganes, M.C. (1995). Quantitation and distribution of altered fatty acids in frying fats. *Journal of the American Oil Chemists' Society* 72: 1171–1176.

Márquez-Ruiz, G., Rodríguez-Pino, V., and de la Fuente, M.A. (2011). Determination of 10-hydroxystearic, 10-ketostearic, 8-hydroxypalmitic, and 8-ketopalmitic acids in milk fat by solid-phase extraction plus gas chromatography-mass spectrometry. *Journal of Dairy Science* 94: 4810–4819.

Martin, F.L. and Ames, J.M. (2001). Formation of Strecker aldehydes and pyrazines in a fried potato model system. *Journal of Agricultural and Food Chemistry* 49: 3885–3892.

Martínez-Yusta, A. and Guillen, M.D. (2014a). Deep-frying food in extra virgin olive oil: a study by (1)H nuclear magnetic resonance of the influence of food nature on the evolving composition of the frying medium. *Food Chemistry* 150: 429–437.

Martínez-Yusta, A. and Guillén, M.D. (2014b). A study by 1H nuclear magnetic resonance of the influence on the frying medium composition of some soybean oil-food combinations in deep-frying. *Food Research International* 55: 347–355.

Matthäus, B. and Haase, N.U. (2014). Acrylamide – still a matter of concern for fried potato food? *European Journal of Lipid Science and Technology* 116: 675–687.

Melton, S.L., Jafar, S., Sykes, D., and Trigiano, M.K. (1994). Review of stability measurements for frying oils and fried food flavor. *Journal of the American Oil Chemists' Society* 71: 1301–1308.

Menéndez-Carreño, M., Ansorena, D., Astiasarán, I. et al. (2010). Determination of non-polar and mid-polar monomeric oxidation products of stigmasterol during thermo-oxidation. *Food Chemistry* 122: 277–284.

Mestdagh, F.J., De Meulenaer, B., Van Poucke, C. et al. (2005). Influence of oil type on the amounts of acrylamide generated in a model system and in French fries. *Journal of Agricultural and Food Chemistry* 53: 6170–6174.

Mestdagh, F., Maertens, J., De Wilde, T. et al. (2007). Chemical pre-treatments of potato products: mechanisms of acrylamide mitigation and effects on the sensorial quality. *Communications in Agricultural and Applied Biological Sciences* 72: 9–12.

Mestre Prates, J.A., Gonçalves Quaresma, M.A., Branquinho Bessa, R.J. et al. (2006). Simultaneous HPLC quantification of total cholesterol, tocopherols and β-carotene in Barrosã-PDO veal. *Food Chemistry* 94: 469–477.

Mitei, Y., Ngila, J., Yeboah, S. et al. (2009). Profiling of phytosterols, tocopherols and tocotrienols in selected seed oils from Botswana by GC-MS and HPLC. *Journal of the American Oil Chemists' Society* 86: 617–625.

Miyashita, K., Frankel, E.N., Neff, W.E., and Awl, R.A. (1990). Autoxidation of polyunsaturated triacylglycerols. III. Synthetic triacylglycerols containing linoleate and linolenate. *Lipids* 25: 48–53.

Miyazawa, T., Fujimoto, K., and Oikawa, S. (1990). Determination of lipid hydroperoxides in low density lipoprotein from human plasma using high performance liquid chromatography with chemiluminescence detection. *Biomedical Chromatography* 4: 131–134.

Miyazawa, T., Kunika, H., Fujimoto, K. et al. (1995). Chemiluminescence detection of mono-, bis-, and tris-hydroperoxy triacylglycerols present in vegetable oils. *Lipids* 30: 1001–1006.

Morales, A., Dobarganes, C., Márquez-Ruiz, G., and Velasco, J. (2010). Quantitation of hydroperoxy-, keto- and hydroxy-dienes during oxidation of FAMEs from high-linoleic and high-oleic sunflower oils. *Journal of the American Oil Chemists' Society* 87: 1271–1279.

Morales, A., Marmesat, S., Dobarganes, M.C. et al. (2012a). Evaporative light scattering detector in normal-phase high-performance liquid chromatography determination of FAME oxidation products. *Journal of Chromatography A* 1254: 62–70.

Morales, A., Marmesat, S., Dobarganes, M.C. et al. (2012b). Quantitative analysis of hydroperoxy-, keto- and hydroxy-dienes in refined vegetable oils. *Journal of Chromatography A* 1229: 190–197.

Morales, A., Marmesat, S., Ruiz-Mendez, V. et al. (2014). Formation of oxidation products in edible oils analyzed as FAME derivatives by HPLC-UV-ELSD. *Food Research International* 62: 1080–1086.

Mubiru, E., Shrestha, K., Papastergiadis, A., and De Meulenaer, B. (2013). Improved gas chromatography-flame ionization detector analytical method for the analysis of epoxy fatty acids. *Journal of Chromatography A* 1318: 217–225.

Mubiru, E., Shrestha, K., Papastergiadis, A., and De Meulenaer, B. (2014). Development and validation of a gas chromatography–flame ionization detection method for the determination of epoxy fatty acids in food matrices. *Journal of Agricultural and Food Chemistry* 62: 2982–2988.

Nakajima, M., Yamato, S., Wakabayashi, H., and Shimada, K. (1995). High-performance liquid chromatographic determination of cholesterol and cholestanol in human serum by precolumn derivatization with 2-[2-(isocyanate)ethyl]-3-methyl-1, 4-naphthoquinone combined with platinum catalyst reduction and electrochemical detection. *Biological & Pharmaceutical Bulletin* 18: 1762–1764.

Narvaez-Rivas, M., Pham, A.J., Schilling, M.W., and Leon-Camacho, M. (2014). A new SPE/GC-fid method for the determination of cholesterol oxidation products. Application to subcutaneous fat from Iberian dry-cured ham. *Talanta* 122: 58–62.

Neff, W.E. and Byrdwell, W.C. (1998). Characterization of model triacylglycerol (triolein, trilinolein and trilinolenin) autoxidation products via high-performance liquid chromatography coupled with atmospheric pressure chemical ionization mass spectrometry. *Journal of Chromatography A* 818: 169–186.

Neff, W.E., Frankel, E.N., and Miyashita, K. (1990). Autoxidation of polyunsaturated triacylglycerols. I. Trilinoleoylglycerol. *Lipids* 25: 33–39.

Negroni, M., D'agostin, A., and Arnoldi, A. (2001). Effects of olive, canola, and sunflower oils on the formation of volatiles from the Maillard reaction of lysine with xylose and glucose. *Journal of Agricultural and Food Chemistry* 49: 439–445.

Noguchi, N., Numano, R., Kaneda, H., and Niki, E. (1998). Oxidation of lipids in low density lipoprotein particles. *Free Radical Research* 29: 43–52.

Oehrl, L.L., Hansen, A.P., Rohrer, C.A. et al. (2001). Oxidation of phytosterols in a test food system. *Journal of the American Oil Chemists' Society* 78: 1073–1078.

Oette, K. (1965). Identification of some lipid peroxides by thin-layer chromatography. *Journal of Lipid Research* 6: 449–454.

Ohshima, T., Hopia, A., German, J.B., and Frankel, E.N. (1996). Determination of hydroperoxides and structures by high-performance liquid chromatography with post-column detection with diphenyl-1-pyrenylphosphine. *Lipids* 31: 1091–1096.

Ontanón, I., Culleré, L., Zapata, J. et al. (2013). Application of a new sampling device for determination of volatile compounds released during heating olive and sunflower oil: sensory evaluation of those identified compounds. *European Food Research and Technology* 236: 1031–1040.

Oracz, J., Nebesny, E., and Zyzelewicz, D. (2011). New trends in quantification of acrylamide in food products. *Talanta* 86: 23–34.

Osawa, C.C., Gonçalves, L.A.G., and Da Silva, M.A.A.P. (2013). Odor significance of the volatiles formed during deep-frying with palm olein. *Journal of the American Oil Chemists' Society* 90: 183–189.

Osório, V.M. and de Lourdes Cardeal, Z. (2011). Determination of acrolein in french fries by solid-phase microextraction gas chromatography and mass spectrometry. *Journal of Chromatography A* 1218: 3332–3336.

Oueslati, I., Taamalli, A., Loubiri, A. et al. (2017). Assessment of conventional and microwave heating effects on the variation of the bioactive compounds of Chétoui VOO using HPLC-DAD-ESI-TOF-MS. *Arabian Journal of Chemistry* https://doi.org/10.1016/j.arabjc.2017.08.011.

Özogul, F., Özogul, Y., and Kuley, E. (2011). Simple extraction and rapid HPLC method for tocopherol analysis in marine and fresh-water fish species. *Food Science and Technology Research* 17: 595–598.

Pagliuca, G., Bozzi, C., Gallo, F.R. et al. (2018). Triacylglycerol 'hand-shape profile' of Argan oil. Rapid and simple UHPLC-PDA-ESI-TOF/MS and HPTLC methods to detect counterfeit Argan oil and Argan-oil-based products. *Journal of Pharmaceutical and Biomedical Analysis* 150: 121–131.

Paul, S. and Mittal, G.S. (1997). Regulating the use of degraded oil/fat in deep-fat/oil food frying. *Critical Reviews in Food Science and Nutrition* 37: 635–662.

Pedreschi, F., Kaack, K., and Granby, K. (2006). Acrylamide content and color development in fried potato strips. *Food Research International* 39: 40–46.

Peers, K.E. and Coxon, D.T. (1983). Controlled synthesis of monohydroperoxides by α-tocopherol inhibited autoxidation of polyunsaturated lipids. *Chemistry and Physics of Lipids* 32: 49–56.

Perkins, E.G. and Anfinsen, J.R. (1971). Characterization of the nonvolatile compounds formed during the thermal oxidation of triolein. *Journal of the American Oil Chemists' Society* 48: 556–562.

Petersen, K.D., Kleeberg, K.K., Jahreis, G., and Fritsche, J. (2012). Assessment of the oxidative stability of conventional and high-oleic sunflower oil by means of solid-phase microextraction-gas chromatography. *International Journal of Food Sciences and Nutrition* 63: 160–169.

Picariello, G., Romano, R., and Addeo, F. (2010). Nitrocellulose film substrate minimizes fragmentation in matrix-assisted laser desorption ionization time-of-flight mass spectrometry analysis of triacylglycerols. *Analytical Chemistry* 82: 5783–5791.

Ponte, P.I.P., Alves, S.P., Bessa, R.J.B. et al. (2008). Influence of pasture intake on the fatty acid composition, and cholesterol, tocopherols, and tocotrienols content in meat from free-range broilers. *Poultry Science* 87: 80–88.

Rafferty, J.L., Zhang, L., Siepmann, J.I., and Schure, M.R. (2007). Retention mechanism in reversed-phase liquid chromatography: a molecular perspective. *Analytical Chemistry* 79: 6551–6558.

Rambaldi, D.C., Reschiglian, P., Zattoni, A., and Johann, C. (2009). Enzymatic determination of cholesterol and triglycerides in serum lipoprotein profiles by asymmetrical flow field-flow fractionation with on-line, dual detection. *Analytica Chimica Acta* 654: 64–70.

Ramesh, A., Walker, S.A., Hood, D.B. et al. (2004). Bioavailability and risk assessment of orally ingested polycyclic aromatic hydrocarbons. *International Journal of Toxicology* 23: 301–333.

Ravi Kiran, C., Sasidharan, I., Soban Kumar, D.R., and Sundaresan, A. (2015). Influence of natural and synthetic antioxidants on the degradation of Soybean oil at frying temperature. *Journal of Food Science and Technology* 52: 5370–5375.

Řezanka, T., Matoulková, D., Kolouchová, I. et al. (2013). Brewer's yeast as a new source of palmitoleic acid – analysis of triacylglycerols by LC–MS. *Journal of the American Oil Chemists' Society* 90: 1327–1342.

Rivera, L., Curto, M.J., Pais, P. et al. (1996). Solid-phase extraction for the selective isolation of polycyclic aromatic hydrocarbons, azaarenes and heterocyclic aromatic amines in charcoal-grilled meat. *Journal of Chromatography A* 731: 85–94.

Romani, S., Bacchiocca, M., Rocculi, P., and Dalla Rosa, M. (2009). Influence of frying conditions on acrylamide content and other quality characteristics of French fries. *Journal of Food Composition and Analysis* 22: 582–588.

Romano, R., Giordano, A., Le Grottaglie, L. et al. (2013). Volatile compounds in intermittent frying by gas chromatography and nuclear magnetic resonance. *European Journal of Lipid Science and Technology* 115: 764–773.

Rudzińska, M., Przybylski, R., and Wąsowicz, E. (2009). Products formed during thermo-oxidative degradation of phytosterols. *Journal of the American Oil Chemists' Society* 86: 651–662.

Rudzińska, M., Przybylski, R., Zhao, Y.Y., and Curtis, J.M. (2010). Sitosterol thermo-oxidative degradation leads to the formation of dimers, trimers and oligomers: a study using combined size exclusion chromatography/mass spectrometry. *Lipids* 45: 549–558.

Ruiz-Gutierrez, V. and Barron, L.J. (1995). Methods for the analysis of triacylglycerols. *Journal of Chromatography B: Biomedical Applications* 671: 133–168.

Saini, R.K. and Keum, Y.-S. (2016). Tocopherols and tocotrienols in plants and their products: a review on methods of extraction, chromatographic separation, and detection. *Food Research International* 82: 59–70.

Saito, E., Tanaka, N., Miyazaki, A., and Tsuzaki, M. (2014). Concentration and particle size distribution of polycyclic aromatic hydrocarbons formed by thermal cooking. *Food Chemistry* 153: 285–291.

Salvatori, G., Filetti, F., Di Cesare, C. et al. (2008). Lipid composition of meat and backfat from Casertana purebred and crossbred pigs reared outdoors. *Meat Science* 80: 623–631.

Sánchez-Muniz, F.J., Cuesta, C., and Garrido-Polonio, C. (1993). Sunflower oil used for frying: combination of column, gas and high-performance size-exclusion chromatography for its evaluation. *Journal of the American Oil Chemists' Society* 70: 235–240.

Schröter, J., Griesinger, H., Reu, E. et al. (2016). Unexpected products of the hypochlorous acid-induced oxidation of oleic acid: a study using high performance thin-layer chromatography electrospray ionization mass spectrometry. *Journal of Chromatography A* 1439: 89–96.

Schwartz, H., Ollilainen, V., Piironen, V., and Lampi, A.-M. (2008). Tocopherol, tocotrienol and plant sterol contents of vegetable oils and industrial fats. *Journal of Food Composition and Analysis* 21: 152–161.

Shammugasamy, B., Ramakrishnan, Y., Manan, F., and Muhammad, K. (2015). Rapid reversed-phase chromatographic method for determination of eight vitamin E isomers and γ-oryzanols in rice bran and rice bran oil. *Food Analytical Methods* 8: 649–655.

Simas, R.C., Catharino, R.R., Cunha, I.B.S. et al. (2010). Instantaneous characterization of vegetable oils via TAG and FFA profiles by easy ambient sonic-spray ionization mass spectrometry. *Analyst* 135: 738–744.

Simsek, U.G., Cerci, I.H., Dalkilic, B. et al. (2009). Impact of stocking density and feeding regimen on broilers: chicken meat composition, fatty acids, and serum cholesterol levels. *Journal of Applied Poultry Research* 18: 514–520.

Sion, B., Grizard, G., and Boucher, D. (2001). Quantitative analysis of desmosterol, cholesterol and cholesterol sulfate in semen by high-performance liquid chromatography. *Journal of Chromatography A* 935: 259–265.

Sjaastad, A.K., Jorgensen, R.B., and Svendsen, K. (2010). Exposure to polycyclic aromatic hydrocarbons (PAHs), mutagenic aldehydes and particulate matter during pan frying of beefsteak. *Occupational and Environmental Medicine* 67: 228–232.

Sjövall, O., Kuksis, A., Marai, L., and Myher, J.J. (1997). Elution factors of synthetic oxotriacylglycerols as an aid in identification of peroxidized natural triacylglycerols by reverse-phase high-performance liquid chromatography with electrospray mass spectrometry. *Lipids* 32: 1211–1218.

Sjövall, O., Kuksis, A., and Kallio, H. (2001a). Analysis of molecular species of peroxide adducts of triacylglycerols following treatment of corn oil with tert-butyl hydroperoxide. *Lipids* 36: 1347–1356.

Sjovall, O., Kuksis, A., and Kallio, H. (2001b). Reversed-phase high-performance liquid chromatographic separation of tert.-butyl hydroperoxide oxidation products of unsaturated triacylglycerols. *Journal of Chromatography A* 905: 119–132.

Sjövall, O., Kuksis, A., and Kallio, H. (2002). Formation of triacylglycerol core aldehydes during rapid oxidation of corn and sunflower oils with tert-butyl hydroperoxide/Fe^{2+}. *Lipids* 37: 81–94.

Sjövall, O., Kuksis, A., and Kallio, H. (2003). Tentative identification and quantification of TAG core aldehydes as dinitrophenylhydrazones in autoxidized sunflowerseed oil using reversed-phase HPLC with electrospray ionization MS. *Lipids* 38: 1179–1190.

Skipski, V.P., Good, J.J., Barclay, M., and Reggio, R.B. (1968). Quantitative analysis of simple lipid classes by thin-layer chromatography. *Biochimica et Biophysica Acta* 152: 10–19.

Solaesa, Á.G., Bucio, S.L., Sanz, M.T. et al. (2014). Characterization of triacylglycerol composition of fish oils by using chromatographic techniques. *Journal of Oleo Science* 63: 449–460.

Sosińska, E., Przybylski, R., Hazendonk, P. et al. (2013). Characterisation of non-polar dimers formed during thermo-oxidative degradation of β-sitosterol. *Food Chemistry* 139: 464–474.

Sosińska, E., Przybylski, R., Aladedunye, F., and Hazendonk, P. (2014). Spectroscopic characterisation of dimeric oxidation products of phytosterols. *Food Chemistry* 151: 404–414.

Soupas, L., Juntunen, L., Lampi, A.-M., and Piironen, V. (2004). Effects of sterol structure, temperature, and lipid medium on phytosterol oxidation. *Journal of Agricultural and Food Chemistry* 52: 6485–6491.

Souza, H.A.L., Mariutti, L.R.B., and Bragagnolo, N. (2017). Microwave assisted direct saponification for the simultaneous determination of cholesterol and cholesterol oxides in shrimp. *Journal of Steroid Biochemistry and Molecular Biology* 169: 88–95.

Srivastava, S., Singh, M., George, J. et al. (2010). Genotoxic and carcinogenic risks associated with the consumption of repeatedly boiled sunflower oil. *Journal of Agricultural and Food Chemistry* 58: 11179–11186.

Stan, H.J. and Scheutwinkel-Reich, M. (1980). Detection of hydroxy fatty acids in biological samples using capillary gas chromatography in combination with positive and negative chemical ionization mass spectrometry. *Lipids* 15: 1044–1050.

Steenhorst-Slikkerveer, L., Louter, A., Janssen, H.G., and Bauer-Plank, C. (2000). Analysis of nonvolatile lipid oxidation products in vegetable oils by normal-phase high-performance liquid chromatography with mass spectrometric detection. *Journal of the American Oil Chemists' Society* 77: 837–845.

Stolyhwo, A. and Privett, O.S. (1973). Studies on the analysis of lipid classes by gradient elution adsorption chromatography. *Journal of Chromatographic Science* 11: 20–25.

Sugino, K. (1999). Simultaneous determination of different classes of lipid hydroperoxides by high-performance liquid chromatography with post column detection by a ferrous/xylenol orange reagent. *Bioscience, Biotechnology, and Biochemistry* 63: 773–775.

Sun, J., Zhao, X.-E., Dang, J. et al. (2017). Rapid and sensitive determination of phytosterols in functional foods and medicinal herbs by using UHPLC–MS/MS with microwave-assisted derivatization combined with dual ultrasound-assisted dispersive liquid–liquid microextraction. *Journal of Separation Science* 40: 725–732.

Sunarić, S., Lalić, J., and Spasić, A. (2017). Simultaneous determination of alpha-tocopherol and alpha-tocopheryl acetate in dairy products, plant milks and health supplements by using SPE and HPLC method. *Food Analytical Methods* 10: 3886–3901.

Suomela, J.P., Ahotupa, M., Sjövall, O. et al. (2004a). Diet and lipoprotein oxidation: analysis of oxidized triacylglycerols in pig lipoproteins. *Lipids* 39: 639–647.

Suomela, J.P., Ahotupa, M., Sjövall, O. et al. (2004b). New approach to the analysis of oxidized triacylglycerols in lipoproteins. *Lipids* 39: 507–512.

Suomela, J.P., Ahotupa, M., and Kallio, H. (2005). Triacylglycerol oxidation in pig lipoproteins after a diet rich in oxidized sunflower seed oil. *Lipids* 40: 437–444.

Suomela, J.P., Leskinen, H., and Kallio, H. (2011). Analysis of isomeric forms of oxidized triacylglycerols using ultra-high-performance liquid chromatography and tandem mass spectrometry. *Journal of Agricultural and Food Chemistry* 59: 8095–8100.

Takeoka, G., Perrino, C., and Buttery, R. (1996). Volatile constituents of used frying oils. *Journal of Agricultural and Food Chemistry* 44: 654–660.

Tamba Sompila, A.W.G., Héron, S., Hmida, D., and Tchapla, A. (2017). Fast non-aqueous reversed-phase liquid chromatography separation of triacylglycerol regioisomers with isocratic mobile phase. Application to different oils and fats. *Journal of Chromatography, B: Analytical Technologies in the Biomedical and Life Sciences* 1041–1042: 151–157.

Tangolar, S.G., Özogul, F., Tangolar, S., and Yağmur, C. (2011). Tocopherol content in fifteen grape varieties obtained using a rapid HPLC method. *Journal of Food Composition and Analysis* 24: 481–486.

Tarvainen, M., Suomela, J.P., Kuksis, A., and Kallio, H. (2010). Liquid chromatography-light scattering detector-mass spectrometric analysis of digested oxidized rapeseed oil. *Lipids* 45: 1061–1079.

Tarvainen, M., Nuora, A., Quirin, K.-W. et al. (2015). Effects of CO2 plant extracts on triacylglycerol oxidation in Atlantic salmon during cooking and storage. *Food Chemistry* 173: 1011–1021.

Tekkeli, S.E., Önal, C., and Önal, A. (2012). A review of current methods for the determination of acrylamide in food products. *Food Analytical Methods* 5: 29–39.

Terao, J., Shibata, S.S., and Matsushita, S. (1988). Selective quantification of arachidonic acid hydroperoxides and their hydroxy derivatives in reverse-phase high performance liquid chromatography. *Analytical Biochemistry* 169: 415–423.

Tonks, D.B. (2006). The estimation of cholesterol in serum: a classification and critical review of methods. *Clinical Biochemistry* 39: 446–463.

Ubhayasekera, S.J.K.A., Verleyen, T., and Dutta, P.C. (2004). Evaluation of GC and GC-MS methods for the analysis of cholesterol oxidation products. *Food Chemistry* 84: 149–157.

Ulberth, F. and Reich, H. (1992). Gas chromatographic determination of cholesterol in processed foods. *Food Chemistry* 43: 387–391.

Vaclavik, L., Cajka, T., Hrbek, V., and Hajslova, J. (2009). Ambient mass spectrometry employing direct analysis in real time (DART) ion source for olive oil quality and authenticity assessment. *Analytica Chimica Acta* 645: 56–63.

Vaclavik, L., Belkova, B., Reblova, Z. et al. (2013). Rapid monitoring of heat-accelerated reactions in vegetable oils using direct analysis in real time ionization coupled with high resolution mass spectrometry. *Food Chemistry* 138: 2312–2320.

Velasco, J., Berdeaux, O., Márquez-Ruiz, G., and Dobarganes, M. (2002). Sensitive and accurate quantitation of monoepoxy fatty acids in thermoxidized oils by gas–liquid chromatography. *Journal of Chromatography A* 982: 145–152.

Velasco, J., Marmesat, S., Bordeaux, O. et al. (2004). Formation and evolution of monoepoxy fatty acids in thermoxidized olive and sunflower oils and quantitation in used frying oils from restaurants and fried-food outlets. *Journal of Agricultural and Food Chemistry* 52: 4438–4443.

Velasco, J., Marmesat, S., Berdeaux, O. et al. (2005). Quantitation of short-chain glycerol-bound compounds in thermoxidized and used frying oils. A monitoring study during thermoxidation of olive and sunflower oils. *Journal of Agricultural and Food Chemistry* 53: 4006–4011.

Verleyen, T., Verhe, R., Huyghebaert, A. et al. (2001). Identification of α-tocopherol oxidation products in triolein at elevated temperatures. *Journal of Agricultural and Food Chemistry* 49: 1508–1511.

Viinanen, E. and Hopia, A. (1994). Reversed-phase high-performance liquid chromatographic analysis of triacylglycerol autoxidation products with ultraviolet and evaporative light-scattering detectors. *Journal of the American Oil Chemists' Society* 71: 537–539.

Wang, H., Lee, A.W.M., Shuang, S., and Choi, M.M.F. (2008). SPE/HPLC/UV studies on acrylamide in deep-fried flour-based indigenous Chinese foods. *Microchemical Journal* 89: 90–97.

Wang, H., Feng, F., Guo, Y. et al. (2013). HPLC-UV quantitative analysis of acrylamide in baked and deep-fried Chinese foods. *Journal of Food Composition and Analysis* 31: 7–11.

Wang, L., Csallany, A.S., Kerr, B.J. et al. (2016). Kinetics of forming aldehydes in frying oils and their distribution in French fries revealed by LC-MS-based chemometrics. *Journal of Agricultural and Food Chemistry* 64: 3881–3889.

Warner, K., Neff, W.E., Byrdwell, W.C., and Gardner, H.W. (2001). Effect of oleic and linoleic acids on the production of deep-fried odor in heated triolein and trilinolein. *Journal of Agricultural and Food Chemistry* 49: 899–905.

Wijesundera, C., Ceccato, C., Watkins, P. et al. (2008). Docosahexaenoic acid is more stable to oxidation when located at the sn-2 position of triacylglycerol compared to sn-1(3). *Journal of the American Oil Chemists' Society* 85: 543–548.

Wilson, R. and Lyall, K. (2002). Simultaneous determination by GC-MS of epoxy and hydroxy FA as their methoxy derivatives. *Lipids* 37: 917–924.

Wilson, R., Smith, R., Wilson, P. et al. (1997). Quantitative gas chromatography–mass spectrometry isomer-specific measurement of hydroxy fatty acids in biological samples and food as a marker of lipid peroxidation. *Analytical Biochemistry* 248: 76–85.

Wong, Y.H., Muhamad, H., Abas, F. et al. (2017). Effects of temperature and NaCl on the formation of 3-MCPD esters and glycidyl esters in refined, bleached and deodorized palm olein during deep-fat frying of potato chips. *Food Chemistry* 219: 126–130.

Xia, W. and Budge, S.M. (2017). Techniques for the analysis of minor lipid oxidation products derived from triacylglycerols: epoxides, alcohols, and ketones. *Comprehensive Reviews in Food Science and Food Safety* 16: 735–758.

Xia, W., Budge, S., and Lumsden, M.D. (2016). 1H-NMR characterization of epoxides derived from polyunsaturated fatty acids. *Journal of the American Oil Chemists' Society* 93: 467–478. https://doi.org/10.1007/s11746-016-2800-2.

Xu, X.M., He, H.L., Zhu, Y. et al. (2013). Simultaneous determination of 3-monochloropropane-1,2-diol and acrylamide in food by gas chromatography-triple quadrupole mass spectrometry with coupled column separation. *Analytica Chimica Acta* 760: 93–99.

Yamada, K., Terao, J., and Matsushita, S. (1987). Electrochemical detection of phospholipid hydroperoxides in reverse-phase high performance liquid chromatography. *Lipids* 22: 125–128.

Yen, T.Y., Inbaraj, B.S., Chien, J.T., and Chen, B.H. (2010). Gas chromatography–mass spectrometry determination of conjugated linoleic acids and cholesterol oxides and their stability in a model system. *Analytical Biochemistry* 400: 130–138.

Yu, T.H., Wu, C.M., and Ho, C.T. (1993). Volatile compounds of deep-oil fried, microwave-heated and oven-baked garlic slices. *Journal of Agricultural and Food Chemistry* 41: 800–805.

Yuan, C., Xie, Y., Jin, R. et al. (2017). Simultaneous analysis of tocopherols, phytosterols, and squalene in vegetable oils by high-performance liquid chromatography. *Food Analytical Methods* 10: 3716–3722.

Zeb, A. (2012). Triacylglycerols composition, oxidation and oxidation compounds in camellia oil using liquid chromatography-mass spectrometry. *Chemistry and Physics of Lipids* 165: 608–614.

Zeb, A. (2015a). Chemistry and liquid chromatography methods for the analyses of primary oxidation products of triacylglycerols. *Free Radical Research* 49: 549–564.

Zeb, A. (2015b). A reversed phase HPLC-DAD method for the determination of phenolic compounds in plant leaves. *Analytical Methods* 7: 7753–7757.

Zeb, A. (2017). A simple, sensitive HPLC-DAD method for simultaneous determination of carotenoids, chlorophylls and α-tocopherol in leafy vegetables. *Journal of Food Measurement and Characterization* 11: 979–986.

Zeb, A. and Haq, I. (2017). Polyphenolic composition, lipid peroxidation and antioxidant properties of chapli kebab during repeated frying process. *Journal of Food Measurement and Characterization* **12**: 555–563. https://doi.org/10.1007/s11694-017-9667-2.

Zeb, A. and Mehmood, A. (2012). Effects of oxidized vanaspati ghee on the serum lipids profile and radical scavenging activity of the in vitro lipids of liver, brain and muscles. *Turkish Journal of Biochemistry* 37: 417–423.

Zeb, A. and Murkovic, M. (2010a). Analysis of triacylglycerols in refined edible oils by isocratic HPLC-ESI-MS. *European Journal of Lipid Science and Technology* 112: 844–851.

Zeb, A. and Murkovic, M. (2010b). Characterization of the effects of beta-carotene on the thermal oxidation of triacylglycerols using HPLC-ESI-MS. *European Journal of Lipid Science and Technology* 112: 1218–1228.

Zeb, A. and Murkovic, M. (2010c). Thin-layer chromatographic analysis of carotenoids in plant and animal samples. *JPC Journal of Planar Chromatography Modern TLC* 23: 94–103.

Zeb, A. and Murkovic, M. (2011). Carotenoids and triacylglycerols interactions during thermal oxidation of refined olive oil. *Food Chemistry* 127: 1584–1593.

Zeb, A. and Murkovic, M. (2013a). Determination of thermal oxidation and oxidation products of β-carotene in corn oil triacylglycerols. *Food Research International* 50: 534–544.

Zeb, A. and Murkovic, M. (2013b). Pro-oxidant effects of β-carotene during thermal oxidation of edible oils. *Journal of the American Oil Chemists Society* 90: 881–889.

Zeng, M., Li, Y., He, Z. et al. (2016). Effect of phenolic compounds from spices consumed in China on heterocyclic amine profiles in roast beef patties by UPLC–MS/MS and multivariate analysis. *Meat Science* 116: 50–57.

Zhang, Q., Qin, W., Li, M. et al. (2015). Application of chromatographic techniques in the detection and identification of constituents formed during food frying: a review. *Comprehensive Reviews in Food Science and Food Safety* 14: 601–633.

Zhang, X., Eigendorf, G., Stebbing, D.W. et al. (2002). Degradation of trilinolein by laccase enzymes. *Archives of Biochemistry and Biophysics* 405: 44–54.

Zhang, Y., Dong, Y., Ren, Y., and Zhang, Y. (2006). Rapid determination of acrylamide contaminant in conventional fried foods by gas chromatography with electron capture detector. *Journal of Chromatography A* 1116: 209–216.

Zommick, D.H., Kumar, G.N., Knowles, L.O., and Knowles, N.R. (2013). Translucent tissue defect in potato (Solanum tuberosum L.) tubers is associated with oxidative stress accompanying an accelerated aging phenotype. *Planta* 238: 1125–1145.

Part III

Biochemistry of Food Frying

8

Digestion and Absorption of Fried Foods

8.1 Introduction

Frying oils and fats are important sources of essential fatty acids, phytosterols, polyphenols, and several other important compounds. Upon ingestion, fat provides more energy than carbohydrates or proteins. Foods and frying fats containing highly unsaturated fatty acids undergo oxidation and produce several compounds with both good and bad characteristics. Bad characteristics include those off-flavour compounds responsible for the rancidity of foods and fats. Fried foods with off-flavour are not acceptable to consumers and are thus wasted. The significant issues in the fried food industry are consumer health and likeness. Fried foods that cause any adverse effects are not accepted.

The adverse effects caused by oxidized lipids are a hot topic in modern food science safety research. The formation of several toxic compounds in foods, such as acrylamide, heterocyclic aromatic amines, and furfural, is a cause of concern for both health scientists and consumers. Thus, it is important for consumers to have some basic knowledge of what happens after they ingest fried foods.

8.2 Acceptability of Fried Foods

The acceptability of fried fats and foods depends on numerous factors, including chemical composition, physical and chemical properties, and human perception. Wayler et al. (1982) showed that a moderate degree of unreplaced tooth loss in humans had no significant effect on perceptual estimates of food acceptability, and unilateral intact dentition can provide sufficient masticatory performance to maintain food acceptance. The hardness and taste of fried foods are thus considered the first line of acceptability. Due to the variation in individual human perception, only trained panellists can be used to determine some parameters of the acceptability. The hardness of fried foods is measured using a texture tester.

Ryan et al. (2008) fried peanuts at 170 °C in fresh sunflower (FP-S), corn (FP-C), soybean (FP-So), peanut (FP-P), and olive (FP-O) oils in a domestic fryer. The observed consumer acceptability by attribute of peanuts fried in different vegetable oils (S, C, So, P, and O) is shown in Table 8.1. Significant differences ($\alpha = 0.05$) in consumer

Food Frying: Chemistry, Biochemistry, and Safety, First Edition. Alam Zeb.

Table 8.1 Average consumer acceptability by attribute of fried salted peanuts in different vegetable oils: sunflower (FP-S), corn (FP-C), soybean (FP-So), peanut (FP-P), and olive (FP-O).

	Consumer acceptability by attribute[a]			
Fried salted peanuts	**Colour**	**Aroma**	**Texture**	**Flavour**
FP-S	5.12 ± 0.86 B	4.90 ± 1.17 B	5.22 ± 1.07 B	5.63 ± 1.01 C
FP-C	4.86 ± 1.05 A	4.43 ± 1.22 A	4.99 ± 1.11 A	5.13 ± 1.31 B
FP-So	5.13 ± 1.06 B	4.33 ± 1.23 A	4.91 ± 1.14 A	4.85 ± 1.36 A
FP-P	5.37 ± 1.02 C	5.36 ± 1.03 C	5.39 ± 1.08 B	5.98 ± 0.93 D
FP-O	5.04 ± 1.11 B	4.33 ± 1.30 A	4.93 ± 1.22 A	5.03 ± 1.54 AB

Source: Reproduced with permission of John Wiley & Sons (Ryan et al. 2008).
[a] Mean ± SD. The same letter within a column indicates acceptability is not significantly different at $\alpha = 0.05$.

acceptance for each attribute were found among the products. In texture, FP-P and FP-S had similar consumer acceptance, which was much higher than in the other products. FP-P had the highest consumer acceptances for colour, aroma, and flavour.

Several techniques are reported in the scientific literature for checking the physical properties and chemical changes of frying fats and foods. Physical methods include density, viscosity, smoke point, colour, refractive index, ultraviolet (UV) absorption, infrared spectroscopy (IRS), and dielectric constant (Gertz 2000). Chemical examinations involve the determination of free fatty acids (FFAs), iodine value (IV), anisidine value (AV), saponification value, nonoxidized monomer fatty acids, polymerized triglycerides, fatty acids insoluble in petrol ether, and total polar compounds. In the production of fried food, the quality control analyst, an inspector, or a member of the health food inspection service looks for a simple, easy technique, preferably not involving the use of chemicals. The results obtained should be in line with official methods, such as polar compounds (Dobarganes et al. 2000).

Fat is one of the principal components of fried food texture and acceptability, but several fat replacers can be used in fried foods. Raben et al. (1997) revealed that sucrose or starch is the best in terms of the diet. Similarly, Archer et al. (2004) showed that inulin and lupin-kernel fibre have potential fat-replacer properties in meat products. In short-dough biscuits, fat replacement by inulin showed similar textural properties to control, with additional health benefits (Rodríguez-García et al. 2013). The effects of 50–56% rice flour (w/w) and gum Arabic (0–5%, w/w) on the physical, sensory, and structural features of fried dough strands were explored by Shanthilal and Bhattacharya (2017). An oil reduction of 25.8% was achieved by the incorporation of gum Arabic. The wide variations in failure force (13.8–25.3 N) and failure strain (11.0–28.6%) indicated the formation of snacks varying in texture from a soft-to-bite brittle to a hard-to-eat less brittle (Figure 8.1). The snacks possessed a porous microstructure comprising air cells, pores, and vacuoles. High moisture content in the dough reduced the sensory acceptance of the fried snacks. The high desirability index of 0.9 was accomplished with a high level of rice flour (56%, w/w), while gum Arabic content was between 3.50 and 3.75% (w/w).

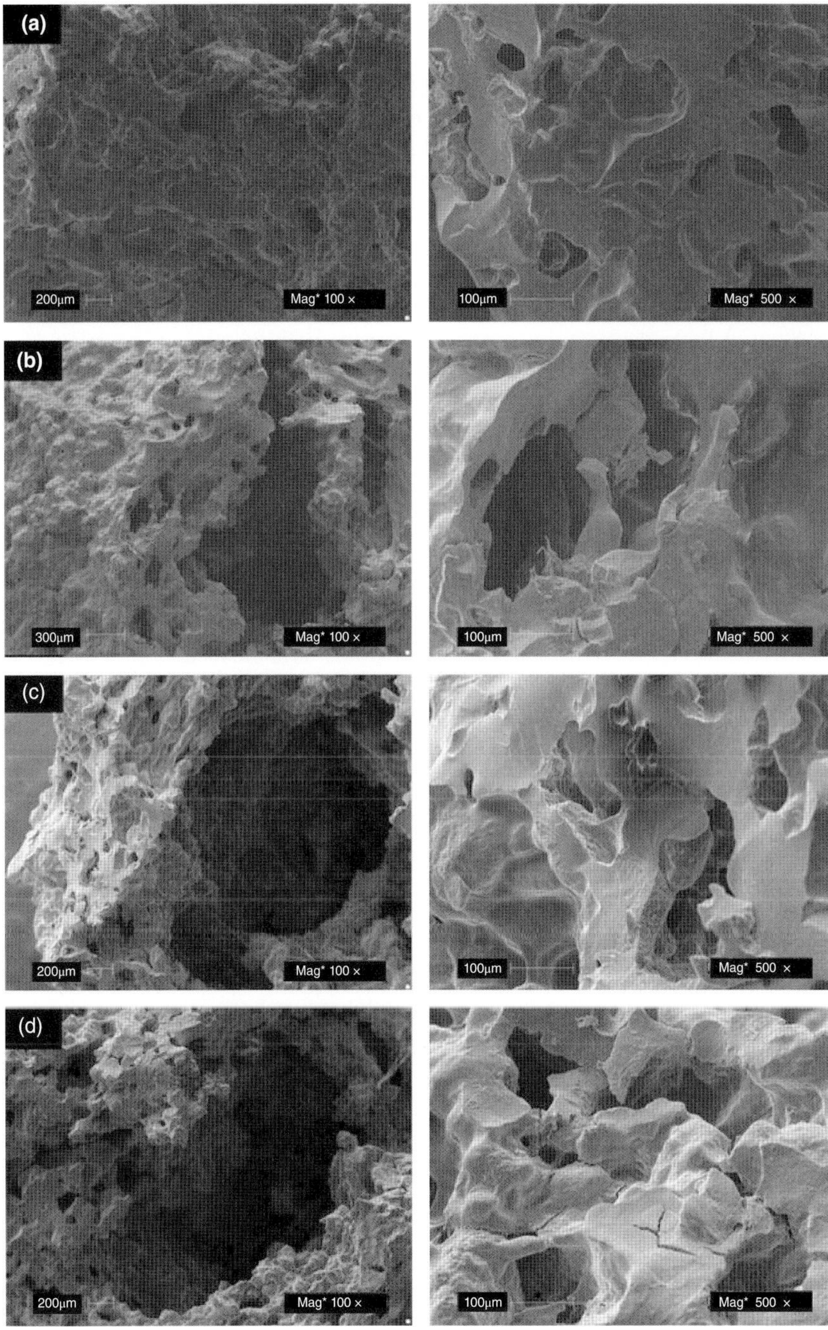

Figure 8.1 Microstructure of fried rice dough strands prepared with 52% rice flour and gum Arabic at (a) 0%, (b) 1%, (c) 3%, and (d) 5% levels. *Left column:* 100× magnification. *Right column:* 500× magnification. *Source:* Reproduced with kind permission of Springer Nature (Shanthilal and Bhattacharya 2017).

8.3 Digestion of Fried Foods

The digestion of food in the gastrointestinal tract involves several steps, starting at the mouth and ending in the large intestine. The digestion of food containing thermally oxidized fats is similar to that of fats and other food components. Digestion-based experiments have been classified into two classes: in vitro and in vivo.

8.3.1 In Vitro Digestion

In vitro digestion involves the use of similar conditions to those found in the human digestive system. A two- or three-step digestion process that includes both gastric and intestinal digestion can be employed (Etcheverry et al. 2012). For gastric digestion, pepsin is added prior to the acidification of the samples to pH 2 or 4. Acidification is important because pepsin denatures itself and thus will lose its activity if pH ≥ 5. Before the start of intestinal digestion, the samples are neutralized to pH 5.5–6.0 prior to the addition of pancreatin and bile salts, and then readjusted to pH 6.5–7.0. The third digestion step, digestion by lingual α-amylase, precedes the gastric phase. This enzyme is used with the aim of breaking apart the glycosidic bonds of starch molecules (i.e. amylose and amylopectin). Once the sample has been digested, bioaccessibility can be measured via either solubility or gastrointestinal models. Carbonell-Capella et al. (2014) review the analytical methods for determining the bioavailability and bioaccessibility of bioactive compounds found in fruits and vegetables.

Garrett et al. (2000) applied an *in vitro* digestion model to a stir-fried meal consisting of fresh spinach, carrot, tomato paste, and vegetable oil. They focused on the bioavailability of carotenoids in fresh stir-fried vegetables, finding that carotenoid transfer from the meal to the aqueous fraction was inhibited when bile extract was omitted from the intestinal phase of digestion. The bioavailability of the micellarized carotenoids was validated using differentiated cultures of Caco-2 human intestinal cells. All four carotenoids were accumulated in a linear manner throughout a 6-hour incubation period. Among the carotenoids, β-carotene content was found to be directly proportional to the *in vitro* digestion of different cassava fruits (Thakkar et al. 2007). After the disruption of food matrix during digestion, an increase in fat amount appears to improve the absorption of carotenes to a larger extent than an increase in xanthophylls (Yonekura and Nagao 2007). Carotenes have been found to be efficiently taken up by enterocytes. However, carotenoid emulsification and micellization steps are greatly affected by the food matrix and dietary components, being the main determinant of carotenoid bioavailability from foodstuffs. Although the intestinal uptake of carotenoids has been thought to occur through simple diffusion, recent studies reported the existence of receptor-mediated transport of carotenoids in enterocytes. The mechanism of absorption of several carotenoids, phytochemicals, and other food components in the intestine is still under investigation. There are, however, several gaps in the reported studies on the bioavailability of phytochemicals, as described by Bohn et al. (2015).

Nieva-Echevarría et al. (2017b) studied the *in vitro* digestion of farmed European sea bass fillets (raw, brine-salted, and dry-salted). They analysed fish lipid extracts before and after digestion using 1H NMR, and examined the headspace composition of the digestates using SPME-GC/MS. They found that during digestion, both lipolysis and fish lipid oxidation took place, the latter evidenced by the generation of conjugated dienes supported on chains with hydroperoxy and hydroxy groups, by the increase of volatile secondary oxidation products, and by the decrease of the antioxidant 2,6-di-tert-butyl-hydroxytoluene (BHT). Similarly, esterification and Maillard-type reactions also occurred. Salting, and especially dry-salting, boosted all of these reactions (except lipolysis) during digestion. Oxidation of polyunsaturated fatty acid and vitamin A was also found to occur during *in vitro* digestion (Nieva-Echevarria et al. 2017a,c).

Meat and meat products have been found to be oxidized, forming several compounds with characteristic harmful properties during digestion (Hecke et al. 2017). The use of a dynamic artificial digestive system for meat digestion has recently been reported by De La Pomélie et al. (2018), who found that both oxidation and nitrozation of meat protein occurred during digestion.

8.3.2 In Vivo Digestion

In vivo digestion studies are very difficult to carry out. They pose several important questions about what happens to each metabolite, and how. The fate of digestion is studied at the end-product level in excreta (e.g. urine, stool) or blood. The latter is considered a principal means of determining the roles of specific food components. *In vivo* digestion studies are mostly based on animal models; human studies are limited.

González-Muñoz et al. (2003) tested the *in vivo* digestibility of the highly polymerized sunflower oil in rats. The digestibility of polymers, dimers, and total oligomers was significantly lower than in unheated oil (Figure 8.2). Moreover, the true digestibility coefficient of the whole heated oil tended to be less than that of the whole unheated oil. The digestibility of triacylglycerol (TAG) monomers was similar for both heated and unheated oils.

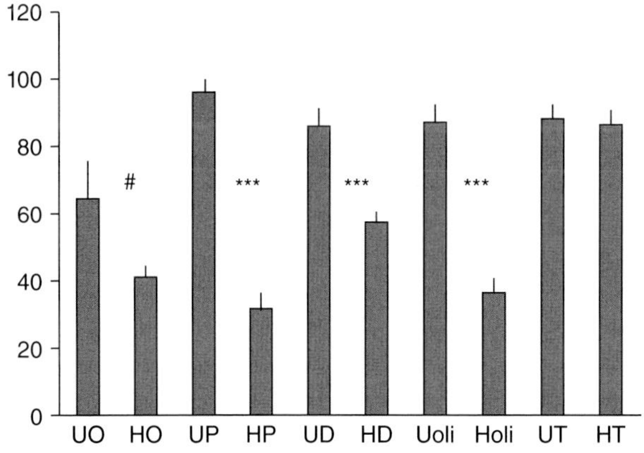

Figure 8.2 True digestibility coefficients (mg fat 100 mg^{-1}) of whole unheated sunflower oil (UO), heated sunflower oil (HO), polymers from unheated sunflower oil (UP), polymers from heated sunflower oil (HP), dimers from unheated sunflower oil (UD), dimers from heated sunflower oil (HD), total oligomers from unheated sunflower oil (Uoli), total oligomers from heated sunflower oil (Holi), triacylglycerol (TAG) monomers from the unheated sunflower oil (UT), and TAG monomers from heated sunflower oil (UH). # $p < 0.1$; *** $p < 0.001$. *Source:* Reproduced with kind permission of John Wiley & Sons (González-Muñoz et al. 2003).

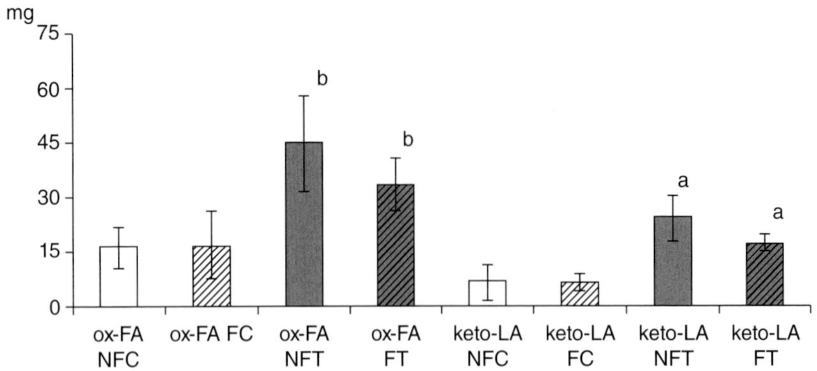

Figure 8.3 Percentage of administrated TAG oligomers, dimers, and polymers found in the small intestine after 4 hours. *Source:* Reproduced with kind permission of Wiley-VCH Verlag GmbH & Co (Olivero-David et al. 2010).

Similarly, Olivero-David et al. (2010) showed the percentages of TAG oligomers, dimers, and polymers found in the small intestine 4 hours after oil ingestion (Figure 8.3). There was no significant difference between the levels of these compounds in rats given oils from different frying operations. Nonetheless, the percentage of oligomers, dimers, and polymers in the small bowel of rats given oil from the 40th frying tended to be higher than that in animals given an unused oil, while that in rats administered oil from the 90th frying tended to be lower (Olivero-David et al. 2010).

The effect of oxidized oil on the digestibility of oxidized fatty acid (Ox-FA) and keto-linoleic acid (keto-LA) in rats reveals that the presence of oxidized species in the small intestine is influenced by oil oxidation (Olivero-David et al. 2011). Nonfasted (NFT) and fasted (FT) animals showed significantly higher levels of ox-FA and keto-LA in intestinal tissue than their control NFC and FC counterparts, as shown in Figure 8.4. These authors showed that total ox-FA and keto-LA were efficiently absorbed during the first 5 hours after sunflower oil administration, but poorly absorbed in the case of fresh oils. It was suggested that oil alteration influenced the digestibility of ox-FA and keto-LA more than fasting, while the digestibility of oxidized oil was significantly affected by fasting.

Feinle et al. (2001) showed that fat digestion is required for the modulation of gastrointestinal sensations, whereas fatty acid chain length plays an important role in digestion. Orozco-Solano et al. (2013) showed that the serum level of essential fatty acids (EFAs) and non- essential fatty acids (NEFAs) is an important biomarker of intervention diets. These authors studied the effect of the intake of breakfasts prepared with four edible oils subjected to a simulated deep-frying protocol in human volunteers and evaluated them in terms of serum fatty acids levels. Statistical analysis has revealed the discrimination capability of omega-6 and omega-3 polyunsaturated fatty acids (PUFAs). Among them, serum levels of arachidonic acid and EPA allowed for an interpretation of the metabolic differences observed in the profiles of fatty acids from individuals after intake of four breakfasts. It is worth noting that gastric digestion of fried or unfried food is greatly influenced by the structural aspects of the food (Bornhorst and Paul Singh 2014). Thus, determining the structural properties of fried food is very important in quality determination.

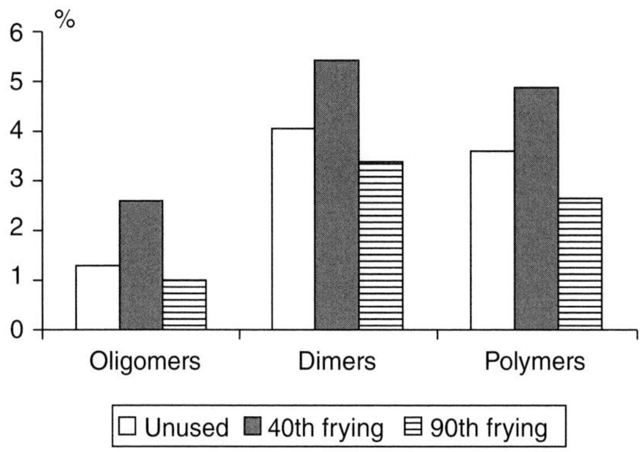

Figure 8.4 Amount (mg) of total oxidized fatty acids (ox-FA) and ketolinoleic acids (keto-LA) present in the intestinal tissue 5 hours after administration of unused sunflower oil and sunflower oil used in 40 frying operations. For ox-FA, fasting conditions effect, nonsignificant; oil effect, $P < 0.001$; fasting conditions × oil alteration interaction, nonsignificant. NFT and FT data bearing a letter (b, $P < 0.01$) were significantly different from NFC and FC values, respectively. For ketolinoleic acids (keto-LA), fasting conditions effect, nonsignificant; oil effect, $P < 0.001$; fasting conditions × oil alteration interaction, nonsignificant. NFT and FT data bearing a letter (a, $P < 0.001$) were significantly different from NFC and FC values, respectively. NF, nonfasted animals; F, fasted animals; NFC, nonfasted animals given control oil; FC, fasted animals given control oil; NFT, nonfasted animals given test oil; F, fasted animals given test oil. Data are mean ± SD of six animals per group. *Source:* Reproduced with permission of American Chemical Society (Olivero-David et al. 2011).

8.4 Absorption of Fried Foods

Digestion of fried food helps in the absorption of food components. Absorptions studies have been reported in animal models. These model studies are based on the absorption of specific compounds or classes of compound. For example, Tomoyori et al. (2004) showed that phytosterol oxidation products are absorbed in the intestinal lymphatic system in rats. Cholesterol oxidation products have been reported to be absorbed and accumulated in the epicardium of male swine, leading to cardiac dysfunction (Shimabukuro et al. 2016). This is due to the absorption of oxidized cholesterol into the bloodstream and, ultimately, the heart through lipoproteins. Thermally oxidized edible oils have been found to increase low-density lipoproteins (LDLs) and decrease high-density lipoproteins (HDLs) in rabbits (Zeb and Ullah 2015; Zeb and Rahman 2017).

In humans, ingestion of fried foods has been found to follow normal absorption. TAGs, sterols (or their ester), and phospholipids are emulsified by the bile acids. Emulsification increases the chances of hydrolysis or of the activity of various lipase enzymes, as depicted in Figure 8.5. About 15–75% of cholesterol is absorbed, while the rest is excreted in faeces. The solubilization of cholesterol is very poor in the aqueous environment of the gut, and thus its digestion and absorption are reliant on separating it into bile micelles. Nonesterified cholesterol can be incorporated into bile acid micelles, whereas esterified cholesterol in the diet (mostly in the range 10–15%) is hydrolysed by cholesterol esterase (Cohn et al. 2010). These reactions result in the formation of FFAs, mono-acylglycerols (MAGs), and free cholesterol (FC), which are taken up by enterocytes through various transporters and transported to the endoplasmic

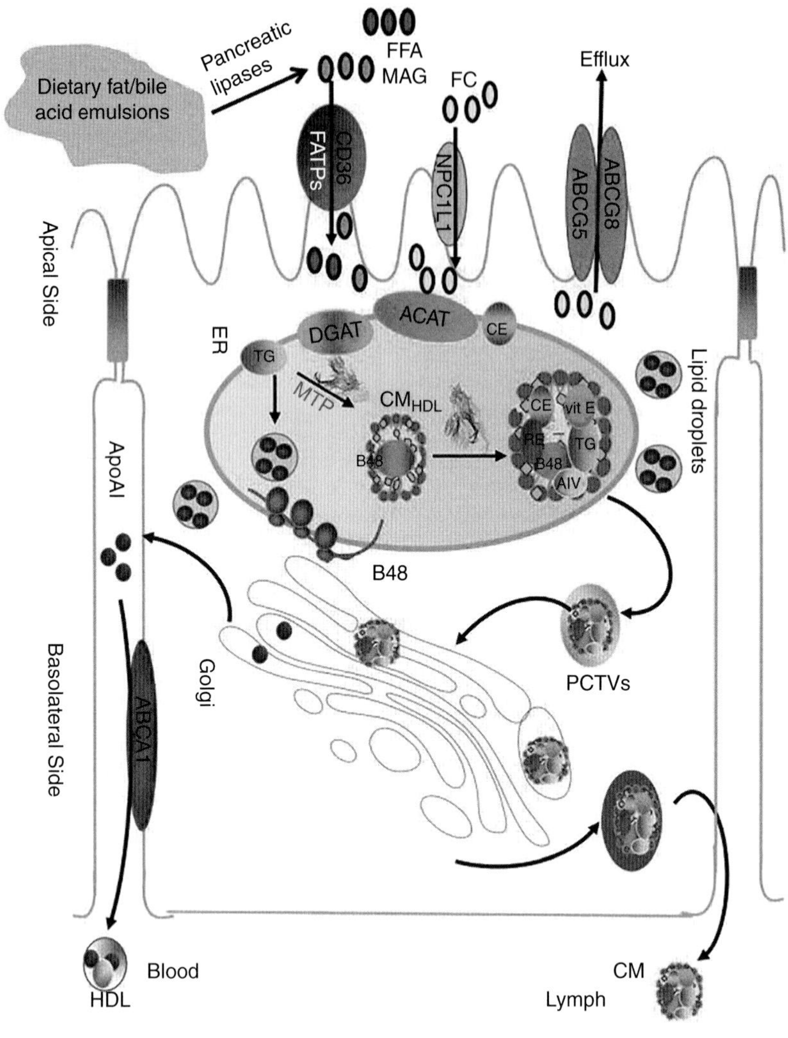

Figure 8.5 Absorption of lipid in the intestinal lumen. *Source:* Reproduced with kind permission of Wolters Kluwer Health (Hussain 2014).

reticulum (ER), where they are used for the synthesis of phospholipids, TAGs, and cholesterol esters. These lipids are then assembled into chylomicron particles using apoB48 as a scaffolding protein with the help of microsomal triglyceride transfer protein (MTP) (Hussain 2014). Alternatively, they can be stored in the cytosol as lipid droplets. FC can be either excreted back into the lumen through ABCAG5/ABCG8 transporters or effluxed into the blood circulation by ABCA1 and apoAI. The Niemann-Pick C1-like 1 (NPC1L1) protein plays a major role in the uptake of cholesterol by enterocytes. The microdomains present in these proteins are then internalized and transported to the endocytic recycling compartment, where cholesterol is released and transported to other organelles, including the ER. The mechanism of these transports is unknown.

Staprans et al. (1994) showed that oxidized lipids in the diet are a source of oxidized lipids in chylomicrons in postprandial human serum. Similarly, Wilson et al. (2002) studied the absorption of fat containing ^{13}C-labelled hydroxyl and dihydroxy TAGs in two groups of healthy women. Postprandial ^{13}C-labelled hydroxy fatty acid concentration increased in a pattern somewhat different from that of plasma triglycerides, with peak levels being reached at between 4 and 6 hours. The ^{13}C monohydroxy triglycerides seemed to be absorbed to a larger degree than the ^{13}C dihydroxy triglycerides.

Goodman (2010) reviewed the digestion and absorption of major nutrients in humans, including carbohydrates, proteins, and fats. Phenolic compounds (PCs) are present in large amounts in plant foods and are effectively absorbed by the body. Dietary oils rich in phenolic compounds are effective in preventing the absorption of toxic substances from fried oils and foods (Zeb and Ahmad 2017). Dominguez-Avila et al. (2017) reviewed the absorption of dietary phenolic compounds. They showed that dietary PCs must pass through gut epithelia before reaching the bloodstream. Passive diffusion (PD), carrier-mediated active (AT), facilitated transport (FT), and para-cellular transport in tight junctions (TJ) are the most widely known absorption mechanisms (see Figure 8.6). Transcellular transport is facilitated by transmembrane pumps, channels, and carriers, while paracellular transport is driven by solute gradients. Transport out of epithelial cells is further assisted by basolateral membrane transporters, which deliver both free and vesicle-contained PCs into the bloodstream and lymphatic system, or else efflux them back to the lumen. Novel transporters and mechanisms for the intestinal handling of PCs and the inhibitory action of PCs on well-known transporters have been elucidated in recent years. Several other parameters and components also affect the absorption of PCs from the diet. Polyphenols plays a significant role in the regulation of tight junctions and, ultimately, the absorption of other metabolites from the diet (Yang et al. 2017).

8.5 Excretion of Fried Foods

Fried food and its components may not be fully absorbed by the body and therefore be excreted. Oxidized or small oxidation products are absorbed and further metabolized to form water-soluble compounds, such as phenolic compounds, which are excreted by the kidneys in the form of urine. Mullen et al. (2006) studied the excretion of several metabolites of quercetin in urine. They revealed that the profile of metabolites excreted in human urine was significantly different from that of corresponding plasma. Several major urinary metabolites, such as quercetin-3-glucuronide, two quercetin glucoside sulfates, and a methyl quercetin diglucuronide, were either absent or present in only trace amounts in the bloodstream. Hydroxymethylfurfural (HMF) (Husøy et al. 2008) and acrylamide (Bjellaas et al. 2007) have been found to be excreted in urine.

In rats, long-term feeding of heated and fried oils showed the excretion of several lipids in faeces (Purushothama et al. 2003). The maximum absorption of fats, irrespective of whether the oils were heated or fried, showed only marginal variations compared to controls. Analysis of faecal fatty acid composition indicated high excretion of oleic acid as compared to palmitic and linoleic acids, as shown in Table 8.2. The fried fats were less absorbed than were heated oil and fresh oils.

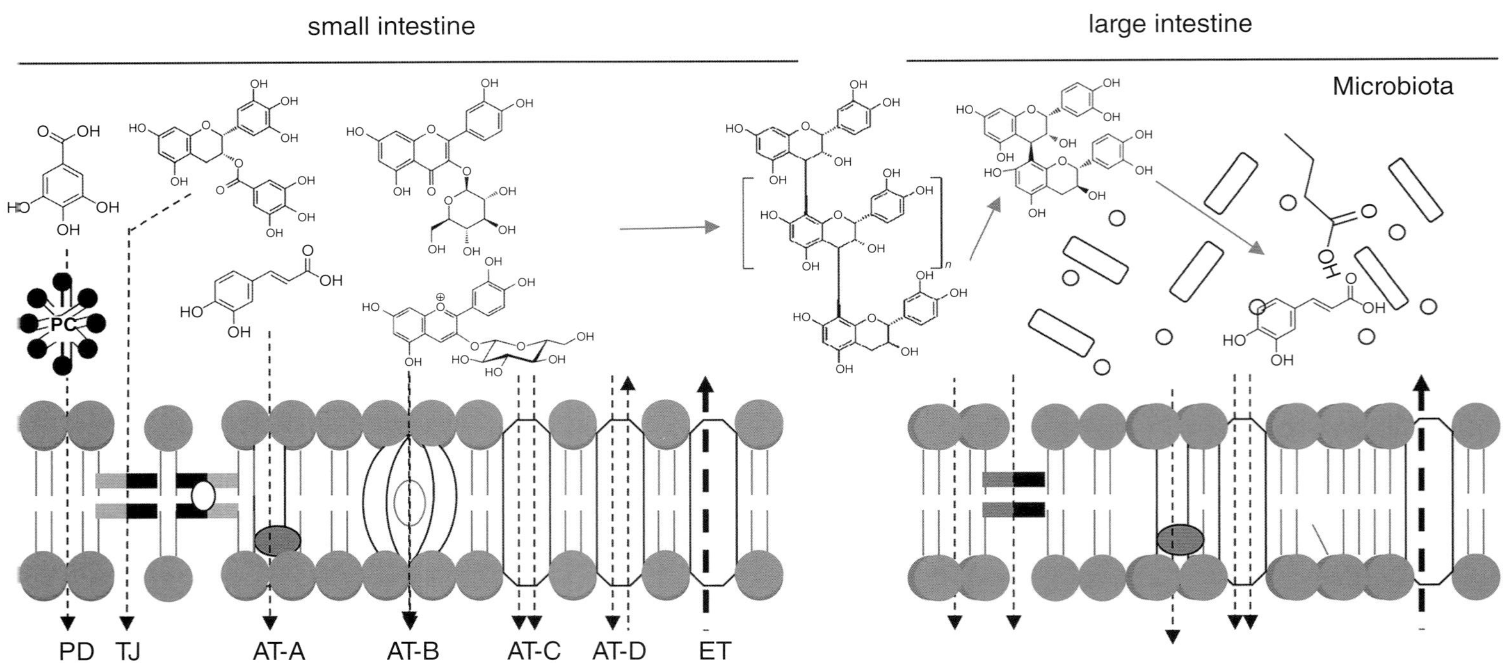

Figure 8.6 Hypothetical transport mechanisms for phenolic compounds (PCs): passive diffusion (PD), para-cellular (TJ), single-solute (AT-A), translocase-type (AT-B), co-solute symporter (AT-C) and antiporter (AT-D) active transporters, and efflux transport (ET). *Source:* Reproduced with kind permission of Royal Society of Chemistry (Dominguez-Avila et al. 2017).

Table 8.2 Faecal fat excretion and fatty acid composition of rats fed fresh, heated, and fried oils.

	Peanut oil (PNO)			Rice bran oil (RBO)			Palm oil (PO)		
Parameters	**FO**	**HO**	**FRO**	**FO**	**HO**	**FRO**	**FO**	**HO**	**FRO**
Total food given (g)	115	115	115	115	115	115	115	115	115
Total food consumed (g)	109	112	108	106	108	112	112	110	113
Fat consumed (g)	21.8	24.4	21.6	21.2	21.6	22.4	22.4	22.0	22.6
Faecal fat excreted (mg)	118.6	178.0	138.0	122.0	162.4	186	108	190	160.5
% absorption	94.6	92.1	90.4	94.3	92.5	91.7	95.2	91.4	93.0
Palmitic acid[a]	32.6	28.4	38.4	40.2	43.4	38.4	40.3	46.4	51.4
Oleic acid[a]	58.2	61.3	52.3	49.4	42.2	45.4	56.4	48.2	41.4
Linoleic acid[a]	8.44	10.2	9.33	12.3	14.3	15.6	3.22	5.44	7.22

Source: Reproduced with kind permission of Springer Nature (Purushothama et al. 2003).
Values are mean of n = 8 rats.
[a] Results of 7 days' collection of faeces. FO, fresh oil; HO, heated oil; FRO, fried oil.

8.6 Key Concepts

- The acceptability of fried fats and foods depends on several physical and chemical parameters.
- Physical methods of determining acceptability include density, viscosity, smoke point, colour, refractive index, UV absorption, IRS, and dielectric constant.
- Chemical parameters for determining acceptability include FFA, IV, AV, saponification value, nonoxidized monomer fatty acids, polymerized triglycerides, fatty acids insoluble in petrol ether, and total polar compounds.
- Digestion of fried foods has been studied using both *in vitro* and *in vivo* methods.
- *In vitro* digestion involves instrumental digestion mimicking stomach and intestinal digestion.
- *In vivo* studies reveal the different means of digestion of fried foods, based on human perception.
- Absorption of fried food components is mediated by a special transport system or mechanism, specific for each class of compounds.
- Undigestible and absorbable food components are excreted in urine or faeces.

References

Archer, B.J., Johnson, S.K., Devereux, H.M., and Baxter, A.L. (2004). Effect of fat replacement by inulin or lupin-kernel fibre on sausage patty acceptability, post-meal perceptions of satiety and food intake in men. *British Journal of Nutrition* 91: 591–599.

Bjellaas, T., Stølen, L.H., Haugen, M. et al. (2007). Urinary acrylamide metabolites as biomarkers for short-term dietary exposure to acrylamide. *Food and Chemical Toxicology* 45: 1020–1026.

Bohn, T., Mcdougall, G.J., Alegria, A. et al. (2015). Mind the gap-deficits in our knowledge of aspects impacting the bioavailability of phytochemicals and their metabolites – a position paper focusing on carotenoids and polyphenols. *Molecular Nutrition & Food Research* 59: 1307–1323.

Bornhorst, G.M. and Paul Singh, R. (2014). Gastric digestion in vivo and in vitro: how the structural aspects of food influence the digestion process. *Annual Review of Food Science and Technology* 5: 111–132.

Carbonell-Capella, J.M., Buniowska, M., Barba, F.J. et al. (2014). Analytical methods for determining bioavailability and bioaccessibility of bioactive compounds from fruits and vegetables: a review. *Comprehensive Reviews in Food Science and Food Safety* 13: 155–171.

Cohn, J., Kamili, A., Wat, E. et al. (2010). Dietary phospholipids and intestinal cholesterol absorption. *Nutrients* 2: 116.

De La Pomélie, D., Santé-Lhoutellier, V., Sayd, T., and Gatellier, P. (2018). Oxidation and nitrosation of meat proteins under gastro-intestinal conditions: consequences in terms of nutritional and health values of meat. *Food Chemistry* 243: 295–304.

Dobarganes, M., Velasco, J., and Dieffenbacher, A. (2000). Determination of polar compounds, polymerized and oxidized triacylglycerols, and diacylglycerols in oils and fats: results of collaborative studies and the standardized method (technical report). *Pure and Applied Chemistry* 72: 1563–1575.

Dominguez-Avila, J.A., Wall-Medrano, A., Velderrain-Rodriguez, G.R. et al. (2017). Gastrointestinal interactions, absorption, splanchnic metabolism and pharmacokinetics of orally ingested phenolic compounds. *Food & Function* 8: 15–38.

Etcheverry, P., Grusak, M., and Fleige, L. (2012). Application of in vitro bioaccessibility and bioavailability methods for calcium, carotenoids, folate, iron, magnesium, polyphenols, zinc, and vitamins B6, B12, D, and E. *Frontiers in Physiology* 3: 317.

Feinle, C., Rades, T., Otto, B., and Fried, M. (2001). Fat digestion modulates gastrointestinal sensations induced by gastric distention and duodenal lipid in humans. *Gastroenterology* 120: 1100–1107.

Garrett, D.A., Failla, M.L., and Sarama, R.J. (2000). Estimation of carotenoid bioavailability from fresh stir-fried vegetables using an in vitro digestion/Caco-2 cell culture model. *Journal of Nutritional Biochemistry* 11: 574–580.

Gertz, C. (2000). Chemical and physical parameters as quality indicators of used frying fats. *European Journal of Lipid Science and Technology* 102: 566–572.

González-Muñoz, M.J., Bastida, S., and Sánchez-Muniz, F.J. (2003). Short-term in vivo digestibility assessment of a highly oxidized and polymerized sunflower oil. *Journal of the Science of Food and Agriculture* 83: 413–418.

Goodman, B.E. (2010). Insights into digestion and absorption of major nutrients in humans. *Advances in Physiology Education* 34: 44–53.

Hecke, T., Camp, J., and Smet, S. (2017). Oxidation during digestion of meat: interactions with the diet and helicobacter pylori gastritis, and implications on human health. *Comprehensive Reviews in Food Science and Food Safety* 16: 214–233.

Husøy, T., Haugen, M., Murkovic, M. et al. (2008). Dietary exposure to 5-hydroxymethylfurfural from Norwegian food and correlations with urine metabolites of short-term exposure. *Food and Chemical Toxicology* 46: 3697–3702.

Hussain, M.M. (2014). Intestinal lipid absorption and lipoprotein formation. *Current Opinion in Lipidology* 25: 200–206.

Mullen, W., Edwards, C.A., and Crozier, A. (2006). Absorption, excretion and metabolite profiling of methyl-, glucuronyl-, glucosyl-and sulpho-conjugates of quercetin in human plasma and urine after ingestion of onions. *British Journal of Nutrition* 96: 107–116.

Nieva-Echevarría, B., Goicoechea, E., and Guillén, M.D. (2017a). Polyunsaturated lipids and vitamin A oxidation during cod liver oil in vitro gastrointestinal digestion. Antioxidant effect of added BHT. *Food Chemistry* 232: 733–743.

Nieva-Echevarría, B., Goicoechea, E., Manzanos, M.J., and Guillén, M.D. (2017b). Fish in vitro digestion: influence of fish salting on the extent of lipolysis, oxidation, and other reactions. *Journal of Agricultural and Food Chemistry* 65: 879–891.

Nieva-Echevarria, B., Goicoechea, E., Manzanos, M.J., and Guillen, M.D. (2017c). 1H NMR and SPME-GC/MS study of hydrolysis, oxidation and other reactions occurring during in vitro digestion of non-oxidized and oxidized sunflower oil. Formation of hydroxy-octadecadienoates. *Food Research International* 91: 171–182.

Olivero-David, R., González-Muñoz, M.J., Benedí, J. et al. (2010). Thermally oxidized palm olein exposure increases triglyceride polymer levels in rat small intestine. *European Journal of Lipid Science and Technology* 112: 970–976.

Olivero-David, R., Paduano, A., Fogliano, V. et al. (2011). Effect of thermally oxidized oil and fasting status on the short-term digestibility of ketolinoleic acids and total oxidized fatty acids in rats. *Journal of Agricultural and Food Chemistry* 59: 4684–4691.

Orozco-Solano, M.I., Priego-Capote, F., and Luque De Castro, M.D. (2013). Analysis of esterified and nonesterified fatty acids in serum from obese individuals after intake of breakfasts prepared with oils heated at frying temperature. *Analytical and Bioanalytical Chemistry* 405: 6117–6129.

Purushothama, S., Ramachandran, H.D., Narasimhamurthy, K., and Raina, P.L. (2003). Long-term feeding effects of heated and fried oils on hepatic antioxidant enzymes, absorption and excretion of fat in rats. *Molecular and Cellular Biochemistry* 247: 95–99.

Raben, A., Macdonald, I., and Astrup, A. (1997). Replacement of dietary fat by sucrose or starch: effects on 14 d ad libitum energy intake, energy expenditure and body weight in formerly obese and never-obese subjects. *International Journal of Obesity* 21: 846–859.

Rodríguez-García, J., Laguna, L., Puig, A. et al. (2013). Effect of fat replacement by inulin on textural and structural properties of short dough biscuits. *Food and Bioprocess Technology* 6: 2739–2750.

Ryan, L.C., Mestrallet, M.G., Nepote, V. et al. (2008). Composition, stability and acceptability of different vegetable oils used for frying peanuts. *International Journal of Food Science and Technology* 43: 193–199.

Shanthilal, J. and Bhattacharya, S. (2017). Frying of rice flour dough strands containing gum Arabic: texture, sensory attributes and microstructure of products. *Journal of Food Science and Technology* 54: 1293–1303.

Shimabukuro, M., Okawa, C., Yamada, H. et al. (2016). The pathophysiological role of oxidized cholesterols in epicardial fat accumulation and cardiac dysfunction: a study in swine fed a high caloric diet with an inhibitor of intestinal cholesterol absorption, ezetimibe. *Journal of Nutritional Biochemistry* 35: 66–73.

Staprans, I., Rapp, J.H., Pan, X.M. et al. (1994). Oxidized lipids in the diet are a source of oxidized lipid in chylomicrons of human serum. *Arteriosclerosis and Thrombosis* 14: 1900–1905.

Thakkar, S.K., Maziya-Dixon, B., Dixon, A.G., and Failla, M.L. (2007). β-Carotene micellarization during in vitro digestion and uptake by Caco-2 cells is directly proportional to β-carotene content in different genotypes of cassava. *Journal of Nutrition* 137: 2229–2233.

Tomoyori, H., Kawata, Y., Higuchi, T. et al. (2004). Phytosterol oxidation products are absorbed in the intestinal lymphatics in rats but do not accelerate atherosclerosis in apolipoprotein E–deficient mice. *Journal of Nutrition* 134: 1690–1696.

Wayler, A.H., Kapur, K.K., Feldman, R.S., and Chauncey, H.H. (1982). Effects of age and dentition status on measures of food acceptability. *Journal of Gerontology* 37: 294–299.

Wilson, R., Lyall, K., Smyth, L. et al. (2002). Dietary hydroxy fatty acids are absorbed in humans: implications for the measurement of 'oxidative stress' in vivo. *Free Radical Biology and Medicine* 32: 162–168.

Yang, G., Bibi, S., Du, M. et al. (2017). Regulation of the intestinal tight junction by natural polyphenols: a mechanistic perspective. *Critical Reviews in Food Science and Nutrition* 57: 3830–3839.

Yonekura, L. and Nagao, A. (2007). Intestinal absorption of dietary carotenoids. *Molecular Nutrition & Food Research* 51: 107–115.

Zeb, A. and Ahmad, S. (2017). Changes in acylglycerols composition, quality characteristics and in vivo effects of dietary pumpkin seed oil upon thermal oxidation. *Frontiers in Chemistry* 5: 55.

Zeb, A. and Rahman, S.U. (2017). Protective effects of dietary glycine and glutamic acid toward the toxic effects of oxidized mustard oil in rabbits. *Food and Function* 8: 429–436.

Zeb, A. and Ullah, S. (2015). Sea buckthorn seed oil protects against the oxidative stress produced by thermally oxidized lipids. *Food Chemistry* 186: 6–12.

9

Nutrition and Metabolism of Fried Foods

9.1 Introduction

The frying medium and fried foods provide several important nutrients that are considered vital for human survival and a balanced healthy life. Dietary fat contains important lipids, which are essential for maintaining several biological functions. It provides more energy than its counterpart biomolecules, such as carbohydrates and proteins. Triacylglycerols (TAGs) in the fat are hydrolysed by lipase enzymes, releasing fatty acids (FAs) and glycerol. Upon reaching cells, the FAs enter their mitochondria through the carnitine transport system and combine with carnitine to form acylcarnitine. Acylcarnitine is then transported to the mitochondrial matrix, where it is converted back to acyl CoA and carnitine. The acyl CoA is metabolized to produce acetyl CoA by several metabolic processes, in a process called β-oxidation. The acetyl CoA is then metabolized in the citric acid cycle to produce reducing equivalents or adenosine triphosphate (ATP) at the substrate level. Heat is released when an ATP is hydrolysed. Acetyl CoA produced by the oxidation of FAs may serve as a precursor for several other lipid metabolites, such as cholesterol.

Carbohydrates are absorbed in the body in the form of monosaccharides such as glucose and fructose. Glucose is oxidized to form pyruvate, and finally acetyl CoA via a glycolytic process. Pyruvate is transported to the cell and metabolized to form acetyl CoA. Condensation of acetyl CoA and oxaloacetate forms the first product of citric acid metabolism, as shown in Figure 9.1. The citric acid cycle plays an amphibolic role in the formation and degradation of metabolites. Thus, metabolism of amino acids is also linked with this cycle. For example, glutamate is produced from the amination reaction of α-ketoglutarate.

Sterols, vitamins, minerals, and polyphenols are taken up by cells according to their need and the transport efficiency of the cell membrane. These metabolites are used by several metabolic reactions. Reducing equivalents such as nicotinamide adenine dinucleotide (NADH) and flavin adenine dinucleotide ($FADH_2$) are metabolized by an electron transport system (ETS) in order to produce ATP. ATP is one of the principal

Food Frying: Chemistry, Biochemistry, and Safety, First Edition. Alam Zeb.

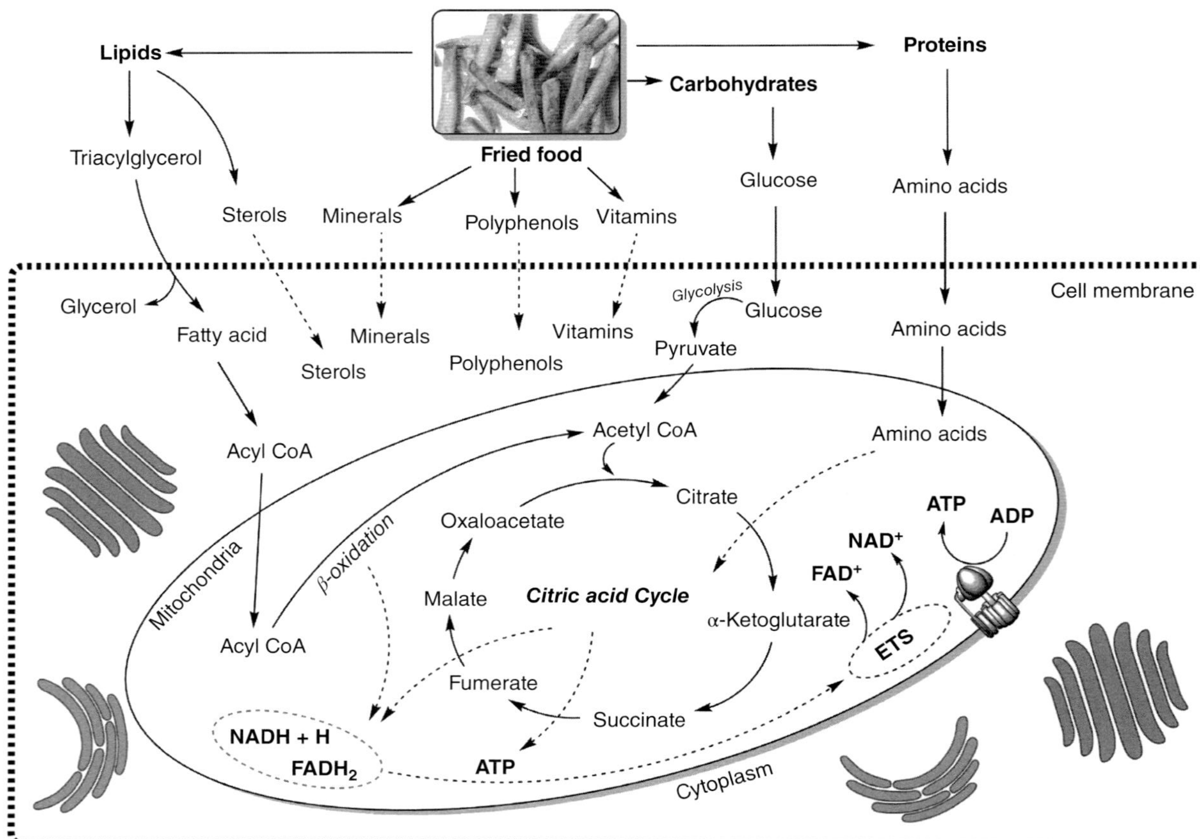

Figure 9.1 Overview of the cellular intake and subsequent metabolism of different biomolecules obtained from fried foods.

sources of energy compounds, which provide enough energy for an endergonic[1] reaction. Upon hydrolysis, proteins yield corresponding amino acids, which take part in either the synthesis of new proteins or the metabolism of other products.

9.2 Metabolism of Fried Lipids

Some lipids are biosynthesized by the human body, but the majority are obtained from the diet. Lipids contained in fried foods are ingested along with other food materials. Frying has been found to contribute to fat accumulation, and high-fat diets (HFDs) affect several metabolic pathways. Raddatz et al. (2011) showed that protein kinase C epsilon (Prkce) deletion ameliorates diet-induced glucose intolerance and provides protection against insulin resistance, which is associated with changes in hepatic lipid partitioning and may reduce the acute inhibitory effects of FA catabolism, such as reactive oxygen species (ROS) generation (Figure 9.2). Similarly, Beyaz et al. (2016) showed that HFD-induced obesity enhances the numbers and functions of mammalian intestinal stem cells and increases the chances of cancer. Analysis of liver mRNAs encoding hepatic genes known to regulate lipid homeostasis revealed that an HFD based on phenolic-rich extra virgin olive oil had a significantly increased expression of the FA transporter after 36 weeks of nutritional intervention in mice (Jurado-Ruiz et al. 2017). These studies show that HFD can severely affect metabolism and, ultimately, health.

Therefore, it is important to discuss the metabolism of dietary lipids rather than endogenously synthesized lipids. The details provided in this section have been acquired

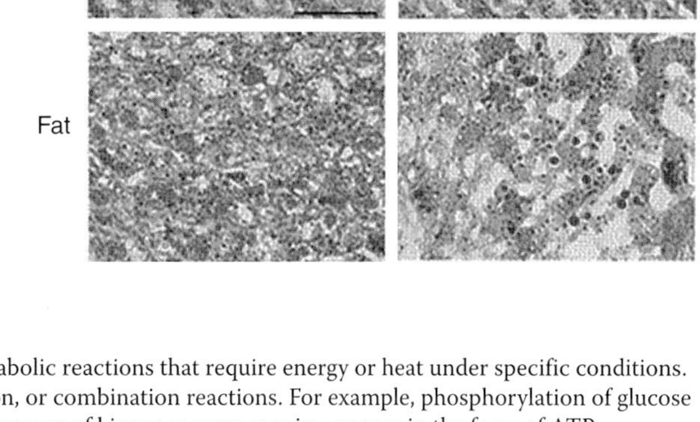

Figure 9.2 Effects of a high-fat diet (HFD) and standard chow diet on the accumulation of neutral lipids in the liver of protein kinase C epsilon (Prkce$^{-/-}$) and wild-type (WT) mice for 1 week. *Source:* Reproduced with kind permission of Springer Nature (Raddatz et al. 2011).

1 Endergonic reactions are those metabolic reactions that require energy or heat under specific conditions. They are mostly addition, condensation, or combination reactions. For example, phosphorylation of glucose to form glucose-6-phosphate in the presence of kinase enzyme requires energy in the form of ATP. Exergonic reactions are those which produce energy or release heat. These reactions may involve hydrolysis of different types of esters, such as that of ATP.

from the review article of Ringseis and Eder (2011), with updated literature. In recent years, the contribution of oxidized fats to total energy intake in industrialized countries has markedly increased. This is mainly due to the rising consumption of deep-fried products, which are very popular because of their desirable flavours, colours, and crispy textures. In a cohort from the European Prospective Investigation into Cancer and Nutrition (EPIC) study, for instance, the percentage of energy intake coming from fried food was more than 23% in the highest quintile of consumption (Guallar-Castillón et al. 2007). During deep frying, several chemical reactions occur within the frying oil, resulting in the formation of a mixture of chemically distinct lipid peroxidation products. Large quantities of the frying oil are absorbed into the fried food during deep frying, and therefore ingested during the food's consumption.

Feeding experiments with rats, guinea pigs, and pigs revealed that ingestion of oxidized fats provokes various effects, such as depletion of antioxidants, an increase in lipid peroxidation (Izaki et al. 1984; Eder et al. 2003a; Keller et al. 2004; Ringseis and Eder 2005), impairment of glucose tolerance, and alteration of thyroid hormone homeostasis (Eder and Stangl 2000; Eder et al. 2002). In addition, a great number of studies show that oxidized fats influence lipid metabolism. In this regard, it is noteworthy that most studies dealing with lipid metabolism alterations of oxidized fats report reduced TAG and cholesterol concentrations in the liver and plasma of animals fed oxidized fat (Huang et al. 1988; Eder and Kirchgessner 1998; Ringseis and Eder 2004; Ringseis et al. 2007a, 2007b, 2007c). Considering that elevated blood lipid concentrations are known risk factors for the development of atherosclerosis, the lipid-lowering effect of oxidized fat indicates that oxidized fats not only induce adverse effects but obviously also exert potentially beneficial ones.

In most of the early studies dealing with oxidized fats (Eder et al. 2004; Zeb and Rahman 2012), strongly oxidized fats were used, which contained concentrations of lipid peroxidation products that were clearly above the limit allowed for use in restaurants by EU legislation. Feeding of such fats not only reduced food intake and growth rates, but also induced pronounced oxidative stress and even provoked toxic effects (Benedetti et al. 1990; Liu and Huang 1995). Moreover, it could be shown that the digestibility of fat from such abused oxidized fats was reduced. It has therefore been argued that the observed lipid-lowering actions of oxidized fats were mainly due to these confounding effects – in particular, the reduced food and energy intake and the decreased digestion and absorption of the fats. To avoid these confounding effects, more recent studies have used only moderately oxidized fats and a controlled feeding system, in which animals in all treatment groups are fed identical amounts of food (Chao et al. 2001). Moreover, in these studies, the loss of antioxidants and polyunsaturated fatty acids (PUFAs) during the thermo-oxidation of the fats was offset by supplementing the oxidized fat diets with synthetic antioxidants (vitamin E) and PUFA-rich fats, respectively, thereby excluding secondary effects due to a reduced intake of antioxidants or PUFAs. Nonetheless, the lipid-lowering actions of oxidized fats were still observed in these studies. These observations strongly suggest that the lipid-lowering effects of oxidized fats are due not to any potential confounding effects (reduced food intake, decreased nutrient digestibility, depletion of antioxidants, essential FA deficiency) but rather to alterations of lipid metabolism directly in the organism.

9.2.1 Phenotypical Alterations of Lipid Metabolism

9.2.1.1 TAG Concentrations in Liver and Plasma

TAG-lowering effects have been consistently observed in numerous independent experiments, most of which have been performed with rats. Table 9.1 summarizes the studies reporting an effect of oxidized fats on liver and plasma TAG concentrations in different species (Ringseis and Eder 2011). Evidently, the TAG-lowering effects of oxidized fats observed were stronger in rats and mice than in guinea pigs and pigs, indicating a species dependency of this effect. In rats and mice, the reported TAG-lowering effects in the liver were in the range of −21 to −85% (compared with the fresh fat group), whereas in pigs the hepatic TAG concentration was only marginally decreased, by approximately 3%. The strongest TAG-lowering effect observed in the liver was reported in a study with lactating rats, at 85% (Ringseis et al. 2007a, 2007b, 2007c). The oxidized fat in this study was prepared by heating sunflower oil at 60 °C for 25 days and was fed to the rats for a total of 21 days, from parturition until weaning. Notably, in all other studies considered in Table 9.1, the oxidized fats were fed to the animals for longer time periods (up to 98 days) (Ringseis et al. 2007a, 2007b, 2007c). This, the strongest TAG-lowering effect was seen in the study with the shortest feeding time. This indicates that the alteration of hepatic TAG content is rapidly induced in the liver and that the duration of administration is relatively unimportant. Also of great interest is the observation from a recent study that feeding of an oxidized fat compared with a fresh fat can prevent hepatic TAG accumulation induced by simultaneous ethanol administration (Ringseis et al. 2007a, 2007b, 2007c). This shows that the TAG-lowering action of the oxidized fat is useful during conditions of disturbed hepatic lipid metabolism, such as fatty liver disease.

In plasma, the degree of TAG lowering induced by oxidized fat was in a similar range as in the liver, at between −11 (Chao et al. 2001) and −76% (Eder et al. 2003b). As in the liver, it is evident that the effect was generally lower in guinea pigs and pigs than in rats and mice (Luci et al. 2007a, 2007b). This again indicates a species dependency of this effect.

In rabbits, a significant increase was seen in serum TAGs (Zeb and Mehmood 2012; Zeb and Rahman 2012, 2017; Zeb and Ahmad 2017; Zeb and Uddin 2017). Liver weight was significantly higher than in controls, as evident from a histological examination of the liver section of the rabbits, which revealed fats accumulated in all oxidized fats, as shown in Figure 9.3. The OG1 liver samples showed hepatic tissue with almost intact architecture; however, the focal area portal tract showed proliferation of bile duct. The hepatocytes were normal in appearance. Sinusoids showed a mild degree of dilation at places of hepatic tissue, and oedematous change of bile duct occurred. The OG2 hepatic tissues revealed necrotic changes and increased lymphocytic infiltration. However, the portal tract showed lymphatic and fat infiltration. Sinusoids showed a mild degree of dilation at places of hepatic tissue and oedematous changes of the bile duct. The OG3 hepatic tissue revealed coagulated necrosis changes. The bile ducts were dilated, and some were filled with crystalloids, while others showed dilation. The OGNG hepatic tissue revealed bile duct dilation. In some places, moderate cholestasis was also found. The NG3 liver had an architecture that appeared to be almost normal, revealing a central vein (CV), hepatic cords, and portal tracts. The results could be explained with the help of Zhou et al. (2016), who showed that deep-fried oil inhibited lipid metabolism by

Table 9.1 Effect on liver and plasma TAG concentrations of oxidized fats, as observed in studies with different species.

Species	Fat source, heating regime	Feeding period (days)	Liver TAGs (%)[a]	Serum TAGs (%)[a]	References
Rat	Soybean oil, 205 °C, 6 hours with wheat flour dough sheets in the oil, 4 consecutive days	56	−36%	−38%	Huang et al. (1988)
Rat	Soybean oil, 80 °C, 5 days	56	−13%	+5%	Hayam et al. (1993)
Rat	Soybean oil, 130 °C, 22 hours	40	−22%	−21%	Eder and Kirchgessner (1998)
Rat	Lard: safflower oil (2:1, w/w), 150 °C, 6 days	51	−21%	−32%	Eder (1999)
Rat	Soybean oil, 205 °C, 6 hours with wheat flour dough sheets in the oil, 4 consecutive days	42	−72%	−11%	Chao et al. (2001)
Rat	Sunflower oil, 98 °C, 48 hours	69	–	−47%	Ammouche et al. (2002)
Rat	Sunflower oil: lard (1:1, w/w), 50 °C, 38 days	63	−41%	−68%	Eder et al. (2003b)
Rat	Sunflower oil: lard (1:1, w/w), 105 °C, 81 hours	63	−24%	−63%	Eder et al. (2003b)
Rat	Sunflower oil: lard (1:1, w/w), 190 °C, 24 hours	63	−32%	−76%	Eder et al. (2003b)
Rat	Sunflower oil, 55 °C, 42 days	56	−21%	−66%	Eder et al. (2003b)
Rat	Soybean oil, 205 °C, 6 hours with wheat flour dough sheets in the oil, 4 consecutive days	42	−71% (male)	−27% (male)	Chao et al. (2004)
Rat	Sunflower oil: lard (1:1, w/w), 50 °C, 38 days	63	−41%	−63%	Sülzle et al. (2004)
Rat	Sunflower oil: linseed oil (80:20, w/w), 50 °C, 16 days	98	−36%	–	Brandsch et al. (2004)
Rat	Sunflower oil, 60 °C, 25 days	32	−32%	−41%	Ringseis et al. (2007a, 2007b, 2007c)
Rat	Sunflower oil, 60 °C, 25 days	21	−85%	−61%	Ringseis et al. (2007a, 2007b, 2007c)
Rat	Sunflower oil, 60 °C, 25 days	22	−59%	–	Ringseis et al. (2007a, 2007b, 2007c)
Rat	Olive oil, 100 °C, 15 hours	30	–	−10%	Khan et al. (2017)

Table 9.1 (Continued)

Species	Fat source, heating regime	Feeding period (days)	Liver TAGs (%)[a]	Serum TAGs (%)[a]	References
Mouse	Heated hydrogenated palm fat : fresh sunflower oil (92 : 8, w/w), 170 °C, 48 hours with repeated frying of French fries for 6 minutes every 30 minutes	112	–	−53%	Kämmerer et al. (2011)
Guinea pig	Soybean oil, 205 °C, 6 hours with potato sticks in the oil, 4 consecutive days	60	–	−21%	Liu and Lee (1998)
Pig	Sunflower oil, 180 °C, 24 hours	28	−3%	−12%	Luci et al. (2007a, 2007b)
pig	Corn, canola, poultry, and tallow, 185 °C, 7 hours	28	−5%	−5%	Liu et al. (2014)
Rabbit	Tallow, 130 °C, 45 hours	07	–	+139%	Zeb and Mehmood (2012)
Rabbit	Vanaspati ghee, 130 °C, 45 hours	07		+76%	Zeb and Rahman (2012)
Rabbit	Desi ghee, 160 °C, 9 hours	07	–	+ 30%	Zeb and Uddin (2017)
Rabbit	Pumpkin seed oil, 160 °C, 9 hours	14	–	+34%	Zeb and Ahmad (2017)
Rabbit	Mustard oil, 170 °C, 9 hours	14	–	+20%	Zeb and Rahman (2017)

Source: Part of the table is reproduced with kind permission of Wiley-VCH Verlag GmbH & Co. (Ringseis and Eder 2011).
[a] Per cent change of liver or plasma TAG concentration as compared with control.

reducing energy metabolism in rats. Thus, an HFD rich in oxidized fat may be responsible for the deposition of dietary fats in the liver.

Given that the heating regime (temperature and time) influences the spectrum of lipid peroxidation products present in oxidized fat (low temperature, long time: predominance of primary lipid peroxidation products; high temperature, short time: predominance of secondary lipid peroxidation products), it has also been studied whether the heating regime influences the degree of TAG lowering. Eder et al. (2003b) investigated the TAG-lowering property of oxidized fats prepared by different heating regimes (either 50 °C for 38 days, 105 °C for 81 hours, or 190 °C for 24 hours). A slightly stronger TAG-lowering effect was observed in the liver of rats fed the oxidized fat prepared at the lowest temperature. However, strong TAG-lowering effects in the liver were also found in rats fed oxidized fats prepared at 205 °C (Chao et al. 2001, 2004). It is therefore likely that both primary and secondary lipid peroxidation products contribute to the TAG-lowering effect of oxidized fats.

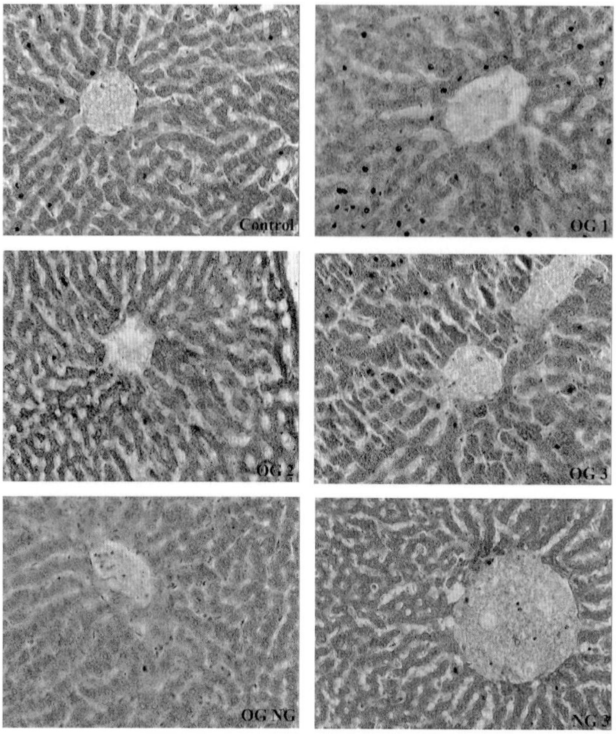

Figure 9.3 Effects of co-administration of unoxidized Desi ghee with oxidized ghee on the liver structure of rabbits. Liver micrographs at ×40, control, OG1 (oxidized ghee 1 g kg^{-1}), OG2 (oxidized ghee 2 g kg^{-1}), OG3 (oxidized ghee 3 g kg^{-1}), OGNG (oxidized ghee 3 g kg^{-1} + normal ghee 3 g kg^{-1}), NG3 (normal ghee 3 g/kg BW^{-1}). CV, central vein; F, fats accumulation (Zeb and Uddin, 2017).

Interestingly, studies in lactating rats revealed that oxidized fats not only reduce hepatic and plasma TAG concentrations but also cause a reduction of TAG in the milk (Brandsch et al. 2004; Ringseis et al. 2007a, 2007b, 2007c). As maternal milk is the only source of nutrients for suckling pups, it is not surprising that the reductions in TAGs and, concomitantly, energy in the milk in these studies were accompanied by a significantly impaired development of litters during lactation. It is also notable that when oxidized fat was fed to pregnant rats, a TAG-lowering effect was observed in the liver of the foetuses (Ringseis et al. 2007a, 2007b, 2007c), suggesting that the components of oxidized fats responsible for TAG lowering pass the placental barrier in sufficient amounts to exert this effect in the foetal liver.

9.2.1.2 Cholesterol Concentrations in Liver and Plasma

Reports on the effects of oxidized fat on cholesterol concentrations in liver and plasma are somewhat less consistent than those on TAG concentrations. Nevertheless, the majority of studies also report reduced cholesterol concentrations in liver and plasma following the administration of oxidized fats, as summarized in Table 9.2 (Ringseis and Eder 2011). No effect on hepatic cholesterol concentration was observed in one study with rats, which received either fresh soybean oil or oxidized soybean oil prepared by

Table 9.2 Effects of dietary oxidized lipids on the liver and plasma total cholesterol (TC) in different species.

Species	Fat source, heating regime	Feeding period (days)	Liver TC (%)[a]	Plasma TC (%)[a]	References
Rat	Soybean oil, 205 °C, 6 hours with wheat flour dough sheets in the oil, 4 consecutive days	56	−22%	−15%	Huang et al. (1988)
Rat	Soybean oil, 130 °C, 22 hours	40	+2%	–	Eder and Kirchgessner (1998)
Rat	Lard: safflower oil (2:1, w/w), 150 °C, 6 days	51	−20%	−13%	Eder (1999)
Rat	Soybean oil, 205 °C, 6 hours with wheat flour dough sheets in the oil, 4r consecutive days	42	−60%	−18%	Chao et al. (2001)
Rat	Sunflower oil, 98 °C, 48 hours	69	n.d.[a]	−23%	Ammouche et al. (2002)
Rat	Sunflower oil: lard (1:1, w/w), 50 °C, 38 days	63	−33%	−9%	Eder et al. (2003a)
Rat	Sunflower oil: lard (1:1, w/w), 105 °C, 81 hours	63	−22%	+1%	Eder et al. (2003a, 2003b)
Rat	Sunflower oil: lard (1 : 1, w/w), 190 °C, 24 hours	63	−25%	−1%	Eder et al. (2003a, 2003b)
Rat	Sunflower oil: lard (1:1, w/w), 55 °C, 49 days	56	–	−15%	Eder et al. (2003a)
Rat	Sunflower oil, 55 °C, 42 days	56	−13%	+2%	Eder et al. (2003a)
Rat	Soybean oil, 205 °C, 6 hours with wheat flour dough sheets in the oil, 4 consecutive days	42	−55% (male)	−14% (male)	Chao et al. (2004)
Rat	Sunflower oil: lard (1:1, w/w), 50 °C, 38 days	63	−29%	−9%	Sülzle et al. (2004)
Rat	Sunflower oil: linseed oil (80:20, w/w), 50 °C, 16 days	98	−55%	–	Brandsch et al. (2004)
Rat	Sunflower oil, 60 °C, 25 days	6	−19%	−19%	Koch et al. (2007b)
Rat	Olive oil, 100 °C, 15 hours	30	–	+35%	Khan et al. (2017)
Mouse	Heated hydrogenated palm fat : fresh sunflower oil (92 : 8, w/w), 170 °C, 48 hours with repeated frying of French fries for 6 minutes every 30 min	112	–	−15%	Kämmerer et al. (2011)
Guinea pig	Soybean oil, 205 °C, 6 hours with potato sticks in the oil, 4 consecutive days	60	–	+30%	Liu and Lee (1998)

(*Continued*)

Table 9.2 (Continued)

Species	Fat source, heating regime	Feeding period (days)	Liver TC (%)[a]	Plasma TC (%)[a]	References
Guinea pig	Sunflower oil: lard (1 : 1, w/w), 55 °C, 49 days	29	−26%	−18%	Eder et al. (2004)
Pig	Sunflower oil, 110 °C, 48 hours	35	–	−30%	Eder and Stangl (2000)
Pig	Sunflower oil, 180 °C, 24 hours	28	+6%	−7%	Luci et al. (2007a, 2007b)
Rabbit	Tallow, 130 °C, 45 hours	07	–	+139%	Zeb & Mehmood (2012)
Rabbit	Vanaspati ghee, 130 °C, 45 hours	07		+108%	Zeb & Rahman (2012)
Rabbit	Desi ghee, 160 °C, 9 hours	07	–	+24%	Zeb & Uddin (2017)
Rabbit	Pumpkin seed oil, 160 °C, 9 hours	14	–	+41%	Zeb & Ahmad (2017)
Rabbit	Mustard oil, 170 °C, 9 hours	14	–	+35%	Zeb & Rahman (2017)

Source: Part of the table is reproduced with kind permission of Wiley-VCH Verlag GmbH & Co. (Ringseis and Eder 2011).

[a] Percent change of liver or plasma TAG concentration as compared with control.

heating at 130 °C (Eder and Kirchgessner 1998). In this study, an effect of oxidized fat on plasma cholesterol concentrations was not reported. In three other experiments with rats (Eder et al. 2003b), only the cholesterol concentration in the liver was decreased (by 22, 25, and 13%, respectively), while plasma cholesterol concentration remained unchanged. In a study with guinea pigs, the concentration of cholesterol in the liver was increased by approximately 30% on feeding with oxidized soybean oil (Liu and Lee 1998). In this study, an effect of the oxidized fat on plasma cholesterol concentrations was not reported. The increase of hepatic cholesterol concentration seems not to be a species-specific phenomenon, however, because in another study with guinea pigs cholesterol concentrations in liver and plasma were reduced by 26 and 18%, respectively (Eder et al. 2004). Furthermore, one study in pigs revealed only negligible effects of oxidized fat on liver and plasma cholesterol concentrations. The lack of effect in the study by Luci et al. (2007a, 2007b) cannot be attributed to a species-specific feature because oxidized fat caused a strong cholesterol-lowering effect in the plasma of pigs in another study (Eder and Stangl 2000).

The feeding periods of the studies considered lasted from 6 to 98 days (Brandsch and Eder 2004; Koch et al. 2007b). Interestingly, the strength of the cholesterol-lowering effect of oxidized fat in the liver and plasma was independent of this; for instance, the per cent changes of cholesterol concentrations in the liver and plasma were similar whether the oxidized fats were administered for 6 (Koch et al. 2007b), 29 (Eder et al. 2004), 51 (Eder 1999), or 56 days (Huang et al. 1988). This indicates that even short-term administration of oxidized fats causes a significant cholesterol-lowering effect and that the duration of oxidized fat administration is of minor importance for this effect. It is, moreover, noteworthy that the type of oxidized fat (fat source and heating regime) also has little influence on the cholesterol-lowering effect. This was shown in a study from Eder et al. (2003b) which investigated the effects of oxidized fats (1 : 1 mixture of sunflower oil and lard) prepared by different heating regimes (either 50 °C for 38 days, 105 °C for 81 hours, or 190 °C for 24 hours). On the other hand, the strength of the cholesterol-lowering effect in the liver differed markedly between two other studies (Huang et al. 1988; Chao et al. 2001) eventhough the oxidized fats administered in these studies were prepared by the same conditions (soybean oil heated for 205 °C with repeated frying of wheat flour dough sheets on 4 consecutive days). Although the reason for these divergent results cannot be definitely explained, the observations from these studies and the study from Eder et al. (2003b) suggest that the spectrum of lipid peroxidation products in the oxidized fat, which is influenced by the heating regime, is also of low significance for the cholesterol-lowering effect.

In contrast to rats, rabbits have a highly organized and developed organ system, which may be one of main reasons for the significant increase in serum cholesterol levels observed in Zeb and Rahman (2012, 2017), Zeb and Mehmood (2012), Zeb and Uddin (2017), and Zeb and Ahmad (2017). These studies showed that liver weight was significantly higher than in controls. A human study showed that metabolic syndrome is not associated with frequent intake of fried foods in young peoples, but adiposity and high blood pressure were clearly evident. It can be easily concluded that repeated fried lipids may not be healthy for humans.

9.2.1.3 Phospholipid Concentrations in Liver and Plasma

Besides cholesterol and TAG concentrations, few studies have additionally reported an effect of oxidized fat on phospholipid concentrations. Three studies revealed that administration of an oxidized fat increased the concentration of total phospholipids in the liver (Eder 1999; Chao et al. 2001; Sulzle et al. 2004). Sulzle et al. (2004), moreover, reported elevations in the hepatic concentrations of specific phospholipid subclasses such as phosphatidylethanolamine, phosphatidylcholine, phosphatidylinositol, and cardiolipin. A further study in rats did not reveal any effect of oxidized fat on liver phospholipid concentration (Chao et al. 2004). Eder (1999) and Chao et al. (2004) reported the phospholipid concentration in plasma to be reduced by approximately 21 and 10%, respectively, by the administration of oxidized fats. No effect on plasma phospholipid concentration was found in the earlier study from Chao et al. (2001).

9.2.1.4 Carnitine Concentrations in the Liver

A further interesting finding with respect to phenotypical alterations of lipid metabolism by oxidized fats is that rats treated with oxidized fat have increased hepatic carnitine concentrations (Koch et al. 2007a). Carnitine is a water-soluble quaternary amine (3-hydroxy-4-*N,N,N*-trimethylaminobutyric acid) which plays an important role in lipid and energy metabolism by acting as a shuttling molecule for the translocation of long-chain FAs (acyl groups) from the cytosol into the mitochondrial matrix, where β-oxidation occurs. Hence, all tissues that use FAs as a fuel source, such as liver, skeletal muscle, heart, and placenta, require carnitine for normal function. Consequently, energy metabolism is critically impaired in these tissues of patients with primary or secondary carnitine deficiency, and such patients develop severe pathological abnormalities such as cardiomyopathy or hepatomegaly. The finding that oxidized fat increases hepatic carnitine concentration is therefore indicative of an increased capacity of the hepatocytes for the import of long-chain FAs into the mitochondria and subsequent oxidative degradation of these FAs within the mitochondria.

9.2.1.5 Free Fatty Acids and Ketone Bodies in Plasma

Feeding experiments with rats (Ringseis et al. 2007a, 2007b, 2007c) and pigs (Luci et al. 2007a, 2007b) revealed that oxidized fats increased plasma concentrations of ketone bodies like β-hydroxybutyrate. Ketogenesis occurs physiologically in the liver during fasting to provide alternative energy substrates to the brain due to the reduced availability of glucose. Inthis state, nonesterified FAs are released from the white adipose tissue by the action of hormone-sensitive lipase, transferred to the tissues via the blood, taken up into the tissues by FA transporters, and degraded via β-oxidation to acetyl-CoA, which is the precursor for ketone bodies. Hence, the findings in rats and pigs indicate that oxidized fats stimulate ketogenesis in the liver.

It is important to note that in the design of most of the recent studies reporting the lipid metabolism alterations of oxidized fats (Ringseis et al. 2007a, 2007b, 2007c), several precautions have been taken (use of moderately oxidized fats, controlled feeding systems, substitution of antioxidant, and PUFA losses during fat oxidation) to avoid potentially confounding effects such as reduced food intake, decreased nutrient digestibility, depletion of antioxidants, and essential FA deficiency which might be causative for the observed phenotypical alterations. It is therefore likely that these alterations are mediated by a direct effect on lipid metabolism in the liver, which will be discussed in subsequent chapters.

9.2.2 Molecular Regulators of Lipid Metabolism

Observations from well-controlled studies (Eder and Kirchgessner 1998; Eder et al. 2004; Sulzle et al. 2004; Ringseis et al. 2007a, 2007b, 2007c; Luci et al. 2007a, 2007b) have led to the suggestion that the pronounced lipid metabolism alterations induced by feeding of oxidized fats are mediated directly by primary and secondary lipid peroxidation products which are absorbed from the ingested fats in the intestine and transported via the blood to the liver. Hepatic lipid metabolism in mammals is regulated critically by two transcription factors acting as master regulators of genes involved in lipid catabolism and lipid synthesis: peroxisome proliferator-activated receptor α (PPARα) and sterol regulatory element-binding protein (SREBP)-1 and -2. Due to their fundamental role in the regulation of lipid metabolism, these transcription factors have been considered in several feeding experiments dealing with oxidized fats, in order to elucidate the lipid metabolism alterations induced by the administration of such fats.

PPARα is a ligand-activated transcription factor that is abundantly expressed in tissues with high rates of FA oxidation such as the liver (Mandard et al. 2004). It predominantly controls genes involved in all aspects of lipid catabolism, including cellular FA uptake, intracellular FA transport, FA transport through the mitochondrial membrane, mitochondrial and peroxisomal FA oxidation, ketogenesis, and gluconeogenesis

(Kersten et al. 1999). In addition, genes involved in carnitine uptake and carnitine biosynthesis have recently been shown to be regulated by PPARα (van Vlies et al. 2007; Koch et al. 2008; Wen et al. 2010). PPARα can be activated by both endogenous and synthetic ligands. Endogenous ligands are FAs and their derivatives (eicosanoids) (Forman et al. 1997). Their activation of PPARα is observed during fasting, since free nonesterified FAs are released from adipose tissue and taken up into tissues at increased levels during this state (Leone et al. 1999). Consequently, in the liver, where PPARα is most abundant, genes involved in FA β-oxidation, ketogenesis, gluconeogenesis, carnitine uptake, and carnitine biosynthesis are strongly upregulated during fasting. PPARα is also activated by a heterogeneous group of synthetic compounds, including the fibrate class of lipid-lowering drugs (clofibrate, fenofibrate, bezafibrate, and gemfibrozil), which have been in clinical use for the treatment of hypertriacylglycerolemia and hypercholesterolemia for about 40 years.

The SREBPs are a family of transcription factors that regulate the coordinated expression of genes involved in lipid synthesis and uptake. Three isoforms of SREBP are known in mammals: SREBP-1a, SREBP-1c, and SREBP-2. Although SREBP-1c, the predominant isoform in adult liver, preferentially activates genes required for FA synthesis and their incorporation into TAGs and phospholipids, SREBP-2 preferentially activates the low-density lipoprotein (LDL) receptor gene and various genes required for cholesterol synthesis, such as 3-hydroxy-3-methylglutaryl-CoA reductase (Horton et al. 2002). SREBP-1a is an activator of both the cholesterol and the FA biosynthetic pathways, but it is present in liver in much lower amounts than the other two forms (Shimomura et al. 1997). SREBPs are synthesized as 120 kDa integral membrane proteins of the endoplasmic reticulum (ER) and form a complex with SREBP cleavage-activating protein (SCAP). When sterol concentrations in cells are high, the SCAP/SREBP complex is retained in the ER. When cells are depleted of sterols, SCAP escorts SREBP to the Golgi for proteolytic processing. In the Golgi, sequential cleavages occur, releasing the mature *N*-terminal domain of SREBP, which then translocates to the nucleus and activates transcription of sterol regulatory element-containing genes (Goldstein et al. 2002). Retention of the SCAP/SREBP complex in the ER is mediated by sterol-dependent binding of the complex to one of the two ER retention proteins, designated insulin-induced genes (Insig)-1 and -2 (Yang et al. 2002).

9.2.3 Evidence for the Activation of PPARα by Oxidized Fat

In a study by Chao et al. (2001), which showed the feeding of oxidized soybean oil prepared under frying conditions to growing rats for 6 weeks, it was reported for the first time that oxidized fat causes activation of PPARα in the liver and the kidney. From their findings, the authors concluded that dietary oxidized fat, by activating PPARα, alters lipid metabolism in rats. In subsequent studies performed by different groups, the PPARα-activating effect of oxidized fat was confirmed (Chao et al. 2005). In all of these studies, activation of PPARα was evaluated by determining the transcript levels of typical PPARα target genes such as acyl-CoA oxidase (ACO), cytochrome P450 isoenzyme 4A1 (CYP4A1), and carnitine-palmitoyltransferase I (CPT I), which are involved in peroxisomal and mitochondrial FA oxidation. Determination of the transcript levels of PPARα target genes is a widely accepted approach to the evaluation of PPARα activation, because they are elevated in response to PPARα activation. On the other hand, expression of the PPARα nuclear receptor itself remains largely unchanged in response to PPARα agonists. Using a transcript profiling approach, Sulzle et al. (2004) provided further evidence for the activation of the PPARα pathway by oxidized fats in showing that oxidized fats upregulate a comprehensive set of PPARα target genes, including ACO, CYP4A1, CPT I, CPT II, medium-chain acyl-CoA dehydrogenase, long-chain acyl-CoA dehydrogenase, and 3-keto-acyl-CoA dehydrogenase. Moreover, liver enlargement (hepatomegaly) due to peroxisome proliferation, which is a typical response to a PPARα agonist (Peters et al. 2005), was observed in experimental animals fed oxidized fats, and is, therefore, a further indication for PPARα activation by oxidized fats.

Table 9.3 provides an overview of the degree of PPARα activation by oxidized fats as evaluated by the upregulation of the PPARα target genes ACO and CYP4A1 in the liver of different species. This overview shows that activation of PPARα by oxidized fat is stronger in the liver of rodents than in that of pigs. This difference is probably explained by variations in the expression level of PPARα in tissues between these species. In rodents, PPARα is extremely highly expressed in tissues, and activation of PPARα not only induces many genes involved in various metabolic pathways but also causes severe peroxisome proliferation in the liver (Peters et al. 2005), with rodents belonging to the proliferating species. In contrast, PPARα agonists like fibrates do not induce peroxisome proliferation in the liver of nonproliferating species, such as guinea pigs, pigs, monkeys, and humans. Nonproliferating species thus have a lower expression of PPARα in the liver, and the response of many genes to PPARα activation is much weaker than in proliferating species (Holden and Tugwood 1999; Peffer et al. 2005). Therefore, effects related to PPARα activation observed in rodents cannot be directly applied to nonproliferating species such as humans. We have recently shown that pigs have a similar mRNA concentration of PPARα in the liver as humans (Luci et al. 2007a, 2007b), which is approximately tenfold lower than in rats. Considering this and the similarity between pig and human liver cells in the gene response to PPARα agonists (Cheon et al. 2005), it is not unlikely that a similar PPARα response to that observed in pigs is induced in the liver of humans consuming oxidized fats.

A further indication for PPARα activation by oxidized fat was provided in a more recent study showing that administration of oxidized fat increases the hepatic transcript level of novel organic cation transporter 2 (OCTN2) (Koch et al. 2007a). This gene has recently been identified as a novel PPARα target gene (Wen et al. 2010). The OCTNs operate on the intestinal absorption and renal reabsorption of carnitine and play a major role in tissue distribution by catalysing the uptake of carnitine into body cells. Due to its high binding affinity for carnitine and its wide expression, OCTN2 is the physiologically most important carnitine transporter, operating for the reabsorption of carnitine from the urine and playing a major role in tissue distribution. Due to these functions, it is likely that upregulation of OCTN2 was responsible for the observation that administration of oxidized fat caused an increase in the carnitine concentration in the liver of rats (Koch et al. 2007a).

Based on the knowledge that oxidized fats markedly activate hepatic PPARα and reduce hepatic TAG concentrations, a study has recently been conducted to investigate whether dietary oxidized fat is useful in the prevention of alcoholic fatty liver, using an established rat model (Ringseis et al. 2007a, 2007b, 2007c). This study clearly revealed that TAG accumulation in response to ethanol feeding is markedly reduced in rats

Table 9.3 Effects of dietary protein oxidation products on specific pathological processes.

Target cell/tissue	Dietary oxidation product	Cellular/tissue response	Signalling pathways/ genes affected	Pathological process	References
IEC-6/intestines	Advanced oxidation protein products	Apoptosis	NADPH oxidase, JNK, PARP-1	Inflammatory bowel disease	Xie et al. (2014)
Intestines	Oxidized casein	Intestinal flora and redox state disturbance	Nrf-2	Inflammatory bowel disease	Fang et al. (2012)
Colon	Advanced oxidation protein products	Cytoskeletal disruption	–	Inflammatory bowel disease	Keshavarzian et al. (2003)
Small intestine	Advanced oxidation protein products	Decreased Ca transport channels	p44/42, MAPK	Inflammatory bowel disease	Wu et al. (2015)
BTC6/isolated murine islets	Amino adipic acid	Oxidative stress, insulin secretion	POX-1, MafA	Diabetes	Wang et al. (2013)
Liver	Oxidized casein	Oxidative stress, apoptosis	Nrf-2, TGF-β, ET-1	Fibrosis	Li et al. (2013)
Kidney, liver	Oxidized casein	Oxidative stress, apoptosis	Nrf-2, TGF-β, ET-2	Fibrosis	Li et al. (2014)
Liver	Tyrosine oxidation products	Oxidative stress	TGF-β1	Fibrosis	Li et al. (2015)

Source: Reproduced with kind permission of Taylor & Francis (Estévez and Luna 2017).

simultaneous administered an oxidized fat, as compared with a fresh one. Moreover, this study showed that, in agreement with previous studies (Fischer et al. 2003; Crabb et al. 2004), the function of PPARα was impaired by ethanol feeding. However, administration of the oxidized – but not the fresh – fat during ethanol feeding resulted in the upregulation of PPARα target genes, indicating that PPARα is activated by oxidized fat even in the presence of ethanol. Thus, these findings strongly suggest that the prevention of alcoholic fatty liver development by oxidized fat is due to the restoration of impaired PPARα function.

The components of oxidized fats which are supposed to be responsible for PPARα activation are oxidized FAs such as hydroxy and hydroperoxy FAs (e.g. 9-hydroxy octadecadienoic acid (9-HODE), 13-HODE, and 13-hydroperoxy octadecadienoic acid (13-HPODE). Fats heated at moderate temperatures of below 100 °C for a long period (several days or weeks) usually contain high amounts of these primary lipid peroxidation products, whereas fats heated at high temperatures have low levels of these compounds because hydroperoxides are relatively unstable and easily decompose at these temperatures. This largely explains why PPARα activation is usually stronger in experimental animals fed oxidized fats heated at moderate temperatures compared with those heated at high temperatures (Table 9.3). Cell-culture studies using different cell types clearly revealed that the addition of pure hydroxy or hydroperoxy FAs to the incubation medium caused an activation of PPARα, as determined in transactivation experiments and by expression analysis of PPARα target genes (König and Eder 2006). Besides hydroxy and hydroperoxy FAs, cyclic fatty acid monomers (CFAMs) have also been suggested to be potential mediators of the activation of PPARα by oxidized fat. CFAMs contain five- and six-membered rings in different positions in the original alkyl chain, with two substituents in either *cis-* or *trans-*conformation relative to the ring. The main precursor FAs of CFAM are the essential FAs linoleic acid and α-linolenic acid, and cyclization occurs when vegetable oils rich in these FAs are heated at temperatures above 200 °C (Sebedio and Grandgirard 1989). Such temperatures are reached when foods are fried and during the refining of oils, especially in the deodorization step. Martin et al. (1997) observed that rats fed TAGs containing purified CFAM generated from linseed oil developed a peroxisome proliferator-like response, including hepatomegaly and elevated activities of ACO, CYP4A, and CPT I. In a subsequent study by the same group, a $5\,g\,kg^{-1}$ diet of an isolated CFAM fraction was administered for 3 weeks to wild-type and PPARα-null mice to elucidate the role of PPARα in mediating these effects (Bretillon et al. 2003). Among other things, the authors found that dietary CFAM enhanced ACO and CYP4A activities two- and threefold, respectively, in the liver of wild-type mice but had no effect in the liver of PPARα-null mice, indicating that CFAM indeed induces a PPARα response.

Interestingly, activation of hepatic PPARα is also observed in the foetal organism when oxidized fat is fed to pregnant rats, suggesting that the previously mentioned components of oxidized fat that are decisive for PPARα activation pass the placental barrier and activate PPARα in the foetal liver. Although no literature data are available on the transplacental transport of oxidized FAs, it is well known that the transplacental transport of FAs, in general, is highly selective for individual FAs (Herrera et al. 2006). The observation that the upregulation of transcript levels of PPARα-responsive genes by oxidized fat in the foetal liver was even more pronounced than in the liver of pregnant rats therefore indicates that the components of oxidized fat responsible for PPARα activation are transported through the placenta with high preference.

Wallert et al. (2015) showed that a long-chain metabolite of α-tocopherol, α-13-carboxychromanol (α-13-COOH), a potent anti-inflammatory molecule, is formed by hepatic α-tocopherol (α-TOH) catabolism and acts as a regulatory metabolite through pathways that are different from its metabolic precursor. Similarly, Schmölz et al. (2016) showed that metabolism of tocopherol initiated with one cycle of CYP4F2/CYP3A4-dependent ω-hydroxylation followed by five cycles of β-oxidation and formed the water-soluble end-product carboxyethylhydroxychroman. All known hepatic metabolites can be conjugated and excreted through either urine or faeces, depending on the length of their side-chain.

9.3 Metabolism of Fried Proteins

Proteins present in fried foods undergo several biochemical changes upon ingestion, the most important of which is oxidation. Lund et al. (2011) reviewed the oxidation of food proteins, one of the most innovative areas of study within the Food Chemistry field. The fact that proteins are targets for ROS was ignored for several decades, while the oxidation of other food components (namely, lipids) was studied in depth. The lack of suitable and specific methodologies to assess protein oxidation (Pox) and the assumption that lipid oxidation was, together with microbial spoilage, the main cause of food deterioration slowed the development of this highly innovative research line. In fact, most of the studies carried out on Pox during the last several decades have been conducted to examine the role played by oxidized proteins in a variety of age-related diseases (e.g. Alzheimer's) (Stadtman and Berlett 1997; Davies 2003), while the occurrence of Pox in food systems has been largely unexplored. Following the discovery that myofibril proteins are affected by ROS during meat maturation and storage (Martinaud et al. 1997), numerous research studies began to look at the occurrence of Pox in muscle foods and to try to shed light on the influence of meat origin, composition, and industrial processing on the development and intensity of muscle Pox. Some recent advances in protein metabolism, such as the basic chemical mechanisms involved in the oxidative modification of muscle proteins, are being clarified, although further studies are still needed. It is known, however, that the reaction of radicals with proteins and peptides in the presence of oxygen gives rise to alterations in both the backbone and the amino acid side chain. These oxidative changes include cleavage of peptide bonds, modification of amino acid side chains, and formation of covalent intermolecular cross-linked protein derivatives (Davies and Dean 1997), as shown in Figure 9.4. Some of the most general amino acid modifications are the formation of protein carbonyl groups and protein hydroperoxides. Cross-linking has mostly been described as the formation of disufide and dityrosine through the loss of cysteine and tyrosine residues (Estévez and Cava 2006; Lund et al. 2011).

Estévez and Luna (2017) showed that dietary Pox is a silent threat to human health, especially that found in meat (Soladoye et al. 2015). It has been shown that a typical post-translational modification of proteins mediated by ROS or oxidized lipids may lead to malfunction and disease (Esterbauer et al. 1991). Other radical and nonradical species, such as reducing sugars that have undergone Maillard-type reactions and reactive nitrogen species (RNS), may be able to induce modifications in proteins such as glycation or nitration, with biological consequences (Gong et al. 2005). The cellular

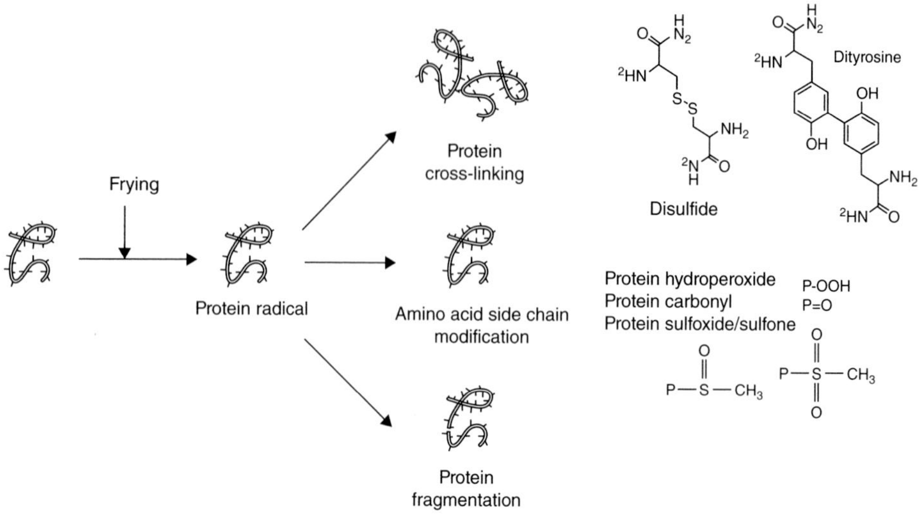

Figure 9.4 The most common consequences of the oxidation of proteins. *Source:* Reproduced with kind permission of Wiley-VCH Verlag, GmbH & Co (Lund et al. 2011).

responses to these molecular changes usually imply the activation of particular signalling pathways that involves gene expression and/or suppression. Numerous studies point to the nuclear transcription factor nuclear factor kappa beta (NF-κB) as a major pathway for the molecular mechanisms that underlie inflammatory, autoimmune, and proliferative disorders (Aw 1999). The fact that dietary oxidized proteins and their products are themselves active executors of specific pathological processes was mostly unknown until recently.

9.3.1 Metabolism of Heterocyclic Aromatic Amines

Most carcinogenic substances formed during the frying of food proteins do not turn mutagenic directly upon their intake. They are only activated in the organism with the help of metabolic enzymes, especially when reaching the liver. The heterocyclic aromatic amines (HAAs) are resorbed in the small intestine, transported to the liver, and activated there, as shown in Figure 9.5 (Lynch et al. 1995). N-oxidation by cytochrome p450 enzymes is the first step in the metabolic activation of HAA compounds, followed by an acetylation reaction. Oxidation of the amino group by means of cytochrome p450 1A2 follows a predicted one-electron transfer, rather than the previously suggested two-electron transfer mechanism. An N-acetoxy amine is formed by O-acetylation from the resulting N-hydroxyl ion. This reaction step is catalysed by N-acetyltransferases (NATs). Alternatively, N-hydroxy-HAA metabolites can be esterified by sulfotransferases (SULTs), L-seryl-tRNA, L-prolyl-tRNA synthetases, and other ATP-dependent enzymes (Turesky and Le Marchand 2011).

The resulting esters are unstable and undergo degradation reaction. The N-acetoxy amine is a very reactive molecule that can form a nitrenium ion of the exocyclic amine group, which is the critical metabolite implicated in toxicity and DNA damage. The nitrenium ion is able to combine with DNA and can cause mutations, which in turn

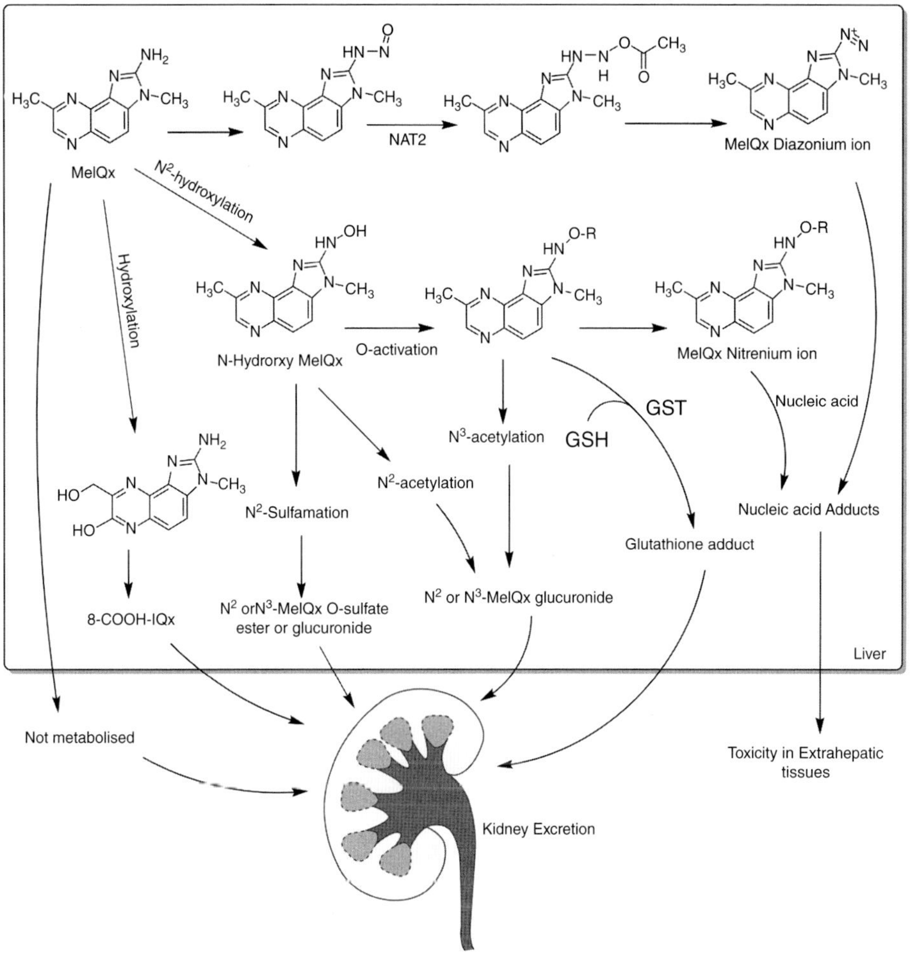

Figure 9.5 Metabolism of heterocyclic aromatic amines (MeIQx). Most of the end products are synthesized in the liver before being excreted by the kidneys with urine. GST, glutathione S-transferase; GSH, glutathione; NAT, *N*-acetyltransferase.

cause tumours. IQ, MeIQ, MeIQx, PhIP, Trp-P-2, Glu-P-1, and MeAαC could demonstrably form adducts with the guanidine bases occurring in DNA (Wakabayashi and Sugimura 1998). Figure 9.5 shows the major pathway of metabolism of MeIQx as a prototype of HAAs, including the activation paths by nitrozation with nitric oxide under inflammatory conditions. Some HAAs, such as IQ and MeIQx, can undergo nitrozation with nitric oxide at neutral pH conditions to form 2-nitrosoamino-3-methylimidazo [4,5-*f*]quinoline and 2-nitrosoamino-3,8-dimethylimidazo-[4,5-*f*]quinoxaline, respectively. These N-nitroso-IQ compounds are transformed to very reactive diazonium compounds that may generate covalent DNA adducts with the DNA bases, such as guanine (Lakshmi et al. 2005).

The mechanism for the N-acetyltransferases (NAT2)-catalysed activation of N-nitroso-MeIQx shown in Figure 9.5 was suggested by Zenser et al. (2009). There are large differences in p450s between humans and other species, such as rodents, in terms of the region-selectivity and catalytic activity of HAA oxidation, which affects its toxicological characteristics (Turesky 2002). Several epidemiological studies have proposed a role for NAT2 activity in the genetic susceptibility of humans to various types of cancer (Le Marchand et al. 2002; Butler et al. 2008). The data support the hypothesis that individuals who smoke and have genetic susceptibility to the rapid phenotypes for both NAT2 and cytochrome (CYP1A2), as well as an increased intake of meat carcinogens, have a higher risk for colorectal cancer (Le Marchand et al. 2002). Human NAT2 catalyses O-acetylation of N-hydroxyl derivatives of IQ, MeIQ, and PhIP, and a number of polymorphic genotypes of human NAT1 and NAT2 are known. In particular, the genotype NAT2 (NAT2-4 wild allele) as a fast acetylator (Sugimura et al. 2004) has often been associated with a higher risk of colorectal cancers (Ishibe et al. 2002). In contrast to carbon hydroxylation is the inactivation step, the resulting compounds undergo different reactions catalysed by phase II enzymes, such as sulfotransferase, glucuronosyltransferase, and glutathione transferase, and the metabolites formed are excreted by the kidney in urine.

9.4 Metabolism of Fried Carbohydrates

Carbohydrates found in fried foods – especially oligosaccharides and polysaccharides – start degrading in the stomach, reaching a maximum rate in the large intestine. The final product of carbohydrate digestion is monosaccharides such as glucose, fructose, and mannose. These compounds are absorbed by the cells and ultimately utilized. These processes are well known to biochemists, so we will not discuss them here. As mentioned in Chapter 5, during frying, carbohydrates formed several toxic compounds, the best-known of which are acrylamide and furfurals. These compounds are ingested in a variety of fried foods, such as potato chips, fried chicken, and fried bread. They are absorbed by the intestine and reach the liver through the blood, where they are metabolized or detoxified and finally excreted in a nontoxic or reduced-toxicity form.

9.4.1 Metabolism of Acrylamide

Acrylamide has two reactive sites: the conjugated double bond and the amide moiety. These groups are responsible for further *in vivo* or *in vitro* reactions. The functional groups in amino acids, such as –SH, and $-NH_2$ groups, are very reactive towards acrylamide. Since the pH of the stomach is very low, conversion of acrylamide to acrylic acid and ammonia is more favourable (Friedman 2003). Several *in vitro* studies involving the reaction of different amino acids and proteins with acrylamide showed promising results, which were later found to be similar to those in *in vivo* studies. Friedman (2003) reviewed the reactions of amino acids and proteins with acrylamide and showed that the –SH group of an amino acid or protein is 100–300 times more reactive with conjugated vinyl compound than is the amino group. The formation of haemoglobin and glycidamide adducts has been reported by several authors (Perez et al. 1999; Paulsson et al. 2002). These adducts have been found to cause toxicity. Studies based on these model systems have thus been extrapolated to animal models and humans.

Figure 9.6 shows the metabolism of acrylamide. In human liver, the concentration of glutathione (GSH) is high, ranging from 3 to 5 $\mu mol\, g^{-1}$; thus, liver is a very good place

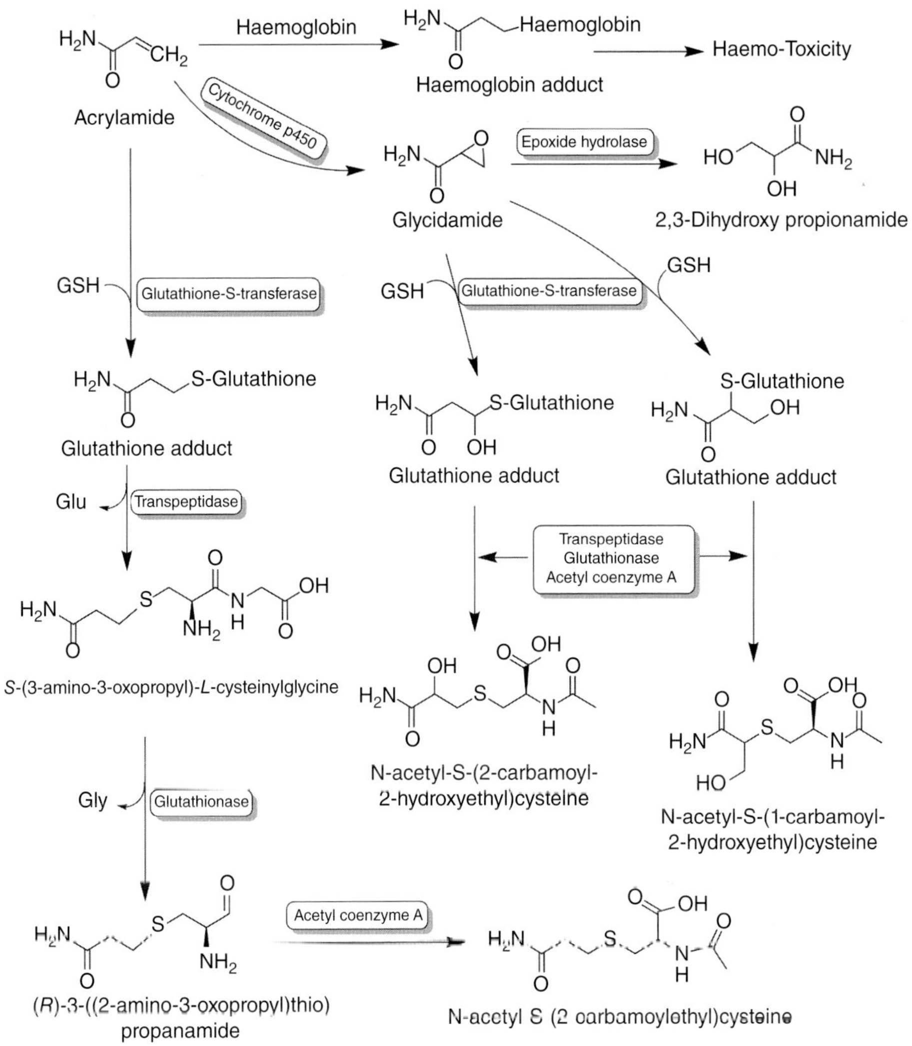

Figure 9.6 Metabolism of acrylamide, indicating the formation of haemoglobin and glutathione (GSH) adducts and of end-products.

for acrylamide detoxification. Upon reaching the liver, acrylamide reacts with GSH in the presence of glutathione S-transferase (GST) enzyme, forming a GSH adduct. This GSH adduct releases glutamyl residue through a transpeptidase enzyme to form S-(3-amino-3-oxopropyl)-cysteinylglycine. Glutaminase enzyme releases glycine to form a thio-propionamide derivative. The end product of this reaction is mercapturic acid (N-acetyl-S-[2-carbamoylethyl]-cysteine), formed by the acetylation. This is excreted in urine. Watzek et al. (2012) showed that mercapturic acid was found in a higher level in human urine after consumption of fried potato chips. A recent human intervention study showed that intake of acrylamide from fried foods, as well as other heated foods, can be monitored by the level of mercapturic acid in the urine (Ruenz et al. 2016).

A decrease in the concentration of GSH may result in oxidative stress and, ultimately, cause toxicity. Toxicity may also result from ageing, liver damage, and cancer. Another parallel pathway has been found to be present, which involves glycidamide formation by cytochrome p450. Three possibilities were reported for the further reaction of glycidamide. The first is the formation of 2,3-dihydroxy-propionamide, which is a water-soluble substance that is excreted as such in urine. This reaction is carried out by epoxide hydrolase in the presence of water (Kirman et al. 2003). The second and third involve the formation of two GSH adducts, which are isomeric in nature; these are metabolized further by transpeptidase, glutathionase, and acetyl coenzyme-A to form N-acetyl-S-(1-carbamoyl-2-hydroxyethyl) cysteine and N-acetyl-S-(2-carbamoyl-2-hydroxyethyl) cysteine as end-products (Friedman 2003), which are removed from the body through the kidneys. The latter reactions consume GSH completely, and may thus contribute to oxidative stress in the body. The oxidative stress induced by fried food may contribute to a variety of diseases, including cancers (Stott-Miller et al. 2013).

9.4.2 Metabolism of Furfurals

Experiments have shown that isotopically labelled HMF administered to rats or mice through oral gavage was rapidly absorbed from the gastrointestinal tract (Germond et al. 1987). *In vitro* studies of the Caco2 cEll line showed a higher absorption and transport of HMF during treatment (Godfrey et al. 1999). The authors suggested that food composition – especially fibre content – may affect HMF uptake. Unabsorbed HMF can be converted to furfural alcohol by enteric bacterial strains (Boopathy et al. 1993). The converted alcohols are then excreted by the kidneys, and have been reported to be present in kidneys, bladder, and, to some extent, the liver. Figure 9.7 shows the metabolism of HMF. The pathways have been found to be the same as those involved in the biotransformation of furfurals in rodents. In the presence of sulfotransferase (SULT) enzyme, HMF is converted to form a more soluble sulfated compound called 5-sulfoxymethyl-furfural (SMF) and sulfo group donor 3-phosphoadenosine-5-phosphosulfate (PAPS) (Capuano and Fogliano 2011).

The sulfate is a good leaving group in SMF, producing a highly reactive intermediate that can react with DNA (Glatt and Sommer 2006) and other macromolecules, giving rise to toxic and mutagenic effects (Surh et al. 1994). Due to the high reactivity of the sulfur group, SMF is very unstable, but it has been detected in the blood of HMF-treated mice, which indicates that HMF is metabolized to SMF *in vivo* (Monien et al. 2009). Another minor pathway is the formation of α-ketoglutaric acid, which may enter the citric acid cycle and produce carbon dioxide (Nomeir et al. 1992), or else produce glutamate by the action of the transaminase enzyme.

The major metabolic pathway contains an early oxidation of HMF to 5-hydroxymethyl-2-furanoic acid (HMFA) and a later conjugation to glycine to yield N-(5-hydroxymethyl-2-furoyl) glycine (HMFG, the glycine conjugate of HMFA), which are readily excreted in the urine (Capuano and Fogliano 2011). In both rats and humans, the ratio HMFA/HMFG decreases as HMF intake increases, suggesting that the availability of free glycine may limit the rate of conjugation, resulting in the excretion of free furoic acid or 2,5-furan dicarboxylic acid (FDCA) through the second pathway. HMFA, HMFG, and FDCA have been identified in the urine of rats and mice (Godfrey et al. 1999). They have also been identified in human urine, and it has been demonstrated

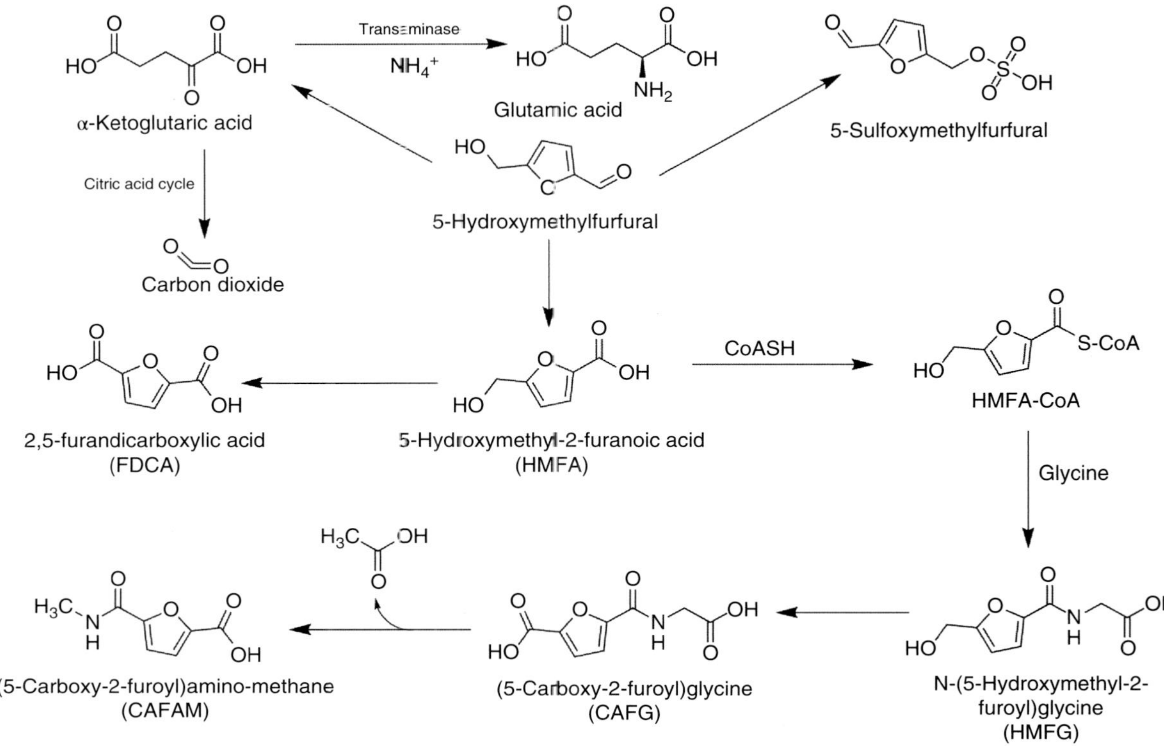

Figure 9.7 Metabolism of hydroxymethyl furfural (HMF).

that they are derived from HMF in food (Husøy et al. 2008). HMFA, HMFG, (5-carboxylic acid-2-furoyl)glycine (CAFG), and (5-carboxylic acid-2-furoyl) amino-methane (CAFAM), but not FDCA, were recently identified in the urine of subjects who had consumed dried plum juice (Prior et al. 2006). Incubation of urine samples with a mixture of aglucuronidase and sulfatase did not significantly change the amount of HMF derivatives detected in them. Therefore, Capuano and Fogliano (2011) concluded that no glucuronide or sulfate conjugates of HMF had formed. In addition, as with furfural and 5-methyl furfural metabolism, it may be inferred that HMF is converted to 5-hydroxymethylfuranacryloyl-CoA and excreted via urine after glycine conjugation as 5-hydroxymethylfuranacryloylglycine (HMFAG).

9.5 Metabolism of Other Metabolites

Polyphenols customarily have been considered antinutrients by animal nutritionists, because of the adverse effect of tannins on protein digestibility. However, interest in food phenolics has increased greatly in recent times, owing to their antioxidant capacity and possible beneficial implications in human health, particularly in relation to cancer, cardiovascular disease, and other pathologies. Much of the literature refers to a single group of plant phenolics: the flavonoids (Bravo 1998). Phenolic glycosides are absorbed in a very small amount in the stomach and moreso in the large intestine. Glycosides are hydrolysed to form free phenolic compounds and aglycones. Phenolic compounds may inhibit several enzymes and increase the antioxidant potential in the body. Because of the diverse nature of phenolic compounds, it is not possible to accommodate the metabolism of all of them. Thus, ellagic acid has been the most widely studied.

Ellagic acid has been reported to produce beneficial effects during oxidative stress by regulating several pathways. These include activation of the antioxidant response via Nrf2 (nuclear erythroid factor 2) (Ding et al. 2014; Baluchnejadmojarad et al. 2017); inhibition of cyclooxygenase 2 (COX-2) (Chatterjee et al. 2012) and cytokines (Anderson and Teuber 2010) with the help of NF-κB (Ahad et al. 2014); modulation of cell survival or apoptosis (Ho et al. 2014); and enhancement of the activity of biological antioxidants and antioxidant enzymes (Anitha et al. 2013). Details regarding these oxidative stress studies are given in Figure 9.8, which shows some important sources of ellagic acid and the oxidative stresses induced by different *in vitro* and *in vivo* stressors. Oxidative stress has been studied in antioxidant enzymes, DNA damage, and oxidized DNA. Such studies are classed as general studies. Mechanistic studies involved the determination of cytokines, transcription factors, kinase enzymes, and cell properties, as described by Zeb (2018). The lack of human studies on the effect of ellagic acid on oxidative stress induced by fried food makes it impossible to determine the exact amount and duration required for treatment.

9.6 Key Concepts

- Most unaltered carbohydrates, lipids, and proteins follow well-known metabolic pathways.
- Fried lipids are mostly oxidized, which significantly alters the metabolism of other metabolites (including lipids), depending upon the amount taken and the extent of oxidation in dietary lipids.

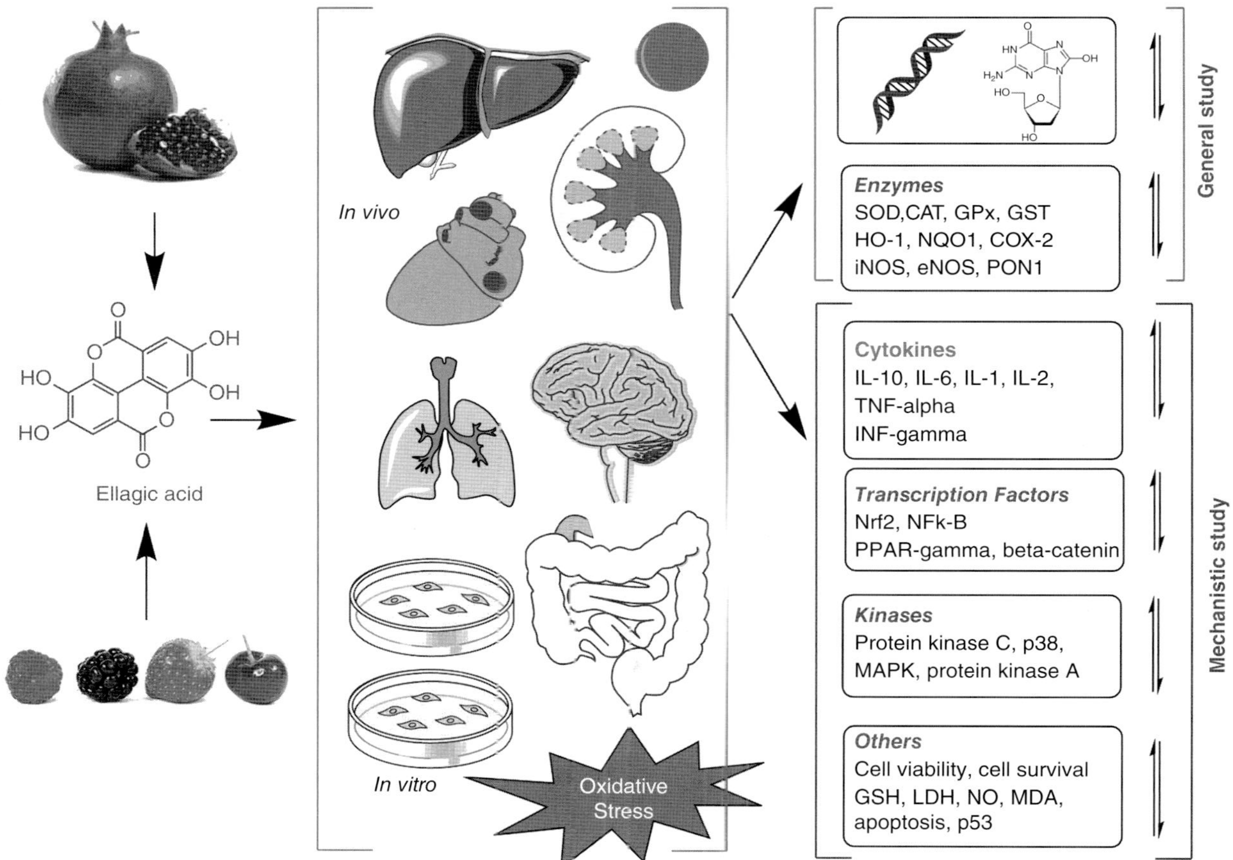

Figure 9.8 The role of ellagic acid in oxidative stress at different organ sites. The parameters examined in mechanistic and general protective studies (*in vitro* and *in vivo*) are shown. *Source:* Reproduced with kind permission of Springer Nature (Zeb 2018).

- Oxidized lipids may decrease or significantly increase serum lipids profile, depending upon the amount taken and the extent of oxidation.
- Recent studies have shown that repeated ingestion of frying oils increases the deposition of fats in the liver and causes necrosis and cancer.
- Proteins are oxidized during frying, but there is little information regarding their subsequent metabolism.
- Metabolites of proteins such as HAAs and polycyclic aromatic hydrocarbons (PAHs) produce negative effects on health and significantly alter metabolism.
- HAAs are metabolized to form soluble compounds, which are excreted from the body.
- Acrylamide and furfurals can also be metabolized to form relatively soluble and low-toxicity compounds, which are excreted from the body.
- GSH and its antioxidant enzymes have been found to play a significant role in the metabolism of acrylamide, furfurals, and HAAs.
- Among the other metabolites, phenolic compounds play a large part in metabolism, in which GSH and its antioxidant enzymes have a significant role.

References

Ahad, A., Ganai, A.A., Mujeeb, M., and Siddiqui, W.A. (2014). Ellagic acid, an NF-kappaB inhibitor, ameliorates renal function in experimental diabetic nephropathy. *Chemico-Biological Interactions* 219: 64–75.

Ammouche, A., Rouaki, F., Bitam, A., and Bellal, M.M. (2002). Effect of ingestion of thermally oxidized sunflower oil on the fatty acid composition and antioxidant enzymes of rat liver and brain in development. *Annals of Nutrition and Metabolism* 46: 268–275.

Anderson, K.C. and Teuber, S.S. (2010). Ellagic acid and polyphenolics present in walnut kernels inhibit in vitro human peripheral blood mononuclear cell proliferation and alter cytokine production. *Annals of the New York Academy of Sciences* 1190: 86–96.

Anitha, P., Priyadarsini, R.V., Kavitha, K. et al. (2013). Ellagic acid coordinately attenuates Wnt/beta-catenin and NF-kappaB signaling pathways to induce intrinsic apoptosis in an animal model of oral oncogenesis. *European Journal of Nutrition* 52: 75–84.

Aw, T.Y. (1999). Molecular and cellular responses to oxidative stress and changes in oxidation-reduction imbalance in the intestine. *American Journal of Clinical Nutrition* 70: 557–565.

Baluchnejadmojarad, T., Rabiee, N., Zabihnejad, S., and Roghani, M. (2017). Ellagic acid exerts protective effect in intrastriatal 6-hydroxydopamine rat model of Parkinson's disease: possible involvement of ERbeta/Nrf2/HO-1 signaling. *Brain Research* 1662: 23–30.

Benedetti, P.C., Di Felice, M., Gentili, V. et al. (1990). Influence of dietary thermally oxidized soybean oil on the oxidative status of rats of different ages. *Annals of Nutrition and Metabolism* 34: 221–231.

Beyaz, S., Mana, M.D., Roper, J. et al. (2016). High-fat diet enhances stemness and tumorigenicity of intestinal progenitors. *Nature* 531: 53.

Boopathy, R., Bokang, H., and Daniels, L. (1993). Biotransformation of furfural and 5-hydroxymethyl furfural by enteric bacteria. *Journal of Industrial Microbiology* 11: 147–150.

Brandsch, C. and Eder, K. (2004). Effects of peroxidation products in thermoxidised dietary oil in female rats during rearing, pregnancy and lactation on their reproductive

performance and the antioxidative status of their offspring. *British Journal of Nutrition* 92: 267–275.
Brandsch, C., Nass, N., and Eder, K. (2004). A thermally oxidized dietary oil does not lower the activities of lipogenic enzymes in mammary glands of lactating rats but reduces the milk triglyceride concentration. *Journal of Nutrition* 134: 631–636.
Bravo, L. (1998). Polyphenols: chemistry, dietary sources, metabolism, and nutritional significance. *Nutrition Reviews* 56: 317–333.
Bretillon, L., Alexson, S.E.H., Joffre, F. et al. (2003). Peroxisome proliferator-activated receptor α is not the exclusive mediator of the effects of dietary cyclic FA in mice. *Lipids* 38: 957–963.
Butler, L.M., Millikan, R.C., Sinha, R. et al. (2008). Modification by N-acetyltransferase 1 genotype on the association between dietary heterocyclic amines and colon cancer in a multiethnic study. *Mutation Research, Fundamental and Molecular Mechanisms of Mutagenesis* 638: 162–174.
Capuano, E. and Fogliano, V. (2011). Acrylamide and 5-hydroxymethylfurfural (HMF): a review on metabolism, toxicity, occurrence in food and mitigation strategies. *LWT – Food Science and Technology* 44: 793–810.
Chao, P.M., Chao, C.Y., Lin, F.J., and Huang, C. (2001). Oxidized frying oil up-regulates hepatic acyl-CoA oxidase and cytochrome P450 4 A1 genes in rats and activates PPARalpha. *Journal of Nutrition* 131: 3166–3174.
Chao, P.M., Hsu, S.C., Lin, F.J. et al. (2004). The up-regulation of hepatic acyl-CoA oxidase and cytochrome P450 4A1 mRNA expression by dietary oxidized frying oil is comparable between male and female rats. *Lipids* 39: 233–238.
Chao, P.M., Yang, M.F., Tseng, Y.N. et al. (2005). Peroxisome proliferation in liver of rats fed oxidized frying oil. *Journal of Nutritional Science and Vitaminology (Tokyo)* 51: 361–368.
Chatterjee, A., Chatterjee, S., Das, S. et al. (2012). Ellagic acid facilitates indomethacin-induced gastric ulcer healing via COX-2 up-regulation. *Acta Biochimica et Biophysica Sinica (Shanghai)* 44: 565–576.
Cheon, Y., Nara, T.Y., Band, M.R. et al. (2005). Induction of overlapping genes by fasting and a peroxisome proliferator in pigs: evidence of functional PPARα in nonproliferating species. *American Journal of Physiology – Regulatory, Integrative and Comparative Physiology* 288: R1525–R1535.
Crabb, D.W., Galli, A., Fischer, M., and You, M. (2004). Molecular mechanisms of alcoholic fatty liver: role of peroxisome proliferator-activated receptor alpha. *Alcohol* 34: 35–38.
Davies, M.J. (2003). Singlet oxygen-mediated damage to proteins and its consequences. *Biochemical and Biophysical Research Communications* 305: 761–770.
Davies, M.J. and Dean, R.T. (1997). *Radical-Mediated Protein Oxidation: From Chemistry to Medicine*. Boca Oxford: Oxford University Press.
Ding, Y., Zhang, B., Zhou, K. et al. (2014). Dietary ellagic acid improves oxidant-induced endothelial dysfunction and atherosclerosis: role of Nrf2 activation. *International Journal of Cardiology* 175: 508–514.
Eder, K. (1999). The effects of a dietary oxidized oil on lipid metabolism in rats. *Lipids* 34: 717–725.
Eder, K. and Kirchgessner, M. (1998). The effect of dietary vitamin E supply and a moderately oxidized oil on activities of hepatic lipogenic enzymes in rats. *Lipids* 33: 277–283.

Eder, K. and Stangl, G.I. (2000). Plasma thyroxine and cholesterol concentrations of miniature pigs are influenced by thermally oxidized dietary lipids. *Journal of Nutrition* 130: 116–121.

Eder, K., Skufca, P., and Brandsch, C. (2002). Thermally oxidized dietary fats increase plasma thyroxine concentrations in rats irrespective of the vitamin E and selenium supply. *Journal of Nutrition* 132: 1275–1281.

Eder, K., Keller, U., Hirche, F., and Brandsch, C. (2003a). Thermally oxidized dietary fats increase the susceptibility of rat LDL to lipid peroxidation but not their uptake by macrophages. *Journal of Nutrition* 133: 2830–2837.

Eder, K., Suelzle, A., Skufca, P. et al. (2003b). Effects of dietary thermoxidized fats on expression and activities of hepatic lipogenic enzymes in rats. *Lipids* 38: 31–38.

Eder, K., Keller, U., and Brandsch, C. (2004). Effects of a dietary oxidized fat on cholesterol in plasma and lipoproteins and the susceptibility of low-density lipoproteins to lipid peroxidation in Guinea pigs fed diets with different concentrations of vitamins E and C. *International Journal for Vitamin and Nutrition Research* 74: 11–20.

Esterbauer, H., Schaur, R.J., and Zollner, H. (1991). Chemistry and biochemistry of 4-hydroxynonenal, malonaldehyde and related aldehydes. *Free Radical Biology and Medicine* 11: 81–128.

Estévez, M. and Cava, R. (2006). Effectiveness of rosemary essential oil as an inhibitor of lipid and protein oxidation: contradictory effects in different types of frankfurters. *Meat Science* 72: 348–355.

Estévez, M. and Luna, C. (2017). Dietary protein oxidation: a silent threat to human health? *Critical Reviews in Food Science and Nutrition* 57: 3781–3793.

Fang, W., Sun, J., Li, Z.L. et al. (2012). Effect of oxidated food protein on mice gut flora and redox state. *Chinese Journal of Microecology* 24: 193–196.

Fischer, M., You, M., Matsumoto, M., and Crabb, D.W. (2003). Peroxisome proliferator-activated receptor α (PPARα) agonist treatment reverses PPARα dysfunction and abnormalities in hepatic lipid metabolism in ethanol-fed mice. *Journal of Biological Chemistry* 278: 27997–28004.

Forman, B.M., Chen, J., and Evans, R.M. (1997). Hypolipidemic drugs, polyunsaturated fatty acids, and eicosanoids are ligands for peroxisome proliferator-activated receptors α and δ. *Proceedings of the National Academy of Sciences of the United States of America* 94: 4312–4317.

Friedman, M. (2003). Chemistry, biochemistry, and safety of acrylamide. A review. *Journal of Agricultural and Food Chemistry* 51: 4504–4526.

Germond, J.E., Philippossian, G., Richli, U. et al. (1987). Rapid and complete urinary elimination of [14C]5-hydroxymethyl-2-furaldehyde administered orally or intravenously to rats. *Journal of Toxicology and Environmental Health* 22: 79–89.

Glatt, H. and Sommer, Y. (2006). Health risks of 5-hydroxymethylfurfural (HMF) and related compounds. In: *Acrylamide and Other Hazardous Compounds in Heat-Treated Foods* (ed. K. Skog and J. Alexander). Cambridge: Woodhead.

Godfrey, V.B., Chen, L.-J., Griffin, R.J. et al. (1999). Distribution and metabolism of (5-hydroxymethyl)furfural in male F344 rats and B6C3F1 mice after oral administration. *Journal of Toxicology and Environmental Health, Part A* 57: 199–210.

Goldstein, J.L., Rawson, R.B., and Brown, M.S. (2002). Mutant mammalian cells as tools to delineate the sterol regulatory element-binding protein pathway for feedback regulation of lipid synthesis. *Archives of Biochemistry and Biophysics* 397: 139–148.

Gong, C.-X., Liu, F., Grundke-Iqbal, I., and Iqbal, K. (2005). Post-translational modifications of tau protein in Alzheimer's disease. *Journal of Neural Transmission* 112: 813–838.

Guallar-Castillón, P., Rodríguez-Artalejo, F., Fornés, N.S. et al. (2007). Intake of fried foods is associated with obesity in the cohort of Spanish adults from the European prospective investigation into cancer and nutrition. *American Journal of Clinical Nutrition* 86: 198–205.

Hayam, I., Cogan, U., and Mokady, S. (1993). Dietary oxidized oil enhance the activity of (Na+ K+) ATPase and acetylcholinesterase and lowers the fluidity of rats erythrocyte membrane. *Journal of Nutrient Biochemistry* 4: 563–568.

Herrera, E., Amusquivar, E., López-Soldado, I., and Ortega, H. (2006). Maternal lipid metabolism and placental lipid transfer. *Hormone Research in Pædiatrics* 65 (Suppl. 3): 59–64.

Ho, C.C., Huang, A.C., Yu, C.S. et al. (2014). Ellagic acid induces apoptosis in TSGH8301 human bladder cancer cells through the endoplasmic reticulum stress- and mitochondria-dependent signaling pathways. *Environmental Toxicology* 29: 1262–1274.

Holden, P. and Tugwood, J. (1999). Peroxisome proliferator-activated receptor alpha: role in rodent liver cancer and species differences. *Journal of Molecular Endocrinology* 22: 1–8.

Horton, J.D., Goldstein, J.L., and Brown, M.S. (2002). SREBPs: activators of the complete program of cholesterol and fatty acid synthesis in the liver. *Journal of Clinical Investigation* 109: 1125–1131.

Huang, C.-J., Cheung, N.-S., and Lu, V.-R. (1988). Effects of deteriorated frying oil and dietary protein levels on liver microsomal enzymes in rats. *Journal of the American Oil Chemists' Society* 65: 1796–1803.

Husøy, T., Haugen, M., Murkovic, M. et al. (2008). Dietary exposure to 5-hydroxymethylfurfural from Norwegian food and correlations with urine metabolites of short-term exposure. *Food and Chemical Toxicology* 46: 3697–3702.

Ishibe, N., Sinha, R., Hein, D.W. et al. (2002). Genetic polymorphisms in heterocyclic amine metabolism and risk of colorectal adenomas. *Pharmacogenetics and Genomics* 12: 145–150.

Izaki, Y., Yoshikawa, S., and Uchiyama, M. (1984). Effect of ingestion of thermally oxidized frying oil on peroxidative criteria in rats. *Lipids* 19: 324–331.

Jurado-Ruiz, E., Varela, L.M., Luque, A. et al. (2017). An extra virgin olive oil rich diet intervention ameliorates the nonalcoholic steatohepatitis induced by a high-fat 'Western-type' diet in mice. *Molecular Nutrition & Food Research* 61: 1600549.

Kämmerer, I., Ringseis, R., and Eder, K. (2011). Feeding a thermally oxidised fat inhibits atherosclerotic plaque formation in the aortic root of LDL receptor-deficient mice. *British Journal of Nutrition* 105: 190–199.

Keller, U., Brandsch, C., and Eder, K. (2004). Supplementation of vitamins C and E increases the vitamin E status but does not prevent the formation of oxysterols in the liver of Guinea pigs fed an oxidised fat. *European Journal of Nutrition* 43: 353–359.

Keshavarzian, A., Banan, A., Farhadi, A. et al. (2003). Increases in free radicals and cytoskeletal protein oxidation and nitration in the colon of patients with inflammatory bowel disease. *Gut* 52: 720–728.

Kersten, S., Seydoux, J., Peters, J.M. et al. (1999). Peroxisome proliferator-activated receptor α mediates the adaptive response to fasting. *Journal of Clinical Investigation* 103: 1489–1498.

Kirman, C.R., Gargas, M.L., Deskin, R. et al. (2003). A physiologically based pharmacokinetic model for acrylamide and its metabolite, glycidamide, in the rat. *Journal of Toxicology and Environmental Health, Part A: Current Issues* 66: 253–274.

Khan, I.T., Nadeem, M., Imran, M. et al. (2017). Antioxidant capacity and fatty acids characterization of heat treated cow and buffalo milk. *Lipids in Health and Disease* 16: 163.

Koch, A., Konig, B., Luci, S. et al. (2007a). Dietary oxidised fat up regulates the expression of organic cation transporters in liver and small intestine and alters carnitine concentrations in liver, muscle and plasma of rats. *The British Journal of Nutrition* 98: 882–889.

Koch, A., König, B., Spielmann, J. et al. (2007b). Thermally oxidized oil increases the expression of insulin-induced genes and inhibits activation of sterol regulatory element-binding protein-2 in rat liver. *Journal of Nutrition* 137: 2018–2023.

Koch, A., König, B., Stangl, G.I., and Eder, K. (2008). PPARα mediates transcriptional upregulation of novel organic cation transporters-2 and -3 and enzymes involved in hepatic carnitine synthesis. *Experimental Biology and Medicine* 233: 356–365.

König, B. and Eder, K. (2006). Differential action of 13-HPODE on PPARα downstream genes in rat Fao and human HepG2 hepatoma cell lines. *Journal of Nutritional Biochemistry* 17: 410–418.

Lakshmi, V.M., Hsu, F.F., and Zenser, T.V. (2005). Nitric oxide-mediated nitrosation of 2-amino-3,8-dimethylimidazo[4,5-f]quinoxaline potentiated by hemin and myeloperoxidase. *Chemical Research in Toxicology* 18: 1038–1047.

Le Marchand, L.C., Hankin, J.H., Pierce, L.M. et al. (2002). Well-done red meat, metabolic phenotypes and colorectal cancer in Hawaii. *Mutation Research, Fundamental and Molecular Mechanisms of Mutagenesis* 506–507: 205–214.

Leone, T.C., Weinheimer, C.J., and Kelly, D.P. (1999). A critical role for the peroxisome proliferator-activated receptor α (PPARα) in the cellular fasting response: the PPARα-null mouse as a model of fatty acid oxidation disorders. *Proceedings of the National Academy of Sciences* 96: 7473–7478.

Li, Z.L., Wu, L., Le, G., and Shi, Y. (2013). Effect of oxidized casein on the oxidative damage of blood and digestive organs in mice. *Acta Nutrimenta Sinica* 35: 39–43.

Li, Z.L., Mo, L., Le, G., and Shi, Y. (2014). Oxidized casein impairs antioxidant defense system and induces hepatic and renal injury in mice. *Food and Chemical Toxicology* 64: 86–93.

Li, Z.L., Shi, Y., Le, G. et al. (2015). 24-Week exposure to oxidized tyrosine induces hepatic fibrosis involving activation of the MAPK/TGF-β1 signaling pathway in Sprague–Dawley rats model. *Oxidative Medicine and Cellular Longevity* art. 763272.

Liu, J.-F. and Huang, C.-J. (1995). Tissue α-tocopherol retention in male rats is compromised by feeding diets containing oxidized frying oil. *Journal of Nutrition* 125: 3071–3080.

Liu, J.F. and Lee, Y.W. (1998). Vitamin C supplementation restores the impaired vitamin E status of Guinea pigs fed oxidized frying oil. *Journal of Nutrition* 128: 116–122.

Liu, M., Bai, J., He, S. et al. (2014). rb10 Promotes lipolysis and thermogenesis by phosphorylation-dependent feedback inhibition of mTORC1. *Cell Metabolism* 19: 967–980.

Luci, S., Giemsa, B., Kluge, H., and Eder, K. (2007a). Clofibrate causes an upregulation of PPAR-α target genes but does not alter expression of SREBP target genes in liver and adipose tissue of pigs. *American Journal of Physiology – Regulatory, Integrative and Comparative Physiology* 293: R70–R77.

Luci, S., König, B., Giemsa, B. et al. (2007b). Feeding of a deep-fried fat causes PPARα activation in the liver of pigs as a non-proliferating species. *British Journal of Nutrition* 97: 872–882.

Lund, M.N., Heinonen, M., Baron, C.P., and Estévez, M. (2011). Protein oxidation in muscle foods: a review. *Molecular Nutrition & Food Research* 55: 83–95.

Lynch, A.M., Murray, S., Gooderham, N.J., and Boobis, A.R. (1995). Exposure to and activation of dietary heterocyclic amines in humans. *Critical Reviews in Oncology/Hematology* 21: 19–31.

Mandard, S., Müller, M., and Kersten, S. (2004). Peroxisome proliferator-activated receptor α target genes. *Cellular and Molecular Life Sciences* 61: 393–416.

Martin, J.C., Caselli, C., Broquet, S. et al. (1997). Effect of cyclic fatty acid monomers on fat absorption and transport depends on their positioning within the ingested triacylglycerols. *Journal of Lipid Research* 38: 1666–1679.

Martinaud, A., Mercier, Y., Marinova, P. et al. (1997). Comparison of oxidative processes on myofibrillar proteins from beef during maturation and by different model oxidation systems. *Journal of Agricultural and Food Chemistry* 45: 2481–2487.

Monien, B.H., Frank, H., Seidel, A., and Glatt, H. (2009). Conversion of the common food constituent 5-hydroxymethylfurfural into a mutagenic and carcinogenic sulfuric acid ester in the mouse in vivo. *Chemical Research in Toxicology* 22: 1123–1128.

Nomeir, A.A., Silveira, D.M., Mccomish, M.F., and Chadwick, M. (1992). Comparative metabolism and disposition of furfural and furfuryl alcohol in rats. *Drug Metabolism and Disposition* 20: 198–204.

Paulsson, B., Grawe, J., and Tornqvist, M. (2002). Hemoglobin adducts and micronucleus frequencies in mouse and rat after acrylamide or N-methylolacrylamide treatment. *Mutation Research* 516: 101–111.

Peffer, P.L., Lin, X., and Odle, J. (2005). Hepatic β-oxidation and carnitine palmitoyltransferase I in neonatal pigs after dietary treatments of clofibric acid, isoproterenol, and medium-chain triglycerides. *American Journal of Physiology – Regulatory, Integrative and Comparative Physiology* 288: R1518–R1524.

Perez, H.L., Cheong, H.K., Yang, J.S., and Osterman-Golkar, S. (1999). Simultaneous analysis of hemoglobin adducts of acrylamide and glycidamide by gas chromatography-mass spectrometry. *Analytical Biochemistry* 274: 59–68.

Peters, J.M., Cheung, C., and Gonzalez, F.J. (2005). Peroxisome proliferator-activated receptor-α and liver cancer: where do we stand? *Journal of Molecular Medicine* 83: 774–785.

Prior, R.L., Wu, X., and Gu, L. (2006). Identification and urinary excretion of metabolites of 5-(Hydroxymethyl)-2-furfural in human subjects following consumption of dried plums or dried plum juice. *Journal of Agricultural and Food Chemistry* 54: 3744–3749.

Raddatz, K., Turner, N., Frangioudakis, G. et al. (2011). Time-dependent effects of Prkce deletion on glucose homeostasis and hepatic lipid metabolism on dietary lipid oversupply in mice. *Diabetologia* 54: 1447–1456.

Ringseis, R. and Eder, K. (2004). Dietary oxidized cholesterol increases expression and activity of antioxidative enzymes and reduces the concentration of glutathione in the liver of rats. *International Journal for Vitamin and Nutrition Research* 74: 86–92.

Ringseis, R. and Eder, K. (2005). Effects of dietary fat and oxidized cholesterol on gene expression in rat liver as assessed by cDNA expression array analysis. *European Journal of Nutrition* 44: 231–241.

Ringseis, R. and Eder, K. (2011). Regulation of genes involved in lipid metabolism by dietary oxidized fat. *Molecular Nutrition & Food Research* 55: 109–121.

Ringseis, R., Gutgesell, A., Dathe, C. et al. (2007a). Feeding oxidized fat during pregnancy up-regulates expression of PPARα-responsive genes in the liver of rat fetuses. *Lipids in Health and Disease* 6: 6.

Ringseis, R., Muschick, A., and Eder, K. (2007b). Dietary oxidized fat prevents ethanol-induced triacylglycerol accumulation and increases expression of PPAR α target genes in rat liver. *Journal of Nutrition* 137: 77–83.

Ringseis, R., Dathe, C., Muschick, A. et al. (2007c). Oxidized fat reduces milk triacylglycerol concentrations by inhibiting gene expression of lipoprotein lipase and fatty acid transporters in the mammary gland of rats. *Journal of Nutrition* 137: 2056–2061.

Ruenz, M., Bakuradze, T., Eisenbrand, G., and Richling, E. (2016). Monitoring urinary mercapturic acids as biomarkers of human dietary exposure to acrylamide in combination with acrylamide uptake assessment based on duplicate diets. *Archives of Toxicology* 90: 873–881.

Schmölz, L., Birringer, M., Lorkowski, S., and Wallert, M. (2016). Complexity of vitamin E metabolism. *World Journal of Biological Chemistry* 7: 14–43.

Sebedio, J.L. and Grandgirard, A. (1989). Cyclic fatty acids: natural sources, formation durin heat treatment, synthesis and biological properties. *Progress in Lipid Research* 28: 303–336.

Shimomura, I., Shimano, H., Horton, J.D. et al. (1997). Differential expression of exons 1a and 1c in mRNAs for sterol regulatory element binding protein-1 in human and mouse organs and cultured cells. *Journal of Clinical Investigation* 99: 838–845.

Soladoye, O.P., Juárez, M.L., Aalhus, J.L. et al. (2015). Protein oxidation in processed meat: mechanisms and potential implications on human health. *Comprehensive Reviews in Food Science and Food Safety* 14: 106–122.

Stadtman, E.R. and Berlett, B.S. (1997). Reactive oxygen-mediated protein oxidation in aging and disease. *Chemical Research in Toxicology* 10: 485–494.

Stott-Miller, M., Neuhouser, M.L., and Stanford, J.L. (2013). Consumption of deep-fried foods and risk of prostate cancer. *The Prostate* 73: 960–969.

Sugimura, T., Wakabayashi, K., Nakagama, H., and Nagao, M. (2004). Heterocyclic amines: mutagens/carcinogens produced during cooking of meat and fish. *Cancer Science* 95: 290–299.

Sulzle, A., Hirche, F., and Eder, K. (2004). Thermally oxidized dietary fat upregulates the expression of target genes of PPAR alpha in rat liver. *Journal of Nutrition* 134: 1375–1383.

Surh, Y.J., Liem, A., Miller, J.A., and Tannenbaum, S.R. (1994). 5-Sulfooxymethylfurfural as a possible ultimate mutagenic and carcinogenic metabolite of the Maillard reaction product, 5-hydroxymethylfurfural. *Carcinogenesis* 15: 2375–2377.

Turesky, R.J. (2002). Heterocyclic aromatic amine metabolism, DNA adduct formation, mutagenesis, and carcinogenesis. *Drug Metabolism Reviews* 34: 625–650.

Turesky, R.J. and LE Marchand, L. (2011). Metabolism and biomarkers of heterocyclic aromatic amines in molecular epidemiology studies: lessons learned from aromatic amines. *Chemical Research in Toxicology* 24: 1169–1214.

van Vlies, N., Ferdinandusse, S., Turkenburg, M. et al. (2007). PPARα-activation results in enhanced carnitine biosynthesis and OCTN2-mediated hepatic carnitine accumulation. *Biochimica et Biophysica Acta (BBA) – Bioenergetics* 1767: 1134–1142.

Wakabayashi, K. and Sugimura, T. (1998). Heterocyclic amines formed in the diet: carcinogenicity and its modulation by dietary factors. *Journal of Nutritional Biochemistry* 9: 604–612.

Wallert, M., Schmölz, L., Koeberle, A. et al. (2015). α-Tocopherol long-chain metabolite α-13'-COOH affects the inflammatory response of lipopolysaccharide-activated murine RAW264.7 macrophages. *Molecular Nutrition & Food Research* 59: 1524–1534.

Wang, T.J., Ngo, D., Psychogios, N. et al. (2013). 2-Aminoadipic acid is a biomarker for diabetes risk. *Journal of Clinical Investigation* 123: 4309–4317.

Watzek, N., Scherbl, D., Feld, J. et al. (2012). Profiling of mercapturic acids of acrolein and acrylamide in human urine after consumption of potato crisps. *Molecular Nutrition & Food Research* 56: 1825–1837.

Wen, G., Ringseis, R., and Eder, K. (2010). Mouse OCTN2 is directly regulated by peroxisome proliferator-activated receptor alpha (PPARalpha) via a PPRE located in the first intron. *Biochemical Pharmacology* 79: 768–776.

Wu, P., Xie, F., Xue, M. et al. (2015). Advanced oxidation protein products decrease the expression of calcium transport channels in small intestinal epithelium via the p44/42 MAPK signaling pathway. *European Journal of Cellular Biology* 94: 190–203.

Xie, F., Sun, S., Xu, A. et al. (2014). Advanced oxidation protein products induce intestine epithelial cell death through a redox-dependent, c-jun N-terminal kinase and poly (ADP-ribose) polymerase-1-mediated pathway. *Cell Death and Disease* 5: e1006.

Yang, T., Espenshade, P.J., Wright, M.E. et al. (2002). Crucial step in cholesterol homeostasis: sterols promote binding of SCAP to INSIG-1, a membrane protein that facilitates retention of SREBPs in ER. *Cell* 110: 489–500.

Zeb, A. (2018). Ellagic acid in suppressing in vivo and in vitro oxidative stresses. *Molecular and Cellular Biochemistry* 448: 27–41. https://doi.org/10.1007/s11010 018 3310-3

Zeb, A. and Ahmad, S. (2017). Changes in acylglycerols composition, quality characteristics and in vivo effects of dietary pumpkin seed oil upon thermal oxidation. *Frontiers in Chemistry* 5: 55.

Zeb, A. and Mehmood, A. (2012). Effects of oxidized vanaspati ghee on the serum lipids profile and radical scavenging activity of the in vitro lipids of liver, brain and muscles. *Turkish Journal of Biochemistry* 37: 417–423.

Zeb, A. and Rahman, W.U. (2012). Effects of oxidized tallow on the rabbit serum lipids and antioxidant activity of the in-vitro lipids. *Toxicology Research* 28: 151–157.

Zeb, A. and Rahman, S.U. (2017). Protective effects of dietary glycine and glutamic acid toward the toxic effects of oxidized mustard oil in rabbits. *Food and Function* 8: 429–436.

Zeb, A. and Uddin, I. (2017). The coadministration of unoxidized and oxidized desi ghee ameliorates the toxic effects of thermally oxidized ghee in rabbits. *Journal of Nutrition and Metabolism* 2017: 4078360.

Zenser, T.V., Lakshmi, V.M., Schut, H.A.J. et al. (2009). Activation of aminoimidazole carcinogens by nitrosation: mutagenicity and nucleotide adducts. *Mutation Research, Genetic Toxicology and Environmental Mutagenesis* 673: 109–115.

Zhou, Z., Wang, Y., Jiang, Y. et al. (2016). Deep-fried oil consumption in rats impairs glycerolipid metabolism, gut histology and microbiota structure. *Lipids in Health and Disease* 15: 86.

10

Fried Foods in Health and Disease

10.1 Introduction

Food frying can help maintain and improve human health. Meat and fish muscle foods are rich in high-quality proteins, vitamins, and minerals (Sobral et al. 2018). Anthropologists believe that regular meat consumption was crucial to human evolution (Smil 2002). Hominids began using fire to cook foods approximately 1.8 million years ago, which played an important role in the improvement of several human-specific characteristics; the energy gained from cooked food contributed to the development of larger brains, as well as to several adaptations in the mouth and intestinal tract (Ferraro et al. 2013).

Today, according to the Food and Agriculture Organization of the United Nations (FAO), the supply of muscle foods represents a strong market. In contrast to bovine meat, which has experienced a slight decrease in consumption in recent years, the consumption of fish, pig, and chicken meat has increased. Red meat has begun to be viewed negatively for its possible carcinogenic effect. In October 2015, the International Agency for Research on Cancer (IARC) released a report that classified processed meat as carcinogenic (group 1) and red meat as probably carcinogenic (group 2A) to humans. Despite this, meat is an important source of essential micronutrients, such as B vitamins and trace elements, as well as fatty acids. Howe et al. (2006) found that 43% of total dietary intake of *n*-3 polyunsaturated fatty acids (PUFAs) could be achieved by consuming meat products, compared to 48% from oily fish; this was assumed for Australian adults, given their high meat intake compared with that of oily fish (Sobral et al. 2018).

Fried food quality is a function of oil quality. If the cooking oil is abused or damaged, it affects the texture, taste, and overall flavour perception of the food (Stier 2013). If these properties are still intact when the product is stored, several preliminary changes may also occur, which are caused by a number of detrimental health effects. Thus, quality control and strict guidelines must be followed in order to safeguard public health.

10.2 Fried Foods and Health

10.2.1 Health Maintenance

Fried foods have been subject to great scrutiny in recent years due to their involvement in a number of health-related issues. However, Davis (2013) has argued that the health risks of fried foods are overblown and require further research, while Bittman (2013)

Food Frying: Chemistry, Biochemistry, and Safety, First Edition. Alam Zeb.

suggests that fried foods may be good for human consumptions. There are a number of parameters according to which frying may be considered healthy:

1) Source of important lipids
2) Quick conversion of indigestible substances to digestible
3) High acceptability compared to other cooking methods (in most cases)
4) Increased nutrient availability
5) Increased taste and sensory characteristics
6) Flavour formation

McDonald and Eskin (2007) showed that fat is an important component of our diet. Frying medium and fried foods contain important fatty acids such as essential fatty acids. The carbohydrate-rich diet of plants usually contains a relatively small amount of essential fatty acids; the frying of such foods contributes to their essential fatty acid content. For example, a modest reduction of EPA and DHA in humpback salmon occurred during frying (Gladyshev et al. 2006). The authors hypothesized that the absence of a significant reduction in the PUFA content of the red flesh of fishes of the *Salmonidae* family during heat treatment may be due to a high level of natural antioxidants in such fish, formed in the course of their evolution. Table 10.1 shows the fat contents of several fried foods.

Frying also contributes several important antioxidants, such as phenolic compounds, tocopherol, and vitamin C. Although water-soluble vitamins are mostly destroyed during frying (Lešková et al. 2006), some are retained in the inner core of the fried foods, as shown in Figure 10.1. It has been observed that the outer coats of such foods are usually rich in fats, whereas the core mostly contains water, vitamins, carbohydrates, and proteins. These components thus act as an important source of nutrients in our daily diet.

The polyphenol content of French fries is largely unaffected by the frying process, and many health-promoting microconstituents migrate to the fries from the frying oil (Chiou et al. 2012). The total phenolic content (TPC) of watercress leaves was evaluated

Table 10.1 Range of fat contents in different fried foods.

Fried Food	Fat Content (%)
Potatoes and potato products	15–36
Cereal products, doughnuts	18–30
Bread slices	35–50
Vegetables	35–75
Mushrooms (breaded)	60–80
Beef, pork, meat balls, patties	10–27
Chicken (battered and breaded)	10–30
Fish, prawns (breaded)	20–42
Sausages and other meat products	22–70
Tomatoes and its products	12–35
Leafy vegetables	10–55

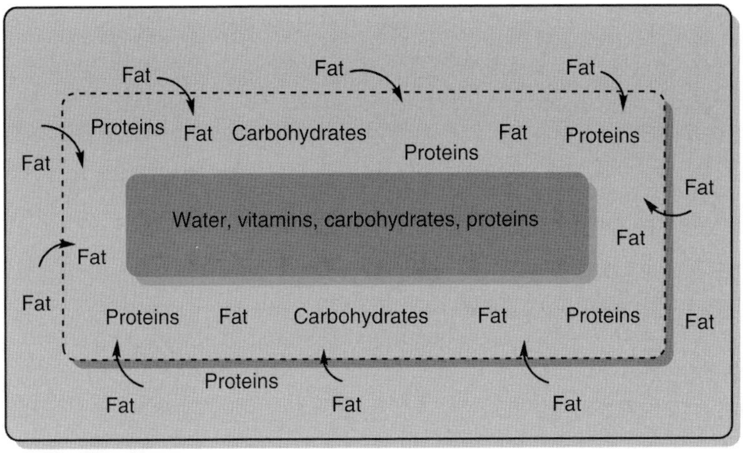

Figure 10.1 Retention and transfer of different nutrients during frying.

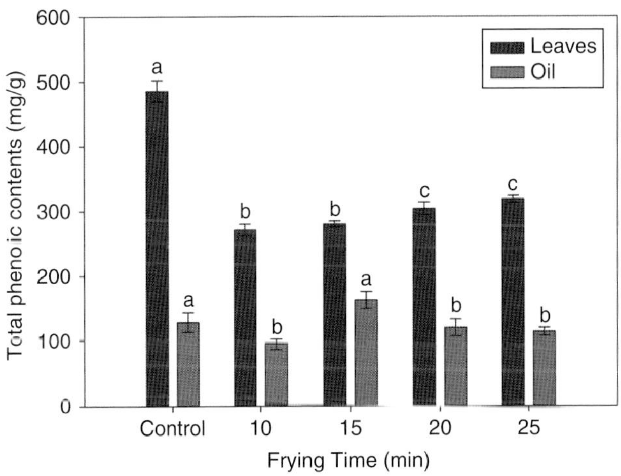

Figure 10.2 Effects of frying on total phenolic contents (TPC) of watercress leaves and sunflower oil. Data are mean with standard deviation (error bars) of replicate measurement (n = 5). Different letter in each case represents significance. *Source*: Reproduced with kind permission of Springer Nature (Zeb and Habib 2018).

to determine the amount of phenolic compounds leached to the oil from the leaves (or vice versa) during frying. Figure 10.2 shows that control oil contained 128.8 mg g^{-1} GAE, which decreased significantly ($p < 0.05$) after 10 minutes of frying, and increased after 15 minutes. There was no significant difference between 10, 20, and 25 minutes of frying on the TPC of sunflower oil. This suggests that phenolic compounds remained stable over these frying times, which could be one reason there was no hydroperoxide production (Zeb and Habib 2018). Phenolic compounds are strong antioxidants that help in preventing the oxidation of edible oils. When phenolic compounds such as gallic acid or its derivatives were added to sunflower oil, the oxidative stability of the oil was improved (Velasco et al. 2009). The presence of these phenolic compounds in fried food provides beneficial properties during ingestion.

Pokorný and Dostálová (2010) reviewed changes in the nutrient contents of frying foods and found that several flavour compounds are formed during frying, which is important for consumer acceptability. The flavour compounds were mostly formed by the frying medium, but could also be produced by chemical reactions in the carbohydrate and protein contents of the food.

10.2.2 Frying and Health Hazards

Márquez-Ruiz and Dobarganes (2007) reviewed the nutritional and physiological effects of heated fats/oils, thermo-oxidized oils and fats, and model oxidized lipids. They found most studies on this topic are based either on a model system or on edible oils oxidized by heat, which mimics frying oils. These studies lack the analytical data required to establish valid associations between the toxic compounds present and the widely evaluated effects on molecular targets, metabolic pathways, and chronic diseases. The findings of Qi et al. (2014) suggest that consumption of fried food could interact with a genetic background in relation to obesity, highlighting the particular importance of reducing fried food consumption in individuals genetically predisposed to obesity. Dobarganes and Marquez-Ruiz (2015) reviewed the adverse effects of frying using vegetable oils, finding that the degradation products of these oils are responsible for adverse health effects. To date, however, such studies have focused on the few known compounds that can be easily synthesized and characterized.

10.3 Fried Foods and Cancer

Knekt et al. (1994) showed that intake of fried foods has little effect on colon cancer in humans. Similarly, Steineck et al. (1993) evaluated a number of epidemiological studies and suggested that no preventive measures are justifiable. In a follow-up study, it was proposed that mutagens in fried meat may be involved in the progression of cancer. Consequently, the relationships between the intake of fried meat and the risk of cancer at different sites were studied among 9990 Finnish men and women aged between 15 and 99 and initially free of cancer (Knekt et al. 1994). During a 24-year follow-up, 853 cancer cases were diagnosed. The intake of fried meat was estimated from a dietary history interview, which covered the total diet in the previous year. The authors found a positive correlation between fried meat intake and the risk of female hormone-related cancers (i.e. cancer of the breast, endometrium, and ovary, combined). In a Swedish epidemiological study, Gerhardsson de Verdier (1995) revealed that the frying of meat could be a potential carcinogenic contributor; they argued that specific animal model studies were necessary.

The carcinogenic effect of heterocyclic aromatic amines (HAAs) was demonstrated in animal trials with rats and mice. In these trials, the animals were fed concentrations of 0.03 or 0.04% HAAs in their food over a period of two years. After several months, feeding with IQ, MeIQ, and MeIQx resulted in the formation of tumours in all of the trial animals (Ohgaki et al. 1987). The tumours appeared mainly in the animals' livers or intestines. Of the control rats that received normal food, only 4% developed tumours. After feeding with 0.06% MeIQx for several weeks, tumours appeared in the rats' livers, lungs, and lymph nodes (Ohgaki et al. 1987). A dose of 10–20 mg of IQ per kg body weight in the food of nonhuman primates similarly caused carcinoma formation after several months (Adamson et al. 1996). Based on animal trials, the IARC has classified IQ as a possible carcinogen (class 2A) and eight other HAAs (MeIQ, MeIQx, PhIP, AαAC, MeAαAC, Trp-P-1, Trp-P-2, and Glu-P-2) as probable carcinogens (class 2b) (IARC 1993).

However, it is difficult to prove a connection between HAA intake and the development of different types of tumours in humans. The quantity of HAAs ingested by humans is much smaller than that fed to trial animals. Thus, the results of long-term animal studies cannot be used to set a specific threshold value for the human intake of HAAs. Nevertheless, epidemiological studies in the last decade have shown an increased risk of incidence of different cancers for humans who eat high amounts of very well-done meat. In a cancer cohort study in Sweden, HAAs were reported to have no association with breast cancer in postmenopausal women (Sonestedt et al. 2008).

Lippi and Mattiuzzi (2015) reviewed the association between fried foods and the risk of prostate cancer in humans. Table 10.2 shows that the risk may differ depending on the food ingested. For example, a large intake of fried fish lowers the risk of prostate cancer (Pawlega et al. 1996), or shows no association at all (Joshi et al. 2012). However, Stott-Miller et al. (2013) showed a significant association between prostate cancer risk and a large intake of French fries, fried fish, and chicken. A similar association was reported by Li et al. (2014) in Chinese population study.

Table 10.2 Studies investigating the association between a large consumption of fried food and prostate cancer risk.

Country	Study design	Study population	Endpoint	Results	References
Poland	Case–control	76 cancer patients and 152 matched controls	Prevalence of prostate cancer according to fried fish intake	Risk of prostate cancer lower in subjects with a large intake of fried fish (OR 0.5; 95% CI 0.2–0.9)	Pawlega et al. (1996)
USA	Case–control	1857 cancer patients (717 localized and 1140 advanced) and 1096 matched controls	Prevalence of prostate cancer according to deep-fried fish intake	No association between large intake of deep-fried fish and risk of localized (OR 1.2; 95% CI 0.9–1.8) or advanced (OR 1.1; 95% CI 0.8–1.5) prostate cancer	Joshi et al. (2012)
USA	Case–control	1549 cancer patients and 1492 matched controls	Prevalence of prostate cancer according to fried food intake	Significant association between prostate cancer risk and large intake of French fries (OR 1.37; 95% CI 1.11–1.69), fried chicken (OR 1.30; 95% CI 1.04–1.62), and fried fish (OR 1.32; 95% CI 1.05–1.66)	Stott-Miller et al. (2013)
China	Case–control	250 patients with prostate cancer and 500 matched controls	Prevalence of prostate cancer according to fried food intake	Significant association between prostate cancer risk and intake of fried foods (OR 2.35; 95% CI 1.24–4.43)	Li et al. (2014)

Source: Reproduced with kind permission of Taylor & Francis (Lippi and Mattiuzzi 2015).
OR, odds ratio; CI, confidence limit.

Ganesan et al. (2017) found that polycyclic aromatic hydrocarbons (PAHs) are highly lipid-soluble and are therefore freely absorbed by the gastrointestinal system of mammals. They are swiftly disseminated in a variety of tissues, with the marked tendency for localization in body fat (Abdel-Shafy and Mansour 2016). They are genotoxic and produce mutations that lead to cancer. Such chemically induced enzymatic progressions produce metabolites that react with DNA. When the DNA sequence is altered in genes that regulate cell replication, cancer can result. PAHs can bind to DNA at specific sites, forming bulky complexes called DNA adducts that can be either stable or unstable (Ganesan et al. 2017). Stable adducts may lead to DNA replication errors, whereas unstable adducts react with the DNA strand, removing a purine base. Such mutations, if not repaired, can transform genes encoding for normal cell signalling proteins into cancer-causing oncogenes (Figure 10.3). Low-molecular-weight PAHs are more potent as co-carcinogens during the promotional stage of cancer, due to the ease with which

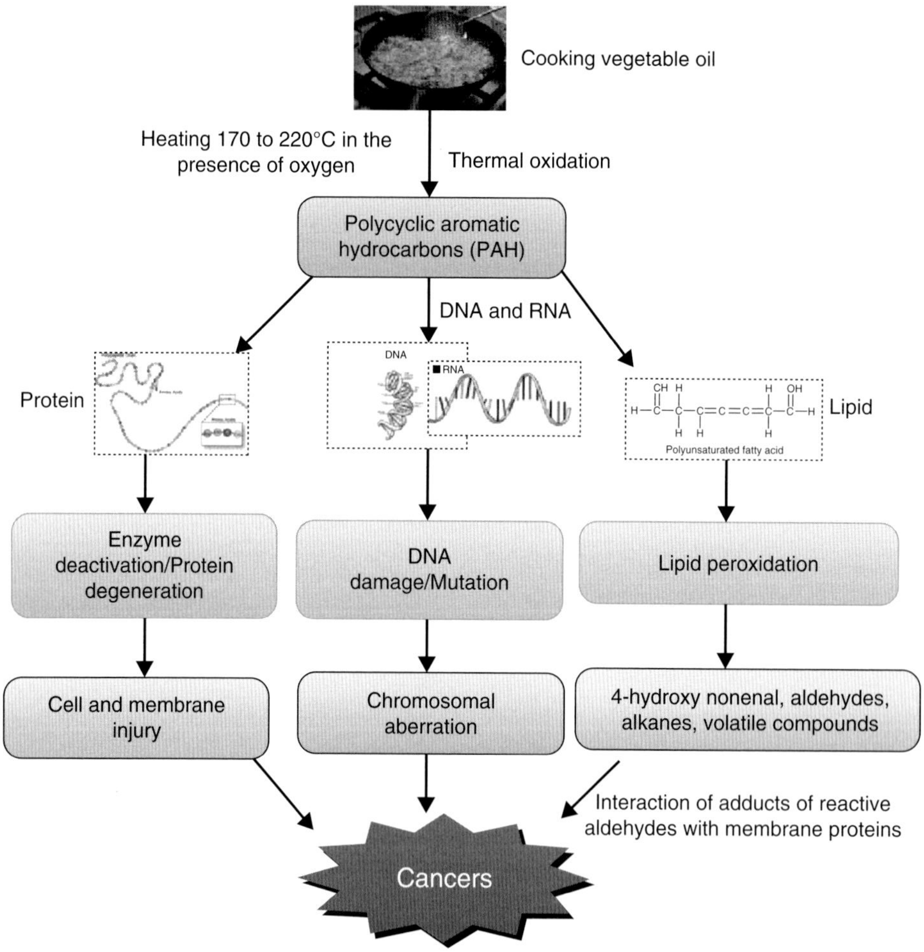

Figure 10.3 Repeatedly heated cooking oil emits polycyclic aromatic hydrocarbons (PAHs) and causes cancer. *Source:* Reproduced with kind permission of Taylor & Francis (Ganesan et al. 2017).

they are absorbed in the intestine after ingestion. In this stage, cells are removed from growth-suppressing signals and begin to clonally replicate. Low–molecular-weight PAHs that have inlet or inlet-like regions can dysregulate gap junction channels, interfering with intracellular communication and affecting mitogen-activated protein kinases (MAPKs), which activate transcription factors involved in cell proliferation (Ganesan et al. 2017).

Most studies on carcinogenic effects are based on heated vegetable oils or their products, such as hydroperoxides and oxidized oils. Table 10.3 lists a number of studies conducted in the last decade showing the effects of heated vegetables oils and fats on various cancers. Recent studies show that feeding oxidized oils or fats to experimental animals causes inflammation, necrosis, and ultimately liver cancer (Goya et al. 2007; Zeb and Haq 2016; Khan et al. 2017; Zeb and Akbar 2018). Lung cancer has also been attributed to the presence of oxidized vegetable oils products (Hung et al. 2007; Cai et al. 2012; Zhang et al. 2012; Lee and Gany 2013; Nie et al. 2014; Duan et al. 2016). Structural questionnaire studies have also confirmed the occurrence of gastric cancer (Mojska et al. 2010; Somi et al. 2015), breast cancer (Cottet et al. 2009; Datta and Biswas 2009; Catsburg et al. 2015; Kojima et al. 2017), pancreatic cancer (Ghorbani et al. 2015), endometrial adenocarcinoma (Takayama et al. 2013), stomach cancer (Pham et al. 2010), and colorectal cancer (Galeone et al. 2007). Agbaria et al. (2015) analysed the effect of deep-fried vegetable oils on the nuclear factor kappa beta (NF-κB) pathway of mouse colon carcinoma (MC38) and Hodgkin's lymphoma (L428) cells. Other human cancers, such as leukaemia (Liu et al. 2015), vascular cancer (Orozco-Solano et al. 2013), cervical cancer (Lee et al. 2010), and pancreatic cancer (Ghorbani et al. 2015), have also been associated with deep-heated frying oils.

These studies revealed that frying generates new compounds, which are mostly carcinogenic if taken in excess or repeatedly. Most come from lipids and proteins, either in fried food or in frying medium. Thus, during frying, standard safety protocols should be followed to avoid or reduce the formation of toxic compounds.

10.4 Fried Foods and Diabetes

Diabetes is a metabolic disorder that is associated with fried food consumption. Animal studies show that administration of fried foods alters serum glucose levels by affecting insulin receptors in the cells, causing type 2 diabetes (T2D) (Chao et al. 2007). Osorio-Yáñez et al. (2017) studied maternal periconceptional fried food intake in a large (n = 3414) prospective birth cohort in Seattle and Tacoma, WA using a food frequency questionnaire. The authors concluded that regular consumption of fried foods such as fried chicken and fish is associated with gestational diabetes mellitus (GDM). Bao et al. (2014) showed that consumption of fried foods, especially away from home, is significantly associated with pre-pregnancy GDM in women. Similar results were also reported by Cahill et al. (2014) in US women and men. The authors concluded that frequent consumption of fried food was significantly associated with risk of incident T2D.

Wallin et al. (2017) reported overall no association between total fish consumption and T2D in a prospective cohort study of Swedish men. The authors indicated that dietary contaminants in fish may influence the relationship. Fried fish and shellfish

Table 10.3 Effects of heated vegetable oils and fats on various cancers.

Model	Frying oil/fat	Investigation	Results	References
Rabbits	Tallow	Serum biochemistry and liver histology, GST, GSH, MDA, RSA	Liver cancer	Zeb and Akbar (2018)
Rats	Olive oil	Liver histology and serum biochemistry	Liver cancer	Khan et al. (2017)
Rabbits	Sunflower oil	Serum biochemistry and liver histology	Liver cancer	Zeb and Haq (2016)
human hepatoma HepG2 cells	Olive oil	GSH, GST, GPx, MDA	Liver cancer	Goya et al. (2007)
Human lung adenocarcinoma A549 cells	Rapeseed oil	Assay of IAP2 and phosphorylated Akt proteins, flow cytometry, and terminal deoxynucleotidyl transferase-mediated dUTP nick end labelling (TUNEL) analysis	Lung cancer	Hung et al. (2007)
Human	Vegetable oils	Analysis of DNA damage	Lung cancer	Lee and Gany (2013)
Human	Vegetable oils	Dietary intake exposure assessment and intra-individual variance	Lung cancer	Cai et al. (2012), Zhang et al. (2012), Nie et al. (2014), Duan et al. (2016)
Human	Vegetable oils	Structured questionnaire	Gastric cancer	Somi et al. (2015)
Human	Vegetable oils	Structured questionnaire	Gastric cancer	Mojska et al. (2010)
Human	Vegetable oils	Structured questionnaire and haematological profile	Breast cancer	Kojima et al. (2017)
Human	Vegetable oils	Structured questionnaire	Breast cancer	Catsburg et al. (2015), Cottet et al. (2009)
Human	Vegetable oils	Structured questionnaire	Breast cancer	Datta and Biswas (2009)
Human	Vegetable oils	Structured questionnaire	Pancreas cancer	Ghorbani et al. (2015)
Human	Vegetable oils	Structured questionnaire	Endometrial adenocarcinoma	Takayama et al. (2013)
Human	Vegetable oils	Structured questionnaire	Stomach cancer	Pham et al. (2010)
Human	Olive oil	Structured questionnaire, BMI	Colorectal cancer	Galeone et al. (2007)

Mouse colon carcinoma (MC38) cells and Hodgkin's lymphoma (L428) cells	Vegetable oils	Analysis of NF-κB pathway	Colon cancer	Agbaria et al. (2015)
Human	Vegetable oils	Structured questionnaire and haematological profile	leukaemia	Liu et al. (2015)
Human obese	Sunflower seed oil and virgin olive oil	Assay of omega-3 and omega-6 EFAs	Vascular cancer	Orozco-Solano et al. (2013)
Human	Vegetable oils	Pap smear screening and biopsy examination, community-based case–control study	Cervical cancer	Lee et al. (2010)
Human	Vegetable oils	Structured questionnaire – gender, age, body mass index (BMI), years of education, diabetes and alcohol history, smoking status, and opium use	Pancreatic cancer	Ghorbani et al. (2015)

consumption was significantly associated with higher T2D incidence. These findings suggest that more specific advice on fish species subtypes and preparation methods may be warranted.

10.5 Fried Foods and Cardiovascular Diseases

Cardiovascular diseases (CVDs) are associated with a high intake of lipids and with a potentially life-threatening disturbance in the lipid metabolism. Fried foods usually contain high amounts of fat, and thus cause an increase in serum lipids. Recent animal studies from our lab showed that dietary administration of thermally oxidized edible oils or fats increased serum lipid contents (Zeb and Rahman 2012; Zeb and Ullah 2015; Zeb and Ahmad 2017; Zeb and Rahman 2017) and was thus responsible for CVDs. Similar observations have been reported in a study of Indonesian civil servants (Sudargo et al. 2017). It was found that a significant correlation was present between fried food consumption and serum lipid profile, which is responsible for hypertension. The risk of hypertension is positively and independently associated with the intake of cooking oil polar compounds and inversely related to blood concentrations of monounsaturated fatty acids (Soriguer et al. 2003). Ng et al. (2014) found that oxidation compounds present in heated oil disturb the physiologically critical aspects of cardiovascular functions, such as blood pressure, inflammation, steady redox state, and lipid profile, setting off a complex, interlinking disorder in the normal homeostasis. The response can potentially lead to CVD, as shown in Figure 10.4.

In pregnant women, studies show an optimistic association between fried food intake and being overweight, waist circumference, and weight gain (Tiwari et al. 2009). Regular fried food ingestion has been associated with a higher risk of incident overweight/obesity in two large prospective studies. The European Prospective Study showed a positive association of fried food consumption with central and general obesity in women (Guallar-Castillón et al. 2012). Sayon-Orea et al. (2013) reported similar results, showing the development of overweight/obesity status with the consumption of fried foods more than four times per week, as compared with fewer than two times a week (Sayon-Orea et al. 2014).

Qi et al. (2014) showed that fried food consumption interacts with the genes responsible for obesity and, ultimately, cardiovascular health. They found that significant interactions of the genetic risk score with total fried food consumption, fried food consumed at home, and fried food consumed away from home on body mass index (BMI) were replicated in the Women's Genome Health Study, as shown in Figure 10.5. The difference in BMI between individuals who consumed fried foods more than four times a week and those who consumed fried foods less than once a week was more pronounced among participants in the highest third than among those in the lowest third of the genetic risk score. In the three cohorts combined, the association between fried food consumption and BMI strengthened across the thirds of the genetic risk score; viewed differently, the association between the genetic risk score and BMI was more pronounced in those who often ate fried foods.

Gadiraju et al. (2015) reviewed the association of fried food consumption with cardiovascular health. They found enough evidence to suggest adverse health effects

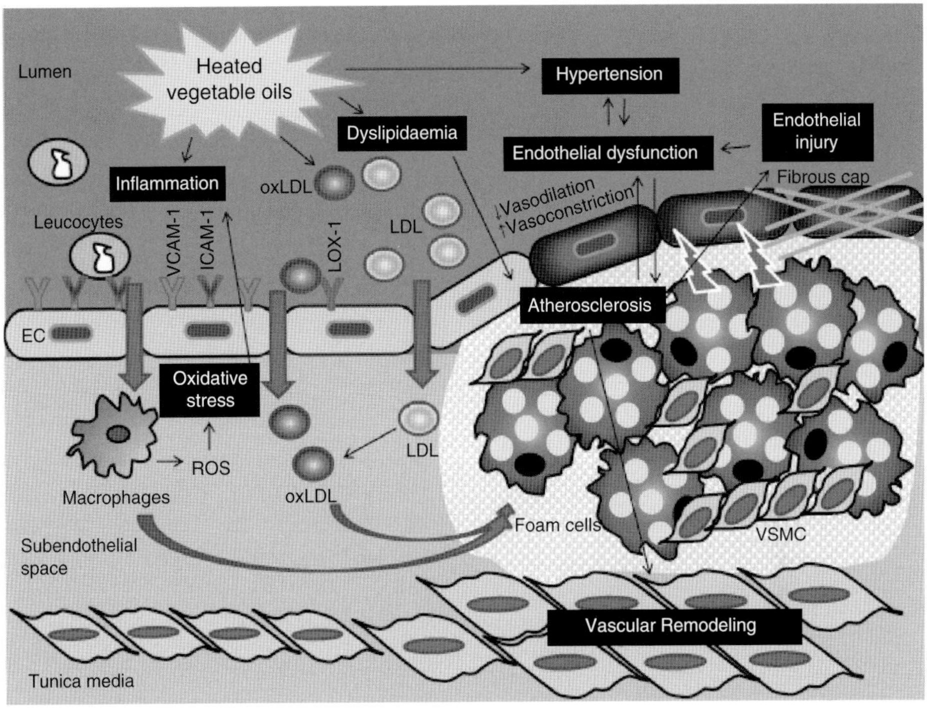

Figure 10.4 Proposed pathological mechanisms responsible for increased cardiovascular disease (CVD) risk following consumption of heated oil. EC, endothelial cell; ICAM-1, intercellular adhesion molecule-1; LDL, low-density lipoprotein; LOX-1, lectin-like oxidized low-density lipoprotein receptor; oxLDL, oxidized low-density lipoprotein; ROS, reactive oxygen species; VCAM-1, vascular cell adhesion molecule-1; VSMC, vascular smooth muscle cell. *Source:* Reproduced with kind permission of Elsevier Inc. (Ng et al. 2014).

with a higher frequency of fried food consumption. Similarly, Lippi and Mattiuzzi (2015) and Lippi et al. (2015) also reviewed fried food consumption and its association with heart diseases. Djoussé et al. (2015) conducted a prospective cohort study of 15 362 participants from the Physicians' Health Study, which showed a positive association of heart failure with fried food intake frequency in male physicians. The recent Costa Rican Heart Study reported that fried food consumed outside of the home is highly associated with cardiovascular health (Hu et al. 2018). While awaiting confirmation from future studies, it may be advisable to the public to consume fried foods in moderation while maintaining an overall healthy diet.

10.5.1 *Trans* Fats and Cardiovascular Health

Dietary fats have an important role in coronary heart disease (CHD) (Willett 2012). Positive associations between intake of *trans* fatty acids and risk of CHD were observed in the Nurses' Health Study cohort (Willett et al. 1993) and confirmed in other large prospective studies (Mozaffarian et al. 2006). In a meta-analysis, the summary relative

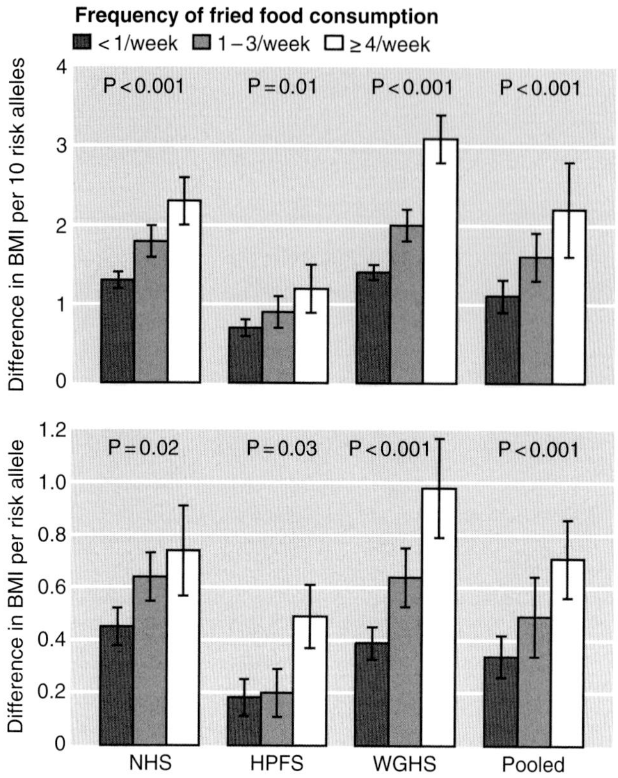

Figure 10.5 Genetic associations with body mass index (BMI) according to frequency of fried food consumption in three cohorts. Data are differences (SE) in BMI per 10 risk alleles of genetic risk score and differences (SE) in BMI per risk allele (A-allele) of the FTO (fat mass and obesity associated) variant rs15581002. In the Nurses' Health Study (NHS) and Health Professionals Follow-Up Study (HPFS), data were adjusted for age, source of genotyping data, physical activity, television watching, smoking, alcohol intake, intake of sugar-sweetened beverages, alternative healthy eating index, and total energy intake. In the Women's Genome Health Study (WGHS), data were adjusted for age, physical activity, smoking, alcohol intake, intake of sugar-sweetened beverages, alternative healthy eating index, and total energy intake. Data from three cohorts, pooled by means of fixed-effects meta-analyses (if $p \geq 0.05$ for heterogeneity between studies) or random-effects meta-analyses (if $p < 0.05$ for heterogeneity between studies). *Source:* Reproduced with kind permission of BMJ Publishing Group Ltd. (Qi et al. 2014).

risk for an increment of 2% of energy from *trans* fat was 1.23 (95% CI 1.11–1.37; $p < 0.001$). More recently, the relationship between *trans* fatty acids in red cell membranes and risk of CHD was examined in the Nurses' Health Study using a nested case–control design; the relative risk for the highest versus lowest quartile was 3.3 (1.5–7.2, p for trend <0.01).

The combination of many metabolic studies demonstrating adverse effects of *trans* fatty acids on blood lipid fractions and prospective epidemiologic studies has provided strong support for an adverse effect on CHD risk. Thus, actions to reduce intake have included banning the use of partially hydrogenated fats in Denmark, Brazil, and Chile, as well as in several cities in the United States. Although much progress has been made in reducing intake in some countries, intake remains high in many parts of the world.

Also, even if the average intake has been greatly reduced in a particular country, a modest percentage of the population may still consume large amounts of *trans* fats through fried foods (Willett 2012).

The number of different *trans* isomers in partially hydrogenated vegetable oils is very large, because the *trans* double-bonds can vary in number (usually 0, 1 or 2) and position in the carbon chain, and the number of carbons atoms can differ. In two studies using analyses of red cells or adipose tissue, 18: 2 *trans* isomers were substantially more strongly associated with risk of CHD than were 18: 1 *trans* isomers (Baylin et al. 2003; Sun et al. 2007). This may have practical implications, because if the degree of partial hydrogenation is reduced but not eliminated, the amount of 18: 2 *trans* isomers can be increased. This light degree of hydrogenation has often been used when the goal was to destroy alpha-linolenic acid (ALA) (which is susceptible to oxidation, producing rancidity and shortened shelf life) rather than to create a hard fat. Although our understanding of the effects of different *trans* isomers is incomplete, the goal should be to eliminate partial hydrogenation, not just reduce it.

Trans isomers from ruminant fats will usually not exceed 1% of fatty acids, even with high intakes of red meat and butterfat. However, ruminant fats have generated increased interest as other sources of *trans* fatty acids have been reduced. The relationship between the intake of *trans* fatty acids from ruminants and the risk of CHD has been examined in four prospective studies; none identified a significant positive association, and in three a nonsignificant inverse trend was observed (Willett et al. 1993; Pietinen et al. 1997; Jakobsen et al. 2008).

Although major progress has been made in understanding the consequences of partial hydrogenation of vegetable oils, not all issues regarding cooking fats are resolved. The sustained heating of fats, as used in the deep frying processes of the fast food industry, can lead to the oxidation of PUFAs (Kanner 2007), but intake is usually low because oxidation creates undesirable flavours. Virtually no epidemiologic data exist on the consequences of oxidized cooking fats.

10.6 Fried Foods and Aging

Oxidative stress is a specific condition of a cell or group of cells under which reactive oxygen-containing compounds exist in higher number than under normal conditions. It can also be defined as the disturbance of the oxidant–antioxidant ratio in favour of oxidant species (Sies 1991). This means that the internal defence system lacks the potential to neutralize the oxidant species; these may be reactive oxygen species (ROS), reactive nitrogen species (RNS) (Pisoschi and Pop 2015), or oxidized lipids (Zeb 2015). Oxidative stress is one of the factors responsible for ageing. A recent animal study showed that thermally oxidized lipids induce oxidative stress (Zeb and Akbar 2018).

Ageing is linked with an increase in oxidation products derived from nucleic acids, sugars, sterols, and lipids, especially from fried foods. Evidence shows that oxidation products induce changes in the cell membrane structure, either by proliferation, ageing, or, especially, wounding (Spiteller 2010). A recent public health study showed that furan or its derivatives, which are commonly present in fried foods (Chain et al. 2017), are a key factor affecting ageing and health. Thus, healthy fried foods – especially with reduced toxicants – will maintain health and reduce ageing.

10.7 Key Concepts

- Food is fried in order to maintain human health.
- Fried foods are considered healthy because they are a source of lipids, provide quick processing, and impart a high acceptability of taste, flavour, and sensory characteristics.
- Fried foods also contain several important nutrients and antioxidants.
- Frying produces nutritional and physiological changes in the frying medium and in the foods being fried.
- Several studies show that fried foods and heated oils cause cancers of various types.
- Few studies show the effects of fried foods on diabetes.
- Fried foods also contribute to the prevalence of CVD.
- *Trans* fats are especially notorious for affecting cardiovascular health.
- Fried foods and oils may cause oxidative stress and ultimately contribute to ageing.

References

Abdel-Shafy, H.I. and Mansour, M.S.M. (2016). A review on polycyclic aromatic hydrocarbons: source, environmental impact, effect on human health and remediation. *Egyptian Journal of Petroleum* 25: 107–123.

Adamson, R.H., Thorgeirsson, U.P., and Sugimura, T. (1996). *Extrapolation of Heterocyclic Amine Carcinogenesis Data from Rodents and Nonhuman Primates to Humans.* Berlin, Heidelberg: Springer.

Agbaria, R., Gabarin, A., Dahan, A., and Ben-Shabat, S. (2015). Anticancer activity of *Nigella sativa* (black seed) and its relationship with the thermal processing and quinone composition of the seed. *Drug Design, Development and Therapy* 9: 3119–3124.

Bao, W., Tobias, D.K., Olsen, S.F., and Zhang, C. (2014). Pre-pregnancy fried food consumption and the risk of gestational diabetes mellitus: a prospective cohort study. *Diabetologia* 57: 2485–2491.

Baylin, A., Kabagambe, E.K., Ascherio, A. et al. (2003). High 18:2 trans-fatty acids in adipose tissue are associated with increased risk of nonfatal acute myocardial infarction in costa rican adults. *Journal of Nutrition* 133: 1186–1191.

Bittman, M. 2013. Deep fried and good for you. *New York Times*, 21 October.

Cahill, L.E., Pan, A., Chiuve, S.E. et al. (2014). Fried-food consumption and risk of type 2 diabetes and coronary artery disease: a prospective study in 2 cohorts of US women and men. *American Journal of Clinical Nutrition* 100: 667–675.

Cai, Y., Lv, J., Zhang, W., and Zhang, L. (2012). Dietary exposure estimates of 16 polycyclic aromatic hydrocarbons (PAHs) in Xuanwei and Fuyuan, counties in a high lung cancer incidence area in China. *Journal of Environmental Monitoring* 14: 886–892.

Catsburg, C., Kim, R.S., Kirsh, V.A. et al. (2015). Dietary patterns and breast cancer risk: a study in 2 cohorts. *American Journal of Clinical Nutrition* 101: 817–823.

Chain, E., Panel, O.C.I.T.F., Knutsen, H.K. et al. (2017). Risks for public health related to the presence of furan and methylfurans in food. *EFSA Journal* 15: e05005.

Chao, P.M., Huang, H.L., Liao, C.H. et al. (2007). A high oxidised frying oil content diet is less adipogenic, but induces glucose intolerance in rodents. *British Journal of Nutrition* 98: 63–71.

Chiou, A., Kalogeropoulos, N., Boskou, G., and Salta, F.N. (2012). Migration of health promoting microconstituents from frying vegetable oils to French fries. *Food Chemistry* 133: 1255–1263.

Cottet, V., Touvier, M., Fournier, A. et al. (2009). Postmenopausal breast cancer risk and dietary patterns in the E3N-EPIC prospective cohort study. *American Journal of Epidemiology* 170: 1257–1267.

Datta, K. and Biswas, J. (2009). Influence of dietary habits, physical activity and affluence factors on breast cancer in East India: a case-control study. *Asian Pacific Journal of Cancer Prevention* 10: 219–222.

Davis, R. J. 2013. Health risks of fried foods may be overblown. *Time*, 24 September.

Djoussé, L., Petrone, A.B., and Gaziano, J.M. (2015). Consumption of fried foods and risk of heart failure in the physicians' health study. *Journal of the American Heart Association* 4 (4): e001740.

Dobarganes, C. and Marquez-Ruiz, G. (2015). Possible adverse effects of frying with vegetable oils. *British Journal of Nutrition* 113 (Suppl. 2): S49–S57.

Duan, X., Shen, G., Yang, H. et al. (2016). Dietary intake polycyclic aromatic hydrocarbons (PAHs) and associated cancer risk in a cohort of Chinese urban adults: inter- and intra-individual variability. *Chemosphere* 144: 2469–2475.

Ferraro, J.V., Plummer, T.W., Pobiner, B.L. et al. (2013). Earliest archaeological evidence of persistent hominin carnivory. *PLoS ONE* 8: e62174.

Gadiraju, T.V., Patel, Y., Gaziano, J.M., and Djousse, L. (2015). Fried food consumption and cardiovascular health: a review of current evidence. *Nutrients* 7: 8424–8430.

Galeone, C., Talamini, R., Levi, F. et al. (2007). Fried foods, olive oil and colorectal cancer. *Annals of Oncology* 18: 36–39.

Ganesan, K., Sukalingam, K., and Xu, B. (2017). Impact of consumption and cooking manners of vegetable oils on cardiovascular diseases – a critical review. *Trends in Food Science & Technology* 71: 132–154.

Gerhardsson De Verdier, M. (1995). Epidemiologic studies on fried foods and cancer in Sweden. *Princess Takamatsu Symposia* 23: 292–298.

Ghorbani, Z., Hekmatdoost, A., Zinab, H.E. et al. (2015). Dietary food groups intake and cooking methods associations with pancreatic cancer: a case-control study. *Indian Journal of Gastroenterology* 34: 225–232.

Gladyshev, M.I., Sushchik, N.N., Gubanenko, G.A. et al. (2006). Effect of way of cooking on content of essential polyunsaturated fatty acids in muscle tissue of humpback salmon (Oncorhynchus gorbuscha). *Food Chemistry* 96: 446–451.

Goya, L., Mateos, R., and Bravo, L. (2007). Effect of the olive oil phenol hydroxytyrosol on human hepatoma HepG2 cells. *European Journal of Nutrition* 46: 70–78.

Guallar-Castillón, P., Rodríguez-Artalejo, F., Lopez-Garcia, E. et al. (2012). Consumption of fried foods and risk of coronary heart disease: Spanish cohort of the European prospective investigation into cancer and nutrition study. *British Medical Journal* 344: e363.

Howe, P., Meyer, B., Record, S., and Baghurst, K. (2006). Dietary intake of long-chain ω-3 polyunsaturated fatty acids: contribution of meat sources. *Nutrition* 22: 47–53.

Hu, P., Li, Y., and Campos, H. (2018). Fried food intake and risk of nonfatal acute myocardial infarction in the Costa Rica heart study. *PLoS ONE* 13: e0192960.

Hung, H.S., Wu, W.J., Cheng, Y.W. et al. (2007). Association of cooking oil fumes exposure with lung cancer: involvement of inhibitor of apoptosis proteins in cell survival and proliferation in vitro. *Mutation Research* 628: 107–116.

IARC (1993). Some naturally occurring substances: food items and constituents, heterocyclic aromatic amines and mycotoxins. In: *World Health Organization. IARC Monographs on the Evaluation of the Carcinogenic Risk of Chemicals to Humans*. Geneva: World Health Organization.

Jakobsen, M.U., Overvad, K., Dyerberg, J., and Heitmann, B.L. (2008). Intake of ruminant trans fatty acids and risk of coronary heart disease. *International Journal of Epidemiology* 37: 173–182.

Joshi, A.D., John, E.M., Koo, J. et al. (2012). Fish intake, cooking practices, and risk of prostate cancer: results from a multi-ethnic case–control study. *Cancer Causes & Control* 23: 405–420.

Kanner, J. (2007). Dietary advanced lipid oxidation endproducts are risk factors to human health. *Molecular Nutrition & Food Research* 51: 1094–1101.

Khan, A.A., Zeb, A., and Sherazi, S.T. (2017). Thermally oxidized olive oil produces biochemical, physiological effects and fatty liver in rats. *Chiang Mai Journal of Science* 44: 847–857.

Knekt, P., Steineck, G., Jarvinen, R. et al. (1994). Intake of fried meat and risk of cancer: a follow-up study in Finland. *International Journal of Cancer* 59: 756–760.

Kojima, R., Okada, E., Ukawa, S. et al. (2017). Dietary patterns and breast cancer risk in a prospective Japanese study. *Breast Cancer* 24: 152–160.

Lee, C.H., Yang, S.F., Peng, C.Y. et al. (2010). The precancerous effect of emitted cooking oil fumes on precursor lesions of cervical cancer. *International Journal of Cancer* 127: 932–941.

Lee, T. and Gany, F. (2013). Cooking oil fumes and lung cancer: a review of the literature in the context of the U.S. population. *Journal of Immigrant and Minority Health* 15: 646–652.

Lešková, E., Kubíková, J., Kováčiková, E. et al. (2006). Vitamin losses: retention during heat treatment and continual changes expressed by mathematical models. *Journal of Food Composition and Analysis* 19: 252–276.

Li, M.L., Ji, L., Hou, J.G. et al. (2014). Environmental and psycho-social factors related to prostate cancer risk in the Chinese population: a case-control study. *Biomedical and Environmental Sciences* 27: 707–717.

Lippi, G. and Mattiuzzi, C. (2015). Fried food and prostate cancer risk: systematic review and meta-analysis. *International Journal of Food Sciences and Nutrition* 66: 587–589.

Lippi, G., Mattiuzzi, C., and Cervellin, G. (2015). Fried food consumption and ischemic heart disease: a systematic literature review. *International Journal of Cardiology* 190: 210–211.

Liu, P., Holman, C.D., Jin, J., and Zhang, M. (2015). Diet and risk of adult leukemia: a multicenter case-control study in China. *Cancer Causes & Control* 26: 1141–1151.

Márquez-Ruiz, G. and Dobarganes, M.C. (2007). Nutritional and physiological effects of used frying oils and fats. In: *Deep Frying*, 2e (ed. M. Erickson). New York: Elsevier.

Mcdonald, B.E. and Eskin, M.N. (2007). Role of fat in the diet. In: *Deep Frying*, 2e (ed. M. Erickson). New York: Elsevier.

Mojska, H., Gielecinska, I., Szponar, L., and Oltarzewski, M. (2010). Estimation of the dietary acrylamide exposure of the Polish population. *Food and Chemical Toxicology* 48: 2090–2096.

Mozaffarian, D., Katan, M.B., Ascherio, A. et al. (2006). Trans fatty acids and cardiovascular disease. *New England Journal of Medicine* 354: 1601–1613.

Ng, C.-Y., Leong, X.-F., Masbah, N. et al. (2014). Heated vegetable oils and cardiovascular disease risk factors. *Vascular Pharmacology* 61: 1–9.

Nie, J., Shi, J., Duan, X. et al. (2014). Health risk assessment of dietary exposure to polycyclic aromatic hydrocarbons in Taiyuan, China. *Journal of Environmental Sciences* 26: 432–439.

Ohgaki, H., Hasegawa, H., Suenaga, M. et al. (1987). Carcinogenicity in mice of a mutagenic compound, 2-amino-3,8-dimethylimidazo[4,5-f]quinoxaline (MeIQx) from cooked foods. *Carcinogenesis* 8: 665–668.

Orozco-Solano, M.I., Priego-Capote, F., and Luque De Castro, M.D. (2013). Analysis of esterified and nonesterified fatty acids in serum from obese individuals after intake of breakfasts prepared with oils heated at frying temperature. *Analytical and Bioanalytical Chemistry* 405: 6117–6129.

Osorio-Yáñez, C., Gelaye, B., Qiu, C. et al. (2017). Maternal intake of fried foods and risk of gestational diabetes mellitus. *Annals of Epidemiology* 27: 384.e1–390.e1.

Pawlega, J., Rachtan, J., and Dyba, T. (1996). Dietary factors and risk of prostate cancer in Poland. Results of case–control study. *Neoplasma* 43: 61–63.

Pham, T.M., Fujino, Y., Kikuchi, S. et al. (2010). Dietary patterns and risk of stomach cancer mortality: the Japan collaborative cohort study. *Annals of Epidemiology* 20: 356–363.

Pietinen, P., Ascherio, A., Korhonen, P. et al. (1997). Intake of fatty acids and risk of coronary heart disease in a cohort of Finnish men: the alpha-tocopherol, beta-carotene cancer prevention study. *American Journal of Epidemiology* 145: 876–887.

Pisoschi, A.M. and Pop, A. (2015). The role of antioxidants in the chemistry of oxidative stress: a review. *European Journal of Medicinal Chemistry* 97: 55–74.

Pokorný, J. and Dostálová, J. (2010). Changes in nutrients, antinutritional factors, and contaminants at frying temperatures. In: *Frying of Food: Oxidation, Nutrient and Non-nutrient Antioxidants, Biologically Active Compounds and High Temperatures* (ed. D. Boskou and I. Elmadfa). Boca Raton, FL: CRC Press.

Qi, Q., Chu, A.Y., Kang, J.H. et al. (2014). Fried food consumption, genetic risk, and body mass index: gene-diet interaction analysis in three US cohort studies. *British Medical Journal* 348: g1610.

Sayon-Orea, C., Bes-Rastrollo, M., Basterra Gortari, F. et al. (2013). Consumption of fried foods and weight gain in a Mediterranean cohort: the SUN project. *Nutrition, Metabolism, and Cardiovascular Diseases* 23: 144–150.

Sayon-Orea, C., Bes-Rastrollo, M., Gea, A. et al. (2014). Reported fried food consumption and the incidence of hypertension in a Mediterranean cohort: the SUN (Seguimiento Universidad de Navarra) project. *British Journal of Nutrition* 112: 984–991.

Sies, H. (1991). Oxidative stress: from basic research to clinical application. *American Journal of Medicine* 91: 31S–38S.

Smil, V. (2002). Eating meat: evolution, patterns, and consequences. *Population and Development Review* 28: 599–639.

Sobral, M.M.C., Cunha, S.C., Faria, M.A., and Ferreira, I.M. (2018). Domestic cooking of muscle foods: impact on composition of nutrients and contaminants. *Comprehensive Reviews in Food Science and Food Safety* https://doi.org/10.1111/1541-4337.12327 **17**: 309–333.

Somi, M.H., Mousavi, S.M., Naghashi, S. et al. (2015). Is there any relationship between food habits in the last two decades and gastric cancer in North-Western Iran. *Asian Pacific Journal of Cancer Prevention* 16: 283–290.

Sonestedt, E., Ericson, U., Gullberg, B. et al. (2008). Do both heterocyclic amines and omega-6 polyunsaturated fatty acids contribute to the incidence of breast cancer in postmenopausal women of the Malmö diet and cancer cohort? *International Journal of Cancer* 123: 1637–1643.

Soriguer, F., Rojo-Martínez, G., Dobarganes, M.C. et al. (2003). Hypertension is related to the degradation of dietary frying oils. *American Journal of Clinical Nutrition* 78: 1092–1097.

Spiteller, G. (2010). Is lipid peroxidation of polyunsaturated acids the only source of free radicals that induce aging and age-related diseases? *Rejuvenation Research* 13: 91–103.

Steineck, G., Gerhardsson De Verdier, M., and Overvik, E. (1993). The epidemiological evidence concerning intake of mutagenic activity from the fried surface and the risk of cancer cannot justify preventive measures. *European Journal of Cancer Prevention* 2: 293–300.

Stier, R.F. (2013). Ensuring the health and safety of fried foods. *European Journal of Lipid Science and Technology* 115: 956–964.

Stott-Miller, M., Neuhouser, M.L., and Stanford, J.L. (2013). Consumption of deep-fried foods and risk of prostate cancer. *The Prostate* 73: 960–969.

Sudargo, T., Sari, F.T., and Naomi, N.D. (2017). The correlation of obesity, smoking, fried foods consumption pattern and food intake with lipid profile in civil servant in Yogyakarta, Indonesia. *International Journal of Community Medicine and Public Health* 4: 690.

Sun, Q., Ma, J., Campos, H. et al. (2007). A prospective study of trans fatty acids in erythrocytes and risk of coronary heart disease. *Circulation* 115: 1858–1865.

Takayama, S., Monma, Y., Tsubota-Utsugi, M. et al. (2013). Food intake and the risk of endometrial endometrioid adenocarcinoma in Japanese women. *Nutrition and Cancer* 65: 954–960.

Tiwari, R., Srivastava, D., and Gour, N. (2009). A cross-sectional study to determine prevalence of obesity in high income group colonies of Gwalior city. *Indian Journal of Community Medicine* 34: 218.

Velasco, J., Holgado, F., Dobarganes, C., and Márquez-Ruiz, G. (2009). Antioxidant activity of added phenolic compounds in freeze-dried microencapsulated sunflower oil. *Journal of the American Oil Chemists' Society* 86: 445–452.

Wallin, A., DI Giuseppe, D., Orsini, N. et al. (2017). Fish consumption and frying of fish in relation to type 2 diabetes incidence: a prospective cohort study of Swedish men. *European Journal of Nutrition* 56: 843–852.

Willett, W.C. (2012). Dietary fats and coronary heart disease. *Journal of Internal Medicine* 272: 13–24.

Willett, W.C., Stampfer, M.J., Manson, J.E. et al. (1993). Intake of trans fatty acids and risk of coronary heart disease among women. *The Lancet* 341: 581–585.

Zeb, A. (2015). Chemistry and liquid chromatography methods for the analyses of primary oxidation products of triacylglycerols. *Free Radical Research* 49: 549–564.

Zeb, A. and Ahmad, S. (2017). Changes in acylglycerols composition, quality characteristics and in vivo effects of dietary pumpkin seed oil upon thermal oxidation. *Frontiers in Chemistry* 5: 55.

Zeb, A. and Akbar, A. (2018). Ellagic acid suppresses the oxidative stress induced by dietary-oxidized tallow. https://doi.org/10.1007/s11010-018-3310-3. *Oxidative Medicine and Cellular Longevity* **2018**: Article ID 7408370.

Zeb, A. and Habib, A. (2018). Lipid oxidation and changes in the phenolic profile of watercress (Nasturtium officinale L.) leaves during frying. *Journal of Food Measurement and Characterization* https://doi.org/10.1007/s11694-018-9885-2

Zeb, A. and Haq, I. (2016). The protective role of tomato powder in the toxicity, fatty infiltration and necrosis induced by oxidized tallow in rabbits. *Journal of Food Biochemistry* 40: 428–435.

Zeb, A. and Rahman, W.U. (2012). Effects of oxidized tallow on the rabbit serum lipids and antioxidant activity of the in-vitro lipids. *Toxicology Research* 28: 151–157.

Zeb, A. and Rahman, S.U. (2017). Protective effects of dietary glycine and glutamic acid toward the toxic effects of oxidized mustard oil in rabbits. *Food & Function* 8: 429–436.

Zeb, A. and Ullah, S. (2015). Sea buckthorn seed oil protects against the oxidative stress produced by thermally oxidized lipids. *Food Chemistry* 186: 6–12.

Zhang, L., Lv, J., and Liao, C. (2012). Dietary exposure estimates of 14 trace elements in Xuanwei and Fuyuan, two high lung cancer incidence areas in China. *Biological Trace Element Research* 146: 287–292.

Part IV

Safety in Food Frying

11

Safety Assessment of Food Frying

11.1 Introduction

Frying fats and oils have been subject to much attention over their health risks. In the early 1970s, the German authorities received several complaints about the quality of fried food served in restaurants, which led to a detailed inspection of the food service industry. This investigation revealed that the quality of the frying medium has a direct impact on the quality of the final fried product. Thus, the German Society for Fat Science (DGF) recommended several regulations for the use of frying oils in restaurants. Later studies revealed several quality parameters that can be used to assess the quality of an oil or fat. These include peroxide value, free fatty acid (FFA) content, anisidine value, iodine value, carbonyl value, colour, and refractive index.

Stier (2001) reviewed some of the important criteria used to assess the quality of frying fats. These are based on the objectives of producing fried foods of high quality and controlling the degradation of oils during frying to achieve maximum economic parameters. The basic criteria were originally obtained from the work of Robertson (1968).

- Proper design, construction, and maintenance of equipment.
- Proper operation of equipment.
- Proper cleaning of equipment.
- Monitoring of chemical indices of oil degradation.
- Minimization of exposure to ultraviolet (UV) light.
- Keeping salt and other sources of metal away from oil.
- Filtering oil regularly.

Once the frying process is initiated, it begins to degrade the frying medium, in a process that cannot be reversed. However, recent research revealed that this degradation can be slowed by various physical processes, in order to achieve maximum use from frying oils.

Food Frying: Chemistry, Biochemistry, and Safety, First Edition. Alam Zeb.

11.2 Guideline for Assessment

The most important issue in the evaluation of deep-frying oil quality is finding the optimal point at which to discard the used oil (Weisshaar 2014). Recommendations from the 3rd International Symposium on Deep-Fat Frying in 2000 define some criteria for the assessment of used frying oils and frying fats (Donner and Ritcher 2000):

- The principle quality index for deep fat frying should be the sensory parameters of the food being fried.
- Analysis of suspect frying fats and oils should utilize two tests to confirm abuse:
  - Total polar materials (<24%).
  - Polymeric triglycerides (<12%).

When oxidative alterations strongly predominate over thermal alterations, serious sensory defects can present before total polar materials (TPM) and polymer triglycerides (PTG) reach the recommended limits. In that case, additional parameters like anisidine value or carbonyl value should be used for evaluation.

In many countries, regulations for abused frying fats and oils were established as guidelines or regulatory limits, addressed to fryer operators and official food control agencies. In most cases, TPM is the critical parameter, with limits from 24 to 27%. A synoptic table of the actual regulations used in different countries was recently presented by Stier (2013). The author stated that regulations are largely in line with those just given, but further harmonization is desirable, at least within the European Union.

For the daily practice of frying, it is reasonable to use rapid tests and to define individual cut-off criteria based on correlation to sensory attributes and the practical experience of the fryer operator. From time to time, individual cut-off values should be rechecked by laboratory analysis.

The European Federation for the Science and Technology of Lipids (Euro Fed Lipid) and the American Oil Chemist's Society (AOCS) hosted the 7th International Symposium on Deep-Fat Frying from February 20 to 22, 2013 in San Francisco, CA. The program was held at the Radisson Fisherman's Wharf and featured 20 speakers. It was attended by more than 80 scientists from all over the world (Gertz and Stier 2013). The aim was to help processors, food service operators, regulators, suppliers, and academics to better understand the frying process in the light of a worldwide emphasis on health and wellness, and concerns about the effect of fried products on obesity. Rather than condemning the frying process, the aim was to improve its efficiency, understand the performance of the new generation of healthy oils (low- and no-*trans*), and look to the past for lessons on how to do things better.

Food processors, restaurant operators, and suppliers to these industries, which include oil producers, food and ingredient suppliers, equipment manufacturers, and the service trade, have to understand that they must produce foods that are not only safe and nutritious but also tasty and healthy. The pressure on the industry is great, but an understanding of the processes and technologies, the markets and demands can help build a basis on which to grow. Like at past International Symposia on Deep-Fat Frying, the participants generated a series of recommendations for enhancing the science and technology of frying, as follows:

1) Support risk-based research and analysis of data to more clearly define safety of potentially hazardous oil degradation products, particularly as oxidized products, and/or those formed in foods during frying. We recommend that these evaluations be expanded to examine the whole food system.

2) Operations working with fried foods at both restaurant and industrial level need to emphasize what they are doing to produce more healthy meal items (holistic approach).
3) Continue to encourage the development, quantification, and validation of rapid test technologies for monitoring fresh and used oil properties.
4) Fryer operators must clearly define specifications for their frying oil(s) with their suppliers and initiate programs to ensure that those specifications are being met.
5) Fryer operators must establish a baseline for their current operations. This serves as a yardstick to properly evaluate any changes to the system, whether that change is a new oil, filter system, or product mix. This baseline should include but need not be limited to oil chemistry, food quality, organoleptic parameters of food, and operational issues.
6) The potential microbiological risks inherent with coated products and coating applicators should be researched and the real risk defined.
7) The group reaffirms their 2000 recommendation that the best indices for assessment of used oils are total polar materials (TPM) and polymeric triacylglycerols using recognized methods. Peroxide value (PV), free fatty acids, and anisidine value should not be used as regulatory indices when it comes to monitoring and comparing the degree of degradation of different frying oils.
8) Scientists from the European Union working in the industry, academia, and the government should meet to evaluate current regulations for frying oils with regards to harmonization across the EU to ensure the production of safe foods. Regulations should be supported by clinical data (Gertz and Stier 2013).

11.3 Quality Indicators for Used Frying Oils

A number of analytical laboratory methods for monitoring frying oil quality have been reported in the last 4 decades. At DGF's 1st Symposium for Used Frying Oils and Fats in 1973, smoke point and petroleum-ether insoluble oxidized fatty acids (OXF) were recommended as suitable parameters for the evaluation of frying fat quality. The determination of petroleum-ether insoluble OXF is extremely laborious and time-consuming, and requires large amounts of organic solvents, so this method found no substantial proliferation. In the following years, TPM and PTG were found to be the most reliable methods for monitoring the changes in fats and oils during the frying process. A number of official methods (e.g. from AOCS and DGF) were published for the determination of TPM and PTG (DGF 2010; Firestone 2013). Determination of TPM in frying oil is usually performed by silica gel column chromatography and gravimetric analysis (Guhr et al. 1981). As an alternative to time- and solvent-consuming classical column chromatography, rapid methods using microcolumns or high-performance liquid chromatographic (HPLC) methods have been reported for the analysis of TPM (Kaufmann et al. 2001; Caldwell et al. 2011). For analysis of PTG, high-performance liquid chromatography size-exclusion chromatography (HPLC-SEC) is used. Separation can be achieved with polymer-based separation phases that can be used with organic solvents. Detection can be achieved with a refractive index detector (RID) or evaporative light scattering detector (ELSD). Several studies have shown that there is a strong correlation between TPM and PTG in used frying oils (Gertz 2000; Farhoosh and

Table 11.1 Quality parameters for evaluation of the characteristics of frying fats and oils.

S. No	Parameter	Method(s) applied	Purpose
1	Peroxide value (PV)	Spectroscopic and titrimetric	Determine the hydroperoxides formed
2	Free fatty acid (FFA)	Titrimetric	Help determine stability
3	Iodine value (IV)	Titrimetric	Determine the amount of unsaturation present in fatty acids
4	Anisidine value (AV)	Spectroscopic	Determine the secondary oxidation
5	Saponification number (SN)	Titrimetric	Determine the soap formation ability of frying fats
6	Fatty acids profile	Chromatographic	Obtain an exact profile of each fatty acid content
7	Total polar materials (TPM)	Chromatographic and quick test	Evaluate the level of polar materials after frying
8	Glyceride contents (GC)	Chromatographic	Determine mono-, di-, and triacylglycerol compositions
9	Total polymeric materials	Chromatographic & quick test	Determine the level of polymerization of triglycerides during frying
10	Metals content (MC)	Spectrometric	Determine the degree of metal contamination during preparation of the frying medium
11	Total phenolic contents (TPC)	Spectroscopic	Specific for plant oils – evaluate TPC

Tavassoli-Kafrani 2011). It can be stated that both parameters – TPM and PTG – are the best indicators to check the quality of used frying oils. Some important quality parameters are given in Table 11.1.

But there is a little fly in the ointment: when oxidative alterations strongly predominate over thermal alterations, sensory defects like train oil or fishy flavour can appear long before TPM and PTG reach critical values. This can be observed for frying oils with higher amounts of linolenic acid (>3%) or for oils in open fryers with an unfavourable surface-to-volume (S/V) ratio, used for the preparation of doughnuts and other yeast-raised baked goods. In these cases, anisidine value (Tompkins and Perkins 1999; Aladedunye and Przybylski 2009), carbonyl value (Farhoosh and Tavassoli-Kafrani 2011; Endo et al. 2003; Farhoosh and Moosavi 2008), or conjugated diene value at 234 nm (Farhoosh et al. 2008) can be helpful tools in evaluating oxygen-stressed frying oils, considering the oil's type.

A huge number of volatile compounds are formed during the frying process, according to the nature of the oil and the conditions of frying. These volatiles cover different classes of compound, such as alkanes, alkenes, alcohols, saturated and unsaturated aldehydes and ketones, and short-chain fatty acids. They are responsible for the typical odour of both good and abused frying oil, and can be used as marker compounds for frying oil quality. Some of them, like acrolein and other α,β-unsaturated aldehydes, are undesirable because of health concerns. For analysis, sophisticated techniques like headspace gas chromatography mass spectrometry (GC-MS) are used (Guillén and Goicoechea 2008; Berdeaux et al. 2012; Petersen et al. 2013).

A novel approach to the evaluation of oxidative altered frying oils might be the detection of triglyceride-bound epoxy fatty acids by gas chromatography with flame ionization detection (GC-FID) (Velasco et al. 2004; Mubiru et al. 2013). Epoxy fatty acids are present in the g/kg range in used frying oils. They are unexpectedly stable end products of oxidation, and like TPM and PTG, they are accumulated during the frying process.

In the laboratory of Dr. Rüdiger Weisshaar, the development of a new 'hydrogen bromide value' (HBV), which correlates to the content of epoxy fatty acids, is under way. In this work, the reaction of HBr with epoxy groups is used to develop a simple analytical method of estimating the content of epoxy acids in used frying oils. The first results look very promising (Weisshaar 2014).

FFAs are very easy to determine by titration and have been frequently applied as a quality parameter for used frying oils (Fritsch 1981; Stevenson et al. 1984). The amount of FFAs covers both FFAs formed by hydrolysis and short-chain fatty acids formed by oxidation. Short-chain fatty acids are volatile and are stripped off from the hot frying oil, so their amount will depend on many accidental individual circumstances. For example, the level of FFAs increases strongly if dimethylpolysiloxane is used as an antifoaming additive, due to surface effects which hinder volatile fatty acids from evaporating. Many researchers have come to the conclusion that there is no direct relationship between FFA content and the quality of a used frying oil, and FFAs are no longer considered a reliable marker for frying fat deterioration (Gertz 2000).

11.4 Physical Assessment

11.4.1 Sensory Evaluation, Colour, and Foaming

The organoleptic examination is the first and most important step in the evaluation of frying fat quality. If the taste and flavour of a frying oil or fat are not acceptable, neither will be any products fried in it. Smell and taste are subjective qualities, but in many cases (e.g. restaurants and small fast food outlets) they will be the only form of quality control applied to an oil or fat used in frying. The results of a sensory analysis can be significantly enhanced if it is performed under scientific conditions by a trained sensory panel (Ravelli et al. 2010).

There are numerous studies aimed at assisting or replacing the human ability to smell through the use of a so-called 'electronic nose' (Mildner-Szkudlarz et al. 2008). This is usually an array of multiple metal-oxide sensors that are sensitive to different volatile compounds formed by oxidative alteration of oils. Although tested successfully for some special applications, electronic noses have not yet found broad distribution because of their high cost and unsuitability for field use.

Changes in colour are the most eye-catching alterations of frying fats during their time of use. There is no correlation between colour and frying fat quality – good oil can be a dark and bad oil can be light in colour (Stier 2004). However, the grade of dark colouring can be helpful if a consistent constellation of frying oil and fried goods is present. If a defined grade of dark colour is reached, the oil is considered to be unfit for use and should be discarded. For this purpose, schemes for colour comparison are provided by different manufacturers.

Altered frying oils begin to foam because of the formation of volatiles like short-chain fatty acids and of different surface-active substances. There is no real correlation between foam height and food quality, but if foaming is too strong, the oil can boil over.

11.4.2 Quick Tests for Viscosity and Dielectricity

Polymer compounds formed during the use of a frying oil lead to an increase in that oil's viscosity. Therefore, viscosity is an excellent tool for monitoring frying fat quality. There is a strange correlation between viscosity, TPM, and PTG (Gertz 2000). Unfortunately, the two instruments designed to measure this, developed by Leatherhead/GEC Marconi (Kress-Rogers et al. 1990) and Gertz (Fri-Check®), are no longer manufactured (Gertz 2000).

During the frying process, the level of polar material increases continuously. With increasing TPM, the dielectric constant of the frying oil increases too. The dielectric constant can be easily measured electronically. The first instrument designed for this purpose was the food oil sensor (FOS) from Northern Instrument Corporation (Lino Lakes, MN, USA). Using this instrument, Hein et al. (1998) conducted basic studies of the correlation between dielectric properties and TPM. They found a very good correlation in the range of 0.8–0.95, as determined by column chromatography and the FOS.

The FOS is again no longer manufactured, but some other devices based on measuring the dielectric constant of frying oil are on the market, such as the CapSens 5000 (Centre for Chemical Information Technology, Switzerland), FOM 310 (Ebro Electronics, Germany), and Testo 265 (Testo AG, Germany). The CapSens instrument is a new version of the FOS; a small sample of frying oil is put into the heated detector and the result is displayed either as an FOS value or as % TPM. FOM 310 and Testo 265 also provide results as % TPM, but in these cases, the sensor probe is immersed directly into the hot oil bath of the fryer. Good accordance with TPM, as determined by column chromatography, was reported for these instruments (Bansal et al. 2010a).

In practice, there are some concerns around the use of instruments that measure the dielectricity constant (Gertz 2000). Traces of water, salt, and minerals may enhance polarity and adversely affect their measurements. Filtration of test samples may be necessary to obtain more reliable results. The disturbing effect of water can be compensated when the oil is measured above 110 °C. Test results are also strongly influenced by the nature and fatty acid composition of the frying oil. Short- and medium-chain fatty acids are more polar than long-chain fatty acids. Therefore, for short- and medium-chain fats like milk fat and coconut oil, significantly higher TPM results will be obtained.

An alternative to electronic tests is a quick test for polar materials, based on chemical test principles. The most important are the 3M PCT 120 oil tester (3M Inc., USA) and the TPM Veri-Fry (Test Kit Technologies Inc., USA) (Xu 1999).

11.5 Chemical Assessment

11.5.1 Chemical Quick Tests

There are several test kits available for the detection of various chemical parameters like FFA and oxidation products using colour reactions, pH indicators, or redox indicators. Several studies of the performance of different test kits and the correlation of their results to other analytical data have been reported in recent decades (Gertz 2000; Muhl et al. 2000; Innawong et al. 2004b; Sanibal and Mancini-Filho 2004; Stier 2004; Marmesat et al. 2007; Bansal et al. 2010a, 2010b; Chen et al. 2013; Gertz and Stier 2013; Osawa et al. 2013; Gertz and Behmer 2014). Kalogianni et al. (2017) recently reviewed and evaluated several quick tests for the evaluation of different quality characteristics, the details of which are given in Table 11.2.

The test strips of the 3M Shortening Monitor (3M, Inc., USA) are used to determine FFA in the range of 0.0–2.5% (lower range) and 0–7% (higher range) (Kalogianni et al. 2017). As already mentioned, there is only a very poor correlation between the FFA level and the frying fat quality measured as by TPM or PTG.

Oxifrit-Test, formerly called Rau-test, can detect OXF using a redox indicator; Fritest can detect carbonyl compounds measuring alkaline colour by comparing with a colour scale. With these tests, frying oils can be classified into four categories from 'good' to 'bad'. Both tests require hazardous and flammable chemicals, so they are unsuitable for use in food production rooms. Another problem is interferences between the basic colour of the frying oil itself and the colour of the reaction. Both tests correlate very poorly with TPM and PTG – they only can detect oxidative deterioration (Marmesat et al. 2007; Bansal et al. 2010b; Weisshaar 2014).

A number of analytical test kits are manufactured under the brand name 'Safe'. These tests are not typical quick tests but fast and portable laboratory tests. Liquid reagents and solvents and a portable spectrophotometer are required. Results can be achieved much faster than with official laboratory methods. For analysis of frying oils, four test kits have been validated (Yee Foo et al. 2006; Bansal et al. 2010b): PeroxySafe for lipid peroxides, AldeSafe for malondialdehyde, FASafe for FFA, and AlkalSafe for alkenals.

11.5.2 Near-Infrared Spectroscopy

Near-infrared spectroscopy (NIRS) is a powerful tool for monitoring the quality of used frying oil. It is a clean, fast, and nondestructive technique that allows the rapid determination of various chemical parameters, such as TPM, PTG, FFA, peroxides, anisidine value, carbonyl value, conjugated dienoic acids (CDA), and *trans* fatty acids (TFAs) (Boor and Speek 1994; Innawong et al. 2004a; Ng et al. 2007, 2011). In most cases, cuvettes are used, but measurement directly in the oil bath is possible with fibre-optic probes. Fourier transformation (FT) of the spectra allows for multiple methods of data analysis. In many cases, calibration data can be transferred from one instrument to another of the same type. The great advantage of NIRS compared to other rapid methods is the fact that typical parameters for thermal alteration (PTG and TPM), typical parameters for oxidative alteration (anisidine value, carbonyl value, CDA), and acid values can be detected simultaneously. Thus, a complete picture of different alterations

Table 11.2 Overview of original and previous independent research on rapid methods of determining frying oil quality.

Test	Manufacturer	In the market	Usable in small establishments	Principle	Research studies	Oil type	Food type	Correlation/ comparison of results
AV-check	Adaptec (Japan)	Yes	Yes	Measurement of acid value – colour comparison	—	—	—	—
Electronic nose	Cyrano Sciences and Alpha (USA and France)	Yes	No	Array of metal-oxide sensors, sensitive to volatile compounds	Muhl et al. (2000), Innawong et al. (2004b)	Soybean peanut oil	French fries, fish fingers, poultry sticks	FFA, dielectric constant
FASafe AldeSafe PeroxySafe	MP Biomedicals (USA)	Yes	No	Measurement of spectrophotometric absorbance; conversion of the absorbance units to FFA	Bansal et al. (2010a, 2010b), Foo et al. (2006)	Refined palm olein; blend of palm olein, sesame oil, and peanut oil; sunflower oil; and partially hydrogenated soybean oil	Frozen French fries, frozen chicken nuggets	TPC, FFA, MDA, PV
FOM 320	EBRO (Germany)	Yes	Yes	Dielectric constant – correlation with TPC	Banasal et al. (2010a, 2010b)	Refined palm olein; blend of palm olein, sesame oil, and peanut oil; sunflower oil	Frozen French fries, frozen chicken nuggets	TPC, FFA
Food oil sensor/ Capsens 5000	Northern Instruments (USA)/C-CIT Sensors AG (Switzerland)	No/Yes	Yes	Dielectric constant – correlation with TPC	Bansal et al. (2010a, 2010b), Marmesat et al. (2007)	Used frying oils from different origins and establishments; refined palm olein; blend of palm olein, sesame oil, and peanut oil; and sunflower oil	Frozen French fries, frozen chicken nuggets	TPC, composition in major fatty acids, PTG, FFA
Fri-check	Fri-check (Belgium)	No	—	Time for a piston-like body dropped in a tube containing the oil – related to viscosity and density	Osawa et al. (2013), Chen et al. (2013)	Disposal samples from commercial establishments	Unknown	TPC

Fritest	Merck (Germany)	Yes	Yes	Measure of carbonyl compounds – colour comparison	Banasal et al. (2010a, 2010b), Marmesat et al. (2007)	Used frying oils from different (unknown) origins and establishments; refined palm olein; blend of palm olein, sesame oil, and peanut oil; and sunflower oil	Frozen French fries, frozen chicken nuggets	TPC, composition in major fatty acids, PTG, FFA
Low-range shortening monitor	3M (USA)	Yes	Yes	Range of FFA – colour comparison	Bansal et al. (2010a, 2010b), Kalogianni et al. (2017)	Refined palm olein; blend of palm olein, sesame oil, and peanut oil; and sunflower oil	Frozen French fries, frozen chicken nuggets	TPC, FFA
Near-infrared	Different manufacturers	Yes	No	Determination of parameters like TPC, PTG, FFA, peroxides, anisidine value, carbonyl value, and conjugated dienoic acids	Gertz et al. (2013)	Unknown	Unknown	TPC, PTG, acid value, *para*-anisidine value
Optifry	Mir-oil (USA)	Yes	Yes	Dielectric constant – correlation with TPC	—	—	—	—
Oxifrit	Merck (Germany)	Yes	Yes	Measurement of oxidized fatty acids – colour comparison	Banasal et al. (2010a, 2010b), Marmesat et al. (2007)	Used frying oils from different (unknown) origins and establishments; refined palm olein; blend of palm olein, sesame oil, and peanut oil; and sunflower oil	Unknown Frozen French fries, frozen chicken nuggets	TPC, composition in major fatty acids, PTG, FFA
PCT 120	3M (France)	Yes	Yes	Correlation of dye travel distance with polar compounds	Kalogianni et al. (2017)	—	—	—

(*Continued*)

Table 11.2 (Continued)

Test	Manufacturer	In the market	Usable in small establishments	Principle	Research studies	Oil type	Food type	Correlation/ comparison of results
Testo 265 Testo 270	Testo (Germany)	Yes	Yes	Dielectric constant – correlation with TPC	Bansal et al. (2010a, 2010b)	Refined palm olein; blend of palm olein, sesame oil, and peanut oil; and sunflower oil	Frozen French fries, frozen chicken nuggets	TPC, FFA
TPM, FFA, Verify	Test Kit Technologies (USA)	Yes	Yes	Reaction of gel with polar compounds to give a specific colour	Bansal et al. (2010a, 2010b)	Refined palm olein; blend of palm olein, sesame oil, and peanut oil; and sunflower oil	Frozen French fries, frozen chicken nuggets	TPC
Viscofrit	Viscofrit (Spain)	Yes	Yes	Measurement of the time required to empty a standard funnel	Osawa et al. (2013), Marmesat et al. (2007), Kalogianni et al. (2017)	Used frying oils from different (unknown) origins and establishments	Unknown	TPC, FFA, PTG

Source: Reproduced with kind permission of John Wiley & Sons (Kalogianni et al. 2017).
TPC, total polar compounds; PTG, polymerized triacyl glycerols; FFA, free fatty acids.

can be achieved rapidly with a single measurement. Recently, the first DGF standard method was developed using NIRS was published (Gertz et al. 2013). A recent study showed that a portable NIRS device can be used to test the acid values of peanuts oils (Yang et al. 2017). Thus, it is not difficult to forecast that NIRS methods will gain strong prominence in the frying industries in coming years.

11.5.3 Mass Spectrometry

Mass spectrometry is one of the most widely accepted techniques for identifying individual components of frying foods and media. Cao et al. (2017) reviewed its use in the quality and safety assessment of cooking oils, highlighting its increasing application in authentication, ageing, and marker detection in used oil.

11.6 Evaluation of Fried Foods

Fried foods are also evaluated for both physical and chemical characteristics. The sensory evaluation of fried foods has received the most attention in recent times. Human sensory panels are usually utilized for this purpose. Colour, taste, crispness, hardness, and other parameters are widely assessed.

Among chemical measures, oil content, water content, and thiobarbituric acid reactive substances (TBARS) have received the greatest attention. Recently, a spectroscopic method was developed to assess TBARS contents in the fast food industry (Zeb and Ullah 2016). This method examines the reaction of malondialdehyde (MDA) and TBA in the glacial acetic acid medium. It was used to determine the TBARS contents in fried fast foods such as Shami kebab, samosa, fried bread, and potato chips. Shami kebab, samosa, and potato chips all had a higher amount of TBARS in a glacial acetic acid–water extraction system than their corresponding pure glacial acetic acid; the reverse was true for fried bread. This method can successfully be used for the determination of TBARS in other food matrices, making it applicable for quality control in the food industry. Several chromatographic methods have been reported to determine different quality indices or toxicants in fried foods.

11.7 The Future of Fried Food Safety

Several countries in the European Union, the Americas, Australasia, and Asia have strict regulations regarding fried food safety. However, a large amount of data is still needed to comprehend the full safety spectrum. Much of the world population may or may not be aware of fried food safety issues, are thus could potentially be exposed to fried food toxicants. Therefore, several steps should be taken to implement and extend already established guidelines to all countries:

- EU and US regulations should be extended to all member countries of the World Health Organization (WHO). This forum could successfully be used to tackle all health concerns in nonregulated countries.
- Extensive work is required to analyze the chemistry of different fried foods across a number countries for which data are not available.

- Risk-assessment programs should be organized by country on an annual basis, focusing on all types of physical and chemical assessment.
- Mass spectrometry-based analysis of quality assessment is an excellent choice for future research on fried food safety.
- Hazard and risk-based approaches should be used to ensure fried food safety.

11.8 Key Concepts

- It is important to follow basic safety criteria when frying.
- Used frying oils or fats must contain <24% TPM and <12% polymeric triglycerides as standard basic indicators.
- Other guidelines must be followed to ensure safe food production.
- Several quality indicators can be analyzed to determine the quality of frying media.
- Physical measures such as sensory quality, colour, foaming, viscosity, and dielectricity should be evaluated.
- Several chemical tests, including instrumental quick tests and laboratory analytical methods, must accompany the physical assessment.
- Existing guidelines and regulations should be extended to nonregulated countries via the WHO.
- Safe fried foods may contribute to the future health of the next generation.

References

Aladedunye, F.A. and Przybylski, R. (2009). Degradation and nutritional quality changes of oil during frying. *Journal of the American Oil Chemists' Society* 86: 149–156.

Bansal, G., Zhou, W., Barlow, P.J. et al. (2010a). Evaluation of commercially available rapid test kits for the determination of oil quality in deep-frying operations. *Food Chemistry* 121: 621–626.

Bansal, G., Zhou, W., Barlow, P.J. et al. (2010b). Review of rapid tests available for measuring the quality changes in frying oils and comparison with standard methods. *Critical Reviews in Food Science and Nutrition* 50: 503–514.

Berdeaux, O., Fontagné, S., Sémon, E. et al. (2012). A detailed identification study on high-temperature degradation products of oleic and linoleic acid methyl esters by GC–MS and GC–FTIR. *Chemistry and Physics of Lipids* 165: 338–347.

Boor, A. and Speek, A.J. (1994). Determination of the sum of dimer and polymer triglycerides and of acid value of used frying fats and oils by near infrared reflectance spectroscopy. *Journal of AOAC International (USA)* 77: 1184–1198.

Caldwell, J.D., Cooke, B.S., and Greer, M.K. (2011). High performance liquid chromatography–size exclusion chromatography for rapid analysis of Total polar compounds in used frying oils. *Journal of the American Oil Chemists' Society* 88: 1669–1674.

Cao, G., Ruan, D., Chen, Z. et al. (2017). Recent developments and applications of mass spectrometry for the quality and safety assessment of cooking oil. *Trends in Analytical Chemistry* 96: 201–211.

Chen, W.-A., Chiu, C.P., Cheng, W.-C. et al. (2013). Total polar compounds and acid values of repeatedly used frying oils measured by standard and rapid methods. *Journal of Food and Drug Analysis* 21: 58–65.

DGF. 2010. Polymerized triacylglycerols – determination in severely heat stressed fats and oils (deep-frying fats) by high-performance size-exclusion chromatography (HPSEC). DGF Standard Methods. German Society for Fat Science.

Donner, M.G. and Ritcher, W.O. (2000). 3rd international symposium on deep-fat frying – optimal operation. *European Journal of Lipid Science and Technology* 102: 306–308.

Endo, Y., Tominaga, M., Tagiri-Endo, M. et al. (2003). A modified method to estimate total carbonyl compounds in frying oils using 1-butanol as a solvent. *Journal of Oleo Science* 52: 353–358.

Farhoosh, R. and Moosavi, S.M. (2008). Carbonyl value in monitoring of the quality of used frying oils. *Analytica Chimica Acta* 617: 18–21.

Farhoosh, R. and Tavassoli-Kafrani, M.H. (2011). Simultaneous monitoring of the conventional qualitative indicators during frying of sunflower oil. *Food Chemistry* 125: 209–213.

Farhoosh, R., Tavakoli, J., and Khodaparast, M.H.H. (2008). Chemical composition and oxidative stability of kernel oils from two current subspecies of Pistacia atlantica in Iran. *Journal of the American Oil Chemists' Society* 85: 723–729.

Firestone, D. (2013). Determination of polar compounds in frying fats, AOCS Official Method Cd 20-91. In: *Official Methods and Recommended Practices of the AOCS*, 6e (ed. D. Firestone). Urbana, IL: American Oil Chemists' Society.

Fritsch, C.W. (1981). Measurements of frying fat deterioration: a brief review. *Journal of the American Oil Chemists' Society* 58: 272–274.

Gertz, C. (2000). Chemical and physical parameters as quality indicators of used frying fats. *European Journal of Lipid Science and Technology* 102: 566–572.

Gertz, C. and Stier, R.F. (2013). 7th International Symposium on Deep-Fat Frying, San Francisco, CA (USA): recommendations to enhance frying. *European Journal of Lipid Science and Technology* 115: 589–590.

Gertz, C. and Behmer, D. (2014). Application of FT-NIR spectroscopy in assessment of used frying fats and oils. *European Journal of Lipid Science and Technology* 116: 756–762.

Gertz, C., Fiebig, H.-J., and Hancock, J.N.S. (2013). FT-near infrared (NIR) spectroscopy – screening analysis of used frying fats and oils for rapid determination of polar compounds, polymerized triacylglycerols, acid value and anisidine value [DGF C-VI 21a (13)]. *European Journal of Lipid Science and Technology* 115: 1193–1197.

Guhr, C.G.G., Waibel, J., and Arens, M. (1981). Bestimmung der polaren Anteile in Fritierfetten. *Fette, Seifen, Anstrichmittel* 83: 373–376.

Guillén, M.D. and Goicoechea, E. (2008). Toxic oxygenated α,β-unsaturated aldehydes and their study in foods: a review. *Critical Reviews in Food Science and Nutrition* 48: 119–136.

Hein, M., Henning, H., and Isengard, H.D. (1998). Determination of total polar parts with new methods for the quality survey of frying fats and oils. *Talanta* 47: 447–454.

Innawong, B., Mallikarjunan, P., Irudayaraj, J., and Marcy, J.E. (2004a). The determination of frying oil quality using Fourier transform infrared attenuated total reflectance. *LWT – Food Science and Technology* 37: 23–28.

Innawong, B., Mallikarjunan, P., and Marcy, J.E. (2004b). The determination of frying oil quality using a chemosensory system. *LWT – Food Science and Technology* 37: 35–41.

Kalogianni, E.P., Georgiou, D., Romaidi, M. et al. (2017). Rapid methods for frying oil quality determination: evaluation with respect to legislation criteria. *Journal of the American Oil Chemists' Society* 94: 19–36.

Kaufmann, A., Ryser, B., and Suter, B. (2001). HPLC with evaporative light scattering detection for the determination of polar compounds in used frying oils. *European Food Research and Technology* 213: 372–376.

Kress-Rogers, E., Gillatt, P.N., and Rossell, J.B. (1990). Development and evaluation of a novel sensor for the in situ assessment of frying oil quality. *Food Control* 1: 163–178.

Marmesat, S., Rodrigues, E., Velasco, J., and Dobarganes, C. (2007). Quality of used frying fats and oils: comparison of rapid tests based on chemical and physical oil properties. *International Journal of Food Science and Technology* 42: 601–608.

Mildner-Szkudlarz, S., Jeleń, H.H., and Zawirska-Wojtasiak, R. (2008). The use of electronic and human nose for monitoring rapeseed oil autoxidation. *European Journal of Lipid Science and Technology* 110: 61–72.

Mubiru, E., Shrestha, K., Papastergiadis, A., and de Meulenaer, B. (2013). Improved gas chromatography-flame ionization detector analytical method for the analysis of epoxy fatty acids. *Journal of Chromatography A* 1318: 217–225.

Muhl, M., Demisch, H.U., Becker, F., and Kohl, C.D. (2000). Electronic nose for detecting the deterioration of frying fat – comparative studies for a new quick test. *European Journal of Lipid Science and Technology* 102: 581–585.

Ng, C.L., Wehling, R.L., and Cuppett, S.L. (2007). Method for determining frying oil degradation by near-infrared spectroscopy. *Journal of Agricultural and Food Chemistry* 55: 593–597.

Ng, C.L., Wehling, R.L., and Cuppett, S.L. (2011). Near-infrared spectroscopic determination of degradation in vegetable oils used to fry various foods. *Journal of Agricultural and Food Chemistry* 59: 12286–12290.

Osawa, C.C., Gonçalves, L.A.G., and da Silva, M.A.A.P. (2013). Odor significance of the volatiles formed during deep-frying with palm Olein. *Journal of the American Oil Chemists' Society* 90: 183–189.

Petersen, K.D., Jahreis, G., Busch-Stockfisch, M., and Fritsche, J. (2013). Chemical and sensory assessment of deep-frying oil alternatives for the processing of French fries. *European Journal of Lipid Science and Technology* 115: 935–945.

Ravelli, D., Matsuoka, C.R., Della Modesta, R.C. et al. (2010). Determination of deep frying soybean oil disposal point by a sensory method. *Journal of the American Oil Chemists' Society* 87: 515–520.

Robertson, C.J. (1968). A review of deep frying current practices point the way to new horizons. *Canadian Institute of Food Technologists Journal* 1: A66–A75.

Sanibal, E.A.A. and Mancini-Filho, J. (2004). Frying oil and fat quality measured by chemical, physical, and test kit analyses. *Journal of the American Oil Chemists' Society* 81: 847–852.

Stevenson, S.G., Vaisey-Genser, M., and Eskin, N.A.M. (1984). Quality control in the use of deep frying oils. *Journal of the American Oil Chemists' Society* 61: 1102–1108.

Stier, R. (2001). The measurement of frying oil quality and authenticity. In: *Frying: Improving Quality* (ed. J.B. Rossell). Cambridge: Woodhead Publishing.

Stier, R.F. (2004). Tests to monitor quality of deep-frying fats and oils. *European Journal of Lipid Science and Technology* 106: 766–771.

Stier, R.F. (2013). Ensuring the health and safety of fried foods. *European Journal of Lipid Science and Technology* 115: 956–964.

Tompkins, C. and Perkins, E.G. (1999). The evaluation of frying oils with the p-Anisidine value. *Journal of the American Oil Chemists' Society* 76: 945–947.

Velasco, J., Marmesat, S., Bordeaux, O. et al. (2004). Formation and evolution of monoepoxy fatty acids in thermoxidized olive and sunflower oils and quantitation in used frying oils from restaurants and fried-food outlets. *Journal of Agricultural and Food Chemistry* 52: 4438–4443.

Weisshaar, R. (2014). Quality control of used deep-frying oils. *European Journal of Lipid Science and Technology* 116: 716–722.

Xu, X.-Q. (1999). A modified VERI-FRY® quick test for measuring total polar compounds in deep-frying oils. *Journal of the American Oil Chemists' Society* 76: 1087–1089.

Yang, M., Chen, Q., Kutsanedzie, F.Y.H. et al. (2017). Portable spectroscopy system determination of acid value in peanut oil based on variables selection algorithms. *Measurement* 103: 179–185.

Yee Foo, S., Cuppett, S., and Schlegel, V. (2006). Evaluation of SafTestTM methods for monitoring frying oil quality. *Journal of the American Oil Chemists' Society* 83: 15.

Zeb, A. and Ullah, F. (2016). A simple spectrophotometric method for the determination of Thiobarbituric acid reactive substances in fried fast foods. *Journal of Analytical Methods in Chemistry* 2016: Article ID 9412767, 5 pages.

12

Toxicity of Food Frying

12.1 Introduction

The main purpose of frying is to produce high-quality foods with good consumer acceptability. However, not all acceptable foods are safe. The early work of Custot (1959) and Nolen et al. (1967) showed that frying fats may not be used repeatedly. Experiments were based on animal models that used dietary frying fats. Changes in serum biochemistry and fat depositions were commonly observed. Food chemists were keenly interested in the new toxic compounds produced during frying. It was found that the initial products of fat oxidation were hydroperoxides, epoxides, and hydroxides. These products caused significant toxic effects in experimental animals. Horton et al. (1987) found that lipid peroxidation products are toxic. Kubow (1990) also found that dietary lipid oxidation products are toxic, increasing atherosclerosis. Marnett (1991) showed that these oxidation products are produced by free radical reactions.

In order to study the mechanisms of certain chronic diseases caused by oxidized lipids, cell culturing and model studies were performed. Marnett (1999) showed that lipid oxidation products react with DNA to form toxic compounds that are unable to proceed with normal metabolic reactions. He reported that the principal compounds involved in the reaction were deoxyguanine and deoxyadenosine, forming adducts such as pyrimidopurinone.

A large variety of oxidized or newly formed compounds found in both frying fats and fried foods have been evaluated for their toxicological properties. The details of some of these products are discussed in this chapter.

12.2 Toxicity of Oxidized Triacylglycerols

The first products of triacylglycerols (TAGs) are hydroperoxides, epoxides, and hydroperoxides. These products have been synthetically produced and fed to experimental animals. It has been found that lipid hydroperoxides from dietary frying fats enhance the growth of hepatocarcinoma cells in rats (Rohr-Udilova et al. 2008), and produce biochemical and physiological changes in them (Khan et al. 2017). In higher animals, such as rabbits, the feeding of fried animals fats led to increased formation of cholesterol and low-density lipoprotein (LDL) (Zeb and Rahman 2012). Similar results have been reported for thermally oxidized vanaspati ghee (Zeb and Mehmood 2012; Zeb and

Food Frying: Chemistry, Biochemistry, and Safety, First Edition. Alam Zeb.

Haq 2016) and desi ghee (Zeb and Uddin 2017). Frying oils produced similar toxic effects in experimental animals (Zeb and Rahman 2017).

Thermally oxidized olive oil was found to significantly enhance the atherogenic indices in rats (Khan et al. 2017). The cardiac risk ratio increased with increasing oxidized olive oil doses. An increase in LDL-cholesterol can contribute to the formation of oxidized LDL, which may be present in the arterial lesions and thus contribute to the increase in the cardiac risk ratio. Similarly, an increase in the atherosclerosis index and atherogenic index of plasma can be attributed to an increase in LDL-cholesterol. Upon ingestion, oxidized TAGs enter chylomicrons in the intestine, where they are interconverted to oxidized phospholipids. The oxidized phospholipids and oxidized TAGs are taken up by LDL and become oxidized LDL. Upon reaching the liver, the oxidized LDL is not metabolized or is metabolized only slowly, due to the increase in reactive oxygen species (ROS). Thus, oxidized TAGs contribute to the development of fatty liver as well as atherogenesis.

12.3 Toxicity of Acrylamide

Acrylamide is a well-known industrial chemical and therefore its toxicological properties have been comprehensively studied (Matthäus and Haase 2014). Mild and reversible peripheral nervous system symptoms have been described for Swedish tunnel workers due to the misuse of an acrylamide-based grouting product in the sealing of tunnel walls (Hagmar et al. 2001). Other symptoms of cerebellar dysfunction followed by neuropathy have been observed in 71 Chinese workers at a factory producing acrylamide and polyacrylamide (Calleman et al. 1994).

In animal studies, primarily with rodents, acrylamide has been shown to be carcinogenic when applied in doses orders of magnitude higher than the typical human dietary exposure. These results led to the classification of acrylamide as probably carcinogenic to humans by the International Agency for Research on Cancer (IARC) in 1994. From animal experiments, it was also found that acrylamide is genotoxic and neurotoxic. Additionally, morphological changes in the nerves (Burek et al. 1980) and reproductive and developmental effects on the testes and male fertility have been observed in rodents (Sakamoto and Hashimoto 1986; Tyl and Friedman 2003), resulting in the definition of no observed adverse effect levels (NOAELs) of 0.2 and 2.0 mg/kg BW day^{-1}, respectively, by the Joint Food and Agriculture Organization/World Health Organization Expert Committee on Food Additives (JECFA) in 2005. Later, in 2011, JECFA concluded that adverse neurological effects are unlikely at the estimated average exposure, but that morphological changes in the nerves cannot be excluded for high acrylamide consumers (Matthäus and Haase 2014).

Acrylamide is metabolized very rapidly after ingestion and moves into the blood and urine, from where it is excreted. Most critical for the genotoxic effect of acrylamide is its epoxidation to glycidamide by Cytochrome P450, a haemo-protein with enzyme activity (Baum et al. 2005; Duale et al. 2009). Glycidamide can react not only with haemoglobin and enzymes, but also with DNA, resulting in point mutations and cancer. Glycidamide is further metabolized to mercapturic acid, and most of it is rapidly excreted in the urine (Settels et al. 2008).

From a 2-year study of male and female B6C3F1 mice and F344/N rats administered 0.0, 0.0875, 0.175, 0.35, or 0.70 mM acrylamide in their drinking water ad libitum, Beland et al. (2013) concluded that the tumours observed in multiple organs from both sexes and species of rodents exposed chronically, coupled with previous mechanistic data, strongly support that acrylamide is a genotoxic carcinogen due to its metabolic activation to glycidamide. However, since the most susceptible organ to acrylamide exposure in the mouse model, the Harderian gland, is not present in humans, transfer of the results is difficult.

Biomarker-based animal studies now exist, which give us an insight into the biological effects of acrylamide at doses approaching those found in normal nutrition (Boettcher et al. 2006; Watzek et al. 2012a, b, 2013), which can significantly help improve the risk assessment of dietary acrylamide exposure. For the calculation of human health risk, the margin of exposure (MOE) should be used, which is defined as the quotient of the lower limit at which adverse effects were observed over the estimated exposure from food. The higher the MOE, the lower the estimated risk to human health. None of the studies conducted to date show an MOE higher than 10 000 for different aged populations; for low-weight humans like babies and consumers of highly contaminated foods, the MOE can be below 200, indicating a possible health risk. JECFA has stated that, for a compound that is both genotoxic and carcinogenic, these MOEs indicate a health concern, which underlines the need for further mitigation strategies (Matthäus and Haase 2014).

As acrylamide has been found in animal studies to have a neurotoxic and carcinogenic effect, resulting from its classification as probably carcinogenic to humans by the IARC, no safety reference values can be established for it. Mitigation of acrylamide during food processing is an ongoing process, and every reasonable measure to reduce the content in the final product has to be taken. The effectiveness and result of the measures must be shown and carefully monitored (Matthäus and Haase 2014).

Since 2002, several studies have been conducted to find a correlation between the uptake of acrylamide by humans and an increased incidence of different types of cancer. Sobel et al. (1986) presented a study on the mortality of 371 employees working with acrylamide or polyacrylamide. They found no correlation between the exposure to acrylamide and overall mortality, total malignant neoplasia, or any specific cancers. A study with a cohort of 8854 men, 2293 exposed to acrylamide by work, was examined for mortality from 1925 to 1983. The results again showed no increased risk of mortality from increased exposure to acrylamide (Collins et al. 1989). Later, after the first announcement of the presence of acrylamide in food, Rice (2005) showed from available human and experimental data that no statistically significantly exposure-related increased risk for cancer in any organ was consistently associated with the presence of acrylamide. Only a doubling of the risk for pancreatic cancer was found for people with the highest exposure. Swaen et al. (2007) also found little evidence for a cancer risk from exposure to acrylamide at production facilities, but they likewise reported a higher risk for pancreatic cancer.

In an assessment of 13 epidemiological studies dealing with the relationship between different types of cancer and exposure to acrylamide conducted in 2011, the German Federal Institute for Risk Assessment (BfR) found no consistent result (Matthäus and Haase 2014). Some studies showed an increased risk, but not a very strong one.

From this assessment, the BfR concluded that a relationship between exposure to acrylamide and cancer can neither be expected nor excluded for humans. It is possible that any such risk is not detectable for the given exposure.

Recently, Hogervorst et al. (2014) showed for the first time that acrylamide may be associated with colorectal cancer with specific somatic mutations, differentially in men and women. They used mutations in KRAS and APC as key markers of colorectal carcinogenesis, and followed a cohort within the Netherlands Cohort Study on diet and cancer. In another study, Hogervorst et al. (2013) found no conclusive associations between acrylamide intake and sex hormones that would lend unequivocal biologic plausibility to the observed increased risks of endometrial, ovarian, and breast cancers. Stott-Miller et al. (2013) found an increased risk of prostate cancer after regular consumption of selected deep-fried foods, but they were not able to determine whether this risk was specific to such foods or represents a risk associated with regular intake of foods exposed to high heat and/or other aspects of the Western lifestyle, such as fast food consumption.

On the same subject, Chen et al. (2012) used a statistical regression model to estimate the additional contribution of acrylamide to overall lifetime cancer risk. They focussed on French fry consumption in a Taiwanese city and looked at different methods of preparation. A Monte Carlo simulation showed that if the level of acrylamide in French fries is greater than 168 $\mu g\,kg^{-1}$, the estimated cancer risk will be higher than the target excess lifetime cancer risk.

More concretely, the European Prospective Investigation into Cancer and Nutrition (EPIC) was used to calculate the association between dietary acrylamide intake and pancreatic cancer risk in a cohort of 477 308 men and women aged 35–75. Acrylamide intake was estimated for each participant by questionnaire-based food consumption data with a harmonized acrylamide database from the EU monitoring database. No overall association between dietary acrylamide intake and risk for exocrine pancreatic cancer could be calculated. Smoking status, sex, diabetes alcohol intake, and geographic region had no influence. Only obese people, classified on the basis of body mass index (BMI), showed an inverse association (Obon-Santacana et al. 2013).

Wang et al. (2017) revealed that occupational exposure to acrylamide showed significant alterations in liver function and normal metabolic biomarkers. They found that acrylamide caused human nerve damage. Baltazar et al. (2017) reviewed the potential risks and effects of acrylamide on humans, and suggested that although acrylamide is toxic in animal models, there are inadequate data concerning human toxicity and carcinogenicity. Thus, epidemiological studies are required to clarify several questions, such as those concerning target organs and admissible daily intake.

12.4 Toxicity of Acrolein

Abraham et al. (2011a) provide a detailed review of how acrolein forms adducts with glutathione and other cellular components and is therefore cytotoxic. A World Health Organization (WHO) working group established a tolerable oral acrolein intake of 7.5 μg per kg BW day^{-1}.

12.4.1 Toxico-Kinetics and Metabolism

Seven days after the administration of a single dose of 2.5 mg radioactively labelled acrolein per kg BW to fasting rats, a total of 52–63% of the radioactivity was excreted in the urine, 12–15% in the faeces, and approximately 30% as CO_2 in respiratory air. Approximately 90% of the dose was excreted within 24 hours. Even with daily dosing (14 days of unlabelled acrolein, followed by a single dose of radioactively labelled acrolein), the excretion profile was similar. The tissue contents (measured as radioactivity) were minimal (<1.2%). However, they increased after intravenous administration (Parent et al. 1996).

After oral administration of a single or repeated dose of 2.5 mg per kg BW or a single high dose of 15 mg per kg BW acrolein to rats, the following metabolites were detected in their urine: oxalic acid, malonic acid, *N*-acetyl-*S*-2-carboxy-2-hydroxyethylcysteine, *N*-acetyl-*S*-3-hydroxypropylcysteine, *S*-(hydroxypropyl)-mercapturic acid (3-HPMA), *N*-acetyl-*S*-2-carboxyethylcysteine, and 3-hydroxy-propionic acid. After oral or intravenous administration of acrolein, 3-HPMA was identified in urine as the main metabolite (Parent et al. 1998). This corresponds to the finding that 78.5% of an oral dose of 13 mg acrolein per kg BW in rats was excreted as 3-HPMA. Other metabolites were not analysed (Sanduja et al. 1989). Oxalic acid was detected only after oral administration. The metabolite pattern was otherwise comparable for the two routes of administration. In faeces, an inert homopolymer of acrolein was identified following oral administration (15% of a dose of 2.5 mg acrolein per kg BW and 30% of a 15 mg per kg BW dose). The authors assumed that the polymer was formed in the intestinal tract (Parent et al. 1998). Other metabolites that have been identified *in vitro* include acrylic acid, glycidaldehyde, and glyceraldehyde (Draminski et al. 1983; Gomes et al. 2001).

According to Parent et al. (1998), the metabolism routes of acrolein consist of (i) epoxidation of the double bond followed by a conjugation with glutathione (GSH), (ii) Michael-type addition of water followed by oxidative degradation, and (iii) GSH addition to the double bond and an oxidative or reductive change of the aldehyde function. The GSH adducts are further metabolized to corresponding mercapturic acids and excreted in the urine (Figure 12.1).

The main metabolic route of acrolein occurs through the formation of GSH adducts. These are formed during the direct reaction of acrolein with GSH, but can also be catalysed by glutathione-*S*-transferases. In different cell lines, a complete GSH depletion was observed from 10 μM acrolein (Grafström et al. 1988; Grafström 1990; Dypbukt et al. 1993). After intraperitoneal administration of 5 mg acrolein per kg BW, the GSH content of rat liver was reduced by approximately 50% (Cooper et al. 1992).

In several studies, the release of acrolein was held responsible for the cytotoxicity of aldehydic mercapturic acid. The results of Hashmi et al. (1992) on LLC-PK1 cElls suggest that the mercapturic acids themselves could be cytotoxic. So far, a noteworthy release of acrolein has been observed only in mercapturic acids that have not been detected *in vivo*. It is unclear whether they could have been detected under the conditions of analysis. Whether acrolein can also be released under physiological conditions is likewise unclear, since the pH of the urine is lower than that in the cited *in vitro* studies. 3-HPMA is relatively stable in urine according to the studies conducted to date, and is not bladder-toxic. Furthermore, *N*-acetyl-*S*-2-carboxyethylcysteine is not cytotoxic.

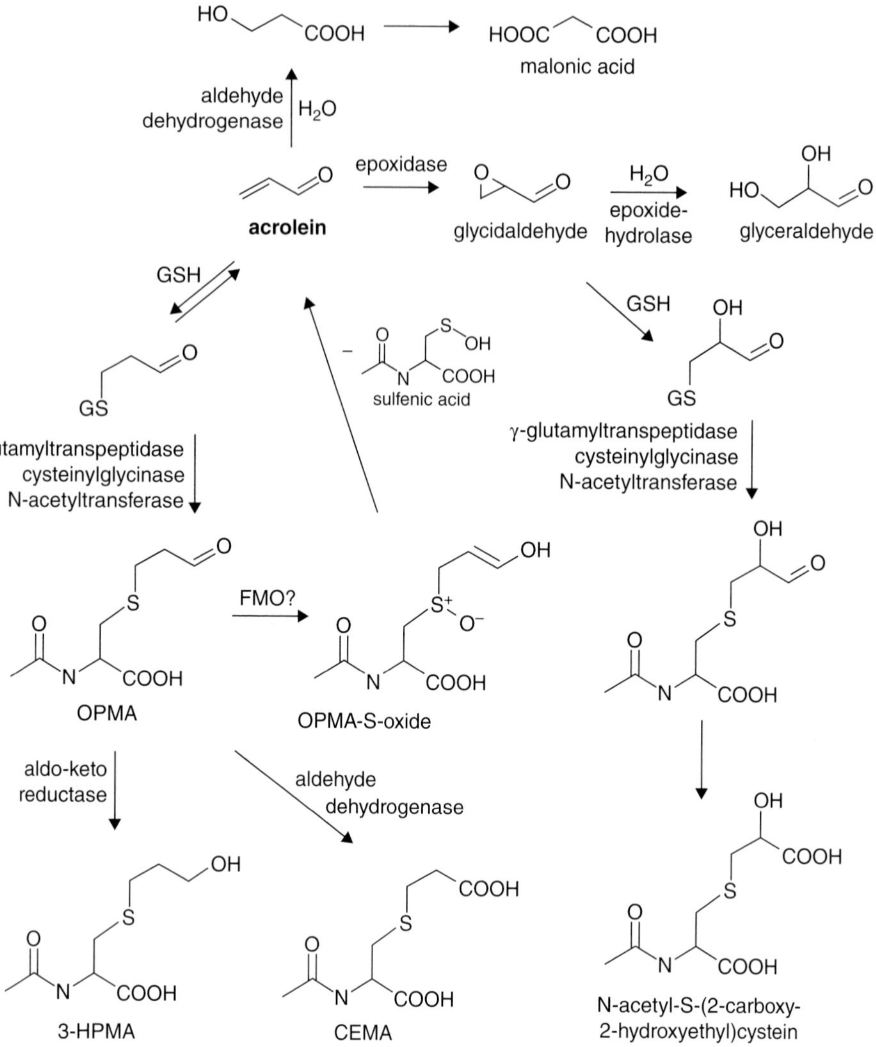

Figure 12.1 Metabolism of acrolein. *Source:* Reproduced with kind permission of John Wiley & Sons (Abraham et al. 2011a).

Overall, there is as yet no evidence to suggest that the oral intake of acrolein could reach levels capable of triggering toxic effects in organs such as the bladder, kidney, and brain (Sanduja et al. 1989).

12.4.2 General Toxic Effects

Acrolein reacts with sulfhydryl (SH) groups and with primary and secondary amines. Acrolein exposure may lead to GSH depletion and the formation of adducts with thiol groups of cysteine residues and amino groups of proteins, nucleic acids, and other cellular components (Kehrer and Biswal 2000; LoPachin et al. 2009). In addition to this adduct formation, acrolein may also cause cross-linking between proteins, as well as between peptides/proteins and DNA (Li et al. 2008; Minko et al. 2008; LoPachin et al. 2009).

Consequently, acrolein is highly cytotoxic both *in vitro* and *in vivo*. In cell culture studies, acrolein (≥10 μM) reduces the viability of numerous cell lines, such as bronchial endothelial and epithelial cells, bronchial and cordial fibroblasts, retinal pigment epithelial cells, and neuronal cells (Kehrer and Biswal 2000; Jia et al. 2007b; Yoshida et al. 2009). Acrolein causes cell death via apoptotic processes (e.g. in human lung cells) (Roy et al. 2010). Furthermore, it modulates several signalling pathways, including those that involve the transcription factors nuclear factor kappa B (NF-κB) and activator protein-1 (Kehrer and Biswal 2000).

Acrolein may cause oxidative stress through the depletion of GSH (e.g. in neuronal cells, bronchial epithelial cells, and endothelial cells of the umbilical cord vein) (Park et al. 2005), which can lead to DNA damage. GSH protects against damaging influences through conjugate formation with acrolein. Although acrolein thus reduces the GSH level in the cell, and hence promotes oxidative stress, normal levels can be restored through a new synthesis of GSH (Horton et al. 1997). A proteomic analysis of acrolein-exposed bronchial epithelial cells revealed that acrolein especially modifies stress proteins, cytoskeletal proteins, and proteins involved in redox signalling (Spiess et al. 2011).

Target organs of toxicity of acrolein are primarily the local tissues affected. After inhalation, irritations of the entire respiratory tract may result in inflammation and in hyperplasias and metaplasias in the respiratory epithelium (Dorman et al. 2008). Oral exposure leads to gastrointestinal symptoms, gastric ulcers, and/or gastric bleeding, the severity of which rises with increasing dose (Auerbach et al. 2008).

In summary, the mechanism of acrolein toxicity includes increased levels of proteins/DNA adducts, decreased GSH levels, and interference with cell signalling pathways such as oxidative stress signalling, NF-κB, and other nuclear factors.

12.4.3 Acute, Subchronic, and Chronic Toxicity

The LD50 of acrolein (single oral dose) has been determined to be between 7 and 46 mg per kg BW for rats, mice, and hamsters (Gomes et al. 2001). In a subchronic National Toxicology Program (NTP) study, hyperplasias of the squamous epithelium were observed in the forestomach of rodents, but no tumours (Abraham et al. 2011a). Rats were exposed to 0.0, 0.75, 1.25, 2.5, 5.0, and 10.0 mg acrolein per kg BW day^{-1} and mice to 0.0, 1.25, 2.5, 5.0, 10.0, and 20.0 mg acrolein per kg BW day^{-1} for 14 weeks by gavage (application 5 days week^{-1}, dissolved in methylcellulose). All mice with the highest acrolein dose died within the first week after the start of the study. Increased mortality rates were also observed in rats and mice in the other higher acrolein dose groups. The incidence of hyperplasias in the forestomach was increased in male rats in the 5 and 10 mg per kg BW acrolein groups, as well as in female rats and mice from the 2.5 mg per kg BW dose group (Auerbach et al. 2008). From the occurrence of hyperplasias in the forestomach, a NOAEL of 1.25 mg per kg BW day^{-1} for female mice and male rats and of 0.75 mg per kg BW day^{-1} for female rats can be derived. For male mice, no NOAEL can be derived from the lowest acrolein dose of 1.25 mg per kg BW, which already results in a slightly higher hyperplasia rate.

Beagle dogs were exposed daily to 0.1, 0.5, and 1.5 mg acrolein per kg BW day^{-1} for 1 year (in gelatine capsules), whereby the highest dose was increased after 4 weeks to 2 mg per kg BW. The main effect was frequent vomiting of the animals after the treatment, the frequency of which decreased after some time, possibly due to an adaptive process. Changes in biochemical parameters such as reduction of calcium, albumin, and total protein levels in the serum were observed. The authors stated that the cause could be related to the vomiting or a functional disorder of the kidney or liver, although there were no pathological indications for the latter (Parent et al. 1992a).

In studies of chronic acrolein toxicity, rats were administered orally (gavage) doses of 0.05, 0.5, and 2.5 mg acrolein (dissolved in water) per kg BW day^{-1} over a period of 2 years. In all treated rats, reduced levels of creatinine phosphokinase were observed in serum, but the underlying cause could not be explained. No increased incidence of microscopically detectable lesions was reported. However, increased mortality rates were observed, which were dose-dependent (Parent et al. 1992c).

12.4.4 Reproduction Toxicity

The reproduction and developmental toxicity of acrolein were examined in rats and rabbits (Parent et al. 1992b, 1993). Rabbits were administered daily doses of 0.0, 0.1, 0.75, and 2.0 mg acrolein per kg BW in water by gavage between days 7 and 19 of gestation. In the highest dose group, an increase in the resorption rates of the foetuses was observed, but this was not statistically significant. Foetal malformations were not observed. At higher doses of 4 and 6 mg per kg BW day^{-1}, there were reports of high maternal mortality, spontaneous miscarriages, increased resorption rates of the foetuses, and gastric ulcers. Acrolein neither induced developmental effects nor exhibited any teratogenic effects (Parent et al. 1993). Furthermore, rats of the parental generation (F0 generation) were exposed to 0, 1, 3, and 6 mg acrolein per kg BW for 70 days, and the females were further treated up to and through the lactation period. A similar treatment was carried out for rats in the F1 generation, in order to examine their offspring (F2 generation). Reproduction parameters such as fertility and viability and the morphology of the offspring remained largely unchanged after acrolein treatment. A reduced weight of the offspring in the F1 generation was, however, observed at the highest acrolein dose. Gastric lesions were established, especially in the high-dose group and in some animals of the F0 and F1 generations receiving the medium dose. In the highest dose group, an increased mortality rate was observed for the F0 and F1 generations. Acrolein was not classified as a reproduction toxin. However, it showed toxic effects at doses of 3 mg per kg BW day^{-1} and higher (Parent et al. 1992b).

12.4.5 Genotoxicity

12.4.5.1 In Vitro Genotoxicity

Studies of the mutagenicity of acrolein in bacterial test systems using *Salmonella typhimurium* strains do not show a uniform picture, since both positive and negative results are observed (Kim et al. 2007b; Li et al. 2008; Liu et al. 2010). In somatic cells, acrolein caused damage to DNA. DNA strand breaks were observed in fibroblasts of the skin, bronchial epithelial cells, and leukaemic cells treated with acrolein (Li et al. 2008). Sister chromatid exchange and chromosomal anomalies were reported in Chinese hamster ovary cells and other cell types (Kim et al. 2007b; Liu et al. 2010). Acrolein showed highly mutagenic effects in human xeroderma pigmentosum cells that were deficient in the nucleotide excision repair systems but not in normal fibroblasts (Curren et al. 1988). Further *in vitro* tests in V79 cells confirmed the mutagenic potential of acrolein (1–2 μM) (Smith et al. 1990). Other mutagenicity analyses with or without activation did not show any mutagenic effects in mammalian cells (Parent et al. 1991a).

12.4.5.2 In Vivo Genotoxicity

In vivo studies on genotoxicity resulting from acrolein exposure are very limited and show negative results (Gomes et al. 2001). After oral administration of up to 10 mg kg BW^{-1} acrolein for 14 weeks, no effects on the formation of micronuclei in normal chromatic erythrocytes in the peripheral blood of male and female mice were observed.

12.4.6 DNA Adducts

Acrolein forms adducts with all DNA bases *in vitro* (Pawlowicz et al. 2006). In cell culture, DNA adducts with desoxyguanosine were detected after acrolein exposure in human lung cells (Feng et al. 2006) and in Chinese hamster ovary cells (Foiles et al. 1990). In addition, acrolein adducts with desoxyadenosine have been detected in rat liver epithelial cells (Kawai et al. 2003).

DNA adducts with desoxyguanosine were detected *ex vivo* in many human tissues as well as in tissues of untreated rodents. Background levels of acrolein DNA adducts were found in the liver of rats and mice (Nath and Chung 1994), whereby another study conducted by the same researchers also detected adducts in the skin of mice as well as in the lung, kidney, brain, colon, prostate gland and mammary gland of rats (Nath et al. 1996). After inhalation exposure to acrolein (1 and 10 ppm over 6 hours), higher acr olein-guanine adduct levels were detected in the abdominal aorta of cockerels. Ten days after the single acrolein exposure, the DNA adduct levels had been repaired and returned to normal values (Penn et al. 2001). Data on DNA adduct levels after an oral acrolein exposure in animal tests are not available.

Background levels of acrolein DNA adducts were also detected in the liver, mammary gland, and leucocytes of humans (Nath et al. 1996). Acrolein-DNA adducts were also detected in brain (Liu et al. 2005) and other human tissues, such as the lung and the colon (Yang et al. 1999). The levels of acrolein DNA adducts of approximately 1 per 10^7 nucleotides found in lung tissue in this study are higher than the levels of some carcinogenic substances, such as polycyclic aromatic hydrocarbons (approximately 0.3 adducts per 10^7 nucleotides) and benzo[a]pyrene-diol-epoxide (non-detectable values of 0.1–0.5 adducts per 10^7 nucleotides) (Zhang et al. 2007).

Other studies have demonstrated mutagenic effects of γ-OH-Acr-dGuo. The controversy regarding the mutagenicity of γ-OH-Acr-dGuo is discussed in detail by Liu et al. (2010). Its ring structure seems to play a role as far as its mutagenicity is concerned. In double-strand DNA, and paired with desoxycytosine, γ-OH-Acr-dGuo occurs with an open ring structure (de los Santos et al. 2001). This can be easily repaired, and hence shows very low mutagenic potential. However, if γ-OH-Acr-dGuo occurs in a closed ring structure, as is the case with free nucleoside in solution or in single-strand DNA, it possesses a rather strong mutagenic potential (Liu et al. 2010). γ-OH-Acr-dGuo can also cause cross-linking of peptides/proteins and DNA, which can lead to DNA damage and mutations (VanderVeen et al. 2001).

In contrast to the rather weak mutagenic potential of γ-OH-Acr-dGuo, α-OH-Acr-dGuo is a potent mutagen, since it can hardly be repaired in the cell (Yang et al. 2002; Sanchez et al. 2003). Although α-OH-Acr-dGuo is formed in smaller amounts than γ-OH-Acr-dGuo after acrolein exposure in different organs, it can accumulate in certain tissues such as the lung, so that similar levels of the two acrolein-guanosine adduct isomers develop.

12.4.7 Carcinogenicity

Given acrolein's cytotoxicity, its possible carcinogenicity is difficult to test because of the necessity of conducting long-term exposure studies (Liu et al. 2010). In female rats that were exposed to 625 ppm acrolein in drinking water (5 days $week^{-1}$) for approximately 2 years, a slightly higher incidence of adenomas of the adrenal cortex was observed: tumours were observed in 5 out of 20 of the exposed rats, compared to 1 out of 20 in the control group (Lijinsky and Reuber 1987).

Parent et al. (1992c) did not observe any influence of acrolein on tumour development in rats that were treated with 0.05, 0.5, and 2.5 mg acrolein (dissolved in water) per kg BW day^{-1} for 2 years by gavage. The increased tumour incidence in the adrenal cortex observed in the first study could not be confirmed (Lijinsky and Reuber 1987). The later authors seem to have had problems with the stability of acrolein, so that the actual acrolein exposure could not be estimated. This is supported by the fact that the acrolein dose calculated by Parent et al. (1992c) and Lijinsky and Reuber (1987) was approximately 50 mg per kg BW day^{-1}, and therefore above the lethal dose for rats.

A further study in mice likewise did not show any influence of acrolein on tumour formation. Animals were exposed daily to 0.5, 2.0, and 4.5 mg acrolein per kg BW (dissolved in water) by gavage over 18 months. The treated animals showed a reduced body weight increase and a higher mortality in the higher dose group (Parent et al. 1991b).

In another study, neonatal mice received 75 or 150 nmol acrolein intraperitoneally on days 8 and 15 after birth. After 12 months, the animals were examined for liver tumours. Acrolein did not lead to an induction of tumours (Von Tungeln et al. 2002).

Sakata et al. (1989) reported on hyperplasias in the urinary bladder after intraperitoneal injection of acrolein. Acrolein (25 mg per kg BW) was administered to male rats intraperitoneally or intragastrically. Many animals died because of the high toxicity. Of the animals that were still alive 2 days after intraperitoneal treatment, 3 out of 12 developed hyperplasias in the bladder. These hyperplasias were not observed after intragastric treatment. In a subsequent experiment, the intraperitoneal injection of 6 mg acrolein per kg BW resulted in a significant increase in proliferation in the mucosa of the bladder. Lower doses did not have any effect.

Cohen et al. (1992) reported a tumorigenic effect of acrolein on the bladder. Acrolein (2 mg/kg BW) was administered to male rats intraperitoneally twice a week for 6 weeks. The rats were then treated for 20 weeks with 3% uracil and another 6 weeks with control diet. The majority of the animals (60%) treated with acrolein and uracil showed papillomas (benign tumours of the epithelium) in the bladder, compared to 27% of rats treated with uracil alone. No tumours were observed in the control animals. This study suggests a possible tumour-initiating activity of acrolein in the urinary bladder if it is applied intraperitoneally.

12.4.7.1 Epidemiology

The possible carcinogenicity of acrolein was examined in an epidemiological study of workers in chemical plants. A relationship between the probable inhalation exposure of acrolein and an increased risk of non-Hodgkin's lymphoma (odds ratio 2.6; two cases), multiple myelomas (odds ratio 1.7; one case), and nonlymphatic leukaemia (odds ratio 2.6; three cases) was observed. However, no statistical significance was determined (Ott et al. 1989). This study presents several deficiencies, such as the small number of cases, an insufficient evaluation of statistical analysis, and a restricted characterization of the exposure, so that its validity is limited (Gomes et al. 2001).

In epidemiological studies, an increased lung cancer incidence was reported for non-smoking Chinese women cooking in a traditional style. The IARC classified emissions from 'high-temperature frying' as 'probably carcinogenic to humans' (2A) (Straif et al. 2006). Compared to the corresponding controls, these women excreted a significantly higher level of mercapturic acids in urine (referred in each case to creatinine), derived from reactions of acrolein, crotonaldehyde, and benzene with GSH (Hecht et al. 2010).

Metabolites such as acrylic acid, glycidaldehyde, and glycerinaldehyde are also formed from acrolein. It is speculated that they could play a role in its toxicological effects (Patel et al. 1980). For glycidaldehyde, carcinogenic effects were observed in laboratory rodents after dermal application. In many recent publications, acrolein and other lipid peroxidation products such as 4-hydroxy-2-*trans*-nonenal, have been associated with neurological diseases. The endogenous level in the brain or medulla is increased in patients suffering from Alzheimer's disease, Parkinson's disease, amyotrophic lateral sclerosis (ALS), and other neurological diseases (Shamoto-Nagai et al. 2007; Butterfield et al. 2010). Increased DNA-acrolein adduct levels were also determined in the brain of Alzheimer's patients (Liu et al. 2005). Whether the oral intake of acrolein via foods is relevant in this connection is rather unlikely, due to its toxicokinetic properties. This aspect should, however, continue to be investigated and clarified.

12.5 Toxicity of Amines and Alcohols

12.5.1 Mutagenicity of Heterocyclic Aromatic Amines

After metabolic activation with S-9 mix, heterocyclic aromatic amines (HAAs) can be assigned to the group of the most strongly mutagenic compounds (Sugimura 1997). The demonstration of mutagenicity was made by means of short-term bacterial tests. These mutagenicity tests also enabled the detection of biologically active substances in complex foods. Substances that are detectable with the Ames test have genotoxic potential (Sugimura et al. 2004). Trp-P-1, Trp-P-2, Glu-P-1, and Glu-P-2 are, after metabolic activation, strong mutagens for the *Salmonella* Typhimurium TA98, TA100, and TA1538 strains used in that test (Sugimura et al. 2004). After the addition of S9 mix, which is necessary for metabolic activation, the IQ compounds MeIQ, IQ, 4,8-DiMeIQx,

MeIQx, Glu-P-1, and Trp-P-1 acted as significantly stronger mutagens than AαC, MeAαC, and PhIP. In addition, the IQ compounds showed a significantly stronger mutagenic activity than aflatoxin B1 and benzo(a)pyrene (Sugimura 1997). The relative mutagenic potentials of MeIQ, MeIQx, and PhIP vary with the test assay. In *Salmonella typhimurium* strain TA98, the relative potencies were MeIQ > MeIQx > PhIP; however, they were PhIP ≥ MeIQ > MeIQx in a Chinese hamster ovary system (US HHS PHS 2002) and IQ > MeIQ > Trp-P-1 ≥ MeIQx >> PhIP in human-derived hepatoma (HepG2) cells (Knasmüller et al. 1999). The relative mutagenicity is therefore difficult to assess, and the effect of intake of certain amounts of HAAs is difficult to predict. Norharman, belonging to the β-carbolines, showed no mutagenic activity in the Ames test (Nagao et al. 1977). However, like harman, it could strengthen the mutagenic activity of the other HAAs and acted co-mutagenically (Wakabayashi et al. 1982). Aniline and norharman (β-carboline) formed aminophenylnorharman via an S9 mix of rat liver or *in vivo* in rodents; hydroxyaminophenylnorharman was generated and converted to an acetoxy derivative, which can form DNA adducts and induce mutations (Totsuka et al. 2004). Co-mutagenic activity can also be perceived with other aromatic amines such as o-toluidine (Hada et al. 2001).

12.5.2 Carcinogenicity of HAAs

The carcinogenic effects of HAAs could be demonstrated in animal trials with rats and mice. In these trials, the animals were fed concentrations of 0.03 or 0.04% HAAs in their food over a period of 2 years. After several months, feeding with IQ, MeIQ, and MeIQx resulted in the formation of tumours in all trial animals, mainly in the liver and instestines (Ohgaki et al. 1987). Among the control rats who received normal food, only 4% developed tumours. After feeding with 0.06% MeIQx for several weeks, tumours appeared in the rats' livers, lungs, and lymph nodes (Ohgaki et al. 1987). A dose of 10–20 mg IQ per kg BW in the food of nonhuman primates similarly caused carcinoma formation after several months (Adamson et al. 1996). Based on animal trials, the IARC has classified IQ as a possible carcinogen (class 2A) and eight other HAAs (MeIQ, MeIQx, PhIP, AαAC, MeAαAC, Trp-P-1, Trp-P-2, and Glu-P-2) as probable carcinogens (class 2b) (Gibis 2016).

However, it is difficult to prove a connection between HAA intake and the development of different types of tumours in humans. The quantity of HAAs ingested by humans is much smaller than that fed to trial animals. In this context, the present results of long-term animal studies cannot be used to set a specific threshold value for the human intake of HAAs. Nevertheless, epidemiological studies in the last decade have shown an increased risk of incidence of different cancers for humans who prefer and eat high amounts of very well-done meat.

12.5.3 Risk Assessment

12.5.3.1 Daily Intake of HAAs

The quantity of mutagenic substances ingested is very difficult to estimate. Daily HAA intake can differ markedly between different epidemiological studies. Alongside different eating habits, the type of preparation and the frequency of meat consumption play important roles in establishing intake. Although the levels observed in heated foods were low, a connection between meat consumption and the development of colorectal carcinomas could be determined in various studies (Le Marchand et al. 2002; Nowell et al. 2002). The results of studies in different countries show that it is difficult to define a standard value for the quantity of ingested HAAs (Table 12.1). In American studies, the HAA intakes calculated were often remarkably higher than those found in, for example, Sweden. One reason for this could be that much more grilled and fried meat is eaten in the United States than in Sweden. In China, the daily HAA intake is estimated at 50 ng day^{-1}. The Spanish intake of 606 ng day^{-1} is clearly higher in comparison, which can be explained by the different types of preparation used there (e.g. grilling, barbecuing, and pan frying), and the higher frequency of meat consumption (Busquets et al. 2004).

In a study from the United States, the authors used consumption data from 3 consecutive days for 3563 participants (Layton et al. 1995), which were collected from a United States Department of Agriculture (USDA)-sponsored random survey of the US population during 1989. It was preferred to investigate the average daily HAA intake with the food in order to be able to estimate the intake of PhIP, MeIQx, DiMeIQx, IQ, and AαC. The HAA concentrations used to calculate the daily HAA intake were taken mainly from databases. In addition, data were based on different heating times and preparation methods (including grilling, frying, and roasting). On this basis, the estimated daily average HAA intake in the United States was determined to be 26 ng per kg BW day^{-1}.

In a New Zealand study, data on prepared meat with known HAA concentrations were taken from the literature and compared with the eating habits of New Zealanders. The estimated intake was around 1000 ng $person^{-1}$ day^{-1} (Thomson et al. 1996). Lower values were obtained in a later New Zealand study that was similarly concerned with the connection between meat consumption and HAA intake. The consumption quantities of different sorts of meat, the modes of preparation, and the degree of browning were established by means of a questionnaire. A total HAA intake of 164 ng $person^{-1}$ day^{-1} was calculated (Norrish et al. 1999). In a Swedish study, 267 women and 277 men in Stockholm aged between 50 and 75 years were asked about their eating habits. The frequency of consumption of grilled or cooked meat and fish with known HAA levels was recorded (Augustsson et al. 1999). In order to take the preferred individual degree of browning into account, the test participants were shown photos of each product with four different degrees of browning. Using this information, the questionnaire produced a mean HAA intake of 160 ng $person^{-1}$ day^{-1} (Augustsson et al. 1997). After several analytical investigations of various meat and fish dishes found in restaurants, HAA intake in Switzerland was calculated to be up to 400 ng $person^{-1}$ day^{-1} (Zimmerli et al. 2001). Data for more than 25 000 US citizens concerning their meat consumption and preparation methods were evaluated in a US cross-sectional study. The calculated daily HAA intake was about 420 ng $person^{-1}$ day^{-1} (Keating and Bogen 2001). The HAA levels in the study were only roughly estimated. In addition, the food was not prepared using the usual household methods. Thus, the value obtained is significantly higher than would be expected for everyday intake.

With the help of a questionnaire, 344 participants in Germany were asked about their frequency of consumption of seven types of meat and fish, taking into account the preparation methods used. In order to record their preferred degree of browning, photos of samples with different degrees from light to very heavy were attached to the questionnaire.

Table 12.1 Estimated daily dietary intakes of major HAAs (ng $person^{-1}$ day^{-1}) reported in various studies.

Country	Subjects	MeIQx	DiMeIQx[a]	PhIP	MeIQ	Other HAAs	Total HAAs	References
United States	3563	182	56	1162		364 AαAc, 21 IQ	1820 (2415)[b]	Layton et al. (1995)[c]
Sweden	544	72	16	72			160	Augustsson et al. (1997)
United States	673	33–45	4	286–458			323–507	Byrne et al. (1998)[d]
United States	374	20–33	1.5–2.2	78–110			100–145	Sinha et al. (2001)[e]
Switzerland	–f	72–85	13–78	124–156	0–39	0–39 IQ,	209–397	Zimmerli et al. (2001)[f]
Japan	39 035	9.8–11	0.7–6.3	39–47	5.6–9.8	2.8–4.9 Trp-P-1	58–77	Kobayashi et al. (2002)[a,g]
United States	537	94–135	6.5–11	161–218			261–364	Nowell et al. (2002)[h]
Germany	385 (344)[i]	34	2	63			103	Rohrmann and Becker (2002)[i]
United States	21 780	98–147	20–33	644–889		176–219 MeIQ and Trp-P-1	938–1288	Keating and Bogen (2004)[j]
Spain	3221	29	14–15	344			606 (934)[k]	Busquets et al. (2004)
Singapore	497	29	13	5.8		2.2 IFP and 7,8-DiMeIQx	50	Wong et al. (2005)[l]
Sweden	15 174	22–256	8–90	29–475			59–821	Ericson et al. (2007)[m]
Germany	25 540	17–22	3–3.8	41–45			61–71	Rohrmann et al. (2009)[n]
Malaysia	600	312	16[a]	195	30		554	Jahurul et al. (2010)[o]
Croatia	94	65–85		164–2968			327–4363	Klapec and Periš (2014)[p]

Source: Reproduced with kind permission of John Wiley & Sons (Gibis 2016).

[a] 4,8-DiMeIQx and 7,8-DiMeIQx.

[b] Includes minor occurring HAAs (Glu-P-1, Glu-P-2, Trp-P-1, Trp-P-2, MeIQ, MeAaC).

[c] Mean intake based on 70 kg body weight.

[d] Questionnaire assessing heterocyclic amine (HCA) intake in each cohort (n = 216, 226, 231 participants) separately.

[e] Case–control study of cases (n = 146), colorectal adenomas.

[f] Sum of IQ, 7,8-DiMeIQx, MeIQ, 4,8-DiMeIQx, MeIQx, and PhIP; total meat (ready to eat) ≈ 130 g person^{-1} day^{-1} (calculated).

[g] Food frequency questionnaire study of 39 035 participants.

[h] Case–control study with 157 cases (colorectal cancer) and 380 controls.

[i] Questionnaire assessing HAA intake of 385 participants (plus repeat 344).

[j] Continuing survey of food intake by individual (CSFII) based on databases (US) with 21 780 records.

[k] Sum of DMIP, Glu-P-2, IQ, MeIQx, MeIQ, Glu-P-1, 7,8-DiMeIQx, 4,8-DiMeIQx, norharman, harman, Trp-P-2, Trp-P-1, PhIP, AαC, and MeAαC.

[l] Food-frequency questionnaires completed by 497 randomly sampled Chinese men and women aged 20–59.

[m] Malmö diet and cancer cohort (n = 15 174) (low intake = 175 °C, high intake = 200 °C).

[n] Prospective cohort study (25 540 participants) (cases of colorectal cancer n = 516 and controls n = 3699).

[o] Food-frequency questionnaire of 600 participants.

[p] Questionnaire of 94 Croatian females (low and high intake); total HAA intake = (Σ intake MeIQx and PhIP) 1.249.

The mean daily HAA intake was found to be 103 ng $person^{-1}$ day^{-1} (Rohrmann and Becker 2002). In a sensory test in another study, some panellists preferred darker and very well-done beef patties, because they liked a well-done appearance and roasted flavour (Gibis and Weiss 2010). There was an increase in HAA levels by factors of 8.5 for MeIQx, 2.0 for PhIP, 4.5 for norharman, and 2.4 for harman between the lightest grilled (for 120 seconds) and darkest grilled (for 180 seconds) products. This shows that preferences concerning colour and roasted flavour play a central role in the personal daily intake of HAAs. As mentioned previously, the frequency of meat consumption and especially the typical methods of preparation contribute to different intake levels of individual HAAs in different countries. Eating habits also play an important role (Rohrmann et al. 2009). Most of the daily intake of HAAs is of PhIP, followed by MeIQx.

In a study obtained from food-frequency questionnaires of 497 randomly sampled Chinese men and women from Singapore, the assessed mean exposure to HAAs was approximately 50 ng day^{-1}; this was 50% higher among younger (20–39 year-olds) compared with older (40–59 year-olds) individuals (Wong et al. 2005). In this assessment, PhIP, MeIQx, and 4,8-DiMeIQx were the most abundant HAAs detected, and concentrations of total HAAs ranged from <0.10 to 6.77 ng g^{-1}. The highest contents were found for Chinese-style roasted pork. Seven types of cooked meat contributed to 90.1% of this intake: pan-fried fish, pork, and chicken; deep-fried chicken and fish; roasted/barbecued pork; and grilled minced beef (Wong et al. 2005). The data in Table 12.1 confirm that there is a large variation in exposure to HAAs between countries (50–1820 ng $person^{-1}$ day^{-1}). One reason for this is the differing frequency of meat consumption; for example, Greece has an approximately 30% lower frequency of consumption than Spain. Another is the different preparation methods employed. Consumer preference for colour and roasted flavour also plays a key role. The food-frequency questionnaires with images of different degrees of browning give additional information concerning HAA intake, which was included in some intake studies (Augustsson et al. 1997; Rohrmann and Becker 2002; Rohrmann et al. 2009).

12.5.4 Epidemiological Data

12.5.4.1 Risk Evaluation

The findings on the mechanisms of action of HAAs and mutagenicity, and the results of carcinogenicity tests, have led international associations concerned with estimating the risks of potentially carcinogenic substances to attempt to classify HAAs according to their risk. Efforts are also being made to issue nutritional, preparation, and other recommendations in order to minimize risks.

Most meat is consumed fried, grilled, or roasted, whereby the HAA intake is related to the intake of meat. Many studies associate high intake of meat – mainly red meat and processed meat – with a high risk of incidence of cancers, particularly colorectal cancer. However, there is evidence that this risk may not be a function of meat as such, but instead reflects a high fat intake and/or the generation of carcinogens through various home and restaurant cooking and processing methods (Ferguson 2010). The cancer risk may be modified by certain genotypes (Barbir et al. 2012). Key to the connection between meat and cancer is not the meat itself, but the intake of carcinogenic substances, such as HAAs, polyaromatic hydrocarbons, and nitrosamines. Physical activity and smoking also contribute to the multifactoriality of cancer development. Further, the combination of HAAs and with other ingredients such as fibres presents evidence that a diet high in dietary fibre sources may reduce the potential carcinogenic effects of HAAs by changing their absorption and excretion (Ferguson 2010). Assessing this risk is exceedingly difficult, however. The carcinogenic effect has been confirmed in animal trials, but significantly higher doses of HAAs were administered than are usually ingested with the human diet. The intake of thermal mutagens with food is low: about 5–10 μg per meal (Skog et al. 1998). Absorption, metabolization, and excretion of HAAs are further important factors in estimating carcinogenic risk (Turesky 2007). On the one hand, other carcinogenic and tumour promoters are ingested alongside HAAs in food, which may enhance the carcinogenic effects of food components such as haem iron and free radicals (Sugimura et al. 2004). On the other, protective substances such as antioxidants are also ingested. These can combine with other food components and provide a positive bioactive effect (Rahman et al. 2014).

It is difficult to establish a connection between food components and various cancers, as carcinogenesis is a very complex process that occurs over long periods of time. Besides HAAs, other carcinogenic compounds such as polycyclic aromatic hydrocarbons and nitrosamines are present in red meat and processed meat. Human case–control studies with large numbers of participants have shown an increasing risk of the incidence of different types of cancer (e.g. colorectal cancer) in many cases, such as with increasing intake of MeIQx, PhIP, and DiMeIQx, increasing consumption of red meat, and increasing mutagenic activity (Table 12.2). Nonetheless, the assumption that a connection exists between the consumption of red and processed meat and an increased risk of cancer is strengthening. It is suggested in a personal recommendation by the World Cancer Research Fund that not more than 500 g of red meat, including meat products, should be consumed per week, and very little if any of this should be processed meat (Gibis 2016).

Red meat refers to beef, pork, lamb, and goat, including that contained in processed foods. The consumption of fish, poultry, and game is preferable to that of pork, beef, and lamb, which is now considered to be connected with an increased risk of various tumours. Above all, no charred meat or meat grilled directly over flames should be consumed. Carcinogenic HAAs are produced through grilling, and their intake should be minimized by minimizing consumption of grilled products. Gentler preparation methods include boiling, stewing, steaming, and microwaving (Felton et al. 1994; Rahman et al. 2014). It has already been determined that the individual genetic disposition for a fast or slow metabolization of HAAs in the body may play a role in the development of colorectal cancer. Colorectal adenoma risk was positively associated with PhIP, MeIQx, and DiMeIQx (p trend = 0.006, 0.022, and 0.045, respectively) intake (Barbir et al. 2012). The phenotypes of sulfotransferases (SULT1A1) also modified the effect of MeIQx on colorectal adenoma risk ($p > 0.01$), such that the association of MeIQx intake with colorectal adenoma was stronger for slow than for normally metabolizing phenotypes (Barbir et al. 2012). It was determined that there is an increased risk of developing breast, lung, and others types of cancer with the consumption of red, strongly roasted meat. However, some researchers could not show a connection between HAA intake and the development of prostate cancer (Norrish et al. 1999). Nevertheless, they determined that there is an increased risk of cancer for people who prefer to eat well-roasted meat. Sinha et al. (2000) investigated the effects of different

Table 12.2 Epidemiological human studies investigating the effect of the consumption of well-done meat or the intake of major HAAs on cancer risk (without indication: $p_{trend} < 0.05$; N-acetyltransferase (NAT1) or (NAT2); sulfotransferase (SULT)).

Study	Material/Content of HAAs	Odds ratio (OR)/relative risk (RR) (95% CI)	Cancer type	Number of subjects (age); cancer cases	References
Case–control	Very well-done meat	OR 4.36 (2.08–9.60)	Colon/rectum	380 controls (20–88) and 157 cases	Nowell et al. (2002)
	High content of MeIQx	OR 4.09 (1.94–9.08)			
Case–control	AhR GA/AA or NAT1 rapid acetylator and intake of red meat	OR 1.60 (1.23–2.09)	Colon/rectum	1002 polyp cases and 1493 controls	Shin et al. (2008)
	AhR GA/AA or NAT1 rapid acetylator and intake of MeIQx	OR 1.48 (1.14–1.92)			
	AhR GA/AA or NAT1 rapid acetylator and intake of PhIP	OR 1.39 (1.11–1.74) $p_{trend} = 0.055$			
	AhR GA/AA or NAT1 rapid acetylator and intake of DiMeIQx	OR 1.37 (1.10–1.71) $p_{trend} = 0.115$			
Case–control	Total HAAs	OR 0.99 (0.21–4.81)	Colon/rectum	117 cases and 238 controls	Kobayashi et al. (2009)
	MeIQx, PhIP, DiMeIQx	No significant OR			
Cohort study	High intake of PhIP	RR 1.47 (1.13–1.93)	Colon/rectum	(25 540 participants) 516 cases and 3966 controls	Rohrmann et al. (2009)
	High intake of MeIQx or DiMeIQx	No significant RR			
Case–control	High red meat consumption	OR 2.02 (1.06–3.83) $p_{trend} = 0.38$	Colon/rectum	158 cases and 649 controls	Ferrucci et al. (2009)
	High intake of MeIQx	OR 1.90 (1.05–3.42) $p_{trend} = 0.07$			
	High intake of pan-fried meat	OR 1.72 (0.96–3.07)			
Case control	High consumption of red meat	OR 1.4 (1.2–1.6)	Colon/rectum	2543 cases (1881 adenomas/ 622 hyperplastic polyp) and 3764 controls	Fu et al. (2012)
	High intake of MeIQx	OR 1.4 (1.2–1.6)			
	High intake of PhIP	OR 1.3 (1.1–1.5)			
	High intake of DiMeIQx	OR 1.3 (1.1–1.5)			

Case–control	High intake of PhIP	OR 1.81 (1.24–2.64)	Colon/rectum	413 cases and 796 controls	Barbir et al. (2012)
	High intake of MeIQx	OR 1.45 (0.99–2.12)			
	High intake of DiMeIQx	OR 1.35 (0.94–1.93)			
	High intake of MeIQx and slow SULT1A1 activity	OR 7.94 (2.20–28.58)			
Case–control	High intake of MeIQx and NAT1 activity (rs6586714)	OR 1.43 (1.11–1.85)	Colon/rectum	1205 cases and 1387 controls	Gilsing et al. (2012)
Case–control	High intake of MeIQx	OR 1.87 (1.44–2.44)	Colon/rectum	1645 controls and 1062 cases	Helmus et al. (2013)
	High intake of DiMeIQx	OR 1.68 (1.29–2.17)			
	Red meat-derived mutagenic activity	OR 1.77, (1.36–2.30)			
Case-control	Fast NAT2 activity and high intake of DiMeIQx	OR 1.70 (1.06–2.75)	Colon/rectum	1016 cases and 1355 controls	Voutsinas et al. (2013)
	Fast NAT2 activity and high intake of MeIQx	OR 1.91 (1.16–3.16)			
	Fast NAT2 activity and high intake of PhIP	OR 2.14 (1.31–3.49)			
Case–control	High consumption of red meat	OR 1.8 (1.5–2.2)	Lung	1903 cases and 2073 controls	Lam et al. (2009)
	High consumption of processed meat	OR 1.7 (1.4–2.1)			
	High intake of PhIP	OR 1.5 (1.2–1.8)			
	High intake of MeIQx	OR 1.4 (1.2–1.7)			
	High intake of DiMeIQx	OR 1.0 (0.8–1.2), $p_{trend} = 0.09$			
	High mutagenic activity	OR 1.4 (1.1–1.7)			
Case–control	High consumption of well-done red meat	OR 1.52 (0.93–2.46)	Prostate	531 advanced and 195 localized cases and 527 controls (40–79)	John et al. (2011)
	Higher consumption of hamburgers	OR 1.79 (1.10–2.92)			
	Medium–high intake of PhIP	OR 1.42 (0.98–2.04)			
	MeIQx, DiMeIQx	No significant OR			

(Continued)

Table 12.2 (Continued)

Study	Material/Content of HAAs	Odds ratio (OR)/relative risk (RR) (95% CI)	Cancer type	Number of subjects (age); cancer cases	References
Case–control	High intake of MeIQx	OR 6.07 (2.24–16.4)	Bladder	884 cases and 878 controls	Lin et al. (2012)
	High intake of PhIP	OR 2.67 (1.2–5.96)			
	High intake of DiMeIQx	OR 3.44 (1.5–7.89)			
	Mutagenic activity	OR 6.32 (2.51–15.94)			
	High red meat consumption and unfavorable genotypes (≥6)	OR 5.09 (2.89–8.96)			
Case control	High consumption of barbecued meat	OR 1.35 (1.01–1.79)	Renal	1192 cases and 1175 controls	Daniel et al. (2011)
	High intake of MeIQx, PhIP, and DiMeIQx	No significant OR			
Case–control	High intake of MeIQx, PhIP, and DiMeIQx	No significant OR	Breast	2686 cases and 3508 controls	Mignone et al. (2009)
Cohort study	High intake of meat mutagens or high consumption of meat	No significant RR	Breast	120 755 post-menopausal women, 3818 cases of invasive breast cancer (over 8 years)	Kabat et al. (2009)

Source: Reproduced with kind permission of John Wiley and Sons (Gibis, 2016).

HAAs on the development of lung cancer. MeIQx was the only one to show an increased effect, with no effect detected for the other HAAs investigated. The authors determined that an increase in MeIQx intake of 10 ng resulted in an increase of 5% in the risk of colorectal adenomas.

A study carried out in the United States on the connection between meat consumption and the development of breast cancer came to the conclusion that HAA intake has no effect on the risk of cancer (Delfino et al. 2000). Although not all studies have led to consistent results, epidemiological studies show divergent results concerning pork and poultry (Joshi et al. 2015). The total composition of a dish, the cooking method used for the meat, and lifestyle factors (e.g. physical activities and fibre intake) are very important to the effects of meats containing HAAs or other carcinogenic compounds (Ferguson 2010). There is now a close consensus that there is an increased risk of cancer with high consumption of red meat and meat products. The evidence that red and processed meats are a cause of colorectal cancer is more convincing today than it was in the mid-1990s (Gibis 2016). For this reason, the IARC has now classified the consumption of red meat as probably carcinogenic to humans (Group 2A), and that of processed meat as carcinogenic to humans (Group 1), based on sufficient evidence in humans that the consumption of processed meat causes colorectal cancer (Bouvard et al. 2015). Experts have concluded that each 50 g portion of processed meat eaten daily increases the risk of colorectal cancer by 18%.

12.6 Toxicity of Aldehydes

12.6.1 Genotoxicity and Mutagenicity of Furfurals

The most important and widely known furfural is 5-HMF, which mainly tested negative in *in vitro* genotoxicity tests; however, it is mutagenic in the presence of the cytosolic fraction of rat liver with 3′-phosphoadenosine-5′-phosphosulphate as the cofactor for SULTs and in genetically modified bacteria and mammalian cells that express the human SULT 1A1 (Janzowski et al. 2000; Abraham et al. 2011b).

Human and rat SULTs bioconvert 5-HMF into the reactive allyl ester SMF. This substance is mutagenic in bacteria and mammalian cells *in vitro* in the absence of an activating system (Surh et al. 1994); the mechanism probably involves reactions of its highly electrophilic allyl carbocation. SMF is inactivated by GSH in the presence of GSH *S*-transferases. A genotoxic effect in *Salmonella typhimurium* TA104 in the absence of a metabolic system could be reduced in the presence of GSH and GSH *S*-transferases (Lee et al. 1995).

5-HMF was negative in the *in vitro* micronucleus test with HepG2 cells that express SULTs (Severin et al. 2010). In the *in vitro* Comet assay, 5-HMF was weakly positive in various cell lines (with different SULT activity), but only at high concentrations (25–100 mM) that were also cytotoxic. In HepG2 cells, 5-HMF also induced DNA damage at somewhat lower concentrations (8–25 mM) in the Comet assay. These concentrations were again comparatively high, since these tests are generally conducted up to a concentration of 10 mM.

The discrepancies between the positive and negative results of *in vitro* tests may be due to the different conditions under which they were performed, including expression of SULT in the cell lines used and the presence or absence of 3′-phosphoadenosine-5′-phosphosulphate in the metabolising system (Durling et al. 2009).

In vivo, no micronuclei were induced in the micronucleus test with the peripheral blood cells of male and female B6C3F1 mice given 5-HMF over 90 days by gavage at doses of 0, 47, 94, 188, 375, and 750 mg per kg BW day^{-1} (Abraham et al. 2011b). The ratio of polychromatic erythrocytes to total erythrocytes was not changed, and thus it is not clear whether the substance reached the bone marrow. The study design deviated from the standard protocol in terms of duration of administration, but is nonetheless acceptable. SMF was positive in the micronucleus test in mice (Abraham et al. 2011b). However, these two studies are not directly comparable because they were conducted with different mouse strains and treatment protocols.

12.6.2 Acute and Subchronic Toxicity

The acute toxicity of 5-HMF is very low. In a study in rats, an acute oral LD50 of 3100 mg per kg BW was observed. For mice, there was an LD50 of 1910 mg per kg BW, and in another study an LD50 of >2000 mg per kg BW (Ulbricht et al. 1984). In rats administered 250 mg 5-HMF per kg BW per day over a 40-week period, there were no significant differences in terms of weight increase, feed consumption, or final weight as compared with control animals. The histological examination of various organs (e.g. heart, liver, and kidneys) did not show any differences between the two groups (Lang et al. 1970).

In a longer-term study, rats were given 0, 40, 80, or 160 mg 5-HMF per kg BW for 6 days a week over a period of 11 months. Protein and lipid metabolism, ascorbic acid levels in the adrenal glands, the activity of the hepatic succinate dehydrogenase, organ morphology, and body weight were unchanged in comparison to controls. At 160 mg per kg BW, however, there were minor changes in the clinical chemical parameters (a temporary increase in the γ-globulin levels in the serum and a tendency to increased activity of the hepatic tributyrase) and an increase in the relative spleen weight; 40 and 80 mg per kg BW did not have any effect (Zaitsev et al. 1975; Ulbricht et al. 1984).

In a study in mice over a 3-month period within the framework of the NTP studies, there was a significantly lower weight increase at the highest dose of 750 mg per kg BW (5 days per week) compared with controls. Effects like minor to mild cytoplasmic changes in the kidneys were significant in the male animals at doses of 188 mg per kg BW (5 days per week) and above. At 94 mg per kg BW (5 days per week) and below, no adverse effects were observed (Abraham et al. 2011b). Hence, the maximum dose providing NOAEL with regards to acute and subacute toxicity in animal experiments is in the range of 80–100 mg per kg BW day^{-1}.

12.6.3 Carcinogenicity in Animal Models

12.6.3.1 Intestinal Micro-Adenomas

A potential carcinogenic effect plays a major role in the assessment of 5-HMF and its reactive allyl ester SMF. Discussions focus above all on the possible formation of intestinal tumours. Corpet et al. (1990) examined a tumour-inducing effect after initiation with azoxymethane in CF1 mice and Fisher344 rats given different dietary components heat-treated at 180 °C. After 100 days, a significantly higher number of large micro-adenomas of the intestines were observed in animals that had been given heat-treated (caramelized) sugar or heat-treated casein and fat (compared with control animals given the same but unheated feed). Five additional heat-treated food components had no significant effects. The authors believed that compounds formed during the heating process were promoters of colon cancer (Corpet et al. 1990). However, this study could not draw any conclusions about which compounds were responsible.

12.6.3.2 Colon Preneoplastic Lesions

The same working group carried out a high-performance liquid chromatography (HPLC) analysis of slightly caramelized sugar in which a level of 1% 5-HMF was identified as the most important reaction product in terms of quantity (Archer et al. 1992). In subsequent animal experiments, Fisher 344 rats were again pretreated with azoxymethane and divided into four groups. The control group was given untreated saccharose, while other groups were given heat-treated sugar, heat-treated sugar without 5-HMF (butanol extraction), or feed with 1% added 5-HMF. The results of the first study (heat-treated saccharose; Corpet et al. 1990) were confirmed, but heat-treated sugar from which 5-HMF had been extracted did not show any effect. In the group given a feed with 1% 5-HMF, there were significantly larger aberrant crypt foci (ACFs, precursors of micro-adenomas) compared to the control group, as well as a higher number of large ACFs. In additional studies which were then carried out with 5-HMF dissolved in water (given twice at an interval of 1 week, doses of 100–300 mg per kg BW, no pretreatment with azoxymethane, evaluation after 30 days), dose-related increases were observed in the numbers of ACFs per colon and in the fraction of animals with ACFs. Hence, it was reported that 5-HMF could have not only a promoting but also an initiating potential in triggering cancer of the colon. However, regarding the evidence, it should be noted that the study design had shortcomings (three separate studies with different numbers of animals; far higher ACF numbers in the control groups of two studies than in the third; unconvincing choice of dose groups), making a more sophisticated overall assessment necessary (Zhang et al. 1993).

12.6.3.3 Studies with Transgenic Mice

Similar results were observed with mice. Svendsen et al. (2009) treated C57BL/6J mice (wt wt^{-1}) shortly after birth with a single subcutaneous injection of 5-HMF (500 mg/kg BW^{-1}). Examination of the animals at 12 weeks identified 'flat' ACFs in 4 out of 18 (22%), which had not been observed in any of the control animals (no treatment). The authors reported this result as not significant but did not provide any information about the general incidence of 'flat' ACFs in untreated animals; however, if this tissue change were shown never to occur in a spontaneous fashion, then the results could be interpreted as relevant.

A genetic variant of the animals (C57BL/6J Min/+) was also examined, which was particularly predisposed to the occurrence of multiple intestinal neoplastic changes using the same study design. Treated animals were found to have a higher number of adenomas in the small intestine compared to untreated animals (mean 119 versus 102). The difference in the number was significant only if the evaluation was undertaken for the middle and distal sections of the small intestine (15–26 cm from the stomach; $p = 0.033$). This method of specifying a section of the intestine only after histological evaluation is deemed to be problematic from a methodological–statistical angle and is primarily suited for generating hypotheses (e.g. that activation of 5-HMF to carcinogenic metabolites might take place only in a specific section of the intestine). A higher number of 'flat' ACFss were found in the large intestine of treated animals than in the control group (mean 32 versus 21); this number was not significant, however. Nor were there any significant differences in the rate of adenomas in the large intestine.

12.6.3.4 NTP Studies

The relevance of the short-term models just described is unclear with regards to the assessment of 5-HMF's potential to cause intestinal cancer. Greater elucidation of this potential could be expected from the NTP studies conducted a few years ago (NTP 2010). In 2-year studies in F344/N rats and B6C3F1 mice, groups of 50 females and 50 males were given 0, 188, 375, or 750 mg of 5-HMF per kg BW 5 days a week by oral gavage.

In the rats, the survival rates of the 188 and 750 mg per kg BW groups were higher than that of the control group. No noticeable differences in body weight gains were observed in treated versus untreated animals. Degenerative effects were observed in the olfactory epithelium of rats (incidences significantly increased in males at 750 mg per kg BW and in females at 188 and 375 mg per kg BW). The incidences of olfactory epithelium respiratory metaplasia and respiratory epithelium squamous metaplasia were increased in the highest dose group in both genders. The only other treatment-related, non-neoplastic effect was a significantly higher incidence of clear cell foci of the liver in male rats in the highest dose group.

In mice of both genders in the highest dose group, there was a significantly lower survival rate compared to the control group. Furthermore, a lower weight increase was observed. Lesions of the olfactory and respiratory epithelium of the nose similar to those observed in the rats were found in mice in the 375 and 750 mg per kg BW dose groups. Starting with month 8, neurological symptoms were observed in the highest dose group in both genders within the first minutes of administration (e.g. decreased exploratory behaviour, piloerection, salivation, catatonia, excitation, dyspnea, clonic–tonic seizures, and unconsciousness). Because of these clinical symptoms and the reduced survival rate in the highest dose group, these animals were excluded from the evaluation of the carcinogenic potential.

The overall result was classified as showing some evidence of a carcinogenic activity of 5-HMF in female B6C3F1 mice, based on elevated incidences of hepatocellular adenomas of 53% in the 188 mg per kg BW dose group and 52% in the 375 mg per kg BW group. In the control group, the incidence was 28%, which is in accordance with historical control data for hepatocellular adenomas in female B6C3F1 mice (incidence on average around 22%). No evidence of a tumourigenic effect was found in rats or in male mice. The indications observed in the short-term studies of a possible potential of 5-HMF to induce intestinal cancer in rodents could therefore not be confirmed in the NTP long-term study.

12.6.3.5 Studies with SMF

Regarding the possible carcinogenic potential of 5-HMF, discussions focus in particular on the reactive metabolite, SMF. When applied topically to the skin of mice, SMF initiated a higher incidence of papillomas than 5-HMF itself (Surh et al. 1994).

In the study by Svendsen et al. (2009), SMF was applied subcutaneously. The dose was a single administration of 25 mg per kg BW (1/20th of the 5-HMF dose). Results were similar to those obtained for 5-HMF: an elevated incidence of 'flat' ACFs (2 out of 15 mice, 13%) in the wild-type animals, which was not significant compared with the controls. With the particularly sensitive genetic variant, the C57BL/6J Min/+ mouse, an elevated number of small intestinal adenomas and 'flat' ACFs in the colon were observed, but with no or only borderline significance.

The high toxicity of SMF was demonstrated when it was administered intraperitoneally to FVB/N mice in a single high dose of 250 mg per kg BW. Most of the animals died within 5–11 days as a consequence of massive damage to the proximal tubules of the kidneys. Here, the uptake of SMF into the proximal tubule cells mediated by the transporters OAT1 and OAT3 may have played a role (Bakhiya et al. 2009; Abraham et al. 2011b). In subsequent studies with various lower doses and observation periods of up to 40 weeks, changes in the kidney tubules were also observed as the main effect. Furthermore, hepatotoxicity and serositis were observed in organs with direct contact to the peritoneum. No ACFs or adenocarcinomas were observed in the animals treated with SMF. In a positive control (20 animals treated with azoxymethane), a total of 1064 ACFs and five adenocarcinomas could be induced, however (Abraham et al. 2011b).

If SMF should prove to be responsible for a possible carcinogenicity of 5-HMF, then attention will focus on the question of the activity of the respective SULT in various species. For humans, it is known that the most important SULT – SULT1A1 – has far higher enzymatic activity in conjunction with 5-HMF than the SULT of rodents examined to date (Glatt and Sommer 2006). Hence, humans could be more sensitive to the toxic effects of 5-HMF, and laboratory rodents would not then be the ideal species to elucidate this issue. To clarify this point, transgenic mice were examined which carried multiple copies of the human SULT1A1-SULT1A2 gene cluster in the middle region of chromosome 9 (FVB/N-hSULT1A1/2 mice). A 3–12 fold higher SULT activity could be achieved in different tissues, including colonic mucosa (Abraham et al. 2011b).

In this study, the FVB/N-hSULT1A1/2 and wild-type (FVB/N) mice were given 5-HMF in drinking water at doses of 0 (control), 134, and 536 mg per kg BW for 12 weeks. The experiment invovled 48 animals treated with 5-HMF and 24 untreated control animals. In the evaluation, only minor histological tissue changes were observed in the kidneys, which surprisingly did not differ between the mice with high SULT activity and the wild type mice. In addition, no ACFs or tumours were observed in either the control animals or the animals treated with high or low SULT activity (Abraham et al. 2011b).

12.6.4 Assessment of the Carcinogenicity of 5-HMF

The incidence of colorectal carcinomas varies between different countries by up to 25-fold. With growing industrialization and urbanization, the rate increases. This implies that lifestyle could play an important role in the onset of intestinal cancer. Results from so-called migrant studies support this (Abraham et al. 2011b). There are, however, a small number of chemicals that induce colon cancer in laboratory rodents, such as HAAs (PhIP). It was therefore surprising when Corpet et al. (1990) found that heated saccharose promoted azoxymethane-initiated preneoplasias in the colon of rat and mice. 5-HMF was identified as the possible active component of heated saccharose (Zhang et al. 1993). Furthermore, it was shown that 5-HMF itself – without prior treatment with azoxymethane – is capable of inducing an elevated number of preneoplastic ACFs (Zhang et al. 1993). This initiation implied the induction of gene mutations in the colon mucosa by 5-HMF. For instance, Femia et al. (2008) detected K-ras mutations up to 100% (14/14) in ACFs, which were induced in the rat colon by 1,2-dimethyl hydrazine.

5-HMF itself is not mutagenic. However, its metabolite SMF has mutagenic potential. The decisive question is whether this metabolite is formed *in vivo* and is responsible for the induction of preneoplastic ACFs in rats treated with 5-HMF. This question was examined in several experiments, although mice were used instead of rats. SMF was detected *in vivo* in the plasma of mice that had been administered 5-HMF intravenously. Transgenic mice were used, which express a high level of human SULT1A1 and 1A2 in many tissues; this genetic modification leads to a clear increase of 5-HMF sulfation to SMF, detected in subcellular preparations of colon mucosa and other tissue (Abraham et al. 2011b). After administration in drinking water, 5-HMF probably reaches the intestinal mucosa of the luminal side and may be directly activated in the target tissue. If absorption takes place, 5-HMF reaches the intestinal mucosa along systemic pathways. In a short-term experiment, however, 5-HMF did not induce any preneoplastic ACFs in either transgenic or wild-type mice.

These negative results for 5-HMF are surprising as the overall exposure in this short-term study was far higher than the exposure that showed positive results in the original rat study by Zhang et al. (1993). It is not clear whether major species differences between mice and rats could play a role in this discrepancy. There are no indications of this. It should, however, be pointed out that the spontaneous rates of ACFs in the Zhang et al. (1993) study varied between experiments and that the number of ACFs in the animals treated with 5-HMF was only slightly higher than the spontaneous rate. Moreover, the report does not clearly describe how the statistical analyses were conducted.

In the only long-term study within the NTP framework conducted in rats and mice, 5-HMF did not demonstrate any neoplastic or other non-neoplastic effects in the intestinal tract of either species. Nonetheless, an elevated incidence of hepatocellular adenomas was observed in female mice. The incidence was not, however, dose-dependent, and the B6C3F1 strain is known for its genetically driven high susceptibility to the formation of liver tumours. No carcinogenic evidence was seen in male mice or in rats of either gender.

In the study with neonate min/+mice, 5-HMF elevated the number of adenoma in the small intestine, whereas SMF increased the number of so-called 'flat ACFs' in the colon (Svendsen et al. 2009). Both effects were indicated as statistically significant in relation to the vehicle control. However, in terms of methodology, this is difficult to interpret, because an evaluation was only undertaken with regard to certain sections of the intestines. These positive effects may also have been caused by the high sensitivity in this specific sensitive model. The min/+mouse is heterozygous for a mutation in the tumour-suppressor gene Apc, which spontaneously leads to the development of numerous adenomas, mainly in the small intestine but also in the colon. Furthermore, the test substances were administered subcutaneously to the neonate animals, and this constitutes artificial intake.

In summary, these results lead to doubts about whether 5-HMF has any carcinogenic potential at all. Importantly, there is to date no reliable evidence that 5-HMF induces tumours in the colon or small intestine. The genotoxic relevance of the low levels of the metabolite SMF detected *in vivo* has still to be clarified. A verification of its detection *in vivo* seems to make sense. Based on the available results, the possible risks of carcinogenic effects are not currently identifiable or can only be estimated as extremely low.

12.7 Pro-Oxidants

A normal antioxidant prevents oxidation of frying medium or frying foods, as described by Márquez-Ruiz et al. (2014). A pro-oxidant, on the other hand, enhances the oxidation during frying. Thus, pro-oxidants are not suitable for use during frying. However, there are several natural antioxidants such as tocopherols and carotenoids present in either edible oils or foods which act as antioxidants at the beginning of the frying process but degrade to form oxidized species as the temperature increases or the duration of frying becomes prolonged. These oxidized antioxidants increase the potential for oxidation of the frying medium or food.

Chapman et al. (2009) showed that the peroxide values of soybean, corn, safflower, and olive oil containing 0, 650, 1300, and 2600 ppm oxidized α-tocopherol increased during storage at 55 °C (Figure 12.2). The addition of oxidized α-tocopherol increased the peroxide value and accelerated the oxidation process during storage. Thus, the oxidized α-tocopherol acted as a pro-oxidant in the oils (Kim et al. 2007a).

Jia et al. (2007a) revealed that un-oxidized soybean oil could increase the stability of β-carotene, whereas oxidized oils acted as pro-oxidants for β-carotene. Peroxide values were significantly increased with increasing exposure time in the presence of β-carotene, as compared with control samples. Recently, it was found that more hydroperoxides were formed in the presence of β-carotene in model TAGs than in olive oil containing β-carotene under similar conditions (Zeb 2012). β-Carotene was stable and protected during the thermal treatment of olive oil. This could be attributed to the presence of antioxidant compounds in olive oils that are not present in model TAGs of similar composition (Zeb and Murkovic 2010, 2011). In corn oil under similar conditions, β-carotene produced more hydroperoxides than in control samples, suggesting a pro-oxidant role of β-carotene (Zeb and Murkovic 2013). In all these cases, β-carotene did not affect the change in peroxide value during the initial hours of thermal treatment. However, with higher exposure times, the oxidation of food oils was enhanced in the presence of β-carotene. Detailed studies of the reactions by liquid chromatography-mass spectrometry (LC-MS) reveal the structures of these primary oxidation products of TAG formed during food oil thermal treatment. The interactions of individual hydroperoxides were also determined. It was found that β-carotene and its degradation products have different effects on the promotion of individual hydroperoxides. Horton et al. (1987) reviewed the mechanism of toxicity of lipid hydroperoxides. Recent results show that dietary lipid hydroperoxides produced during frying are toxic in animal models (Zeb and Mehmood 2012; Zeb and Rahman 2017).

12.8 Disposal of Fried Foods

Fried foods are usually served at soon as possible after preparation to avoid rancidity and off-flavour. However, they are sometimes stored for later use. If they are not kept at freezing temperature, they develop off-flavour and thus become unacceptable to consumers. Such foods are wasted and must be disposed of. There is no proper guideline for the disposal of fried foods, so procedures for general solid foods are followed.

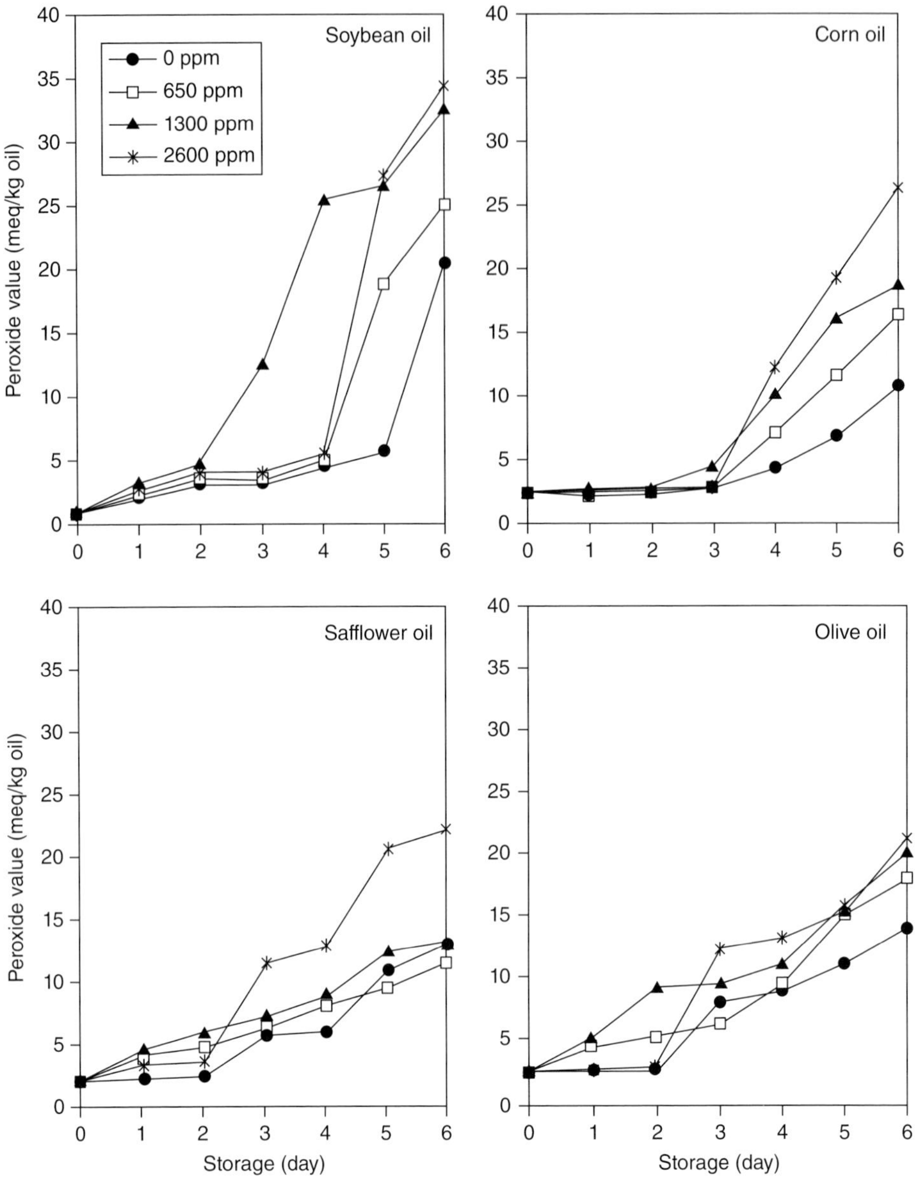

Figure 12.2 Effects of 0, 650, 1300, and 2600 ppm oxidized α-tocopherol on the peroxide value of soybean, corn, safflower, and olive oils stored at 55 °C for 6 days. *Source:* Reproduced with kind permission of John Wiley & Sons (Chapman et al. 2009).

The US Environmental Protection Agency (EPA) suggests reducing food waste at home by following some basic guidelines, which can be found online at https://www.epa.gov/recycle/reducing-wasted-food-home. These guidelines include how to plan for the purchase of foods that may be fried, tips for storage, instructions for the preparation of foods, and some thriftiness tips.

12.9 Disposal and Use of the Frying Medium

Frying oils or fats should be disposed of when they are no longer suitable for repeat use. Some important points should be kept in mind when disposing of frying oils:

- Frying oils that have reached their disposal point should not be reused.
- Frying oils should not be drained or poured down the toilet. They may block sanitary pipes and the city sewer mains and pollute water used for recycling.
- Frying oils should not be added to compost bins or piles. Waste fats may interact with the compost and contribute to pollution.

There are several possible and applied procedures for utilizing waste frying oils; most are based on collecting oils from industry or from restaurants. Specific technologies have been developed to collect and dispose of frying oils. These technologies are designed to be cost-effective and beneficial.

12.9.1 Regeneration of Oils

Edible oils can be regenerated through adsorption on solid adsorbents (Miyagi and Nakajima 2003). Bhattacharya et al. (2008) evaluated the efficacy of activated charcoal powder (ACH), aluminum oxide (ALO), aluminum hydroxide (ALH), activated clay (ACL), celite (CE) and silica gel (SG), activated carbon (ACA), frypowder (FP), britesorb (BR), and magnesol (MA) in regenerating oils. They found showed that treatment with a quaternary combination comprising 3.75% MA, 1.25% SG, 3.75% ACA, and 1.25% ALH was most effective in reducing the decomposition products of fried palm and cottonseed oils. Among the diatomaceous earths, zeolite and lime were most beneficial in reducing free fatty acids (FFAs), viscosity, and colour value in fried oils (Bulut and Yılmaz 2010).

Turan and Yalcuk (2013) conducted purification procedures in a packed column using different amounts of silica gel (SG), aluminium oxide (AO), activated charcoal (AC), bentonite (B), magnesol XL (M), calcium carbonate (CC), zeolite (Z), bleaching earth (BE), and/or their binary, triple, and quaternary combinations. They found that increasing the amount of adsorbent improved the chemical and physical characteristics of the oil. The most effective adsorbents for treating oil were SG and AC, as shown in Table 12.3.

Modified zeolites of Tween 80, β-cyclodextrin, and olive mill waste were also used as adsorbents to test their effectiveness in oil regeneration (Dülger and Yılmaz 2013). The best results were achieved with Tween 80-modified zeolite for TPM, free acidity, instrumental colour values, smoke point, and conjugated dienoic acid measurements. Similarly, the best result for turbidity was achieved with olive mill waste-modified zeolite, and that for viscosity was with hydrochloric acid-modified zeolite. Ali and El Anany (2014) used sugar cane bagasse ash for the regeneration of used frying oils. They found that a filtration system made from sugarcane bagasse ash had a higher regeneration potential than synthetic filter aid. Ogata and Kawasaki (2016) used virgin and calcined magnesium hydroxide for the regeneration of waste edible oils (soybean, rapeseed, and olive). They showed that calcined magnesium hydroxide was useful in regenerating these oils. Similarly, the efficiency of coffee silverskin (CS), CS ash (CSA),

Table 12.3 Effects of adsorbent amount on the physical and chemical characteristics of used oil.

Amount of adsorbent (g)	Amount of oil (g)	FFAs (%)		TPM (%)		CD (%)		AV (mmol/kg oil)		Viscosity (mPa s^{-1}, 30 °C)		Absorbance, 460 nm		Tocopherol (mg kg^{-1})
		Value	PI	Value	PI	Value	PI	Value	PI	Value	PI	Value	PI	Value
SG, 3.75	UO1, 30	0.73 ± 0.01^{a}	26	1.3 ± 0.2^{a}	85	0.47 ± 0.03^{a}	38	21.1 ± 0.9^{a}	49	56.1 ± 0.1^{a}	12	0.391 ± 0.001^{a}	52	382.9 ± 0.8^{b}
SG, 7.5	UO1, 30	0.62 ± 0.01^{b}	40	0.0 ± 0.0^{b}	100	0.39 ± 0.01^{b}	63	10.7 ± 0.4^{b}	86	55.5 ± 0.1^{b}	19	0.222 ± 0.001^{c}	77	389.6 ± 5.8^{b}
SG, 15	UO1, 30	0.27 ± 0.05^{c}	84	0.0 ± 0.0^{b}	100	0.32 ± 0.03^{c}	84	5.8 ± 0.3^{c}	104	54.6 ± 0.1^{c}	30	0.262 ± 0.001^{b}	71	391.9 ± 4.7^{a}
SG, 22.5	UO1, 30	0.16 ± 0.02^{d}	98	0.0 ± 0.0^{b}	100	0.27 ± 0.00^{c}	100	2.2 ± 0.5^{d}	117	52.3 ± 0.0^{d}	58	0.091 ± 0.000^{d}	97	386.6 ± 4.4^{b}
AO, 3.75	UO1, 30	0.59 ± 0.03^{a}	44	3.9 ± 0.2^{a}	55	0.59 ± 0.01^{a}	0	34.0 ± 0.3^{a}	4	54.4 ± 0.1^{b}	32	0.689 ± 0.001^{a}	8	408.5 ± 2.0^{a}
AO, 7.5	UO1, 30	0.18 ± 0.02^{b}	95	1.7 ± 0.3^{b}	80	0.49 ± 0.01^{b}	31	27.7 ± 0.9^{b}	26	55.0 ± 0.1^{a}	25	0.453 ± 0.001^{b}	43	395.3 ± 2.2^{b}
AO, 15 g	UO1, 30	0.12 ± 0.02^{b}	103	0.6 ± 0.0^{c}	93	0.46 ± 0.00^{c}	41	23.1 ± 0.5^{c}	42	54.8 ± 0.3^{a}	27	0.340 ± 0.001^{c}	60	400.8 ± 0.8^{b}
AC, 3.75	UO1, 30	0.60 ± 0.03^{a}	43	3.0 ± 0.2^{a}	66	0.57 ± 0.00^{a}	6	20.9 ± 0.1^{a}	50	54.2 ± 0.1^{a}	34	0.399 ± 0.001^{a}	51	324.7 ± 5.1^{a}
AC, 7.5	UO1, 30	0.20 ± 0.04^{b}	93	0.0 ± 0.0^{b}	100	0.54 ± 0.02^{b}	16	6.0 ± 0.3^{b}	103	54.3 ± 0.1^{a}	33	0.095 ± 0.000^{b}	96	236.6 ± 6.4^{b}
AC, 15	UO1, 30	0.16 ± 0.02^{b}	98	0.0 ± 0.0^{b}	100	0.36 ± 0.01^{c}	72	3.2 ± 0.9^{c}	113	54.3 ± 0.0^{a}	34	0.069 ± 0.000^{c}	100	8.6 ± 0.6^{c}
B, 15	UO1, 30	0.75 ± 0.03^{b}	24	1.6 ± 0.2^{d}	82	$0.52 \pm 0.01^{b,c}$	22	24.1 ± 0.2^{b}	39	53.2 ± 0.0^{c}	47	0.363 ± 0.001^{d}	56	331.3 ± 4.8^{c}
Z, 15	UO1, 30	$0.81 \pm 0.05^{a,b}$	16	7.3 ± 0.2^{a}	16	0.54 ± 0.01^{b}	16	34.6 ± 1.3^{a}	1	57.1 ± 0.1^{a}	−1	0.725 ± 0.001^{a}	2	402.9 ± 0.2^{b}
BE 15	UO1, 30	$0.79 \pm 0.05^{a,b}$	19	2.7 ± 0.4^{c}	100	$0.50 \pm 0.01^{b,c}$	28	20.8 ± 6.2^{b}	51	55.3 ± 0.1^{b}	21	0.452 ± 0.001^{c}	43	313.5 ± 2.5^{d}
M, 15	UO1, 30	0.10 ± 0.02^{c}	105	0.0 ± 0.0^{e}	100	0.46 ± 0.03^{c}	41	8.6 ± 0.1^{c}	94	53.2 ± 0.1^{c}	47	0.141 ± 0.001^{e}	90	409.4 ± 0.4^{b}
CC, 15	UO1, 30	0.86 ± 0.00^{a}	10	6.6 ± 0.5^{b}	24	0.6 ± 0.03^{a}	−3	35.0 ± 0.3^{a}	0	57.2 ± 0.1^{a}	−2	0.691 ± 0.001^{b}	7	425.7 ± 5.4^{a}

Source: Reproduced with kind permission of Springer Nature (Turan and Yalcuk 2013).
Results are given as means with standard deviation.
UO1, used sunflower oil; SG, silica gel; AO, aluminium oxide; AC, activated charcoal; B, bentonite; Z, zeolite; BE, bleaching earth; M, Magnesol XL; CC, calcium carbonate; FFAs, free fatty acids; TPM, change in total polar material; CD, conjugated diene; AV P-anisidine value; PI, percentage improvement.
Means followed by different superscripts in the same column indicate different superscripts ($p < 0.05$).

and nanoparticles of CS (NCS) in regenerating used frying palm oil (UFPO) were studied by Ismail et al. (2017). These authors showed that filtration treatment of UFPO with 4% CS, CSA, and NCS markedly lowered its viscosity and colour values. Nayak et al. (2018) reported that elephant apple powder can be used to improve the quality of fried soybean oil. These studies show that frying oils can be recovered through the adsorption process, but that there are limitations in terms of large numbers of oils, highly oxidized oils, and effectiveness in hot water.

12.9.2 Biodiesel Production

Biodiesel is a type of biofuel used to reduce toxic gases produced during combustion. Used frying oils have been found to possess significant potential for the production of biodiesel (Ouanji et al. 2016). Different methanol and oil ratios are used to determine the best possible fuel for diesel engines. For this purpose, *trans*-esterification is carried out with the help of a catalyst. Encinar et al. (2005) performed *trans*-esterification reactions of used frying oils by means of methanol, sodium, or potassium hydroxides and their methoxides. The biodiesel with the best properties was obtained using a methanol/oil molar ratio of 6 : 1, potassium hydroxide as catalyst (1%), and 65 °C temperature. Similarly, Felizardo et al. (2006) reported that for oils with an acid value of 0.42 mg KOH g^{-1}, a methanol/fried oil ratio of 4.8 and a catalyst/fried oil ratio of 0.6% gave the highest yield of methyl esters. The process of formation of biodiesel is usually catalysed by several substances. Singh et al. (2016) used β-potassium dizirconate as a novel catalyst for the production of biodiesel from used frying oils. They achieved 96.85% direct conversion of biodiesel from waste frying oil with the help of this catalyst.

12.9.3 Other Uses

Frying oils can be used to produce oil water emulsions and paints, electricity, soap, and surfactants. The latter is a very common process utilization in the developing world. The production of biosurfactants on waste frying oils has been reported by Raza et al. (2006), who showed that waste frying oils are good substrates for rhamnolipid production using *Pseudomonas aeruginosa* mutant EBN-8 in the presence or absence of precursor lipid under single-/batch-fed conditions. Similarly, Singh-Ackbarali et al. (2017) evaluated the potential of used frying oils as performance-enhancing additives for road paving applications, suggesting environmental benefits.

12.10 Key Concepts

- During frying, several newly produced compounds are nutritionally significant, but most possess toxic properties.
- Oxidized TAGs such as hydroperoxides have been reported to cause hyperlipidaemia, fatty liver increases in serum lipids, and the growth of hepatocarcinoma in experimental animals.
- Acrylamide is a toxic compound produced during frying that is shown to be associated with nervous system disorders and to possess genotoxic properties.
- Acrolein is detoxified by the liver via glutathione, but higher amounts can cause acute to chronic cytotoxic and genotoxic effects and cancer.

- HAAs produced during frying are strong carcinogenic substances both *in vitro* and *in vivo*.
- Furfurals cause acute to chronic toxicity and cancers in animal models.
- Pro-oxidants have been found to increase the deterioration of fried foods and frying oils, thus contributing to their toxic properties.
- Frying oils should be disposed of with care in order to avoid toxicity and environmental pollution.

References

Abraham, K., Andres, S., Palavinskas, R. et al. (2011a). Toxicology and risk assessment of acrolein in food. *Molecular Nutrition & Food Research* 55: 1277–1290.

Abraham, K., Gurtler, R., Berg, K. et al. (2011b). Toxicology and risk assessment of 5-Hydroxymethylfurfural in food. *Molecular Nutrition & Food Research* 55: 667–678.

Adamson, R.H., Thorgeirsson, U.P., and Sugimura, T. (1996). *Extrapolation of Heterocyclic Amine Carcinogenesis Data from Rodents and Nonhuman Primates to Humans*. Berlin, Heidelberg: Springer.

Ali, R.F. and El Anany, A.M. (2014). Recovery of used frying sunflower oil with sugar cane industry waste and hot water. *Journal of Food Science and Technology* 51: 3002–3013.

Archer, M.C., Bruce, W.R., Chan, C.C. et al. (1992). Aberrant crypt foci and microadenoma as markers for colon cancer. *Environmental Health Perspectives* 98: 195.

Auerbach, S.S., Mahler, J., Travlos, G.S., and Irwin, R.D. (2008). A comparative 90-day toxicity study of allyl acetate, allyl alcohol and acrolein. *Toxicology* 253: 79–88.

Augustsson, K., Skog, K., Jägerstad, M., and Steineck, G. (1997). Assessment of the human exposure to heterocyclic amines. *Carcinogenesis* 18: 1931–1935.

Augustsson, K., Lindblad, J., Overvik, E., and Steineck, G. (1999). A population-based dietary inventory of cooked meat and assessment of the daily intake of food mutagens. *Food Additives and Contaminants* 16: 215–225.

Bakhiya, N., Monien, B., Frank, H. et al. (2009). Renal organic anion transporters OAT1 and OAT3 mediate the cellular accumulation of 5-sulfooxymethylfurfural, a reactive, nephrotoxic metabolite of the Maillard product 5-hydroxymethylfurfural. *Biochemical Pharmacology* 78: 414–419.

Baltazar, A., Joaquim, J.J., and Matos, C. (2017). Potential risks of acrylamide for humans – review. *Clinical Therapeutics* 39: e93.

Barbir, A., Linseisen, J., Hermann, S. et al. (2012). Effects of phenotypes in heterocyclic aromatic amine (HCA) metabolism-related genes on the association of HCA intake with the risk of colorectal adenomas. *Cancer Causes & Control* 23: 1429–1442.

Baum, M., Fauth, E., Fritzen, S. et al. (2005). Acrylamide and glycidamide: approach towards risk assessment based on biomarker guided dosimetry of genotoxic/mutagenic effects in human blood. *Advances in Experimental Medicine and Biology* 561: 77–88.

Beland, F.A., Mellick, P.W., Olson, G.R. et al. (2013). Carcinogenicity of acrylamide in B6C3F(1) mice and F344/N rats from a 2-year drinking water exposure. *Food and Chemical Toxicology* 51: 149–159.

Bhattacharya, A.B., Sajilata, M.G., Tiwari, S.R., and Singhal, R.S. (2008). Regeneration of thermally polymerized frying oils with adsorbents. *Food Chemistry* 110: 562–570.

Boettcher, M.I., Bolt, H.M., Drexler, H., and Angerer, J. (2006). Excretion of mercapturic acids of acrylamide and glycidamide in human urine after single oral administration of deuterium-labelled acrylamide. *Archives of Toxicology* 80: 55–61.

Bouvard, V., Loomis, D., Guyton, K.Z. et al. (2015). Carcinogenicity of consumption of red and processed meat. *The Lancet Oncology* 16: 1599–1600.

Bulut, E. and Yilmaz, E. (2010). Comparison of the frying stability of sunflower and refined olive pomace oils with/without adsorbent treatment. *Journal of the American Oil Chemists' Society* 87: 1145–1153.

Burek, J.D., Albee, R.R., Beyer, J.E. et al. (1980). Subchronic toxicity of acrylamide administered to rats in the drinking water followed by up to 144 days of recovery. *Journal of Environmental Pathology and Toxicology* 4: 157–182.

Busquets, R., Bordas, M., Toribio, F. et al. (2004). Occurrence of heterocyclic amines in several home-cooked meat dishes of the Spanish diet. *Journal of Chromatography B: Analytical Technologies in the Biomedical and Life Sciences* 802: 79–86.

Butterfield, D.A., Bader Lange, M.L., and Sultana, R. (2010). Involvements of the lipid peroxidation product, HNE, in the pathogenesis and progression of Alzheimer's disease. *Biochimica et Biophysica Acta* 1801: 924–929.

Byrne, C., Sinha, R., Platz, E.A. et al. (1998). Predictors of dietary heterocyclic amine intake in three prospective cohorts. *Cancer Epidemiology Biomarkers Previews* 7: 523–529.

Calleman, C.J., Wu, Y., He, F. et al. (1994). Relationships between biomarkers of exposure and neurological effects in a group of workers exposed to acrylamide. *Toxicology and Applied Pharmacology* 126: 361–371.

Chapman, T.M., Kim, H.J., and Min, D.B. (2009). Prooxidant activity of oxidized α-tocopherol in vegetable oils. *Journal of Food Science* 74: C536–C542.

Chen, M.J., Hsu, H.T., Lin, C.L., and Ju, W.Y. (2012). A statistical regression model for the estimation of acrylamide concentrations in French fries for excess lifetime cancer risk assessment. *Food and Chemical Toxicology* 50: 3867–3876.

Cohen, S.M., Garland, E.M., St John, M. et al. (1992). Acrolein initiates rat urinary bladder carcinogenesis. *Cancer Research* 52: 3577–3581.

Collins, J.J., Swaen, G.M., Marsh, G.M. et al. (1989). Mortality patterns among workers exposed to acrylamide. *Journal of Occupational Medicine* 31: 614–617.

Cooper, K.O., Witz, G., and Witmer, C. (1992). The effects of α, β-unsaturated aldehydes on hepatic thiols and thiol-containing enzymes. *Fundamental and Applied Toxicology* 19: 343–349.

Corpet, D.E., Stamp, D., Medline, A. et al. (1990). Promotion of colonic microadenoma growth in mice and rats fed cooked sugar or cooked casein and fat. *Cancer Research* 50: 6955–6958.

Curren, R.D., Yang, L.L., Conklin, P.M. et al. (1988). Mutagenesis of xeroderma pigmentosum fibroblasts by acrolein. *Mutation Research* 209: 17–22.

Custot, F. (1959). Toxicity of heated fats; the problem of frying oils. *Annales de la Nutrition et de l'Alimentation* 13: A417–A448.

Daniel, C.R., Schwartz, K.L., Colt, J.S. et al. (2011). Meat-cooking mutagens and risk of renal cell carcinoma. *British Journal of Cancer* 105 (7): 1096–1104.

Delfino, R.J., Sinha, R., Smith, C. et al. (2000). Breast cancer, heterocyclic aromatic amines from meat and N-acetyltransferase 2 genotype. *Carcinogenesis* 21: 607–615.

de los Santos, C., Zaliznyak, T., and Johnson, F. (2001). NMR characterization of a DNA duplex containing the major acrolein-derived deoxyguanosine adduct gamma-OH-1, -N2-propano-2′-deoxyguanosine. *Journal of Biological Chemistry* 276: 9077–9082.

Dorman, D.C., Struve, M.F., Wong, B.A. et al. (2008). Respiratory tract responses in male rats following subchronic acrolein inhalation. *Inhalation Toxicology* 20: 205–216.

Draminski, W., Eder, E., and Henschler, D. (1983). A new pathway of acrolein metabolism in rats. *Archives of Toxicology* 52: 243–247.

Duale, N., Bjellaas, T., Alexander, J. et al. (2009). Biomarkers of human exposure to acrylamide and relation to polymorphisms in metabolizing genes. *Toxicological Sciences* 108: 90–99.

Dülger, A. and Yilmaz, E. (2013). Effectiveness of modified zeolites as adsorbent materials for frying oils. *European Journal of Lipid Science and Technology* 115: 668–675.

Durling, L.J., Busk, L., and Hellman, B.E. (2009). Evaluation of the DNA damaging effect of the heat-induced food toxicant 5-hydroxymethylfurfural (HMF) in various cell lines with different activities of sulfotransferases. *Food and Chemical Toxicology* 47: 880–884.

Dypbukt, J.M., Atzori, L., Edman, C.C., and Grafström, R.C. (1993). Thiol status and cytopathological effects of acrolein in normal and xeroderma pigmentosum skin fibroblasts. *Carcinogenesis* 14: 975–980.

Encinar, J.M., González, J.F., and Rodríguez-Reinares, A. (2005). Biodiesel from used frying oil. Variables affecting the yields and characteristics of the biodiesel. *Industrial and Engineering Chemistry Research* 44: 5491–5499.

Ericson, U., Wirfält, E., Mattisson, I. et al. (2007). Dietary intake of heterocyclic amines in relation to socio-economic, lifestyle and other dietary factors: estimates in a Swedish population. *Public Health Nutrition* 10: 616–627.

Felizardo, P., Correia, M.J., Raposo, I. et al. (2006). Production of biodiesel from waste frying oils. *Waste Management* 26: 487–494.

Felton, J.S., Fultz, E., Dolbeare, F.A., and Knize, M.G. (1994). Effect of microwave pretreatment on heterocyclic aromatic amine mutagens/carcinogens in fried beef patties. *Food and Chemical Toxicology* 32: 897–903.

Femia, A.P., Tarquini, E., Salvadori, M. et al. (2008). K-ras mutations and mucin profile in preneoplastic lesions and colon tumors induced in rats by 1, 2-dimethylhydrazine. *International Journal of Cancer* 122: 117–123.

Feng, Z., Hu, W., Hu, Y., and Tang, M.S. (2006). Acrolein is a major cigarette-related lung cancer agent: preferential binding at p53 mutational hotspots and inhibition of DNA repair. *Proceedings of the National Academy of Sciences of the United States of America* 103: 15404–15409.

Ferguson, L.R. (2010). Meat and cancer. *Meat Science* 84: 308–313.

Ferrucci, L.M., Cross, A.J., Graubard, B.I. et al. (2009). Intake of meat, meat mutagens, and iron and the risk of breast cancer in the Prostate, Lung, Colorectal, and Ovarian Cancer Screening Trial. *British Journal of Cancer* 101: 178–184.

Foiles, P.G., Akerkar, S.A., Miglietta, L.M., and Chung, F.L. (1990). Formation of cyclic deoxyguanosine adducts in Chinese hamster ovary cells by acrolein and crotonaldehyde. *Carcinogenesis* 11: 2059–2061.

Fu, Z., Shrubsole, M.J., Li, G. et al. (2012). Using gene-environment interaction analyses to clarify the role of well-done meat and heterocyclic amine exposure in the etiology of colorectal polyps. *American Journal of Clinical Nutrition* 96: 1119–1128.

Gibis, M. (2016). Heterocyclic aromatic amines in cooked meat products: causes, formation, occurrence, and risk assessment. *Comprehensive Reviews in Food Science and Food Safety* 15: 269–302.

Gibis, M. and Weiss, J. (2010). Inhibitory effect of marinades with hibiscus extract on formation of heterocyclic aromatic amines and sensory quality of fried beef patties. *Meat Science* 85: 735–742.

Gilsing, A.M., Berndt, S.I., Ruder, E.H. et al. (2012). Meat-related mutagen exposure, xenobiotic metabolizing gene polymorphisms and the risk of advanced colorectal adenoma and cancer. *Carcinogenesis* 33: 1332–1339.

Glatt, H. and Sommer, Y. (2006). Health risks of 5-hydroxymethylfurfural (HMF) and related compounds. In: *Acrylamide and Other Hazardous Compounds in Heat-Treated Foods* (ed. K. Skog and J. Alexander). Cambridge: Woodhead.

Gomes, R., Liteplo, R., and Meek, M. (2001). Acrolein: hazard characterization and exposure – response analysis. *Journal of Environmental Science and Health. Part C, Environmental Carcinogenesis & Ecotoxicology Reviews* 19: 23–43.

Grafström, R.C. (1990). In vitro studies of aldehyde effects related to human respiratory carcinogenesis. *Mutation Research* 238: 175–184.

Grafström, R.C., Dypbukt, J.M., Willey, J.C. et al. (1988). Pathobiological effects of acrolein in cultured human bronchial epithelial cells. *Cancer Research* 48: 1717–1721.

Hada, N., Totsuka, Y., Enya, T. et al. (2001). Structures of mutagens produced by the co-mutagen norharman with o-and m-toluidine isomers. *Mutation Research* 493: 115–126.

Hagmar, L., Törnqvist, M., Nordander, C. et al. (2001). Health effects of occupational exposure to acrylamide using hemoglobin adducts as biomarkers of internal dose. *Scandinavian Journal of Work, Environment & Health* 27: 219–226.

Hashmi, M., Vamvakas, S., and Anders, M. (1992). Bioactivation mechanism of S-(3-oxopropyl)-N-acetyl-L-cysteine, the mercapturic acid of acrolein. *Chemical Research in Toxicology* 5: 360–365.

Hecht, S.S., Seow, A., Wang, M. et al. (2010). Elevated levels of volatile organic carcinogen and toxicant biomarkers in Chinese women who regularly cook at home. *Cancer Epidemiology, Biomarkers & Prevention* 19: 1185–1192.

Helmus, D.S., Thompson, C.L., Zelenskiy, S. et al. (2013). Red meat-derived heterocyclic amines increase risk of colon cancer: a population-based case-control study. *Nutrition and Cancer* 65: 1141–1150.

Hogervorst, J.G., Fortner, R.T., Mucci, L.A. et al. (2013). Associations between dietary acrylamide intake and plasma sex hormone levels. *Cancer Epidemiology, Biomarkers & Prevention* 22: 2024–2036.

Hogervorst, J.G., de Bruijn Geraets, D., Schouten, L.J. et al. (2014). Dietary acrylamide intake and the risk of colorectal cancer with specific mutations in KRAS and APC. *Carcinogenesis* 35: 1032–1038.

Horton, A., Fairhurst, S., and Bus, J.S. (1987). Lipid peroxidation and mechanisms of toxicity. *CRC Critical Reviews in Toxicology* 18: 27–79.

Horton, N.D., Mamiya, B.M., and Kehrer, J.P. (1997). Relationships between cell density, glutathione and proliferation of A549 human lung adenocarcinoma cells treated with acrolein. *Toxicology* 122: 111–122.

Ismail, S.A., Mohammed, E.A., and Ali, R.F.M. (2017). Regeneration of used frying palm oil with coffee silverskin (CS), CS ash (CSA) and nanoparticles of CS (NCS). *Journal of Oleo Science* 66: 897–905.

Jahurul, M.H.A., Jinap, S., Ang, S.J. et al. (2010). Dietary exposure to heterocyclic amines in high-temperature cooked meat and fish in Malaysia. *Food Additives & Contaminants. Part A, Chemistry, Analysis, Control, Exposure & Risk Assessment* 27: 1060–1071.

Janzowski, C., Glaab, V., Samimi, E. et al. (2000). 5-Hydroxymethylfurfural: assessment of mutagenicity, DNA-damaging potential and reactivity towards cellular glutathione. *Food and Chemical Toxicology* 38: 801–809.

Jia, M., Kim, H.J., and Min, D.B. (2007a). Effects of soybean oil and oxidized soybean oil on the stability of β-carotene. *Food Chemistry* 103: 695–700.

Jia, L., Liu, Z., Sun, L. et al. (2007b). Acrolein, a toxicant in cigarette smoke, causes oxidative damage and mitochondrial dysfunction in RPE cells: protection by (R)-α-lipoic acid. *Investigative Ophthalmology & Visual Science* 48: 339–348.

John, E.M., Stern, M.C., Sinha, R., and Koo, J. (2011). Meat consumption, cooking practices, meat mutagens and risk of prostate cancer. *Nutrition and Cancer* 63: 525–537.

Joshi, A.D., Kim, A., Lewinger, J.P. et al. (2015). Meat intake, cooking methods, dietary carcinogens, and colorectal cancer risk: findings from the colorectal cancer family registry. *Cancer Medicine* 4: 936–952.

Kabat, G.C., Cross, A.J., Park, Y. et al. (2009). Meat intake and meat preparation in relation to risk of postmenopausal breast cancer in the NIH-AARP Diet and Health Study. *International Journal of Cancer* 124: 2430–2435.

Kawai, Y., Furuhata, A., Toyokuni, S. et al. (2003). Formation of acrolein-derived 2′-deoxyadenosine adduct in an iron-induced carcinogenesis model. *Journal of Biological Chemistry* 278: 50346–50354.

Keating, G. and Bogen, K. (2001). Methods for estimating heterocyclic amine concentrations in cooked meats in the US diet. *Food and Chemical Toxicology* 39: 29–43.

Keating, G. and Bogen, K. (2004). Estimates of heterocyclic amine intake in the US population. *Journal of Chromatography B: Analytical Technologies in the Biomedical and Life Sciences* 802: 127–133.

Kehrer, J.P. and Biswal, S.S. (2000). The molecular effects of Acrolein. *Toxicological Sciences* 57: 6–15.

Khan, A.A., Zeb, A., and Sherazi, S.T. (2017). Thermally oxidized olive oil produces biochemical, physiological effects and fatty liver in rats. *Chiang Mai Journal of Science* 44: 847–857.

Kim, H.J., Lee, H.O., and Min, D.B. (2007a). Effects and prooxidant mechanisms of oxidized α-tocopherol on the oxidative stability of soybean oil. *Journal of Food Science* 72: C223–C230.

Kim, S.I., Pfeifer, G.P., and Besaratinia, A. (2007b). Lack of mutagenicity of acrolein-induced DNA adducts in mouse and human cells. *Cancer Research* 67: 11640–11647.

Klapec, T. & Periš, D. 2014. Quick estimation of dietary exposure to heterocyclic aromatic amines and acrylamide in a Croatian female population. University Josip Juraj Strossmayer of Osijeck, Croatia.

Knasmüller, S., Schwab, C.E., Land, S.J. et al. (1999). Genotoxic effects of heterocyclic aromatic amines in human derived hepatoma (HepG2) cells. *Mutagenesis* 14: 533–540.

Kobayashi, M., Tsubono, Y., Sasazuki, S. et al. (2002). Vegetables, fruit and risk of gastric cancer in Japan: a 10-year follow-up of the JPHC Study Cohort I. *International Journal of Cancer* 102: 39–44.

Kobayashi, M., Otani, T., Iwasaki, M. et al. (2009). Association between dietary heterocyclic amine levels, genetic polymorphisms of NAT2, CYP1A1, and CYP1A2 and risk of stomach cancer: a hospital-based case-control study in Japan. *Gastric Cancer* 12: 198–205.

Kubow, S. (1990). Toxicity of dietary lipid peroxidation products. *Trends in Food Science and Technology* 1: 67–71.

Lam, T.K., Cross, A.J., Consonni, D. et al. (2009). Intakes of red meat, processed meat, and meat-mutagens increase lung cancer risk. *Cancer Research* 69: 932–939.

Lang, K., Kieckebusch, W., Bassler, K.H. et al. (1970). Tolerance of 5-hydroxymethylfurfural (HMF). 1st communication: chronic administration of HMF and metabolites of the substance. *Zeitschrift für Ernährungswissenschaft* 10: 97–101.

Layton, D.W., Bogen, K.T., Knize, M.G. et al. (1995). Cancer risk of heterocyclic amines in cooked foods: an analysis and implications for research. *Carcinogenesis* 16: 39–52.

Lee, Y.C., Shlyankevich, M., Jeong, H.K. et al. (1995). Bioactivation of 5-Hydroxymethyl-2-Furaldehyde to an electrophilic and mutagenic allylic sulfuric acid ester. *Biochemical and Biophysical Research Communications* 209: 996–1002.

Le Marchand, L.C., Hankin, J.H., Pierce, L.M. et al. (2002). Well-done red meat, metabolic phenotypes and colorectal cancer in Hawaii. *Mutation Research, Fundamental and Molecular Mechanisms of Mutagenesis* 506–507: 205–214.

Li, L., Jiang, L., Geng, C. et al. (2008). The role of oxidative stress in acrolein-induced DNA damage in HepG2 cells. *Free Radical Research* 42: 354–361.

Lijinsky, W. and Reuber, M.D. (1987). Chronic carcinogenesis studies of acrolein and related compounds. *Toxicology and Industrial Health* 3: 337–345.

Lin, J., Forman, M.R., Wang, J. et al. (2012). Intake of red meat and heterocyclic amines, metabolic pathway genes and bladder cancer risk. *International Journal of Cancer* 131: 1892–1903.

Liu, X., Lovell, M.A., and Lynn, B.C. (2005). Development of a method for quantification of acrolein-deoxyguanosine adducts in DNA using isotope dilution-capillary LC/MS/MS and its application to human brain tissue. *Analytical Chemistry* 77: 5982–5989.

Liu, X.Y., Zhu, M.X., and Xie, J.P. (2010). Mutagenicity of acrolein and acrolein-induced DNA adducts. *Toxicology Mechanisms and Methods* 20: 36–44.

Lopachin, R.M., Gavin, T., Petersen, D.R., and Barber, D.S. (2009). Molecular mechanisms of 4-Hydroxy-2-nonenal and Acrolein toxicity: nucleophilic targets and adduct formation. *Chemical Research in Toxicology* 22: 1499–1508.

Marnett, L.J. (1991). Peroxyl free radicals: biological reactive intermediates produced during lipid oxidation. *Advances in Experimental Medicine and Biology* 283: 65–70.

Marnett, L.J. (1999). Lipid peroxidation – DNA damage by malondialdehyde. *Mutation Research, Fundamental and Molecular Mechanisms of Mutagenesis* 424: 83–95.

Márquez-Ruiz, G., Ruiz-Méndez, M.V., and Velasco, J. (2014). Antioxidants in frying: analysis and evaluation of efficacy. *European Journal of Lipid Science and Technology* 116: 1441–1450.

Matthäus, B. and Haase, N.U. (2014). Acrylamide – still a matter of concern for fried potato food?*. *European Journal of Lipid Science and Technology* 116: 675–687.

Mignone, L.I., Giovannucci, E., Newcomb, P.A. et al. (2009). Meat consumption, heterocyclic amines, NAT2 and the risk of breast cancer. *Nutrition and Cancer* 61: 36–46.

Minko, I.G., Kozekov, I.D., Kozekova, A. et al. (2008). Mutagenic potential of DNA–peptide crosslinks mediated by acrolein-derived DNA adducts. *Mutation Research, Fundamental and Molecular Mechanisms of Mutagenesis* 637: 161–172.

Miyagi, A. and Nakajima, M. (2003). Regeneration of used frying oils using adsorption processing. *Journal of the American Oil Chemists' Society* 80: 91.

Nagao, M., Yahagi, T., Honda, M. et al. (1977). Comutagenic actions of norharman derivatives with 4-dimethylaminoazobenzene and related compounds. *Cancer Letters* 3: 339–346.

Nath, R.G. and Chung, F.L. (1994). Detection of exocyclic 1,N2-propanodeoxyguanosine adducts as common DNA lesions in rodents and humans. *Proceedings of the National Academy of Sciences* 91: 7491–7495.

Nath, R.G., Ocando, J.E., and Chung, F.-L. (1996). Detection of 1,N-2-propanodeoxyguanosine adducts as potential endogenous DNA lesions in rodent and human tissues. *Cancer Research* 56: 452–456.

Nayak, P.K., Dash, U., Rayaguru, K., and Krishnan, K.R. (2018). Influence of elephant apple powder as an adsorbent in the regeneration of fried soybean oil: process optimization studies. *Journal of Food Process Engineering* 41: e12637.

Nolen, G.A., Alexander, J.C., and Artman, N.R. (1967). Long-term rat feeding study with used frying fats. *Journal of Nutrition* 93: 337–348.

Norrish, A.E., Ferguson, L.R., Knize, M.G. et al. (1999). Heterocyclic amine content of cooked meat and risk of prostate cancer. *Journal of the National Cancer Institute* 91: 2038–2044.

Nowell, S., Coles, B., Sinha, R. et al. (2002). Analysis of total meat intake and exposure to individual heterocyclic amines in a case-control study of colorectal cancer: contribution of metabolic variation to risk. *Mutation Research, Fundamental and Molecular Mechanisms of Mutagenesis* 506: 175–185.

NTP. 2010. NTP toxicology and carcinogenesis studies of 5-(Hydroxymethyl)-2-furfural (CAS No. 67–47-0) in F344/N rats and B6C3F1 mice (gavage studies). National Toxicology Program Technical Report Series 7.

Obon-Santacana, M., Slimani, N., Lujan-Barroso, L. et al. (2013). Dietary intake of acrylamide and pancreatic cancer risk in the European prospective investigation into cancer and nutrition (EPIC) cohort. *Annals of Oncology* 24: 2645–2651.

Ogata, F. and Kawasaki, N. (2016). Regeneration of waste edible oil by the use of virgin and calcined magnesium hydroxide as adsorbents. *Journal of Oleo Science* 65: 941–948.

Ohgaki, H., Hasegawa, H., Suenaga, M. et al. (1987). Carcinogenicity in mice of a mutagenic compound, 2-amino-3,8-dimethylimidazo[4,5-f]quinoxaline (MeIQx) from cooked foods. *Carcinogenesis* 8: 665–668.

Ott, M.G., Teta, M.J., and Greenberg, H.L. (1989). Lymphatic and hematopoietic tissue cancer in a chemical manufacturing environment. *American Journal of Industrial Medicine* 16: 631–643.

Ouanji, F., Khachani, M., Boualag, M. et al. (2016). Large-scale biodiesel production from Moroccan used frying oil. *International Journal of Hydrogen Energy* 41: 21022–21029.

Parent, R.A., Caravello, H.E., and Harbell, J.W. (1991a). Gene mutation assay of acrolein in the CHO/HGPRT test system. *Journal of Applied Toxicology* 11: 91–95.

Parent, R.A., Caravello, H.E., and Long, J.E. (1991b). Oncogenicity study of acrolein in mice. *Journal of the American College of Toxicology* 10: 647–659.

Parent, R.A., Caravello, H.E., Balmer, M.F. et al. (1992a). One-year toxicity of orally administered acrolein to the beagle dog. *Journal of Applied Toxicology* 12: 311–316.

Parent, R.A., Caravello, H.E., and Hoberman, A.M. (1992b). Reproductive study of acrolein on two generations of rats. *Fundamental and Applied Toxicology* 19: 228–237.

Parent, R.A., Caravello, H.E., and Long, J.E. (1992c). Two-year toxicity and carcinogenicity study of acrolein in rats. *Journal of Applied Toxicology* 12: 131–139.

Parent, R.A., Caravello, H.E., Christian, M.S., and Hoberman, A.M. (1993). Developmental toxicity of acrolein in New Zealand white rabbits. *Fundamental and Applied Toxicology* 20: 248–256.

Parent, R.A., Caravello, H.A., and Sharp, D.E. (1996). Metabolism and distribution of [2,3-14C]acrolein in Sprague-Dawley rats. *Journal of Applied Toxicology* 16: 449–457.

Parent, R.A., Paust, D.E., Schrimpf, M.K. et al. (1998). Metabolism and distribution of [2,3-14C]acrolein in Sprague-Dawley rats: II. Identification of urinary and fecal metabolites. *Toxicological Sciences* 43: 110–120.

Park, Y.S., Misonou, Y., Fujiwara, N. et al. (2005). Induction of thioredoxin reductase as an adaptive response to acrolein in human umbilical vein endothelial cells. *Biochemical and Biophysical Research Communications* 327: 1058–1065.

Patel, J.M., Wood, J.C., and Leibman, K.C. (1980). The biotransformation of allyl alcohol and acrolein in rat liver and lung preparations. *Drug Metabolism and Disposition* 8: 305–308.

Pawlowicz, A.J., Munter, T., Zhao, Y., and Kronberg, L. (2006). Formation of acrolein adducts with 2′-deoxyadenosine in calf thymus DNA. *Chemical Research in Toxicology* 19: 571–576.

Penn, A., Nath, R., Pan, J. et al. (2001). 1,N(2)-propanodeoxyguanosine adduct formation in aortic DNA following inhalation of acrolein. *Environmental Health Perspectives* 109: 219–224.

Rahman, U., Sahar, A., Khan, M.I., and Nadeem, M. (2014). Production of heterocyclic aromatic amines in meat: chemistry, health risks and inhibition. A review. *LWT – Food Science and Technology* 59: 229–233.

Raza, Z.A., Khan, M.S., Khalid, Z.M., and Rehman, A. (2006). Production kinetics and tensioactive characteristics of biosurfactant from a Pseudomonas aeruginosa mutant grown on waste frying oils. *Biotechnology Letters* 28: 1623–1631.

Rice, J.M. (2005). The carcinogenicity of acrylamide. *Mutation Research* 580: 3–20.

Rohrmann, S. and Becker, N. (2002). Development of a short questionnaire to assess the dietary intake of heterocyclic aromatic amines. *Public Health Nutrition* 5: 699–705.

Rohrmann, S., Hermann, S., and Linseisen, J. (2009). Heterocyclic aromatic amine intake increases colorectal adenoma risk: findings from a prospective European cohort study. *American Journal of Clinical Nutrition* 89: 1418–1424.

Rohr-Udilova, N.V., Stolze, K., Sagmeister, S. et al. (2008). Lipid hydroperoxides from processed dietary oils enhance growth of hepatocarcinoma cells. *Molecular Nutrition & Food Research* 52: 352–359.

Roy, J., Pallepati, P., Bettaieb, A., and Averill-Bates, D.A. (2010). Acrolein induces apoptosis through the death receptor pathway in A549 lung cells: role of p53This review is one of a selection of papers published in a special issue on oxidative stress in health and disease. *Canadian Journal of Physiology and Pharmacology* 88: 353–368.

Sakamoto, J. and Hashimoto, K. (1986). Reproductive toxicity of acrylamide and related compounds in mice – effects on fertility and sperm morphology. *Archives of Toxicology* 59: 201–205.

Sakata, T., Smith, R.A., Garland, E.M., and Cohen, S.M. (1989). Rat urinary bladder epithelial lesions induced by acrolein. *Journal of Environmental Pathology, Toxicology and Oncology* 9: 159–169.

Sanchez, A.M., Minko, I.G., Kurtz, A.J. et al. (2003). Comparative evaluation of the bioreactivity and mutagenic spectra of acrolein-derived alpha-HOPdG and gamma-HOPdG regioisomeric deoxyguanosine adducts. *Chemical Research in Toxicology* 16: 1019–1028.

Sanduja, R., Ansari, G., and Boor, P.J. (1989). 3-Hydroxypropylmercapturic acid: a biologic marker of exposure to allylic and related compounds. *Journal of Applied Toxicology* 9: 235–238.

Settels, E., Bernauer, U., Palavinskas, R. et al. (2008). Human CYP2E1 mediates the formation of glycidamide from acrylamide. *Archives of Toxicology* 82: 717–727.

Severin, I., Dumont, C., Jondeau-Cabaton, A. et al. (2010). Genotoxic activities of the food contaminant 5-hydroxymethylfurfural using different in vitro bioassays. *Toxicology Letters* 192: 189–194.

Shamoto-Nagai, M., Maruyama, W., Hashizume, Y. et al. (2007). In parkinsonian substantia nigra, alpha-synuclein is modified by acrolein, a lipid-peroxidation product, and accumulates in the dopamine neurons with inhibition of proteasome activity. *Journal of Neural Transmission (Vienna)* 114: 1559–1567.

Shin, A., Shrubsole, M.J., Rice, J.M. et al. (2008). Meat intake, heterocyclic amine exposure, and metabolizing enzyme polymorphisms in relation to colorectal polyp risk. *Cancer Epidemiology, Biomarkers and Prevention* 17: 320–329.

Singh, V., Bux, F., and Sharma, Y.C. (2016). A low cost one pot synthesis of biodiesel from waste frying oil (WFO) using a novel material, β-potassium dizirconate (β-K2Zr2O5). *Applied Energy* 172: 23–33.

Singh-Ackbarali, D., Maharaj, R., Mohamed, N., and Ramjattan-Harry, V. (2017). Potential of used frying oil in paving material: solution to environmental pollution problem. *Environmental Science and Pollution Research International* 24: 12220–12226.

Sinha, R., Kulldorff, M., Swanson, C.A. et al. (2000). Dietary heterocyclic amines and the risk of lung cancer among Missouri women. *Cancer Research* 60: 3753–3756.

Sinha, R., Kulldorff, M., Chow, W.H. et al. (2001). Dietary intake of heterocyclic amines, meat-derived mutagenic activity, and risk of colerectal adenomas. *Cancer Epidemiology Biomarkers Preview* 10: 559–562.

Skog, K., Johansson, M., and Jägerstad, M. (1998). Carcinogenic heterocyclic amines in model systems and cooked foods: a review on formation, occurrence and intake. *Food and Chemical Toxicology* 36: 879–896.

Smith, R.A., Cohen, S.M., and Lawson, T.A. (1990). Acrolein mutagenicity in the V79 assay. *Carcinogenesis* 11: 497–498.

Sobel, W., Bond, G., Parsons, T., and Brenner, F. (1986). Acrylamide cohort mortality study. *Occupational and Environmental Medicine* 43: 785–788.

Spiess, P.C., Deng, B., Hondal, R.J. et al. (2011). Proteomic profiling of acrolein adducts in human lung epithelial cells. *Journal of Proteomics* 74: 2380–2394.

Stott-Miller, M., Neuhouser, M.L., and Stanford, J.L. (2013). Consumption of deep-fried foods and risk of prostate cancer. *Prostate* 73: 960–969.

Straif, K., Baan, R., Grosse, Y. et al. (2006). Carcinogenicity of household solid fuel combustion and of high-temperature frying. *The Lancet Oncology* 7: 977–978.

Sugimura, T. (1997). Overview of carcinogenic heterocyclic amines. *Mutation Research, Fundamental and Molecular Mechanisms of Mutagenesis* 376: 211–219.

Sugimura, T., Wakabayashi, K., Nakagama, H., and Nagao, M. (2004). Heterocyclic amines: mutagens/carcinogens produced during cooking of meat and fish. *Cancer Science* 95: 290–299.

Surh, Y.-J., Liem, A., Miller, J.A., and Tannenbaum, S.R. (1994). 5-Sulfooxymethylfurfural as a possible ultimate mutagenic and carcinogenic metabolite of the Maillard reaction product, 5-hydroxymethylfurfural. *Carcinogenesis* 15: 2375–2377.

Svendsen, C., Husoy, T., Glatt, H. et al. (2009). 5-Hydroxymethylfurfural and 5-sulfooxymethylfurfural increase adenoma and flat ACF number in the intestine of Min/+ mice. *Anticancer Research* 29: 1921–1926.

Swaen, G.M., Haidar, S., Burns, C.J. et al. (2007). Mortality study update of acrylamide workers. *Occupational and Environmental Medicine* 64: 396–401.

Thomson, B., Lake, R., Cressey, P., and Knize, M. (1996). Estimated cancer risk from heterocyclic amines in cooked meat-a New Zealand perspective. *Proceedings of the Nutrition Society of New Zealand* 21: 106–115.

Totsuka, Y., Takamura-Enya, T., Nishigaki, R. et al. (2004). Mutagens formed from β-carbolines with aromatic amines. *Journal of Chromatography. B, Analytical Technologies in the Biomedical and Life Sciences* 802: 135–141.

Turan, S. and Yalcuk, A. (2013). Regeneration of used frying oil. *Journal of the American Oil Chemists' Society* 90: 1761–1771.

Turesky, R.J. (2007). Formation and biochemistry of carcinogenic heterocyclic aromatic amines in cooked meats. *Toxicology Letters* 168: 219–227.

Tyl, R.W. and Friedman, M.A. (2003). Effects of acrylamide on rodent reproductive performance. *Reproductive Toxicology* 17: 1–13.

Ulbricht, R.J., Northup, S.J., and Thomas, J.A. (1984). A review of 5-hydroxymethylfurfural (HMF) in parenteral solutions. *Fundamental and Applied Toxicology* 4: 843–853.

Vanderveen, L.A., Hashim, M.F., Nechev, L.V. et al. (2001). Evaluation of the mutagenic potential of the principal DNA adduct of acrolein. *Journal of Biological Chemistry* 276: 9066–9070.

Von Tungeln, L.S., Yi, P., Bucci, T.J. et al. (2002). Tumorigenicity of chloral hydrate, trichloroacetic acid, trichloroethanol, malondialdehyde, 4-hydroxy-2-nonenal, crotonaldehyde, and acrolein in the B6C3F(1) neonatal mouse. *Cancer Letters* 185: 13–19.

Voutsinas, J., Wilkens, L.R., Franke, A. et al. (2013). Heterocyclic amine intake, smoking, cytochrome P450 1A2 and *N*-acetylation phenotypes, and risk of colorectal adenoma in a multiethnic population. *Gut* 62: 416–422.

Wakabayashi, K., Yahagi, T., Nagao, M., and Sugimura, T. (1982). Comutagenic effect of norharman with aminopyridine derivatives. *Mutation Research Letters* 105: 205–210.

Wang, S.Y., Yu, C.P., Pan, Y.L. et al. (2017). Metabolomics analysis of serum from subjects after occupational exposure to acrylamide using UPLC-MS. *Molecular and Cellular Endocrinology* 444: 67–75.

Watzek, N., Bohm, N., Feld, J. et al. (2012a). N7-glycidamide guanine DNA adduct formation by orally ingested acrylamide in rats: a dose-response study encompassing human diet-related exposure levels. *Chemical Research in Toxicology* 25: 381–390.

Watzek, N., Scherbl, D., Feld, J. et al. (2012b). Profiling of mercapturic acids of acrolein and acrylamide in human urine after consumption of potato crisps. *Molecular Nutrition & Food Research* 56: 1825–1837.

Watzek, N., Scherbl, D., Schug, M. et al. (2013). Toxicokinetics of acrylamide in primary rat hepatocytes: coupling to glutathione is faster than conversion to glycidamide. *Archives of Toxicology* 87: 1545–1556.

Wong, K.-Y., Su, J., Knize, M.G. et al. (2005). Dietary exposure to heterocyclic amines in a Chinese population. *Nutrition and Cancer* 52: 147–155.

Yang, I.Y., Chan, G., Miller, H. et al. (2002). Mutagenesis by acrolein-derived propanodeoxyguanosine adducts in human cells. *Biochemistry* 41: 13826–13832.

Yang, K., Fang, J., Li, D. et al. (1999). ^{32}P-postlabelling with high-performance liquid chromatography for analysis of abundant DNA adducts in human tissues. *IARC Scientific Publications* 150: 205–217.

Yoshida, M., Tomitori, H., Machi, Y. et al. (2009). Acrolein toxicity: comparison with reactive oxygen species. *Biochemical and Biophysical Research Communications* 378: 313–318.

Zaitsev, A.N., Simonian, T.A., and Pozdniakov, A.L. (1975). Hygienic standards for hydroxymethylfurfural in food products. *Voprosy Pitaniia* 1: 52–55.

Zeb, A. (2012). Thermal degradation of β-carotene in food oils. In: *Vitamin A and Carotenoids: Chemistry* (ed. V.R. Preedy). London: The Royal Society of Chemistry.

Zeb, A. and Haq, I. (2016). The protective role of tomato powder in the toxicity, fatty infiltration and necrosis induced by oxidized tallow in rabbits. *Journal of Food Biochemistry* 40: 428–435.

Zeb, A. and Mehmood, A. (2012). Effects of oxidized vanaspati ghee on the serum lipids profile and radical scavenging activity of the in vitro lipids of liver, brain and muscles. *Turkish Journal of Biology* 37: 417–423.

Zeb, A. and Murkovic, M. (2010). Characterization of the effects of β-carotene on the thermal oxidation of triacylglycerols using HPLC-ESI-MS. *European Journal of Lipid Science and Technology* 112: 1218–1228.

Zeb, A. and Murkovic, M. (2011). Carotenoids and triacylglycerols interactions during thermal oxidation of refined olive oil. *Food Chemistry* 127: 1584–1593.

Zeb, A. and Murkovic, M. (2013). Determination of thermal oxidation and oxidation products of β-carotene in corn oil triacylglycerols. *Food Research International* 50: 534–544.

Zeb, A. and Rahman, W.U. (2012). Effects of oxidized tallow on the rabbit serum lipids and antioxidant activity of the in-vitro lipids. *Toxicology Research* 28: 151–157.

Zeb, A. and Rahman, S.U. (2017). Protective effects of dietary glycine and glutamic acid toward the toxic effects of oxidized mustard oil in rabbits. *Food & Function* 8: 429–436.

Zeb, A. and Uddin, I. (2017). The coadministration of unoxidized and oxidized desi ghee ameliorates the toxic effects of thermally oxidized ghee in rabbits. *Journal of Nutrition and Metabolism* 2017: 4078360.

Zhang, X.-M., Chan, C.C., Stamp, D. et al. (1993). Initiation and promotion of colonic aberrant crypt foci in rats by 5-hydroxymethy1-2-furaldehyde in thermolyzed sucrose. *Carcinogenesis* 14: 773–775.

Zhang, S., Villalta, P.W., Wang, M., and Hecht, S.S. (2007). Detection and quantitation of acrolein-derived 1,N2-propanodeoxyguanosine adducts in human lung by liquid chromatography-electrospray ionization-tandem mass spectrometry. *Chemical Research in Toxicology* 20: 565–571.

Zimmerli, B., Rhyn, P., Zoller, O., and Schlatter, J. (2001). Occurrence of heterocyclic aromatic amines in the Swiss diet: analytical method, exposure estimation and risk assessment. *Food Additives & Contaminants* 18: 533–551.

13

Improving the Quality of Fried Foods

13.1 Introduction

The main objective in improving food frying is to control the process by which hazardous compounds develop in the system. Process control is considered to play a significant role in maintaining the quality of fried foods and of the frying medium. Several factors, such as the nature of the frying medium, temperature, oil uptake, coating, and food ingredients, play important roles here. Food quality and safety can be determined by through a Hazard Analysis Critical Control Point (HACCP) system, which aims to determine and remedy potential contamination areas and toxicants throughout the frying process. Such contaminants may come from the frying medium, food ingredients, the food preparation method, surface treatments, the frying process, management of the frying medium, storage, and preservation. These processes have been explained in detail by Rossell (2001).

Significant progress has been made in recent times in the technological processes of frying. This chapter will therefore focus more on chemical and process control measures. All studies revolve around the frying medium and the food, with most recent studies aiming to control toxicant production in the final fried food product.

13.2 Improving the Quality of Fried Foods

13.2.1 Pretreatment of Food

Oladejo et al. (2018) reviewed the application of pretreatment methods in food frying. Frying involves heat and mass transfer between a food sample and the surrounding oil, which leads to physical and chemical changes in the food (Farkas et al. 1996). The qualities that consumers usually look for in fried foods include low moisture content, low oil uptake, crispness, flavour, and golden colour (Krokida et al. 2001a). The frying process has been described as involving inner vaporization of highly available water, which can be transferred from inside the material at a temperature close to boiling point, and the inner vaporization of bound water at a higher temperature, depending on the equilibrium of water desorption (Vitrac et al. 2002). In the initial stage of frying, drying rate is influenced by both diffusive and convective thermal resistances; the final stage is influenced by a decrease in temperature gradients and core temperature. According to Farkas et al. (1996), the frying process can be divided into four stages: initial heating,

Food Frying: Chemistry, Biochemistry, and Safety, First Edition. Alam Zeb.

surface boiling, falling rate, and bubble endpoint. During frying, water evaporates from the structure of the sample, while oil migrates into the structure through the pores created by evaporation. As frying progresses, the surface of the food sample becomes hardened (crust formation) due to the increase in frying temperature, and this tends to slow the rate of mass transfer. According to Bouchon et al. (2003), the oil absorption mechanism can be divided into three parts: structural oil (oil absorbed during frying), penetrated surface oil (oil suctioned into the food during cooling after removal from fryer), and surface oil (leftover oil on the surface of the food).

The demand for fried snacks has traditionally been high, due to their convenience and organoleptic properties (Dueik and Bouchon 2011). However, the side effects of consuming fat-fried snacks, such as heart disease, cancer, and obesity (Andrés-Bello et al. 2011), have led consumers to prefer low-fat snacks in recent times (Bouchon et al. 2003). There has been a search for improved methods of frying in order to lower the acrylamide (a known carcinogen) content in fried foods (Andrés-Bello et al. 2011). In recent times, vacuum frying has been used to improve the quality of fried foods in terms of their moisture, oil content, texture, colour, and flavour (Troncoso et al. 2009; Dueik et al. 2010; Andrés-Bello et al. 2011). Vacuum frying is the frying of food samples at a pressure (6.65 kPa) below atmospheric pressure in a confined system, which results in the lowering of the boiling point for both moisture and oil (Andrés-Bello et al. 2011; Dueik and Bouchon 2011). Some reports show that the advantages of vacuum frying include a reduction in oil content, preservation of colour, texture, and nutritional contents, and a reduction in acrylamide content (Da Silva and Moreira 2008; Mariscal and Bouchon 2008; Andrés-Bello et al. 2010).

Different pretreatment methods have been applied to food samples in a bid to reduce oil uptake and improve the overall quality of the final products. Food samples like potato, apple (Mariscal and Bouchon 2008), sweet potato (Taiwo and Baik 2007), chicken nuggets (Ngadi et al. 2007), snacks (Rahimi and Ngadi 2014), carrot (Shyi-Liang et al. 2005), and fish nuggets (Chen et al. 2009) that have been pretreated have shown improved quality in terms of oil content, crispness, flavour, colour, and nutritional contents. Oladejo et al. (2018) provided a review and information on the various pretreatment methods used in the food industry to improve the quality of fried foods (Table 13.1).

13.2.2 Classification of Frying Pretreatment Methods

The various pretreatment methods can be classified into two major categories: (i) conventional pretreatment methods and (ii) novel pretreatment methods (Onwude et al. 2016). The conventional pretreatment methods are old and common methods employed as standard in the food industry. These include oven/hot-air drying (Moyano and Pedreschi 2006), blanching (Taiwo and Baik 2007), coating (Chen et al. 2009), osmotic dehydration (Krokida et al. 2001a), and the freezing technique (Maity et al. 2012). Novel pretreatment methods are newly emerging technologies that offer advantages (ecofriendly, fast, low energy usage, cost-effective, etc.) over the conventional methods (Awad et al. 2012; Onwude et al. 2016). These include superheated steam drying (van Loon et al. 2005), infrared (IR) technology (Bingol et al. 2012), ultrasound (Dehghannya et al. 2015; Dehghannya and Abedpour 2018), microwave technology (Adedeji et al. 2009), and Pulsed electric field (PEF) (Janositz et al. 2011). The classification of various pretreatment methods is summarized in Figure 13.1.

Table 13.1 Summary of pretreatment methods applied to different agricultural products.

Pretreatment method	Pretreatment conditions	Frying conditions	Food types	Remarks	References
Hot air drying	Optimized drying temperature at 63.4 °C	Sunflower oil; frying temperature at 152 °C for 207 seconds	Carrot slices	50% reduction in oil content of fried carrot slices	Karacabey et al. (2016)
Hot air drying	Drying at 70 °C for 20, 40, 60, 90, and 120 minutes	Cotton seed oil; frying temperature at 170 °C for 0.3, 0.6, 1, 3, 5, 7, 10, 13, 15, and 20 minutes	Potato	Decrease in oil uptake and moisture content; increase in density, specific volume, and porosity; decreased lightness and yellowness, increased redness (negative effects on colour)	Krokida et al. (2001c)
Hot air drying	Drying at 65 °C (inlet) and 57 °C (outlet) for 20, 40, 60, and 90 minutes	Sunflower oil; frying temperature at 150, 175, and 200 °C for 10, 20, 30, 45, 60, 90, and 120 seconds	Chickpea flour snacks	54% reduction in oil content	Debnath et al. (2003)
Superheated steam drying	Superheated steam drying at 130, 150, and 180 °C	Soybean oil; frying temperature at 160 °C for 2.0–5.5 minutes	Potato	15–27% reduction in fat content; significant alteration in microstructure (compact structure); increased lightness, redness, and yellowness compared to control	Zielinska et al. (2015)
Hot air drying	Drying at 61 °C	Soybean oil; frying temperature at 160 °C for 90 seconds	Chicken nuggets	42–50% reduction in oil content; increased hardness and decreased lightness	Martínez-Ávila et al. (2010)
Blanching	Hot water at 85 °C for 3 minutes	Frying temperature at 120, 150, and 180 °C for 5 minutes	Potato	Oil uptake by 15–19% compared to control; no significant effect of blanching on final texture of fried samples	Pedreschi and Moyano (2005)
Blanching	Hot water at 90 °C for 1 minute	Frying temperature at 180 °C at 2-minute intervals for 16 minutes	Potato	Oil uptake during the first stage of frying	Andrés Grau et al. (2013)
Infrared (IR)	IR at 11 080 $W m^{-2}$ for 30, 60, 90, 120, 150, and 180 seconds	Frying temperature at 146, 160, and 170 °C for 1, 3, 5, and 7 minutes	Potato	37.5, 32.0, and 30.0% less total oil in IR blanched samples at frying temperatures of 146, 160, and 174 °C, respectively, as compared to blanched; on sensory analysis, IR blanched samples mostly preferred in terms of texture, taste, appearance, and colour	Bingol et al. (2012)

(*Continued*)

Table 13.1 (Continued)

Pretreatment method	Pretreatment conditions	Frying conditions	Food types	Remarks	References
Blanching	70 °C for 10 minutes	Canola oil; frying temperature at 170 °C for 0.5, 1, 2, 3 and 5 minutes	Sweet potato	Lower porosity in fried blanched sample compared to control; higher shrinkage compared to control	Taiwo and Baik (2007)
Coating	Gellan gum and guar gum	Soy bean oil; frying temperature at 170 °C for 100 seconds	Potato	Reduced heat transfer coefficients and oil uptake	Kim et al. (2011)
Coating	Methylcellulose (MC)	Sunflower oil; frying temperature at 150, 170, and 190 °C for 0.5, 1, 2, 3, and 4 minutes	Chicken nuggets	Highest level of oil reduction achieved with MC	Soorgi et al. (2012)
Coating	Hydroxypropyl methylcellulose (HPMC), guar gum, xanthan gum, and guar–xanthan gum combination	Sunflower oil; frying temperature at 170 °C for 2, 3, and 4 minutes	Carrot	Maximum 53% reduction in oil content with guar–xanthan gum, as compared to control	Akdeniz et al. (2006)
Coating	Methylcellulose (MC)	Sunflower oil; frying temperature at 150–170 °C for 5–15 minutes	Wheat flour dough	30% reduction in oil uptake	Suárez et al. (2008)
Coating	Pectin, whey protein, and soy protein isolate	Hydrogenated vegetable fat; frying temperature at 180 °C for 1.5 minutes	Cassava	Best fat reduction for cassava puree with whey protein, at 27%	Freitas et al. (2009)
Osmotic dehydration	15% sodium chloride/50% sucrose solution for 90 minutes	Sunflower oil; frying temperature at 170 °C for 2, 4, and 6 minutes	Potato	Reduction of oil and better colour compared to unreated	Karizaki et al. (2013)
Osmotic dehydration	Maltodextrin concentrations (40, 50, and 65, w/v), osmotic dehydration times (45, 60, and 70 minutes)	Canola oil; vacuum fried at 120, 130, and 138 °C for 60, 75, 98, 109, 105, 115, 120, 128, 130, 140, and 150 seconds	Mango	Lowest oil content and best colour, flavour, and carotenoid content obtained in samples pretreated by OD concentration of 65 (w/v) at 40 °C for 60 minutes	Nunes and Moreira (2009)
Osmotic dehydration	Sugar, NaCl, maltodextrine 12, and maltodextrine 21	Cotton seed oil; frying temperature at 170 °C for 0.3, 0.6, 1,3, 5, 7, 10, 13, and 15 minutes	Potato	Decrease in oil and moisture content and increase in porosity and colour	Krokida et al. (2001b)

Osmotic dehydration	Sugar solution (23.07 wt%) for 2 seconds	Canola oil; frying temperature at 180 °C for 6 minutes	Potato	30% less oil and more shrinkage in treated versus untreated samples	Mai Tran et al. (2007)
Osmotic dehydration	Maltodextrin solution 30 and 70% for 1 hour	Canola oil; vacuum frying at 100 °C for 72.5 minutes	Apricots	Acceptable moisture, oil, and beta carotene contents	Diamante et al. (2012)
Osmotic dehydration	NaCl solution at 40 °C for 4 hours	Canola oil; frying temperature at 170 1 °C for 0.5, 1.0 2.0, 3.0, and 5.0 minutes	Sweet potato	Less texture springiness	Taiwo and Baik (2007)
Microwave	Microwave at 14.1 W g^{-1} for 1 and 2 minutes	Canola oil; frying at 160 °C for 0–300 seconds	Chicken nuggets	Less oil uptake, 16.66% reduction in oil compared to control	Ngadi et al. (2009)
Microwave	Microwave at 5.23 W g^{-1}10.47 and 15.70 W g^{-1} for 30 seconds	Sunflower oil; frying at 135, 150, and 160 °C for 0–135 seconds	Ostrich meat plates	Minimum fat uptake obtained in sample micro waved at 5.23 W g^{-1} and fried at 135 °C	Amiryousefi et al. (2012)
Microwave	Microwave power densities at 6.7, 11.0, and 14.5 W g^{-1} for 60 seconds	Canola oil; frying temperature at 170, 180, and 190 °C	Chicken nuggets	Maximum reduction in fat uptake in samples precooked at 6.7 W g^{-1} and fried at 170 °C	Adedeji et al. (2009)
Freezing	Blast freezing at −20 °C for 2 hours and maintained at −20 °C overnight	Sunflower oil; frying temperature at 106 °C, 80 mmHg, for 9 minutes	Carrot	Significant improvement in organoleptic and healthy properties with freezing pretreatment	Albertos et al. (2016)
Freezing	Blast freezing at −40 °C for 3 hours and storage at −40 °C prior to frying	Sunflower oil; frying at 130 °C, 150 °C, and 170 °C for 0–6 minutes	Snacks	Increase in activation energies compared to unfrozen samples	Maity et al. (2012)
Ultrasound	Ultrasound probe of 20 kHz, 1500 W, and 90% amplitude level	Canola oil; frying temperature at 190 °C for 45 seconds	Tortilla chips	Higher breaking force gradient (BFG) and peak fracture force (PFF) in ultrasound pretreated samples	Janve et al. (2015)
Ultrasound	Ultrasound probe of 20 kHz and duty cycle pulses of 50 and 90%	Frying temperature at 170 °C for 2, 4, and 6 minutes	Potato	Reduction of oil by 12.5% and improved colour compared to untreated samples	Karizaki et al. (2013)

(*Continued*)

Table 13.1 (Continued)

Pretreatment method	Pretreatment conditions	Frying conditions	Food types	Remarks	References
Ultrasound	Ultrasound bath of 28 and 40 kHz and 160 W	Mixture of sunflower, soybean, and cottonseed oil; frying temperature at 150, 170, and 190 °C for 60, 120, 180, and 240 seconds	Potato	Less oil uptake in ultrasound pretreated versus untreated samples	Dehghannya et al. (2015)
Pulsed electric field (PEF)	Total energy of 18.9 kJ kg^{-1}; 9000 pulses at 0.75 kV cm^{-1} electric field or 810 pulses at 2.50 kV cm^{-1} electric field	Sunflower oil; frying temperature at 190 °C for 4 minutes	Potato	Lower oil absorption and browning tendency during frying among PEF pretreated samples	Ignat et al. (2015)
Pulsed electric field (PEF)	Specific energy levels between 0.5 and 22.0 kJ kg^{-1}; electric field strengths ranging from 0.3 to 1.2 kV cm^{-1}	Canola oil; frying temperature at 150, 170, and 190 °C for 1 minute	Sweet potato	18% lower oil content in PEF pretreated samples at 1.2 kV cm^{-1} fried at 190 °C versus untreated samples	Liu et al. (2017)

Source: Reproduced with kind permission of Society of Chemical Industry (Oladejo et al. 2017).

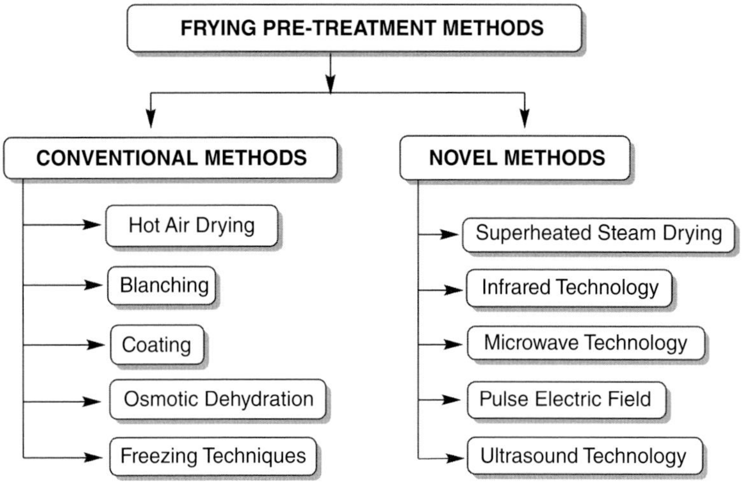

Figure 13.1 Classification of frying pretreatment methods. *Source:* Reproduced with kind permission of the Society of Chemical Industry (Oladejo et al. 2017).

13.2.3 Conventional Pretreatment Methods

13.2.3.1 Predrying (Hot-Air Drying)

Predrying of food samples in a hot-air dryer prior to frying has been reported to be an efficient way of reducing moisture and oil content during the frying process. The drying temperature usually used for predrying of agricultural products in the laboratory and in the food industry ranges between 60 and 70 °C (Krokida et al. 2001c; Debnath et al. 2003; Pedreschi and Moyano 2005). The initial moisture content of a food depends on the drying time prior to frying, and significantly affects the moisture and oil contents of the final fried product. Krokida et al. (2001c) reported that an increase in drying time led to a decrease in the moisture and oil contents of French fries after frying. However, they also found that predrying (hot-air drying) had negative effects on the colour of the fries, causing a decrease in lightness and yellowness and an increase in redness. Debnath et al. (2003) similarly stated that predrying of a chickpea flour-based snack food for 40 minutes led to decreased moisture and oil contents during frying at 175 °C for 45 seconds. The reduction in oil uptake caused by predrying could be a result of the improved initial solid content and hardening of the structure prior to frying (Krokida et al. 2000). Martínez-Ávila et al. (2010) discovered that pre- and post-fry drying of chicken nuggets decreased their oil content by 42 and 50%, respectively, compared to control, at a frying temperature of 160 °C for 3 minutes. Rahimi and Ngadi (2014) observed a decreased moisture content when the predrying time was increased, and found that predrying time significantly affected the oil distribution fractions in fried batter at 180 °C. Karacabey et al. (2016) found the optimal conditions for achieving a 50% reduction of oil content in fried carrot slices to be a predrying temperature of 63.4 °C, drying weight loss of 16%, frying temperature of 152 °C, and frying time of 207 seconds. High texture and colour quality were also obtained at these optimal values.

According to Debnath et al. (2003), the pretreatment of a chickpea flour-based snack in a convective dryer for 0–90 minutes before frying at 175 °C led to a decrease in moisture loss kinetic from 0.045 to 0.096 s^{-1} and a decrease in oil uptake kinetic from 0.056 to 0.082 s^{-1}. They explained that the reduction in the moisture loss and oil uptake transfer rates was due to the formation of a compact structure of potato as a result of predrying. In a similar experiment, Oyedeji et al. (2017) modelled and compared the kinetics of mass transfer during deep fat and vacuum frying of yellow-fleshed cassava slices. The samples were divided into three groups: vacuum frying, atmospheric frying, and predried (to 70% of the initial moisture content) prior to atmospheric frying. Their results showed that the moisture contents were lowest in the predried fried samples, but that these samples also had the slowest rate of moisture loss. Obviously, this was due to an initial reduction in moisture content and the development of a compact structure during convective drying. Similarly, the oil uptake in predried fried samples was lower compared to that of fresh samples subjected to vacuum and atmospheric frying.

13.2.3.2 Blanching

Blanching is the immersion of food samples in hot water for a very short time in order to preserve some of their quality parameters. During blanching, the starch in the food gelatinizes, which alters the structure of the sample. Blanching has also been used as a pretreatment to improve the quality of fried foods in the food industry. Pedreschi and Moyano (2005) assessed the effect of blanching on the oil uptake and texture of potato slices during deep fat frying at 180 °C. They observed that, contrary to their expectations, the blanched sample absorbed more oil than the control. This could be due to surface wetting, which is a crucial mechanism for oil absorption (Bouchon et al. 2003). When they combined blanching and drying together as a single pretreatment, they obtained a lower oil content compared to control at frying temperatures of 120, 150, and 180 °C (Moyano and Pedreschi 2006).

Blanching pretreatment of food samples helps to improve the colour of the fried product. Moyano and Pedreschi (2006) reported that the a^* colour parameter (redness) of blanched potato samples was significantly lower than that of unblanched samples at frying temperatures of 120, 150, and 180 °C. They attributed this to the fact that there was a decrease in the amount of reducing sugar (necessary to the formation of the Maillard reaction) due to leaching caused by blanching. They further reported that the acrylamide content in fried sweet potato pretreated by blanching was reduced by 68, 75, and 49% at temperatures of 120, 150, and 180 °C, respectively. The amount of reducing sugar in a food material determines the level of acrylamide in the final product (Wicklund et al. 2006). Blanching food materials prior to frying helps to reduce the concentration of asparagines and reducing sugar, thus providing low acrylamide content in fried foods (Wicklund et al. 2006; Mariotti et al. 2011).

Blanching (70 °C for 10 minutes) was one of the pretreatments that Taiwo and Baik (2007) applied to sweet potato slices prior to deep fat frying at 170 °C for 0.5–5.0 minutes. They reported that blanched sweet potatoes had a lower porosity and greater change in volume compared to controls. They also reported that blanching led to a 50% reduction in the initial hardness of the sweet potato prior to deep fat frying, which they explained as a result of the dissolution, depolymerization, and degradation of cell-wall pectin. However, on frying, blanched samples were harder compared to other pretreatments.

Pedreschi and Moyano (2005) used normalized textural parameter (MF*) to determine the textural kinetics of blanched and unblanched potato during deep fat frying at 120, 150, and 180 °C. They indicated that both control and blanching pretreatments had similar effects on the texture of the potato during frying. They observed for both blanched and control samples that there was a softening of the texture after a short frying time (<166 seconds), but that hardness developed as the frying progressed (>166 seconds). During the initial stage of frying, there was solubilization of middle lamella and gelatinization of starch. But as the frying progressed, there was compaction of potato structure, which led to the hardening of the crust.

13.2.3.3 Coating

Coating is the addition of edible ingredients to the food material, and this method has been used in the food industry to lower the oil and moisture content in fried foods (Mallikarjunan et al. 1997; Kilincceker and Hepsag 2012). A new way of producing battered fried foods in the food industry was developed by Sanz et al. (2004), which can eliminate the prefrying stage.

In an experiment, Lalam et al. (2013) treated chicken nuggets (crust and core) with 5 g/100 g methylcellulose (MC) prior to frying at 175 and 190 °C for 0, 30, 60, 120 and 240 seconds. They found that MC-coated chicken nuggets had a lower oil content compared to that of controls. However, they also observed lower moisture loss in MC-coated chicken nuggets compared to controls. One of the pretreatment methods used by Durán et al. (2007) was the coating of potato slices with 16 g L^{-1} of edible film solution of hydroxypropylmethylcellulose (HPMC) prior to frying. Contrary to expectations, the potato samples coated with HPMC absorbed more oil than controls. The authors explained that the presence of hydroxypropyl groups could have limited the coating capacity forming of the film. Chen et al. (2009) developed a fish nugget immersed in different batter formulas containing 1% carboxyl methyl cellulose (CMC) or 1% HPMC. They reported that these batters had a higher moisture content and lower oil content compared to batters without 1% CMC or 1% HPMC when fried in a conventional deep fat fryer or microwave fryer. They explained that the gel structure of CMC and HPMC enhances retention of moisture and lowers absorption of oil. They further reported that crusts containing CMC or HPMC had a higher L^* colour value and softer texture compared to other batters without CMC and HPMC.

The coating has also been reported to have an effect on the heat transfer coefficient during the frying of foods. Kim et al. (2011) found potato strips coated with hydrocolloids (gellan gum and guar gum) and fried in fresh soybean oil at 170 °C for 100 seconds significantly reduced the heat transfer coefficient compared to controls. They explained that hydrocolloids produced water vapour bubbles during frying, which inhibited heat transfer from the oil to the surface of the food, thus reducing the heat transfer coefficient. They also investigated the effects of the hydrocolloids on the oil content of potato strips and reported that there was a 41% reduction in oil content compared to controls when the potato strips were coated with 0.9% guar gum. They further discovered a correlation between the heat transfer coefficient and the oil content of the potato strips ($R^2 > 0.99$). They stated that the decreased heat transfer coefficient led to reduced oil uptake during frying.

Akdeniz et al. (2006) investigated the effects of the addition of HPMC, guar gum, xanthan gum, and a guar–xanthan gum combination to the batter of carrot slices deep fat-fried at 170 °C for 2, 3 and 4 minutes on coating pick up, moisture content, oil content, texture, porosity, and sample colour. They observed that coating pick-up was the highest in the batter containing the guar–xanthan gum combination, at about 66.5% higher than in controls. They observed this due to the high degree of synergism. All the gums had a significant effect on moisture loss, but guar gum and the guar–xanthan combination showed the highest retention of moisture. The authors explained that this might be due to the water-binding capacities of the gums, and that the synergistic effect of the combined gums might also provide a covering layer on the surface of the samples, inhibiting moisture loss. The retention of moisture by the gums in the batter formulations also led to a low uptake of oil during frying. Lower oil uptake was observed in a batter formulation coated with HPMC compared to control and xanthan gum because of its thermal gelation property. However, the guar–xanthan gum combination gave a 53% reduction in oil content compared to control. The authors further reported that the gum-coated batter formulation showed an increase in crispness and porosity and a decrease in colour changes (ΔE) compared to controls. In a similar work, Suárez et al. (2008) compared the deep fat frying process in a coated and uncoated dough system. Dough coated with MC and a control was fried at 150–170 °C. The coated dough had a 30% oil reduction compared to the control, but there was no significant difference between the coated and uncoated doughs in terms of moisture content, texture, and colour attributes. The type of hydrocolloid used influences the quality attributes of fried foods. In a bid to reduce the fat uptake in cassava product fried in an electric fryer, Freitas et al. (2009) used three different hydrocolloids (pectin, whey protein, and soy protein isolate) to coat the cassava product. They stated that whey protein-coated cassava product had the highest reduction in oil content compared to other hydrocolloids (pectin and soy protein isolate).

Coating potato chips prior to frying with almond gum improves their sensory properties (colour, appearance, crispiness, taste, odour, and appetence) compared to uncoated samples (Freitas et al. 2009). The use of almond gum as a coating agent can also significantly decrease the hardness of fried foods (Bouaziz et al. 2016). This is due to gel film formation between the moisture content of the food and the almond gum, which contributes to a stronger but more brittle coating (Mellema 2003).

Amaral et al. (2017) showed that coating potato strips with alginate ($20\,g\,L^{-1}$) prior to frying at 180 °C for 6 minutes had no significant effect on the acrylamide content of the fried strips. However, a recent study showed that the addition of a low concentration of chitosan (0.27%) to batters can reduce the acrylamide content of the fried batter by 59% (Sansano et al. 2016). The amino group of chitosan significantly competes with the amino group of asparagines in binding with the carbonyl group of the reducing sugar during frying, leading to a reduction in the acrylamide content of the fried batter.

13.2.3.4 Osmotic Dehydration

Osmotic dehydration is the simultaneous removal of water from a food sample and uptake of solute from the surrounding osmotic solution (Fernandes et al. 2006; Oladejo and Ma 2016). Osmotic dehydration has been widely used in the food industry for the partial removal of water from agricultural products, because of its ease of operation, preservation of nutrients, and maintenance of structure (Taiwo et al. 2003; Chenlo et al. 2006).

Osmotic dehydration has been tested as a way of reducing the oil uptake and improving the quality attributes of fried foods. Krokida et al. (2001b) pretreated French fries in four different osmotic solutions (sucrose solution, NaCl solution, maltodextrin 12, and maltodextrin 21) prior to frying at 170 °C for 0.3, 0.6, 1.0, 3.0, 5.0, 7.0, 10.0, 13.0, and 15.0 minutes. They reported that French fries pretreated in sucrose solution had the lowest moisture and oil content, followed by NaCl solution, maltodextrin 21, and maltodextrin 12. The decrease in initial moisture content of pretreated samples continues during frying, and this also led to a low uptake of oil because of the reduced initial moisture content. According to Mai Tran et al. (2007), a combination of predrying and the addition of sugar solution prior to frying led to about a 30% reduction in the oil content of potato chips compared to controls. Taiwo and Baik (2007) investigated osmotic dehydration as a pretreatment for sweet potato chips, where samples were immersed in NaCl solution at 40 °C for 4 hours. They found that osmotic pretreated samples had less chewiness and adhesiveness compared to controls when fried at 170 °C for 0.5–5.0 minutes. Bunger et al. (2003) soaked potato slices in NaCl solution at an optimized concentration value of 3% for 50 minutes prior to frying at 180 °C for 6 minutes. They reported a significant reduction in oil content, from 0.13 to 1.10 g oil g^{-1} dry matter: about a 22.2% difference compared to controls. The texture (hardness) of the NaCl-soaked slices increased and the colour was preserved compared to untreated samples. Similarly, Santis et al. (2007) reported a paler colour of blanched potato chips treated in NaCl solution prior to frying as compared to blanched potato chips without NaCl solution. Osmotic dehydration has been reported to have a significant effect on the visco-elastic property of potato strips. According to Krokida et al. (2001d), osmotic pretreatment of potato strips prior to frying at 170 °C for 1–20 minutes increases the maximum stress and maximum strain of the fried potatoes, and decreases their elasticity. Osmotic dehydration pretreatment can also be applied to food samples prior to vacuum frying. Nunes and Moreira (2009) investigated the effects of osmotic dehydration on the product qualities of vacuum-fried mango chips. They pretreated mango chips in different maltodextrin concentrations (40, 50, and 65 w/v) for 45, 60, and 70 minutes at osmotic solution temperature of 22 and 40 °C. They reported that mango chips pretreated with 65 w/v maltodextrin concentration for 40 minutes and vacuum fried at 120 °C gave the lowest oil uptake, best colour, best flavour, and best carotenoid content.

Pedreschi et al. (2007b) investigated the colour of fried potato samples pretreated by blanching and soaked in 3% NaCl solution for 5 minutes at 200 rpm agitation at 25 °C. The samples were fried at 120, 140, 160, and 180 °C. Their results showed that blanched potato samples immersed in 3% NaCl solution had improved colour appearance compared to controls (blanching only), as the total colour difference (ΔE) of samples immersed in 3% NaCl was reduced by 6, 16, 9, and 23% with respect to controls at 120, 140, 160, and 180 °C, respectively. Similar observations were reported for the normalized maximum texture (MF), where blanched potato samples soaked in 3% NaCl solution had a crisper texture than controls by 15, 13, 11, and 11% at frying temperatures of 120, 140, 160, and 180 °C, respectively.

One way to reduce the acrylamide content in potato slices is to soak them in NaCl solution prior to frying (Pedreschi et al. 2007a). This method gives a 90% reduction compared to untreated samples. Similarly, soaking potato chips in citric acid solutions can reduce acrylamide formation by 70%, when fried at 150 °C (Pedreschi et al. 2004). Recently, Kalita and Jayanty (2013) showed that soaking potato chips in vanadium sulfate ($VOSO_4$) solutions of 0.001, 0.01, and 0.1 mol L^{-1} before frying reduced the acrylamide content by 57.7, 71.4, and 92.5%, respectively, when compared with untreated samples. They explained that the binding of VO^{2+} to asparagine and the drop in pH of potato samples led to a significant reduction in the acrylamide content of fried potato chips.

13.2.3.5 Freezing

Freezing techniques include freezing and freeze drying of foods commonly used in the food industry (Ceballos et al. 2012; Nowak et al. 2016). Both methods aim at reducing the initial moisture content of a food before further processing (Albertos et al. 2016). Freezing is the formation of crystals within the structure of a food at low temperature; the size of the crystals depends on the rate of freezing (slow versus fast) (Degner et al. 2013; Nowak et al. 2016). Freeze drying is the removal of water from the food substance by sublimation at low pressure and temperature (Jiang et al. 2011; Nakagawa and Ochiai 2015). The possibility of freezing of food samples prior to frying in order to improve the quality of fried foods has been investigated. Frying of frozen snacks helps to remove the foams and smoke that normally occur during industrial frying (Maity et al. 2012).

Freeze drying was one of the pretreatments used by Gamble and Rice (1987) in order to determine the effect of predrying on the oil content and distribution of potato crisp prior to frying at 165 °C. These authors stated that fried freeze-dried potato crisps absorbed more oil than untreated crisps. This observation was confirmed by Kawas and Moreira (2001) when they reported that fried freeze-dried tortilla chips had a higher oil content, softer texture, and lower porosity compared to controls. Pretreatment of food samples with freeze drying results in a low gelatinization of starch, which causes less inhibition of absorption of oil during frying due to the creation of several small pores within the structure.

Freezing pretreatment prior to vacuum frying has been reported to significantly reduce the hardness of fried carrot snacks compared to high-pressure processing and controls (Albertos et al. 2016). The authors explained that the high ice crystal formation during freezing pretreatment led to an increased moisture evaporation during frying, which gave a more degraded structure and lower moisture content in the samples.

Maity et al. (2012) determined the textural kinetics of fried snacks as affected by freezing pretreatment prior to frying at 130–170 °C for 0–6 minutes. They reported an increase of 68% in the hardness of frozen samples compared to unfrozen samples. They further reported that freezing pretreatment increased the activation energies of textural parameters of fried snacks compared to unfrozen samples and that the overall sensory acceptability of the frozen fried snacks was not significantly different from that of the unfrozen samples. Similarly, Taiwo and Baik (2007) indicated that frozen samples of sweet potato required greater force to cut prior to frying, and on frying had 40% higher values for chewiness and the highest values for negative adhesive (stickiness) compared to sweet potato subjected to other pretreatment methods.

13.2.4 Novel Pretreatment Methods

13.2.4.1 Superheated Steam Drying

Superheated steam drying is a novel technology that has recently been applied to the drying of agricultural crops (Zielinska and Cenkowski 2012; Sehrawat et al. 2016). It works by both heating the food products and carrying away the evaporated water from the product, thus acting as a drying medium (van Loon et al. 2005). The final moisture content can be adjusted by altering the pressure, steam temperature, and flow (Kingcam et al. 2008). A schematic diagram of the experimental setup used by Kingcam et al. (2008) is shown in Figure 13.2.

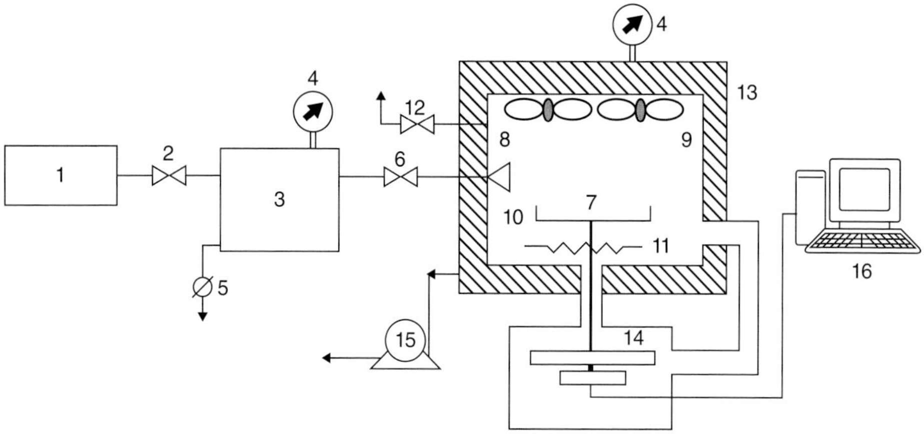

Figure 13.2 Schematic diagram of a low-pressure superheated steam dryer and associated units used by Kingcam et al. (2008): (1) boiler; (2) steam valve; (3) steam reservoir; (4) pressure gauge; (5) steam trap; (6) steam regulator; (7) drying chamber; (8) steam inlet and distributor; (9) two electric fans; (10) sample holder; (11) electric heater; (12) vacuum break-up valve; (13) insulator; (14) on-line mass indicator and logger; (15) vacuum pump; (16) PC with installed data acquisition card.
Source: Reproduced with kind permission of Elsevier Ltd.

The possibility of using superheated steam drying as a pretreatment prior to frying was investigated by van Loon et al. (2005). They stated that it produced a satisfactory product quality when compared with air drying. Zielinska and Cenkowski (2012) reported that potato chips pretreated by superheated steam drying had a 15–27% reduction in fat content and were lighter, redder, and more yellowish than potato chips produced without pretreatment. In a similar experiment, Primo-Martín and van Deventer (2011) used superheated steam drying to treat a battered snack prior to frying in peanut oil at 180 °C for 3 minutes. They indicated that the use of superheated steam pretreatment led to a low total oil content after frying. This could be attributed to the fact that the crust surface was very rough (as a result of starch gelatinization) and to the presence of amylose film on the surface of the crust, which reduced its permeability to oil. The authors concluded that final frying of samples pretreated by superheated steam positively affected their colour. The texture analysis carried out by Primo-Martín and van Deventer (2011) also indicated that battered snacks pretreated by superheated steam had crusts with a high initial crispness and a good crispness retention after frying. The authors attributed this to the fact that the low oil content of the fried snacks enhanced their crispy sound. They suggested that superheated steam pretreatment could be successfully applied in the potato chip industry.

13.2.4.2 Infrared Technology

Infrared (IR) is a thermal radiant energy produced through a catalytic reaction with a catalyst pad built into a catalytic infrared (CIR) emitter powered by natural gas or propene (Wu et al. 2014). It has been used in the food industry for the drying (Wang et al. 2017), peeling (Li et al. 2014), microorganism inactivation (Huang 2004), and roasting (Bagheri et al. 2016) of agricultural products. Huang (2004) designed an IR heating system consisting of two ceramic emitters arranged in parallel. The two aluminium reflectors

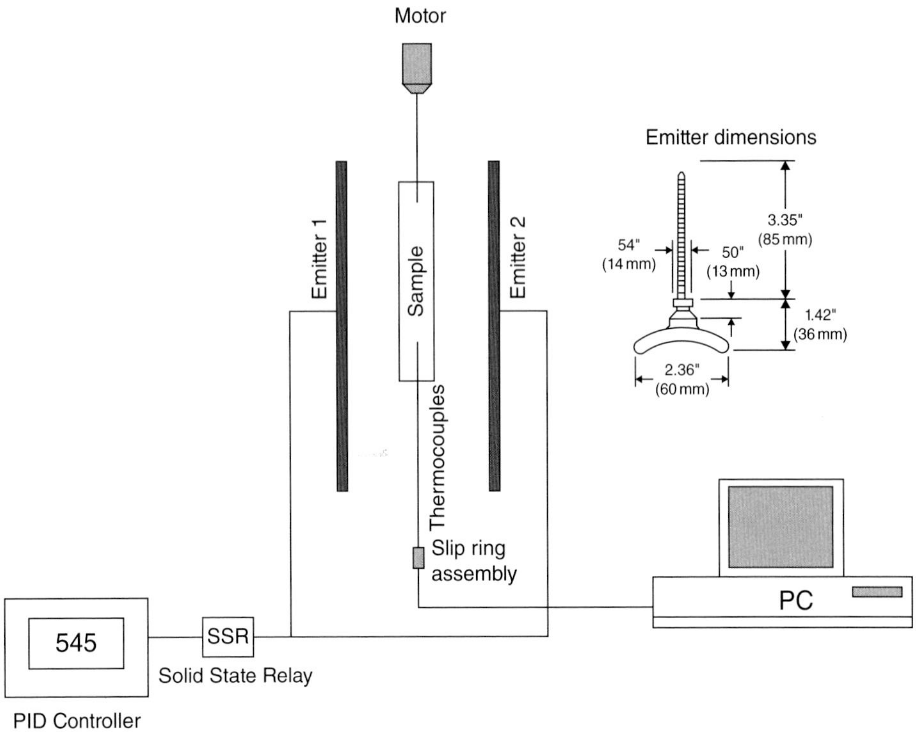

Figure 13.3 Infrared (IR) heating device used by Huang (2004). *Source:* Reproduced with kind permission of Elsevier Ltd.

cover the emitters to redirect the IR waves and enhance the efficiency of the IR radiation. Two thermocouples are inserted into the emitters and connected to a proportional–integral–derivative (PID) controller. The emitter temperature is controlled through a solid-state relay, while the temperature of the IR emitters is controlled at 545 ± 1 °C by the PID process controller, as shown in Figure 13.3.

The suitability of using IR as a pretreatment to improve the quality of fried foods has recently been investigated. Bingol et al. (2014) used a pilot catalytic IR device to pretreat potato strips. During the pretreatment, the strips were heated from above and below for 30, 60, 90, 120, 150, and 180 seconds, then subjected to frying at 146, 160, and 174 °C for 1, 3, 5, and 7 minutes. The authors reported that IR blanched potato strips fried at temperature of 146, 160, and 174 °C produced 37.5, 32.0, and 30.0% less oil, respectively, compared to unblanched strips. This was due to starch gelatinization of the core interior and formation of a surface crust, which inhibited oil uptake, and the formation of a whitish elastic skin, which reduced the diffusion rate of the oil (Bingol et al. 2012). Colour analysis of the IR blanched and unblanched strips showed that the L^*, a^*, and b^* values decreased initially but increased as frying progressed. A sensory analysis showed that IR blanched potato strips were more preferred compared to unblanched strips in terms of texture, taste, colour, and appearance. In a related experiment, Bingol et al. (2014) compared the effect of IR blanching and water blanching pretreatments on the product quality of French fries. They reported that IR pretreated French fries had lower moisture and oil contents than water blanched fries, and that IR blanched fries developed a^* and b^* colour values faster. The operational cost (energy use) of IR blanching was reported to be equal to that of water blanching.

Melito and Farkas (2012) reported the optimal IR parameters for fried doughnuts, in terms of oil content and product quality. They indicated that IR emitters positioned 60 mm above the doughnuts with a residence time of 45 seconds would give a 25.6–30.6% reduction in oil content compared to controls. Sensory analysis showed no significant difference between IR-finished doughnuts and control.

13.2.4.3 Microwave Technology

Microwave technology involves the absorption of energy by foods, leading to the internal generation of heat and the subsequent development of a pressure gradient within the food, which results in an increase in the moisture loss rate (Barutcu et al. 2009). In industry and the laboratory, application of microwave technology in food frying can be done in two major ways: either the frying process taking place directly inside the microwave device, where the source of heat for frying is generated by the microwave, or the food material is precooked by microwave for some time before undergoing deep fat frying.

Chen et al. (2009) stated that microwave frying shortens the frying time and enhances the frying process. According to Barutcu et al. (2009), microwave-fried batter had a higher specific volume, porosity, and inner surface smoothness compared to conventionally fried samples when different flour types were fried in a microwave oven using 365 W at 180 °C for 1.5 minutes versus a conventional deep fat fryer at 180 °C for 5 minutes. They further reported that the hardness of the microwave-fried batter was similar or lower than that of the conventionally fried samples. In a similar experiment, Sensoy et al. (2013) modelled and compared the microwave frying with the conventional frying of chicken breast meat at 180 °C using numerical simulation. They found that the processing time was shorter for microwave frying compared to conventional frying and that the heat transfer coefficient during microwave frying was higher than that in conventional frying. They explained that this could be due to the higher turbulence experienced during microwave frying. They also indicated that it could lead to low oil absorption in samples fried in the microwave. Recently, Su et al. (2016) investigated the application of microwave-assisted vacuum frying (MVF) in a bid to reduce the oil content and improve the quality of potato chips. They discovered that potato chips fried in MVF at a microwave power density of 12, 16, and 20 $W\,g^{-1}$, frying temperature of 100, 110, and 120 °C, and vacuum degree of 0.065, 0.075, and 0.085 MPa had a lower oil content, faster moisture evaporation rate, better crispiness, reduced colour change, and better preserved cell integrity compared to vacuum frying. Microwave frying of potato strips (at 400 W for 1 minutes) has also been demonstrated to reduce acrylamide content by 87.85% compared to conventionally frying at 170 °C for 4.5 minutes (Sahin et al. 2007).

Adedeji et al. (2009) investigated the kinetics of mass transfer in microwave precooked and deep-fat-fried chicken nuggets. They reported that the highest oil reduction was observed in chicken nuggets precooked at 6.7 $W\,g^{-1}$ and fried at 170 °C. They stated that the values of effective moisture diffusivity for the breading coating and the core portion ranged from 6.39×10^{-10} to 14.7×10^{-10} and from 1.77×10^{-10} to $14.0 \times 10^{-10}\,m^2\,s^{-1}$, respectively, with R^2 from 0.86 to 0.99, while the oil mass transfer rate constant for the breading region ranged from 0.04 to 40.96 s^{-1} with R^2 from 0.67 to 0.97. The activation energy was found to range from 13.65 to 54.93 kJ mol^{-1}. In a similar experiment, Amiryousefi et al. (2012) evaluated the kinetics of mass transfer in microwave-precooked and deep-fat-fried ostrich meat plates. According to their report, the minimal fat content was obtained at 5.23 W g^{-1} and 135 °C, the effective moisture diffusivity ranged from 1.47×10^{-8} to $4.17 \times 10^{-8}\, m^2 s^{-1}$, and the fat transfer rate constant ranged from 0.024 to 19.708 s^{-1}. They stated the activation energy to be in the range of 38.84–51.07 kJ mol^{-1}. Ngadi et al. (2009) reported the effect of microwave pretreatment on mass transfer during the deep fat frying of chicken nuggets. Coated chicken nuggets were pretreated in a microwave for 1–2 minutes and then fried at 160 °C for 0–300 seconds. The authors observed that microwave pretreatment reduced the initial moisture content from 1.22 to 0.82 kg kg^{-1} and from 2.48 to 2.01 kg kg^{-1} for the coating and core layer, respectively. After frying at 300 seconds, the moisture content of the coating fell from 1.22, 0.99, and 0.82 kg kg^{-1} to 0.40, 0.37, and 0.35 kg kg^{-1} for microwave pretreatment times of 0, 1, and 2 minutes, respectively, while the moisture content of the core fell from 2.48, 2.26, and 2.01 kg kg^{-1} to 1.80, 1.77, and 1.71 kg kg^{-1}, respectively. For the fat content, microwave pretreatment of the coating for 1 and 2 minutes caused a 26.22 and 33.51% reduction, respectively, after frying for 300 seconds, while the core had a fat content reduction of 10.95 and 16.66%, respectively, compared to controls.

Microwave precooking of food strips at 850 W for 10–30 seconds prior to frying has been reported to reduce acrylamide content by 36, 41, and 60% in potato strips fried at 150, 170, and 190 °C, respectively, compared to untreated samples (Erdoğdu et al. 2007).

13.2.4.4 Ultrasound Technology

Ultrasound refers to sound waves of 20 kHz and above. It produces cavitation, which brings about mechanical, physical, and chemical changes in foods during processing, such as dehydration, mixing, cutting, hydrolysis, extraction, and emulsification (Gallego-Juárez et al. 2007; Awad et al. 2012). Ultrasound technology has been employed in the food industries because of its numerous advantages, which include convenience, low cost, fast processing time, minimized use of materials, and preservation of nutritional content (Siucińska and Konopacka 2014). Karizaki et al. (2013) investigated the effect of ultrasound-assisted osmotic dehydration (UAOD) pretreatment on the quality of deep-fat-fried potato. The quality parameters that were determined included moisture content, oil uptake, colour, texture, and microstructure. Potato slabs were pretreated by UAOD at an ultrasound frequency of 20 kHz (probe type) for 30 minutes using 15% sodium chloride/50% sucrose solution, prior to frying at 170 °C for 2, 4, and 6 minutes. The authors reported that fried potato slabs pretreated by UAOD had a 12.5% (DB) reduction in oil uptake compared to controls. They further stated that pretreated samples had improved colour and texture qualities, but more damaged cell structure, compared with controls. This was due to the cavitation effect on the structure of potato caused by the application of ultrasound. In a related experiment, Dehghannya et al. (2015) determined the effects of ultrasound-assisted air drying pretreatment on the oil content of fried potato strips. The pretreatment was carried out using an ultrasound water bath at frequencies of 28 and 40 kHz and in combination with drying pretreatments at 80 °C for 8 and 15 minutes prior to frying at 150, 170, and 190 °C. The authors observed that the ultrasound pretreated samples had a lower oil content compared with controls (without ultrasound pretreatment). Recently, Amaral et al. (2016) investigated the possibility of using ultrasound as an alternative to blanching in vacuum-packaged potato strips. They concluded that ultrasound preserved the colour and improved the texture (firmness) of vacuum-packaged potato strips. However, they also showed that ultrasound pretreatment had no significant effect on the acrylamide content of fried potato strips compared to other pretreatment methods (blanching and coating).

Ultrasound can reduce the pretreatment (processing) time of food materials before frying. In the preparation of tortilla chips prior to frying, an ultrasound power of 1500 W, frequency of 20 kHz, and amplitude of 90% were applied. This led to a reduction in the traditional corn steeping time from 18.0 to 1.5 hours (Janve et al. 2015). Not only did ultrasound reduce the processing time of the tortilla chips, but it also significantly affected their texture and colour. The texture (breaking force gradient and peak fracture force) of fried tortilla chips pretreated by ultrasound was significantly higher than that of traditionally processed (without ultrasound) chips (Janve et al. 2015). Colour analysis of ultrasound pretreated tortilla chips by machine vision showed that their lightness (L^*) and hue ($°h$) were significantly higher and their colour difference (ΔE) significantly lower compared to untreated chips. The action of the ultrasound during processing may have altered their structure, causing the observed colour and texture changes (Oladejo et al. 2017). A sensory analysis showed no significant difference between the ultrasound processed and traditionally processed fried tortilla chips (Janve et al. 2015).

13.2.4.5 Pulsed Electric Field

Pulsed electric field (PEF) is the application of a high-voltage electric field for a short duration (microseconds to milliseconds) to a sample product placed between the electrodes (Ignat et al. 2015; Liu et al. 2017). It has been widely used in the food industry for drying (Gachovska et al. 2009), freezing (Jalté et al. 2009), and microorganism inactivation (Amiali et al. 2004). Recently, it has also been used as a pretreatment to improve the quality of fried foods. Ignat et al. (2015) investigated the effect of PEF pretreatment on the quality of fried potatoes. Potato cubes were subjected to PEF treatment at 0.75 kV cm^{-1} electric field for 9000 pulses or at 2.50 kV cm^{-1} electric field for 810 pulses. They observed that cubes pretreated by PEF had a lower oil content compared to blanched and water-dipped controls. Application of PEF to potato samples favoured the leaching of reducing sugar, which eventually led to a reduced browning index during frying. In a similar work, Liu et al. (2017) studied the effect of PEF pretreatment on the structure and quality of fried sweet potato. Whole tubers were subjected to PEF treatments ranging from 0.3 to 1.2 kV cm^{-1}. The authors reported that samples pretreated at 1.2 kV cm^{-1} and fried at 190 °C had an 18% lower oil content compared to untreated samples. They stated that their findings could be used in the food industry to design PEF processing parameters suitable for the pretreatment of solid materials prior to frying (Liu et al. 2017). Janositz et al. (2011) showed that the pretreatment of food samples by PEF enhanced the diffusion of water from the interior part of the food to the surface during frying. As the water reached the surface of the food, the vapour pressure difference between the food surface moisture and the oil increased, and the water vapour formed a layer on the surface of the food, reducing the fat uptake. The smooth surface of the PEF-treated food samples also contributed to a lower fat uptake compared to non-pretreated samples, which had rougher surfaces (Bouchon et al. 2003).

13.3 Mitigation Strategies for Acrylamide

13.3.1 Agronomical Methods

Potato products are highly susceptible to acrylamide formation because they contain acrylamide precursors and because traditional processing methods such as frying and roasting favour the Maillard reaction. The two approaches to preventing acrylamide formation are thus the removal of acrylamide precursors and interference with the Maillard reaction. The major challenge in reducing the acrylamide content of potato products lies in maintaining the product specification and final food quality (Vinci et al. 2012). Acrylamide formation is greatly influenced by agronomic conditions such as cultivation, soil properties, and climatic and storage conditions. Studies have shown that starch-derived products are prone to acrylamide concentrations of 5 mg kg^{-1} (Pedreschi et al. 2014). Systematic analysis revealed that acrylamide reductions are attained by focusing on the raw materials and processing conditions. Amrein et al. (2003) found the level of free asparagine in potato to be higher (0.2–0.4% dry weight, representing 20–60% of total amino acids) than that of reducing sugars.

Large, long, oval tubers containing high dry matter and low reducing sugar contents are preferred for the production of French fries, whereas for crisp production, moderate and oval tubers are more liked (Vinci et al. 2010). Reduction of sugar content in potato tuber (*Solanum tuberosum*) is achieved by choosing potato cultivators with low reducing sugar contents. Nitrogen fertilization plays a vital role in the formation of acrylamide, as it has a direct impact on asparagine and reducing sugar concentrations. A decrease in nitrogen fertilization has shown enhancement of reducing sugars, greatly reducing the quality of the potato during frying. An appropriate balance should be maintained in order to obtain a lower acrylamide content in tubers with respect to environmental impact and legal fertilization limits (Gerendás et al. 2007). Climatic conditions also influence precursors such as reducing sugars and asparagine. Potatoes grown in dry summers tend to show lower reducing sugar contents than those grown at temperatures above 25 °C. This is due to the elevated levels of reducing sugars found during respiration. Other variables adopted for the reduction of precursors include the storage of foods in dark conditions, as light activates the sugar metabolism. Finally, low storage temperatures and low oxygen levels inhibit cold-induced enzymes and consequently decrease the level of reducing sugars in the final product (Kerstin et al. 2004).

Muttucumaru et al. (2006) found that the level of acrylamide was greater in wheat grown under depleted sulfate conditions (2600–5200 μg kg^{-1}) than in that grown under normal conditions (600–900 μg kg^{-1}). They further reported that acrylamide formation was sevenfold greater (723–5286 μg kg^{-1}) than in pretreated samples. In bread, nitrogen fertilization of the soil induced amino acid and protein contents with increased levels of acrylamide, in the range of 10.6–55.6 μg kg^{-1} (Pedreschi et al. 2014). The higher asparagine content (137–485 mg kg^{-1}) resulted in higher acrylamide formation (115 μg kg^{-1}) as the ash content was found to be higher with increased activity of the protease. Varietal selection, environmental factors, and the effects of sulfur differ for wheat and rye grains, allowing a better understanding of the factors that affect the potential for acrylamide formation.

The exposure to acrylamide on coffee consumption is 8% in the United States, 13% in the Netherlands, and 39% in Sweden. Acrylamide concentration did not tend to show any significant changes when coffee was stored in a dark place for 3–6 months at 10–12 °C, whereas there was a 40–65% decreases in concentration when it was stored ground at room temperature for 6 months. Hoenicke and Gatermann (2005) reported acrylamide reduction during the storage of cocoa powder, due to the interaction of SH and NH_2 in the solid state. Mizukami et al. (2006) showed that control of roasting temperature in green tea solids lowered acrylamide concentration; an increase in temperature and longer heating times resulted in a decrease in acrylamide formation in the range of 4.8–41.12 mg mL^{-1}.

13.3.2 Biological Methods

One possible method of mitigating acrylamide in foods is to bake bread with lower concentrations of free asparagine and reducing sugar. Data obtained from the US Office of Environmental Health Hazard Assessment (OEHHA) show that cereals are the major dietary source of acrylamide, including in foods such as bread, cereal, cookies, and cakes. The percentage contribution to different population records is around 24% in Swedish adults, 44% in Belgian adolescents, and 40% in all Americans. In the case of cereal-based products, asparagine content is higher than that of reducing sugar. A comprehensive understanding of the role of asparagine and the genetic variation of food products is highly significant in acrylamide mitigation (Curtis et al. 2014).

The reduction of precursors by the adoption of genetic engineering has brought significant changes in acrylamide formation as various genes have been identified as regulating sugar control. RNAi technology with silenced α-glucan water dikinase, starch phosphorylase, or vascular acid invertase was reported to have a significant role in the reduction of acrylamide formation (Rommens et al. 2008). The same authors also showed that overexpressed *SnRK1* potato tuber contains reduced glucose and starch, by 83 and 30%, respectively. Byrne et al. (2012) investigated the G and G × E effects of Spark × Rialto on free asparagine content in bread. They found it had doubled haploid lines and significantly lowered asparagine content, proving that breading could mitigate the risk of acrylamide. The overexpression of *TaGCN2* in wheat significantly reduced free asparagine and acrylamide formation. In transgenic wheat plants overexpressing the *TaGCN2* gene, asparagine synthetase genes were reduced and could not be induced by sulfur deprivation. Asparagine synthetase in *Arabidopsis* is influenced by the regulatory protein kinase *SnRK1*, which controls the gene *ASN1*, affecting light and sugars. Quantitative trait loci (QTL) tend to reduce acrylamide formation through the development of genetic tools facilitating a reduction of asparagine and sugar levels in food crops (Xu et al. 2014).

13.3.3 Physical Methods

Physical methods have mainly focused on temperature, time, and moisture adjustments, as these directly influence the formation of acrylamide. Blanching is an important process in the production of French fries, as it limits oil absorption and improves the texture of the food. It also helps in maintaining the uniform colour of the product, as the precursors are leached, giving a reduced concentration of acrylamide. Reductions of 65 and 96%, respectively, were reported in French fries and potato crisps under blanching condition of 70 °C for 10–15 minutes. The concentration of sugar also influences the reduction of acrylamide. A decrease in sugar extraction of about 10% was noted during extraction in blanching water compared with fresh water. Drying of blanched or soaked potatoes helps in reducing fat absorption and oil hydrolysis prior to frying. It also reduces the concentration of acrylamide, with shorter times required to obtain the desired product quality (Franke et al. 2005).

Frying is the last step in the production of French fries and potato crisps. The frying process tends to form acrylamide under intense conditions (time and temperature). Frying at lower temperatures increases the time of frying, leading to enhanced fat uptake. Hence, frying time and temperature should be optimized carefully (Foot et al. 2007). The other processing factors are the type of oil and the model potato system. Some studies indicate that sunflower and rapeseed oils tend to show higher acrylamide content, but others have stated that the oil has no influence on acrylamide formation. Frying under vacuum conditions helps in the reduction of acrylamide – up to 94% at lower temperatures. Researchers comparing the effects of frying and baking reported that baking of potato chips at 170 °C resulted in a higher acrylamide content than frying (Palazoglu et al. 2010).

Reduction of acrylamide in bread is a major area of research. The quality of a bread product depends upon its surface colour, with greater colour intensity being seen when bread is baked at 200 °C. Temperature, air humidity, and moisture are important parameters noted for the effective reduction of acrylamide. In order to study crust temperature, colour formation, and water content, Ahrné et al. (2007) subjected food products to traditional baking without air, steam baking, and falling-temperature baking. The acrylamide concentration of the outer crust layer was found to be higher ($270\,\mu g\,kg^{-1}$) than that in crumb ($128\,\mu g\,kg^{-1}$), due to lower temperature exposure. Steam injection was better than other techniques, as it reduced the concentration of acrylamide to 50%. Steam and falling temperature were reported to lower acrylamide levels by 40% in the final bread crust. Alternative techniques used for the reduction of acrylamide are IR radiation and air jet impingement baking, with both showing 60% acrylamide reduction (Keramat et al. 2011). The intensity of flour milling and the baking temperature also reduce the concentration of acrylamide during baking. Fermentation plays a significant role, with shorter fermentation time reducing the acrylamide concentration by 87% in bread made with whole-grain wheat and by 77% in bread made with rye bran. In addition to temperature, heat transfer is another processing parameter to be controlled for the minimization of acrylamide content. The baking of bread in a conventional oven resulted in higher acrylamide content due to forced air convection. This conventional process increases the intensity of drying, showing decreased moisture content in cereals. Deck ovens were used to reduce the acrylamide levels formed during the Maillard reaction (Claus et al. 2008).

13.3.4 Chemical Methods

Several researchers have reported the reduction of acrylamide by the addition of additives such as amino acids, acid solutions, asparagine enzyme solutions, salts, cations, and certain antioxidants. The use of low-molecular additives such as organic acid and amino acids has gained attention due to its effect on food matrices. In the case of bakery products, the addition of organic acids such as citric acid at 0.1 or 0.2% decreased the level of acrylamide due to hydrolysis of carboxamide to aspartic acid (Jung et al. 2003).

A reduction of acrylamide formation by 50% was achieved through the addition of a non-asparagine amino acid such as glycine, lysine, or cysteine (Kim et al. 2005). In potato, soaking of crisps prior to frying in acetic solution at 20 °C for 20 minutes reduced acrylamide formation by 90%. The immersion of potato strips in distilled water, followed by a citric acid solution with a longer blanching time of about 80 minutes, also effectively reduced the acrylamide level (Kita et al. 2004).

Antioxidants play a vital role in the reduction of acrylamide by eliminating the free electron produced during the Maillard reaction. The antioxidants present in bamboo leaves and other plant extracts significantly reduced the acrylamide concentration by 69–73% (Zhu et al. 2009). Another promising solution is the use of asparaginase enzyme, as it has the potential to reduce the initial acrylamide content of 1400 $\mu g\, kg^{-1}$ in fried pastry products by 60–90%. As the asparaginase competes with asparagine, the nucleophilic compound helps in the reduction of acrylamide content. The effect of various amino acids in homogenized potatoes fried at 180 °C for 25 minutes with the addition of glycine and glutamine reduced acrylamide formation by 42–70% (Rydberg et al. 2003). Metal ions also play an important role in the reduction of acrylamide, where $CaCl_2$ and NaCl show positive effects on reducing acrylamide levels. The dipping of potatoes in NaCl and $CaCl_2$ inhibited acrylamide formation during frying in heated glucose and fructose systems due to Schiff base intermediate formation (Gökmen and Şenyuva 2007).

13.4 Reducing Oil Uptake

Oil or fat uptake is one of the principal factors that influence the hazard risk of fried foods. Minimal oil uptake should therefore be aimed at in all frying process. Mellema (2003) reviewed the mechanism and reduction of fat uptake by fried foods in detail. Following are some important factors that should be considered when aiming to improve frying food quality:

- The oil or frying medium should be of the best quality. The surface materials produced by the degradation process reduce the surface tension. This increase oil absorption, as degraded oil sticks easily to the food matrix.
- A low temperature and frying time should be used, as a high temperature and high duration may increase oil uptake. Thus, an optimum time and temperature (160–180 °C) should be set for each type of food in order to produce a final product of good quality.
- The surface area of a food is directly correlated with its oil uptake. By reducing surface area, food with less oil can be obtained.
- A food's moisture content is replaced by oil content. Thus, initial moisture content should be kept optimal in order to obtain an optimum oil uptake.
- Some food components enhance oil uptake (i.e. dough can take up more oil than can oily substances), while some ingredients, such as salt, can reduce it.
- Pretreatment processes such as drying can significantly reduce oil uptake.
- Coating of foods with starchy substances such as breadcrumbs, batter, or flour produces a fried crust, which forms a barrier between the food and the oil, thus reducing oil uptake.

13.5 Fortification

Tallow, vanaspati ghee, and other ready-to-use frying media are criticized for lacking basic nutrients such as vitamin A, D, E, and K. Therefore, these media can be fortified with vitamins and other substance. For example, fried potato chips were fortified with antioxidant vitamins such as retinyl palmitate and ascorbyl palmitate. It was found that these antioxidants were helpful in increasing the chips' shelf-life (Allam and Edress El-Sayed 2004). Similarly, potato chips were fortified with grape seed extract (GSE), green tea extract (GTE), and butylated hydroxytoluene (BHT) in order to study lipid oxidation during storage (Rababah et al. 2012). A significant reduction in off-flavour development was observed with the addition of GTE and GSE at a concentration range of 750–1000 ppm. Chiou et al. (2013) showed that olive leaf extract rich in oleuropein is an effective fortification substance in the deep frying of French fries.

13.6 The Role of Natural Antioxidants

Antioxidants (both naturally occurring and artificially added) play a prominent role in protecting frying oils against thermal and oxidative deterioration, and hence in the manufacture of high-quality fried products (Márquez-Ruiz et al. 2014). Primary or chain-breaking antioxidants mainly act as hydrogen donors, reacting with alkyl and alkyl-peroxyl radicals. While, at low temperatures, the reaction of primary antioxidants with alkyl-peroxyl radicals is predominant, under the low-oxygen-pressure conditions found at frying temperatures, the reaction with the abundant alkyl radicals is also significant. Phenolic compounds, naturally occurring tocopherols, tocotrienols, and carotenoids are good antioxidants. Compounds of synergic action, such as citric acid and ascorbyl palmitate, are normally regenerators of primary antioxidants. The action of antioxidants is well known under the room-temperature conditions found during storage and at the moderate temperatures used in accelerated tests measuring oil stability. However, under frying conditions involving oxidative and thermal reactions, evaluation of antioxidant efficacy is far more complicated (Boskou 2003; Choe and Min 2007). This section describes the main antioxidants found in frying oils.

13.6.1 Tocochromanols

Tocochromanols, or, more generally, vitamin E, represent a family of natural, structurally related compounds known as tocopherols and tocotrienols. Tocopherols are the main naturally occurring antioxidants found in edible oils, and are also used as additives in frying oils to improve their oxidative stability. In the European Union, α-, γ-, and δ-tocopherol, coded as E307, E308, and E309, respectively, and a tocopherol-rich extract, coded as E306, can be used as additives *quantum satis* in all fats and oils except virgin and olive oils (Márquez-Ruiz et al. 2014). For refined olive oils, including olive pomace oil, the addition of 200 mg kg^{-1} α-tocopherol is permitted.

A mechanism dependent on temperature seems to be involved in the action of tocopherols (Marinova and Yanishlieva 1992), since the loss of tocopherol under frying conditions is, in general, relative to oil degradation, becoming more rapid as the degree of oil unsaturation decreases (Simonne and Eitenmiller 1998; Wagner et al. 2001); the opposite is found at lower temperatures, from ambient up to 100 °C (Martín-Polvillo et al. 2004; Márquez-Ruiz et al. 2008). Therefore, accelerated tests at moderate temperatures or under the conditions of the Rancimat test may be useful in predicting the relative oxidation behaviour of different oils and fats, although they may not be indicative of their frying performance.

Additionally, it has been found that the amounts of polymers and polar compounds at the point of tocopherol exhaustion are significantly lower in monounsaturated oils as compared to polyunsaturated oils (Barrera-Arellano et al. 2002). These results have been reported in studies in both model systems (Verleyen et al. 2001) and oils (Normand et al. 2001; Verleyen et al. 2002). Since monounsaturated oils become unprotected at levels of polar compounds much lower than the limit established in official regulations, products fried in monounsaturated oils may be more susceptible to oxidation during storage (Márquez-Ruiz et al. 1999).

Figure 13.4 shows the different behaviours of tocopherols at frying temperature in conventional or high linoleic sunflower oil (HLSO) versus high oleic sunflower oil (HOSO) with similar tocopherol contents and compositions (where α-tocopherol makes up more than 95% of total tocopherols) (Barrera-Arellano et al. 2002). Apart from the expected higher polymerization of the tocopherol-stripped oils as compared to their counterparts, the most important finding from this figure is that not only was the loss of tocopherols more rapid in the less unsaturated oil (HOSO), but tocopherols were exhausted at lower oil degradations. It has been reported that tocopherols show a rapid loss at frying temperatures due to their degradation. In this respect, the formation of γ-tocopherolquinone, 4,5-epoxy-γ-tocopherolquinone, and 7,8-epoxy-γ-tocopherolquinone has been demonstrated, and their antioxidant efficacy has been suggested (Verleyen et al. 2001).

Concerning the type of tocopherol, it has been shown that α-tocopherol is less stable at high temperature than δ-tocopherol, while β- and γ-tocopherols degrade at an

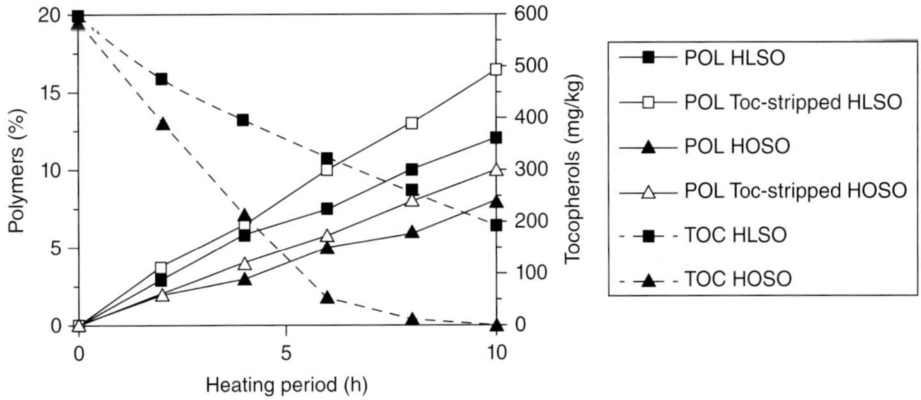

Figure 13.4 Formation of polymers (POL) and loss of tocopherols (TOC) in high linoleic sunflower oil (HLSO), high oleic sunflower oil (HOSO), tocopherol-stripped high linoleic sunflower oil (TOC-stripped HLSO), and tocopherol-stripped high oleic sunflower oil (Toc-stripped HOSO) heated at 180 °C. *Source:* Reproduced with kind permission of John Wiley & Sons (Márquez-Ruiz et al. 2014).

intermediate rate (Gordon and Kourkimskå 1995; Barrera-Arellano et al. 2002; Warner and Moser 2009; Aladedunye and Przybylski 2013). As a result, new sunflower lines have been developed to increase levels of γ-tocopherol and decrease the relative content of linolenic acid (Demurin et al. 1996). In particular, a sunflower oil from a line of seeds characterized by increased levels of both palmitic acid and oleic acid and a modified tocopherol profile made up of γ-tocopherol (>95% of total tocopherols) has been found to be highly stable (Susana et al. 2008).

With respect to tocotrienols, their action is still poorly known. In general, the stability of α-tocotrienol has been found to be similar to that of α-tocopherol (Romero et al. 2007), but the order of stability of the different tocotrienol homologues is different from that of their tocopherol counterparts, with γ-tocotrienol the least stable (Rossi et al. 2007).

13.6.2 Polyphenolic Compounds

The main group of phenolic and polyphenolic compounds showing protective action during frying is present in virgin olive oil (VOO) and mainly comprises hydroxytyrosol and tyrosol and their derivatives, and lignans (Bendini et al. 2007). The other two groups with a reported antioxidant effect in frying are γ-oryzanol (i.e. sterol esters of ferulic acid) in rice bran oil (Lerma-García et al. 2009), and lignans (sesamol, sesaminol, sesamolin, and others) in sesame oil (Moazzami et al. 2007). Interestingly, studies analyzing the losses of tocopherols and polyphenols during frying report a high rate of degradation for hydroxytyrosol and its derivatives, a similar or lower rate for α-tocopherol, and a low degradation rate for tyrosol, its derivatives, and lignans (Andrikopoulos et al. 2002). The rapid loss of hydroxytyrosol suggests a substrate-independent degradation is a more important reaction than its antioxidant effect. Hence, the effective protection found is probably more related to the group of less active antioxidants (i.e. lignans and tyrosol and its derivatives), which remain longer at frying temperatures.

There have been few reports on the good frying performance of sesame and rice bran oils. Most such studies, however, have shown a significant increase in oil thermostability and a high retention of natural tocopherols in blends of different oils that include either sesame or rice bran oil (Chung et al. 2006; Farhoosh and Esmaeilzadeh 2009). In sesame oil, excellent results have been obtained for sesamol at frying temperatures (Hwang et al. 2012, 2013), but neither sesamin nor sesamolin shows appreciable antioxidant activity. However, sesamolin may be hydrolysed during frying to yield sesamol and sesaminol, both of which show high stability and a synergistic effect with tocopherols (Gopala et al. 2005).

Some ingredients based on extracts rich in polyphenols obtained from oils or their byproducts have been commercialized (Kochhar 2000; Gertz 2004). Besides improving the thermostability of oils during frying, the addition of these natural products to oils is aimed at increasing antioxidant intake (Chiou et al. 2009). Also, the addition of individual flavonoids (i.e. pelargonidin, cyanidin, quercetin, and myricetin) has been shown to be very effective in increasing oil thermostability (Naz et al. 2008).

13.6.3 Phytosterols

Phytosterols such as 4-desmethylsterol and 4-methylsterol with an ethylidene group in the side chain (e.g. Δ5-avenasterol, Δ7 avenasterol, fucosterol, vernosterol, and citrostadienol, as well as their esters) have been found to contribute to delaying oil degradation at frying temperature (Boskou 2003), although some authors report a non-existent or only slight effect (Winkler and Warner 2008). They are normally referred as to antipolymerization agents even though they seem to act as antioxidants (i.e. through the formation of allyl radicals in a primary carbon atom followed by isomerization to a relatively stable tertiary radical).

13.6.4 Squalene

Squalene is a highly unsaturated hydrocarbon with important health effects that is found in large quantities in shark liver oil and in small amounts in olive and wheat germ oil. Although there is rather scant information on its efficacy at frying temperatures, it seems to show only low or moderate effect (Małecka 1994). Naziri et al. (2016) showed a loss of squalene during lipid oxidation.

13.6.5 Carotenoids

Carotenoids are natural antioxidants found in a variety of foods. In plant oils, they can be present in low to high amounts, depending upon the variety and type of plant. For example, red palm and sea buckthorn oils are rich in carotenoids (Zeb 2012). These pigments play a significant role during frying. Small amounts of carotenoids are usually used; high amounts added to the frying medium will play a pro-oxidative role (Zeb and Murkovic 2011).

13.6.6 Plant Extracts

In the last decade, a significant number of studies have focused on the positive effects of the addition of plant extracts – mainly herbs and spices – to frying oils (Márquez-Ruiz et al. 2014). These follow earlier studies on the antioxidative properties of such extracts at room temperature (Yanishlieva and Marinova 2001; Yanishlieva et al. 2006). Their addition is directed not only towards delaying deterioration by protecting the natural antioxidants found in these oils (i.e. tocopherols), but also at increasing the consumption of natural, healthy compounds in response to consumer demand.

In general, the action found is positive because plant extracts contain potent antioxidants, as detailed in an excellent review by Yanishlieva et al. (2006). However, the protective effect is highly variable during frying even for the same extract, because of the inherent characteristics of natural products and the complexity of the frying process. Unfortunately, the composition of an extract is often poorly defined and varies greatly depending on the plant variety, extraction method, and concentration. When studies include the characterization of extracts, it is usually based on the analysis of polyphenols (Ajila et al. 2011; Motilva et al. 2013), but their pro- or antioxidative action may also depend on the content and chemical structure of other minor compounds.

13.7 Packaging of Fried Foods

Packaging plays a critical role in the food supply chain. The primary function of packaging is to serve as a container for the food, enabling efficient transport within the whole supply chain, preventing any physical damage, and protecting against manipulation and theft.

The packaging also meets the fundamental need to maintain food quality and safety from production to final consumption by preventing any unwanted chemical and biological changes. Hence, the packaging acts as a barrier protecting the food from environmental influences such as oxygen, moisture, light, dust, pests, volatiles, and both chemical and microbiological contaminants (Arvanitoyannis and Oikonomou 2012; Yildirim et al. 2018).

The protective role of packaging is primarily passive, acting as a barrier between the food, the atmosphere, and the external environment. However, there are some exceptions, such as with fresh produce, for which highly gas-permeable or perforated packaging materials are used to allow gas exchange with the atmosphere (Lee and Paik 1997; Hussein et al. 2015). Such packaging systems are limited in their ability to extend the shelf-life of the packaged food. In recent decades, consumer concern about the safety and additive content of food has received much attention. There is an increasing trend towards natural high-quality foods that are nonprocessed or minimally processed, do not contain preservatives, but offer an acceptable shelf-life (Singh et al. 2011). In response, the protective function of packaging has been refined and improved, leading to the development of new packaging technologies, such as modified atmosphere packaging (MAP) (Rodriguez-Aguilera and Oliveira 2009), active packaging (Singh et al. 2011; Yildirim et al. 2018), smart and intelligent packaging (SP/IP) (Kerry and Butler 2008; Brockgreitens and Abbas 2016), and nanomaterials (Imran et al. 2010; Kim et al. 2014).

Active packaging is an innovative approach to maintaining or prolonging the shelf-life of food products while ensuring their quality, safety, and integrity. As defined in European Regulation (EC) No. 450/2009, active packaging comprises packaging systems that interact with food products in such a way as to 'deliberately incorporate components that would release or absorb substances into or from the packaged food or the environment surrounding the food' (European Union 2009). Active packaging systems can be divided into active scavenging systems (absorbers) and active releasing systems (emitters). The former remove undesired compounds from the food or its environment (e.g. moisture, carbon dioxide, oxygen, ethylene, odour), while the latter add compounds to the food or into the headspace (e.g. antimicrobial compounds, carbon dioxide, antioxidants, flavours, ethylene, ethanol). Table 13.2 provides an overview of the primary active packaging technologies and their potential benefits in food applications.

The addition of active substances, such as antimicrobials and antioxidants, through the use of active packaging instead of a direct addition to the food, may decrease the number of such substances required. Traditionally, active substances are added to the bulk of the food, whereas for most fresh and processed food, food degradation or microbial growth occurs at the surface. Furthermore, the activity of the active substances when directly added to the food may be reduced or inhibited as a result of the interaction between the active substances and the food components, and/or during the processing of the food. Therefore, the addition of active substances via active packaging may be more effective than their addition to the bulk of the food.

A large variety of active packaging systems have been developed, and, to date, numerous reviews have emphasized the potential of active packaging technologies to supply safer, 'healthier', and higher-quality foods to the consumer (Kuorwel et al. 2015; Brockgreitens and Abbas 2016). However, the number of reviews presenting the benefits of active packaging technologies applied to specific food products is limited

Table 13.2 Potential active packaging technologies for food applications.

Type of active packaging	Type of food	Potential benefits
Active scavenging systems (absorbers)		
Oxygen scavenger	(Sliced) cooked meat products	Prevention of discolouration
Moisture scavenger	Grated cheese, (par-baked) bakery products	Prevention of mould growth
Ethylene absorber	Fruit and vegetable juices	Retention of vitamin C content, prevention of browning
	Seeds, nuts, and oils; fat-containing instant powders; fried snacks; dried meat products	Prevention of rancidity
	Mushrooms, tomatoes, strawberries, maize, grains, seeds, fresh fish, and meat	Extension of shelf-life through maintenance of moisture content, decrease in moisture condensation in the packaging, positive impact on appearance, and reduction in browning or discolouration
	Climacteric fruits and vegetables	Reduction in ripening and senescence, thereby enhancing quality and prolonging shelf-life
Active releasing systems (emitters)		
Antioxidant releaser	Fresh fatty fish and meat; fat-containing instant powders; seeds, nuts, and oils; fried products	Improvement of oxidative stability
Carbon dioxide emitter	Fresh fish and meat	Extension of microbiological shelf-life, reduction in head-space volume of modified atmosphere packaging
Antimicrobial packaging systems	Fresh and processed meat, fresh and smoked fish, fresh seafood, dairy products, fresh and processed fruits and vegetables, grain, cereals and bakery products, and ready-to-eat meals	Inhibition or retardation of bacterial growth, extension of shelf-life

Source: Reproduced with kind permission of John Wiley & Sons (Yildirim et al. 2018).

(Llorens et al. 2012; Cichello 2015). Active packaging technologies that have been evaluated in model systems may not always behave in the same way in real food applications. The complex structure of a food product may influence the activity of the packaging. For example, the release rates, absorption rates, or diffusion rates of active substances may be affected. Moreover, active substances or carriers may react with food components or bind to them, thereby inhibiting the desired activity. It is important to critically review active packaging studies pertaining to specific foods in order to enable food and packaging scientists to better understand the benefits of such systems and potentially to accelerate their adoption in commercial applications.

13.8 Quality Control in Frying

In addition to technical quality-control monitoring, the quality of fresh foods and oils, frying oil, and fried foods has received considerable attention in recent times. Fresh oil is analysed for quality characteristics, while frying oil must be critically monitored to ensure high-quality fried foods are obtained. The characteristics of oils include fatty acid (FA) composition, volatile components, polar components, organoleptic properties, smoke point, viscosity, free fatty acids (FFAs), conjugated dienes and trienes, dimers and polymeric triglycerides, moisture, oxidized fatty acids, *trans* fatty acids (TFAs), peroxide value, iodine value, cyclic monomers, antioxidant activity, and sterol contents (Ghidurus et al. 2010; Hosseini et al. 2016).

Fresh or fried foods are also evaluated for quality-control characteristics such as oil uptake, moisture content, crust, and colour. These parameters are strictly monitored to obtain acceptable foods for consumers. Various regulations usually force food manufacturers to follow standard guidelines when frying. Thus, the quality of fried foods can be improved by using quality-control measurements.

13.9 Key Concepts

- Several pretreatment methods are available to improve the quality of fried foods.
- These methods are classified as conventional and novel methods.
- Conventional methods include hot-air drying, blanching, coating, osmotic dehydration, and freezing techniques.
- Novel methods include superheated steam drying, IR technology, microwave technology, ultrasound technology, and PEF technology.
- The acrylamide content of fried foods can be reduced by agronomical, biological, physical, and chemical methods.
- Reduction in oil uptake is one of the key quality-control measures.
- Several natural compounds, such as ascorbyl palmitate, retinyl palmitate, plant extracts, and vitamins, have been used as fortifying agents to improve the shelf-life and quality of fried foods and oils.
- Natural substances such as tocochromanols, polyphenols, phytosterols, squalenes, carotenoids, and plant extracts have potential antioxidant activities during frying.
- Fried foods are packaged with the aim of protecting them from deterioration and contamination.
- Quality-control measures are key to our understanding of frying and consumer acceptability.

References

Adedeji, A.A., Ngadi, M.O., and Raghavan, G.S.V. (2009). Kinetics of mass transfer in microwave precooked and deep-fat fried chicken nuggets. *Journal of Food Engineering* 91: 146–153.

Ahrné, L., Andersson, C.-G., Floberg, P. et al. (2007). Effect of crust temperature and water content on acrylamide formation during baking of white bread: steam and falling temperature baking. *LWT – Food Science and Technology* 40: 1708–1715.

Ajila, C.M., Brar, S.K., Verma, M. et al. (2011). Extraction and analysis of polyphenols: recent trends. *Critical Reviews in Biotechnology* 31: 227–249.

Akdeniz, N., Sahin, S., and Sumnu, G. (2006). Functionality of batters containing different gums for deep-fat frying of carrot slices. *Journal of Food Engineering* 75: 522–526.

Aladedunye, F. and Przybylski, R. (2013). Frying stability of high oleic sunflower oils as affected by composition of tocopherol isomers and linoleic acid content. *Food Chemistry* 141: 2373–2378.

Albertos, I., Martin-Diana, A.B., Sanz, M.A. et al. (2016). Effect of high pressure processing or freezing technologies as pretreatment in vacuum fried carrot snacks. *Innovative Food Science & Emerging Technologies* 33: 115–122.

Allam, S.S.M. and Edress El-Sayed, F. (2004). Fortification of fried potato chips with antioxidant vitamins to enhance their nutritional value and storage ability. *Grasas y Aceites* 55: 10.

Amaral, R.D.A., Benedetti, B.C., Pujolà, M. et al. (2016). A first approach of using ultrasound as an alternative for blanching in vacuum-packaged potato strips. *Food and Bioprocess Technology* 9: 1794–1801.

Amaral, R.D.A., Achaerandio, I., Benedetti, B.C., and Pujolà, M. (2017). The influence of edible coatings, blanching and ultrasound treatments on quality attributes and shelf-life of vacuum packaged potato strips. *LWT – Food Science and Technology* 85: 449–455.

Amiali, M., Ngadi, M.O., Raghavan, V.G.S., and Smith, J.P. (2004). Inactivation of Escherichia coli O157:H7 in liquid dialyzed egg using pulsed electric fields. *Food and Bioproducts Processing* 82: 151–156.

Amiryousefi, M.R., Mohebbi, M., and Khodaiyan, F. (2012). Kinetics of mass transfer in microwave precooked and deep-fat fried ostrich meat plates. *Food and Bioprocess Technology* 5: 939–946.

Amrein, T.M., Bachmann, S., Noti, A. et al. (2003). Potential of acrylamide formation, sugars, and free asparagine in potatoes: a comparison of cultivars and farming systems. *Journal of Agricultural and Food Chemistry* 51: 5556–5560.

Andrés-Bello, A., García-Segovia, P., and Martínez-Monzó, J. (2010). Vacuum frying process of gilthead sea bream (Sparus aurata) fillets. *Innovative Food Science & Emerging Technologies* 11: 630–636.

Andrés-Bello, A., García-Segovia, P., and Martínez-Monzó, J. (2011). Vacuum frying: an alternative to obtain high-quality dried products. *Food Engineering Reviews* 3: 63.

Andrés Grau, A.M., Argüelles Foix, A.L., Castelló, G., and Heredia Gutiérrez, A.B. (2013). Mass transfer and volume changes in French fries during air frying. *Food and Bioprocess Technology* 6: 1917–1924.

Andrikopoulos, N.K., Dedoussis, G.V.Z., Falirea, A. et al. (2002). Deterioration of natural antioxidant species of vegetable edible oils during the domestic deep-frying and pan-frying of potatoes. *International Journal of Food Sciences and Nutrition* 53: 351–363.

Arvanitoyannis, I.S. and Oikonomou, G. (2012). Active and intelligent packaging. In: *Modified Atmosphere and Active Packaging Technologies* (ed. I.S. Arvanitoyannis). Boca Raton, FL: CRC Press.

Awad, T.S., Moharram, H.A., Shaltout, O.E. et al. (2012). Applications of ultrasound in analysis, processing and quality control of food: a review. *Food Research International* 48: 410–427.

Bagheri, H., Kashaninejad, M., Ziaiifar, A.M., and Aalami, M. (2016). Novel hybridized infrared-hot air method for roasting of peanut kernels. *Innovative Food Science & Emerging Technologies* 37: 106–114.

Barrera-Arellano, D., Ruiz-Méndez, V., Velasco, J. et al. (2002). Loss of tocopherols and formation of degradation compounds at frying temperatures in oils differing in degree of unsaturation and natural antioxidant content. *Journal of the Science of Food and Agriculture* 82: 1696–1702.

Barutcu, I., Sahin, S., and Sumnu, G. (2009). Effects of microwave frying and different flour types addition on the microstructure of batter coatings. *Journal of Food Engineering* 95: 684–692.

Bendini, A., Cerretani, L., Carrasco-Pancorbo, A. et al. (2007). phenolic molecules in virgin olive oils: a survey of their sensory properties, health effects, antioxidant activity and analytical methods. An overview of the last decade Alessandra. *Molecules* 12: 1679.

Bingol, G., Zhang, A., Pan, Z., and Mchugh, T.H. (2012). Producing lower-calorie deep fat fried French fries using infrared dry-blanching as pretreatment. *Food Chemistry* 132: 686–692.

Bingol, G., Wang, B., Zhang, A. et al. (2014). Comparison of water and infrared blanching methods for processing performance and final product quality of French fries. *Journal of Food Engineering* 121: 135–142.

Boskou, D. (2003). Losses of natural antioxidants and vitamins during deep-fat frying. *Forum of Nutrition* 56: 343–345.

Bouaziz, F., Koubaa, M., Neifar, M. et al. (2016). Feasibility of using almond gum as coating agent to improve the quality of fried potato chips: evaluation of sensorial properties. *LWT – Food Science and Technology* 65: 800–807.

Bouchon, P., Aguilera, J., and Pyle, D. (2003). Structure oil-absorption relationships during deep-fat frying. *Journal of Food science* 68: 2711–2716.

Brockgreitens, J. and Abbas, A. (2016). Responsive food packaging: recent progress and technological prospects. *Comprehensive Reviews in Food Science and Food Safety* 15: 3–15.

Bunger, A., Moyano, P., and Rioseco, V. (2003). NaCl soaking treatment for improving the quality of french-fried potatoes. *Food Research International* 36: 161–166.

Byrne, E.H., Prosser, I., Muttucumaru, N. et al. (2012). Overexpression of GCN2-type protein kinase in wheat has profound effects on free amino acid concentration and gene expression. *Plant Biotechnology Journal* 10: 328–340.

Ceballos, A.M., Giraldo, G.I., and Orrego, C.E. (2012). Effect of freezing rate on quality parameters of freeze dried soursop fruit pulp. *Journal of Food Engineering* 111: 360–365.

Chen, S.-D., Chen, H.-H., Chao, Y.-C., and Lin, R.-S. (2009). Effect of batter formula on qualities of deep-fat and microwave fried fish nuggets. *Journal of Food Engineering* 95: 359–364.

Chenlo, F., Chaguri, L., Santos, F., and Moreira, R. (2006). Osmotic dehydration/impregnation kinetics of padrón pepper (*Capsicum annuum* L. Longum) with sodium chloride solutions: process modelling and colour analysis. *Food Science and Technology International* 12: 221–227.

Chiou, A., Kalogeropoulos, N., Salta, F.N. et al. (2009). Pan-frying of French fries in three different edible oils enriched with olive leaf extract: oxidative stability and fate of microconstituents. *LWT – Food Science and Technology* 42: 1090–1097.

Chiou, A., Kalogeropoulos, N., Efstathiou, P. et al. (2013). French Fries oleuropein content during the successive deep frying in oils enriched with an olive leaf extract. *International Journal of Food Science & Technology* 48: 1165–1171.

Choe, E. and Min, D.B. (2007). Chemistry of deep-fat frying oils. *Journal of Food Science* 72: R77–R86

Chung, J., Lee, Y., and Choe, E. (2006). effects of sesame oil addition to soybean oil during frying on the lipid oxidative stability and antioxidants contents of the fried products during storage in the dark. *Journal of Food Science* 71: C222–C226.

Cichello, S.A. (2015). Oxygen absorbers in food preservation: a review. *Journal of Food Science and Technology* 52: 1889–1895.

Claus, A., Carle, R., and Schieber, A. (2008). Acrylamide in cereal products: a review. *Journal of Cereal Science* 47: 118–133.

Curtis, T.Y., Postles, J., and Halford, N.G. (2014). Reducing the potential for processing contaminant formation in cereal products. *Journal of Cereal Science* 59: 382–392.

Da Silva, P.F. and Moreira, R.G. (2008). Vacuum frying of high-quality fruit and vegetable-based snacks. *LWT – Food Science and Technology* 41: 1758–1767.

Debnath, S., Bhat, K.K., and Rastogi, N.K. (2003). Effect of pre-drying on kinetics of moisture loss and oil uptake during deep fat frying of chickpea flour-based snack food. *LWT – Food Science and Technology* 36: 91–98.

Degner, B.M., Olson, K.M., Rose, D. et al. (2013). Influence of freezing rate variation on the microstructure and physicochemical properties of food emulsions. *Journal of Food Engineering* 119: 244–253.

Dehghannya, J. and Abedpour, L. (2018). Influence of a three stage hybrid ultrasound-osmotic-frying process on production of low-fat fried potato strips. *Journal of the Science of Food and Agriculture* 98: 1485–1491.

Dehghannya, J., Naghavi, E.A., and Ghanbarzadeh, B. (2015). Frying of potato strips pretreated by ultrasound-assisted air-drying. *Journal of Food Processing and Preservation* 40: 583–592.

Demurin, Y., Skoric, D., and Karlovic, D. (1996). Genetic variability of tocopherol composition in sunflower seeds as a basis of breeding for improved oil quality. *Plant Breeding* 115: 33–36.

Diamante, L.M., Savage, G.P., Vanhanen, L., and Ihns, R. (2012). Vacuum-frying of apricot slices: effects of frying temperature, time and maltodextrin levels on the moisture, color and texture properties. *Journal of Food Processing and Preservation* 36: 320–328.

Dueik, V. and Bouchon, P. (2011). Vacuum frying as a route to produce novel snacks with desired quality attributes according to new health trends. *Journal of Food Science* 76: E188–E195.

Dueik, V., Robert, P., and Bouchon, P. (2010). Vacuum frying reduces oil uptake and improves the quality parameters of carrot crisps. *Food Chemistry* 119: 1143–1149.

Durán, M., Pedreschi, F., Moyano, P., and Troncoso, E. (2007). Oil partition in pre-treated potato slices during frying and cooling. *Journal of Food Engineering* 81: 257–265.

Erdoğdu, S., Palazoğlu, T.K., Gökmen, V. et al. (2007). Reduction of acrylamide formation in French fries by microwave pre-cooking of potato strips. *Journal of the Science of Food and Agriculture* 87: 133–137.

European Union. 2009. EU Guidance to the Commission Regulation (EC) No. 450/2009 of 29 May 2009 on active and intelligent materials and articles intended to come into the contact with food (version 1.0). European Commission.

Farhoosh, R. and Esmaeilzadeh, K.R. (2009). Anti-rancidity effects of sesame and rice bran oils on canola oil during deep frying. *Journal of the American Oil Chemists' Society* 86: 539–544.

Farkas, B.E., Singh, R.P., and Rumsey, T.R. (1996). Modeling heat and mass transfer in immersion frying. I, model development. *Journal of Food Engineering* 29: 211–226.

Fernandes, F.A.N., Rodrigues, S., Gaspareto, O.C.P., and Oliveira, E.L. (2006). Optimization of osmotic dehydration of papaya followed by air-drying. *Food Research International* 39: 492–498.

Foot, R.J., Haase, N.U., Grob, K., and Gonde, P. (2007). Acrylamide in fried and roasted potato products: a review on progress in mitigation. *Food Additives and Contaminants* 24 (Suppl. 1): 37–46.

Franke, K., Sell, M., and Reimerdes, E.H. (2005). Quality related minimization of acrylamide formation – an integrated approach. *Advances in Experimental Medicine and Biology* 561: 357–369.

Freitas, D.D.G.C., Berbari, S.A.G., Prati, P. et al. (2009). Reducing fat uptake in cassava product during deep-fat frying. *Journal of Food Engineering* 94: 390–394.

Gachovska, T.K., Simpson, M.V., Ngadi, M.O., and Raghavan, G. (2009). Pulsed electric field treatment of carrots before drying and rehydration. *Journal of the Science of Food and Agriculture* 89: 2372–2376.

Gallego-Juárez, J.A., Riera, E., de la Fuente Blanco, S. et al. (2007). Application of high-power ultrasound for dehydration of vegetables: processes and devices. *Drying Technology* 25: 1893–1901.

Gamble, M.H. and Rice, P. (1987). Effect of pre-fry drying of oil uptake and distribution in potato crisp manufacture. *International Journal of Food Science & Technology* 22: 535–548.

Gerendás, J., Heuser, F., and Sattelmacher, B. (2007). Influence of nitrogen and potassium supply on contents of acrylamide precursors in potato tubers and on acrylamide accumulation in French fries. *Journal of Plant Nutrition* 30: 1499–1516.

Gertz, C. (2004). Optimising the baking and frying process using oil-improving agents. *European Journal of Lipid Science and Technology* 106: 736–745.

Ghidurus, M., Turtoi, M., Boskou, G. et al. (2010). Nutritional and health aspects related to frying (I). *Romanian Biotechnological Letters* 15: 5675–5682.

Gökmen, V. and Şenyuva, H.Z. (2007). Effects of some cations on the formation of acrylamide and furfurals in glucose–asparagine model system. *European Food Research and Technology* 225: 815–820.

Gopala, K.A.G., Khatoon, S., and Babylatha, R. (2005). Frying performance of processed rice bran oils. *Journal of Food Lipids* 12: 1–11.

Gordon, M.H. and Kourkimskå, L. (1995). The effects of antioxidants on changes in oils during heating and deep frying. *Journal of the Science of Food and Agriculture* 68: 347–353.

Hoenicke, K. and Gatermann, R. (2005). Studies on the stability of acrylamide in food during storage. *Journal of AOAC International* 88: 268–273.

Hosseini, H., Ghorbani, M., Meshginfar, N., and Mahoonak, A.S. (2016). A review on frying: procedure, fat, deterioration progress and health hazards. *Journal of the American Oil Chemists' Society* 93: 445–466.

Huang, L. (2004). Infrared surface pasteurization of Turkey frankfurters. *Innovative Food Science & Emerging Technologies* 5: 345–351.

Hussein, Z., Caleb, O.J., and Opara, U.L. (2015). Perforation-mediated modified atmosphere packaging of fresh and minimally processed produce – a review. *Food Packaging and Shelf Life* 6: 7–20.

Hwang, H.S., Winkler-Moser, J.K., and Liu, S.X. (2012). Structural effect of lignans and sesamol on polymerization of soybean oil at frying temperature. *Journal of the American Oil Chemists' Society* 89: 1067–1076.

Hwang, H.-S., Winkler-Moser, J.K., Bakota, E.L. et al. (2013). Antioxidant activity of sesamol in soybean oil under frying conditions. *Journal of the American Oil Chemists' Society* 90: 659–666.

Ignat, A., Manzocco, L., Brunton, N.P. et al. (2015). The effect of pulsed electric field pre-treatments prior to deep-fat frying on quality aspects of potato fries. *Innovative Food Science & Emerging Technologies* 29: 65–69.

Imran, M., Revol-Junelles, A.-M., Martyn, A. et al. (2010). Active food packaging evolution: transformation from micro-to nanotechnology. *Critical Reviews in Food Science and Nutrition* 50: 799–821.

Jalté, M., Lanoisellé, J.-L., Lebovka, N.I., and Vorobiev, E. (2009). Freezing of potato tissue pre-treated by pulsed electric fields. *LWT – Food Science and Technology* 42: 576–580.

Janositz, A., Noack, A.-K., and Knorr, D. (2011). Pulsed electric fields and their impact on the diffusion characteristics of potato slices. *LWT-Food Science and Technology* 44: 1939–1945.

Janve, B., Yang, W., and Sims, C. (2015). Sensory and quality evaluation of traditional compared with power ultrasound processed corn (*Zea mays*) tortilla chips. *Journal of Food Science* 80: S1368–S1376.

Jiang, H., Zhang, M., Mujumdar Arun, S., and Lim, R.X. (2011). Comparison of the effect of microwave freeze drying and microwave vacuum drying upon the process and quality characteristics of potato/banana re-structured chips. *International Journal of Food Science & Technology* 46: 570–576.

Jung, M.Y., Choi, D.S., and Ju, J. W. (2003). A novel technique for limitation of acrylamide formation in fried and baked corn chips and in French fries. *Journal of Food Science* 68: 1287–1290.

Kalita, D. and Jayanty, S.S. (2013). Reduction of acrylamide formation by vanadium salt in potato French fries and chips. *Food Chemistry* 138: 644–649.

Karacabey, E., Turan, M.S., Ozcelik, S.G. et al. (2016). Optimisation of pre-drying and deep-fat-frying conditions for production of low-fat fried carrot slices. *Journal of the Science of Food and Agriculture* 96: 4603–4612.

Karizaki, V.M., Sahin, S., Sumnu, G. et al. (2013). Effect of ultrasound-assisted osmotic dehydration as a pretreatment on deep fat frying of potatoes. *Food and Bioprocess Technology* 6: 3554–3563.

Kawas, M.L. and Moreira, R.G. (2001). Effect of degree of starch gelatinization on quality attributes of fried tortilla chips. *Journal of Food Science* 66: 300–306.

Keramat, J., Lebail, A., Prost, C., and Jafari, M. (2011). Acrylamide in baking products: a review article. *Food and Bioprocess Technology* 4: 530–543.

Kerry, J. and Butler, P. (2008). *Smart Packaging Technologies for Fast Moving Consumer Goods*. Chichester: John Wiley.

Kerstin, O., Rita, S., and Carl-Axel, R. (2004). Tuber components affecting acrylamide formation and colour in fried potato: variation by variety, year, storage temperature and storage time. *Journal of the Science of Food and Agriculture* 84: 447–458.

Kilincceker, O. and Hepsag, F. (2012). Edible coating effects on fried potato balls. *Food and Bioprocess Technology* 5: 1349–1354.

Kim, C.T., Hwang, E.S., and Lee, H.J. (2005). reducing acrylamide in fried snack products by adding amino acids. *Journal of Food Science* 70: C354–C358.

Kim, D.N., Lim, J., Bae, I.Y. et al. (2011). Effect of hydrocolloid coatings on the heat transfer and oil uptake during frying of potato strips. *Journal of Food Engineering* 102: 317–320.

Kim, Y., Min, B., and Kim, K. (2014). Biopolymer-based composite packaging materials with nanoparticles. In: *Innovations in Food Packaging* (ed. J. Han). San Diego, CA: Academic Press.

Kingcam, R., Devahastin, S., and Chiewchan, N. (2008). Effect of starch retrogradation on texture of potato chips produced by low-pressure superheated steam drying. *Journal of Food Engineering* 89: 72–79.

Kita, A., Brathen, E., Knutsen, S.H., and Wicklund, T. (2004). Effective ways of decreasing acrylamide content in potato crisps during processing. *Journal of Agricultural and Food Chemistry* 52: 7011–7016.

Kochhar, S.P. (2000). Stabilisation of frying oils with natural antioxidative components. *European Journal of Lipid Science and Technology* 102: 552–559.

Krokida, M.K., Oreopoulou, V., and Maroulis, Z.B. (2000). Water loss and oil uptake as a function of frying time. *Journal of Food Engineering* 44: 39–46.

Krokida, M.K., Oreopoulou, V., Maroulis, Z.B., and Marinos-Kouris, D. (2001a). Colour changes during deep fat frying. *Journal of Food Engineering* 48: 219–225.

Krokida, M.K., Oreopoulou, V., Maroulis, Z.B., and Marinos-Kouris, D. (2001b). Effect of osmotic dedydration pretreatment on quality of French fries. *Journal of Food Engineering* 49: 339–345.

Krokida, M.K., Oreopoulou, V., Maroulis, Z.B., and Marinos-Kouris, D. (2001c). Effect of pre-drying on quality of French fries. *Journal of Food Engineering* 49: 347–354.

Krokida, M.K., Oreopoulou, V., Maroulis, Z.B., and Marinos-Kouris, D. (2001d). Effect of pre-treatment on viscoelastic behaviour of potato strips. *Journal of Food Engineering* 50: 11–17.

Kuorwel, K.K., Cran, M.J., Orbell, J.D. et al. (2015). Review of mechanical properties, migration, and potential applications in active food packaging systems containing nanoclays and nanosilver. *Comprehensive Reviews in Food Science and Food Safety* 14: 411–430.

Lalam, S., Sandhu, J.S., Takhar, P.S. et al. (2013). Experimental study on transport mechanisms during deep fat frying of chicken nuggets. *LWT – Food Science and Technology* 50: 110–119.

Lee, D.S. and Paik, H.D. (1997). Use of a pinhole to develop an active packaging system for kimchi, a Korean fermented vegetable. *Packaging Technology and Science* 10: 33–43.

Lerma-García, M.J., Herrero-Martínez, J.M., Simó-Alfonso, E.F. et al. (2009). Composition, industrial processing and applications of rice bran γ-oryzanol. *Food Chemistry* 115: 389–404.

Li, X., Zhang, A., Atungulu, G.G. et al. (2014). Effects of infrared radiation heating on peeling performance and quality attributes of clingstone peaches. *LWT – Food Science and Technology* 55: 34–42.

Liu, T., Dodds, E., Leong, S.Y. et al. (2017). Effect of pulsed electric fields on the structure and frying quality of 'kumara' sweet potato tubers. *Innovative Food Science & Emerging Technologies* 39: 197–208.

Llorens, A., Lloret, E., Picouet, P.A. et al. (2012). Metallic-based micro and nanocomposites in food contact materials and active food packaging. *Trends in Food Science & Technology* 24: 19–29.

Mai Tran, T.T., Chen, X.D., and Southern, C. (2007). Reducing oil content of fried potato crisps considerably using a 'sweet' pre-treatment technique. *Journal of Food Engineering* 80: 719–726.

Maity, T., Raju, P.S., and Bawa, A.S. (2012). Effect of freezing on textural kinetics in snacks during frying. *Food and Bioprocess Technology* 5: 155–165.

Małecka, M. (1994). The effect of squalene on the thermostability of rapeseed oil. *Food/Nahrung* 38: 135–140.

Mallikarjunan, P., Chinnan, M., Balasubramaniam, V., and Phillips, R. (1997). Edible coatings for deep-fat frying of starchy products. *LWT – Food Science and Technology* 30: 709–714.

Marinova, E.M. and Yanishlieva, N.V. (1992). Effect of temperature on the antioxidative action of inhibitors in lipid autoxidation. *Journal of the Science of Food and Agriculture* 60: 313–318.

Mariotti, S., Pedreschi, F., Carrasco, J.A., and Granby, K. (2011). Patented techniques for acrylamide mitigation in high-temperature processed foods. *Recent Patents on Food, Nutrition & Agriculture* 3: 158–171.

Mariscal, M. and Bouchon, P. (2008). Comparison between atmospheric and vacuum frying of apple slices. *Food chemistry* 107: 1561–1569.

Márquez-Ruiz, G., Martín-Polvillo, M., Jorge, N. et al. (1999). Influence of used frying oil quality and natural tocopherol content on oxidative stability of fried potatoes. *Journal of the American Oil Chemists' Society* 76: 421–425.

Márquez-Ruiz, G., Martín-Polvillo, M., Velasco, J., and Dobarganes, C. (2008). Formation of oxidation compounds in sunflower and olive oils under oxidative stability index conditions. *European Journal of Lipid Science and Technology* 110: 465–471.

Márquez-Ruiz, G., Ruiz-Méndez, M.V., and Velasco, J. (2014). Antioxidants in frying: analysis and evaluation of efficacy. *European Journal of Lipid Science and Technology* 116: 1441–1450.

Martínez Ávila, M., Vélez-Ruiz, J.F., and Sosa-Morales, M.E. (2010). Combination of hot air drying and deep-fat frying to reduce the oil content in chicken nuggets. *International Journal of Food Science & Technology* 45: 2101–2107.

Martín-Polvillo, M., Márquez Ruiz, G., and Dobarganes, M.C. (2004). Oxidative stability of sunflower oils differing in unsaturation degree during long-term storage at room temperature. *Journal of the American Oil Chemists' Society* 81: 577–583.

Melito, H.S. and Farkas, B.E. (2012). Impact of infrared finishing on the mechanical and sensorial properties of wheat donuts. *Journal of Food Science* 77: E224–E230.

Mellema, M. (2003). Mechanism and reduction of fat uptake in deep-fat fried foods. *Trends in Food Science & Technology* 14: 364–373.

Mizukami, Y., Kohata, K., Yamaguchi, Y. et al. (2006). Analysis of acrylamide in green tea by gas chromatography-mass spectrometry. *Journal of Agricultural and Food Chemistry* 54: 7370–7377.

Moazzami, A.A., Haese, S.L., and Kamal-Eldin, A. (2007). Lignan contents in sesame seeds and products. *European Journal of Lipid Science and Technology* 109: 1022–1027.

Motilva, M.-J., Serra, A., and Macià, A. (2013). Analysis of food polyphenols by ultra high-performance liquid chromatography coupled to mass spectrometry: an overview. *Journal of Chromatography A* 1292: 66–82.

Moyano, P.C. and Pedreschi, F. (2006). Kinetics of oil uptake during frying of potato slices: effect of pre-treatments. *LWT – Food Science and Technology* 39: 285–291.

Muttucumaru, N., Halford, N.G., Elmore, J.S. et al. (2006). Formation of high levels of acrylamide during the processing of flour derived from sulfate-deprived wheat. *Journal of Agricultural and Food Chemistry* 54: 8951–8955.

Nakagawa, K. and Ochiai, T. (2015). A mathematical model of multi-dimensional freeze-drying for food products. *Journal of Food Engineering* 161: 55–67.

Naz, S., Siddiqi, R., and Asad Sayeed, S. (2008). Effect of flavonoids on the oxidative stability of corn oil during deep frying. *International Journal of Food Science & Technology* 43: 1850–1854.

Naziri, E., Mitić, M.N., and Tsimidou, M.Z. (2016). Contribution of tocopherols and squalene to the oxidative stability of cold-pressed pumkin seed oil (*Cucurbita pepo* L.). *European Journal of Lipid Science and Technology* 118: 898–905.

Ngadi, M., Li, Y., and Oluka, S. (2007). Quality changes in chicken nuggets fried in oils with different degrees of hydrogenatation. *LWT – Food Science and Technology* 40: 1784–1791.

Ngadi, M.O., Wang, Y., Adedeji, A.A., and Raghavan, G.S.V. (2009). Effect of microwave pretreatment on mass transfer during deep-fat frying of chicken nugget. *LWT – Food Science and Technology* 42: 438–440.

Normand, L., Eskin, N., and Przybylski, R. (2001). Effect of tocopherols on the frying stability of regular and modified canola oils. *Journal of the American Oil Chemists' Society* 78: 369–373.

Nowak, D., Piechucka, P., Witrowa-Rajchert, D., and Wiktor, A. (2016). Impact of material structure on the course of freezing and freeze-drying and on the properties of dried substance, as exemplified by celery. *Journal of Food Engineering* 180: 22–28.

Nunes, Y. and Moreira, R.G. (2009). Effect of osmotic dehydration and vacuum-frying parameters to produce high-quality mango chips. *Journal of Food Science* 74: E355–E362.

Oladejo, A.O. and Ma, H. (2016). Optimisation of ultrasound-assisted osmotic dehydration of sweet potato (*Ipomea batatas*) using response surface methodology. *Journal of the Science of Food and Agriculture* 96: 3688–3693.

Oladejo, A.O., Ma, H., Qu, W. et al. (2017). Influence of ultrasound pretreatments on diffusion coefficients, texture and colour of osmodehydrated sweet potato (*Ipomea batatas*). *International Journal of Food Science & Technology* 52: 888–896.

Oladejo, A.O., Ma, H., Qu, W. et al. (2018). Application of pretreatment methods on agricultural products prior to frying: a review. *Journal of the Science of Food and Agriculture* 98: 456–466.

Onwude, D.I., Hashim, N., and Chen, G. (2016). Recent advances of novel thermal combined hot air drying of agricultural crops. *Trends in Food Science & Technology* 57: 132–145.

Oyedeji, A., Sobukola, O., Henshaw, F. et al. (2017). Effect of frying treatments on texture and colour parameters of deep fat fried yellow fleshed cassava chips. *Journal of Food Quality* https://doi.org/10.1155/2017/8373801 2017: Article ID 8373801.

Palazoglu, T.K., Savran, D., and Gokmen, V. (2010). Effect of cooking method (baking compared with frying) on acrylamide level of potato chips. *Journal of Food Science* 75: E25–E29.

Pedreschi, F. and Moyano, P. (2005). Effect of pre-drying on texture and oil uptake of potato chips. *LWT – Food Science and Technology* 38: 599–604.

Pedreschi, F., Kaack, K., and Granby, K. (2004). Reduction of acrylamide formation in potato slices during frying. *LWT – Food Science and Technology* 37: 679–685.

Pedreschi, F., Bustos, O., Mery, D. et al. (2007a). Color kinetics and acrylamide formation in NaCl soaked potato chips. *Journal of Food Engineering* 79: 989–997

Pedreschi, F., Moyano, P., Santis, N., and Pedreschi, R. (2007b). Physical properties of pre-treated potato chips. *Journal of Food Engineering* 79: 1474–1482.

Pedreschi, F., Mariotti, M.S., and Granby, K. (2014). Current issues in dietary acrylamide: formation, mitigation and risk assessment. *Journal of the Science of Food and Agriculture* 94: 9–20.

Primo-Martín, C. and van Deventer, H. (2011). Deep-fat fried battered snacks prepared using super heated steam (SHS): Crispness and low oil content. *Food Research International* 44: 442–448.

Rababah, T.M., Feng, H., Yang, W., and Yücel, S. (2012). Fortification of potato chips with natural plant extracts to enhance their sensory properties and storage stability. *Journal of the American Oil Chemists' Society* 89: 1419–1425.

Rahimi, J. and Ngadi, M.O. (2014). Effect of batter formulation and pre-drying time on oil distribution fractions in fried batter. *LWT – Food Science and Technology* 59: 820–826.

Rodriguez-Aguilera, R. and Oliveira, J.C. (2009). Review of design engineering methods and applications of active and modified atmosphere packaging systems. *Food Engineering Reviews* 1: 66–83.

Romero, N., Robert, P., Masson, L. et al. (2007). Effect of α-tocopherol, α-tocotrienol and Rosa mosqueta shell extract on the performance of antioxidant-stripped canola oil (*Brassica* sp.) at high temperature. *Food Chemistry* 104: 383–389.

Rommens, C.M., Yan, H., Swords, K. et al. (2008). Low-acrylamide French fries and potato chips. *Plant Biotechnology Journal* 6: 843–853.

Rossell, J. (2001). *Frying: Improving Quality*. Chichester: Woodhead Publishing.

Rossi, M., Alamprese, C., and Ratti, S. (2007). Tocopherols and tocotrienols as free radical-scavengers in refined vegetable oils and their stability during deep-fat frying. *Food Chemistry* 102: 812–817.

Rydberg, P., Eriksson, S., Tareke, E. et al. (2003). Investigations of factors that influence the acrylamide content of heated foodstuffs. *Journal of Agricultural and Food Chemistry* 51: 7012–7018.

Sahin, S., Sumnu, G., and Oztop, M.H. (2007). Effect of osmotic pretreatment and microwave frying on acrylamide formation in potato strips. *Journal of the Science of Food and Agriculture* 87: 2830–2836.

Sansano, M., Castelló, M.L., Heredia, A., and Andrés, A. (2016). Protective effect of chitosan on acrylamide formation in model and batter systems. *Food Hydrocolloids* 60: 1–6.

Santis, N., Mendoza, F., Moyano, P. et al. (2007). Soaking in a NaCl solution produce paler potato chips. *LWT – Food Science and Technology* 40: 307–312.

Sanz, T., Salvador, A., and Fiszman, S.M. (2004). Innovative method for preparing a frozen, battered food without a prefrying step. *Food Hydrocolloids* 18: 227–231.

Sehrawat, R., Nema, P.K., and Kaur, B.P. (2016). Effect of superheated steam drying on properties of foodstuffs and kinetic modeling. *Innovative Food Science & Emerging Technologies* 34: 285–301.

Sensoy, I., Sahin, S., and Sumnu, G. (2013). Microwave frying compared with conventional frying via numerical simulation. *Food and Bioprocess Technology* 6: 1414–1419.

Shyi-Liang, S., Lung-Bin, H., and Sun, H.L. (2005). Effects of processing conditions on the quality of vacuum-fried carrot chips. *Journal of the Science of Food and Agriculture* 85: 1903–1908.

Simonne, A.H. and Eitenmiller, R.R. (1998). Retention of Vitamin E and added retinyl palmitate in selected vegetable oils during deep-fat frying and in fried breaded products. *Journal of Agricultural and Food Chemistry* 46: 5273–5277.

Singh, P., Abas Wani, A., and Saengerlaub, S. (2011). Active packaging of food products: recent trends. *Nutrition & Food Science* 41: 249–260.

Siucińska, K. and Konopacka, D. (2014). Application of ultrasound to modify and improve dried fruit and vegetable tissue: a review. *Drying Technology* 32: 1360–1368.

Soorgi, M., Mohebbi, M., Mousavi, S.M., and Shahidi, F. (2012). The effect of methylcellulose, temperature, and microwave pretreatment on kinetic of mass transfer during deep fat frying of chicken nuggets. *Food and Bioprocess Technology* 5: 1521–1530.

Su, Y., Zhang, M., Zhang, W. et al. (2016). Application of novel microwave-assisted vacuum frying to reduce the oil uptake and improve the quality of potato chips. *LWT – Food Science and Technology* 73: 490–497.

Suárez, R.B., Campañone, L.A., García, M.A., and Zaritzky, N.E. (2008). Comparison of the deep frying process in coated and uncoated dough systems. *Journal of Food Engineering* 84: 383–393.

Susana, M., Leonardo, V., Victoria, R.M.M. et al. (2008). Thermostability of genetically modified sunflower oils differing in fatty acid and tocopherol compositions. *European Journal of Lipid Science and Technology* 110: 776–782.

Taiwo, K.A. and Baik, O.D. (2007). Effects of pre-treatments on the shrinkage and textural properties of fried sweet potatoes. *LWT – Food Science and Technology* 40: 661–668.

Taiwo, K.A., Eshtiaghi, M.N., Ade-Omowaye, B.I.O., and Knorr, D. (2003). Osmotic dehydration of strawberry halves: influence of osmotic agents and pretreatment methods on mass transfer and product characteristics. *International Journal of Food Science & Technology* 38: 693–707.

Troncoso, E., Pedreschi, F., and Zúñiga, R.N. (2009). Comparative study of physical and sensory properties of pre-treated potato slices during vacuum and atmospheric frying. *LWT – Food Science and Technology* 42: 187–195.

Van Loon, W.A.M., Linssen, J.P.H., Legger, A. et al. (2005). Study of a new energy efficient process for French fries production. *European Food Research and Technology* 221: 779–786.

Verleyen, T., Kamal-Eldin, A., Dobarganes, C. et al. (2001). Modeling of α-tocopherol loss and oxidation products formed during thermoxidation in triolein and tripalmitin mixtures. *Lipids* 36: 719–726.

Verleyen, T., Kamal-Eldin, A., Mozuraityte, R. et al. (2002). Oxidation at elevated temperatures: competition between α-tocopherol and unsaturated triacylglycerols. *European Journal of Lipid Science and Technology* 104: 228–233.

Vinci, R.M., Mestdagh, F., De Muer, N. et al. (2010). Effective quality control of incoming potatoes as an acrylamide mitigation strategy for the French fries industry. *Food Additives & Contaminants. Part A, Chemistry, Analysis, Control, Exposure & Risk Assessment* 27: 417–425.

Vinci, R.M., Mestdagh, F., and De Meulenaer, B. (2012). Acrylamide formation in fried potato products – present and future, a critical review on mitigation strategies. *Food Chemistry* 133: 1138–1154.

Vitrac, O., Dufour, D., Trystram, G., and Raoult-Wack, A.-L. (2002). Characterization of heat and mass transfer during deep-fat frying and its effect on cassava chip quality. *Journal of Food Engineering* 53: 161–176.

Wagner, K.H., Wotruba, F., and Elmadfa, I. (2001). Antioxidative potential of tocotrienols and tocopherols in coconut fat at different oxidation temperatures. *European Journal of Lipid Science and Technology* 103: 746–751.

Wang, T., Khir, R., Pan, Z., and Yuan, Q. (2017). Simultaneous rough rice drying and rice bran stabilization using infrared radiation heating. *LWT – Food Science and Technology* 78: 281–288.

Warner, K. and Moser, J. (2009). Frying stability of purified mid-oleic sunflower oil triacylglycerols with added pure tocopherols and tocopherol mixtures. *Journal of the American Oil Chemists' Society* 86: 1199.

Wicklund, T., Østlie, H., Lothe, O. et al. (2006). Acrylamide in potato crisp—the effect of raw material and processing. *LWT – Food Science and Technology* 39: 571–575.

Winkler, J.K. and Warner, K. (2008). Effect of phytosterol structure on thermal polymerization of heated soybean oil. *European Journal of Lipid Science and Technology* 110: 1068–1077.

Wu, B., Pan, Z., Qu, W. et al. (2014). Effect of simultaneous infrared dry-blanching and dehydration on quality characteristics of carrot slices. *LWT – Food Science and Technology* 57: 90–98.

Xu, Y., Cui, B., Ran, R. et al. (2014). Risk assessment, formation, and mitigation of dietary acrylamide: current status and future prospects. *Food and Chemical Toxicology* 69: 1–12.

Yanishlieva, N.V. and Marinova, E.M. (2001). Stabilisation of edible oils with natural antioxidants. *European Journal of Lipid Science and Technology* 103: 752–767.

Yanishlieva, N.V., Marinova, E., and Pokorný, J. (2006). Natural antioxidants from herbs and spices. *European Journal of Lipid Science and Technology* 108: 776–793.

Yildirim, S., Röcker, B., Pettersen, M.K. et al. (2018). Active packaging applications for food. *Comprehensive Reviews in Food Science and Food Safety* 17: 165–199.

Zeb, A. (2012). Thermal degradation of β-carotene in food oils. In: *Vitamin A and Carotenoids: Chemistry* (ed. V.R. Preedy). London: The Royal Society of Chemistry.

Zeb, A. and Murkovic, M. (2011). Carotenoids and triacylglycerols interactions during thermal oxidation of refined olive oil. *Food Chemistry* 127: 1584–1593.

Zhu, F., Yi-Zhong, C., Jinxia, K., and Harold, C. (2009). Evaluation of the effect of plant extracts and phenolic compounds on reduction of acrylamide in an asparagine/glucose model system by RP-HPLC-DAD. *Journal of the Science of Food and Agriculture* 89: 1674–1681.

Zielinska, M. and Cenkowski, S. (2012). Superheated steam drying characteristic and moisture diffusivity of distillers' wet grains and condensed distillers' solubles. *Journal of Food Engineering* 109: 627–634.

Zielinska, M., Błaszczak, W., and Devahastin, S. (2015). Effect of superheated steam prefrying treatment on the quality of potato chips. *International Journal of Food Science & Technology* 50: 158–168.

14

The Future of Food Frying

14.1 Introduction

Fried foods are the most popular food items consumed across the globe, due to their specific taste, sensory characteristics, colour, and acceptability. Consumption of fried foods is increasing by the day. It has been observed that since 2011, the consumption of fried chicken has increased significantly in the United States (Statista 2018). This shows that fried foods are still the item of choice both at home and in restaurants. Restaurants use specialized fryer systems to prepare foods of acceptable quality. During frying, newly formed oxidized compounds enter the food matrix as part of oil uptake, where they are likely to be consumed.

14.2 Current Strategies

The science of frying has been subject to severe scrutiny in recent years due to the discovery of hazardous compounds affecting the quality of foods and consumer health. Since fried food consumption is associated with several chronic diseases and health conditions, including obesity, diabetes, cancers, and cardiovascular disease, the main purpose of current research is to obtain fried foods of the highest quality with the fewest negative effects.

14.2.1 Improving Frying

There has been tremendous growth in research into the improvement of food frying. Frying should be safe, easy, cost-effective, and healthy. Rossell (2001) showed us how to improve the quality of fried foods. However, a lot of research is still needed to resolve the important issues pertaining to the quality of foods fried both in industry and in restaurants.

14.2.2 Obesity and Fried Foods

Wainwright and Lampert (2007) showed that consumption of fried foods is one of the reasons leading to the current obesity epidemic. These authors were of the view that fried foods, sweets, and carbonated beverages are the most likely causes of obesity.

Food Frying: Chemistry, Biochemistry, and Safety, First Edition. Alam Zeb.

In experimental model animals such as rabbits, the consumption of oxidized sunflower oils showed an increase in the deposition of visceral fats (Zeb and Hussain 2014). The same is also evident in humans. Prentice and Jebb (2003) showed that the energy density of foods is a key factor affecting energy intake. Foods with high energy densities do not sate human appetites, and thus an excess of energy is consumed, which promotes weight gain and obesity. This has been confirmed by Guallar-Castillon et al. (2007) in a Spanish population consuming fried foods. The authors showed that fried food was clearly associated with general and central obesity among Spanish subjects in the highest quintile of energy intake from fried foods.

In a long-term Mediterranean prospective cohort study of 9850 men and women with a mean age of 38.1 years, more frequent fried food consumption at baseline was associated with subsequent obesity during follow-up at 6.1 years (Sayon-Orea et al. 2013). Fried foods could interact with a genetic background in relation to obesity (Qi et al. 2014). Work to date shows that fried food consumption cannot be ignored in looking for a solution to the obesity problem. The World Health Organization (WHO) presented some key facts relating to worldwide obesity for the year 2016:

- Worldwide obesity has nearly tripled since 1975.
- In 2016, more than 1.9 billion adults, 18 years and older, were overweight. Of these, over 650 million were obese.
- 39% of adults aged 18 years and over were overweight in 2016, and 13% were obese.
- Most of the world's population live in countries where overweight and obesity kills more people than underweight.
- 41 million children under the age of 5 were overweight or obese in 2016.
- Over 340 million children and adolescents aged 5–19 were overweight or obese in 2016.
- Obesity is avoidable (WHO 2018).

14.2.3 Diabetes and Cardiovascular Diseases

Diabetes is a chronic disease affecting individual life significantly. It may lead to cardiovascular diseases. As mentioned in Chapter 10, fried food consumption is associated with an increased incidence of diabetes and cardiovascular diseases. In a recent study, Cahill et al. (2014) assessed fried food consumption using a survey of 70 842 women from the Nurses' Health Study (1984–2010) and 40 789 men from the follow-up the study of health professionals (1986–2010). The subjects of the study had no diabetes, cardiovascular disease, or cancer at baseline. The authors concluded that frequent fried food consumption was significantly linked with risk of incident type-2 diabetes and moderately linked with occurrence of cardiovascular diseases. They also found that links were largely mediated by body weight and comorbid hypertension and hypercholesterolaemia. In another Spanish prospective cohort study, 35 583 men aged 45–79 years were followed from 1998 to 2012 (Wallin et al. 2017). It was concluded that consumption of fried fish was associated with incidence of type-2 diabetes. Similarly, in Costa Rica, Hu et al. (2018) showed that fried food consumption outside of the home could have adverse effects in terms of promoting cardiovascular disease. Gadiraju et al. (2015) reviewed the consumption of fried foods and their association with cardiovascular health. They provide enough evidence to suggest adverse effects of frequent fried food consumption. These studies confirm that consumption of fried food is one of the reasons for the high incidence of diabetes and cardiovascular diseases today.

14.2.4 Cancer and Fried Foods

Fried foods are also associated with an increased incidence of cancer (see Chapter 10). A recent French cohort study showed that ultra-processed food consumption is significantly associated with the incidence of breast cancer (Fiolet et al. 2018). Similar results were reported by Kojima et al. (2017) for breast cancer in the Japanese population. Gastric cancer (Guo et al. 2018), colorectal cancer (Tayyem et al. 2018), and other cancers (Ganesan et al. 2017) are associated with the consumption of fried foods. Thus, our future health is at risk from fried foods.

14.3 Future Scenarios

14.3.1 Frying Medium and Characterization

Frying media are obtained from animal and, mostly, plant sources. There is significant focus on obtaining the highest quality of frying medium with high nutritional value and good thermal stability. For this purpose, genetic modifications of the oils seeds are carried out of on already known oilseed crops. Discovery of new sources of edible oils is a hot topic. For example, Hashemi et al. (2017) characterized novel edible oil sources, including Mashhadi melon (*Cucumis melo* var. Iranians cv. Mashhadi), Iranian watermelon (*Citrullus lanatus* cv. Fire Fon), pumpkin (*Cucurbita pepo subsp. pepo* var. Styriaca), and yellow apple (*Malus domestica* cv. Golden Delicious) seed oils. They found the order of oxidative stability of the analysed oils was pumpkin > Mashhadi melon > Iranian watermelon > yellow apple.

14.3.2 *Trans*-Esterification and Oil Blends

Trans-esterification is a process in which fatty acid (FA) moiety is exchanged or replaced with a new desirable FA. This process is very helpful in converting oils of lower thermal stability to high stability. Oil blends represent a far better and easier way of increasing the nutritional value and thermal properties of edible oils than *trans*-esterification. Blends are now widely used for food frying. However, it is hard to know what types of oil a blend is made of, and adulteration can easily be introduced. Therefore, there has been much focus on the adulteration, determination, and authentication of different oil contents. Jiménez-Carvelo et al. (2017) identified the composition of a blend using normal phase liquid chromatography (NP-LC). They showed that oils blends can be characterized in terms of their constituent oils.

14.3.3 Food Availability and Characterization

Food is a basic component of frying science, and its availability is important in producing fried products of desirable characteristics. Food fryers need fresh foods of the highest quality. Foods are produced by farmers. Scientists have introduced significant advances in food production, processing, preservation, and characterization. To date, however, not all types of food that are usually fried have been characterized. Therefore, continuous food availability is required to produce food of the highest quality.

14.3.4 After-Frying Characterization

Frying media and fried foods are characterized in terms of their quality and consumer acceptability. Food chemists and sensory scientists are working on the quality of fried foods and frying media, especially the storage quality. Novel portable analytical characterization equipments have been developed with wide applicability.

14.3.5 Innovations and Emerging Technology

In restaurants and food industries, frying is carried out using automatic or semiautomatic mechanical fryer systems. Food engineers are working on designing new technologies for the food frying process. These technologies are based on optimum frying conditions set by the user, which helps in preparing high-quality fried foods.

14.3.6 Mitigation Strategies

Several mitigation strategies have been devised to minimize the impact of hazardous substances formed during food frying. These include acrylamide, furans, heterocyclic aromatic amines (HAAs), polycyclic aromatic hydrocarbons (PAHs), and other hazardous compounds (Zuo et al. 2015; Chang et al. 2016; Rannou et al. 2016; Qi et al. 2018).

14.3.7 Quality Control Agencies

Maintaining the highest quality of fried food products can only be done if quality control procedures are strictly followed. This requires a quality control agency in every country. Countries where food quality control agencies are active contribute much to the safety of fried food production.

14.3.8 Curbing Obesity and Health Effects

Eating fried foods induces obesity in animals as well as humans. However, fried foods are unavoidable for humans. Thus, strategies are needed to find alternative ways to be safe while eating fried foods rich in oxidized oils. Experiments show that animals develop obesity upon consumption of dietary oxidized fats, as shown in the Figure 14.1 (Zeb and Hussain 2014). When oxidized sunflower oil was fortified with sea buckthorn seed powder, the fat deposition was significantly reduced. Similarly, sea buckthorn seed oil (Zeb and Ullah 2015), tomato powder (Zeb and Haq 2016), and amino acids such as glycine and glutamic acid (Zeb and Rahman 2017) were effective in ameliorating the toxic effects of fried oils. Thus, significant attention is still needed to curb obesity and its disease contribution by fried foods.

Obesity is a multifactorial disease, and fast food companies are certainly not alone in its causation. Prentice and Jebb (2003) suggested some constructive measures which could be taken by fast food vendors to help stem the rise in obesity:

1) Provide and promote a wider range of healthy options with low energy density.
2) Take steps to reduce the energy density of menu items across the board (usually achievable by reducing the fat content).
3) Provide clear listings of nutritional composition (especially energy and fat) at point of purchase and on packaging.

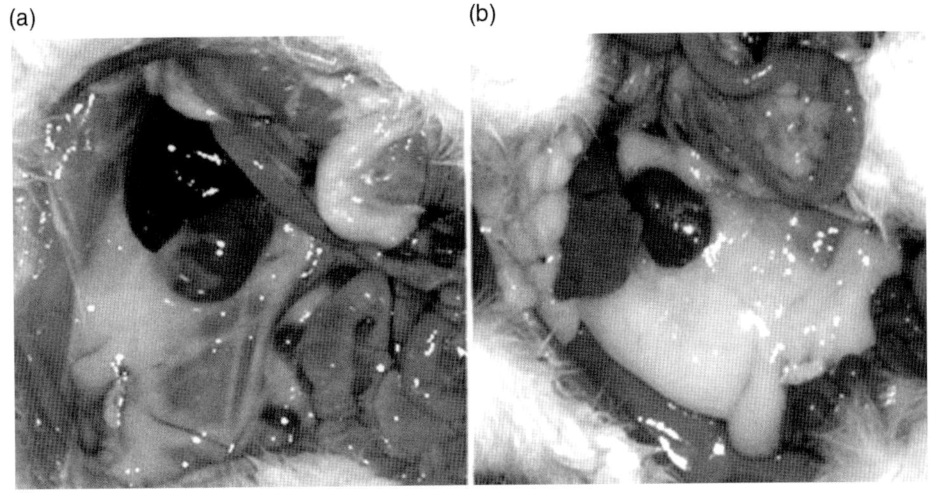

Figure 14.1 Effects of dietary oxidized lipids on the fat deposition in rabbits: (a) visceral fat deposition becomes depleted due to intake of 10% fortified sunflower oil with SBT powder; (b) intake of oxidized sunflower oil (OSFO) increases the visceral fat, affecting liver morphology and physiology (Zeb and Hussain 2014).

4) Stop encouraging inappropriately large portion sizes through advertising, manipulative price structures, and pressure selling of additional or larger menu items.

14.4 Hurdles

Hurdles form part of our daily life, and are also a part of our diet. The more we eat fried foods, the more we are at risk of developing certain chronic diseases, such as obesity. Increasing consumption will result in obese nations with poor health. Buttriss (2013) suggested that isolation and reformulation is unlikely to provide a complete solution to the challenge of improving eating patterns and nutrient provision.

According to a UN report (United Nations 2015), the world population will be 8.5 billion in 2030. The report also shows that the size of the middle class will be 4.9 billion in the same year. This will increase the size of cities, and lead to the establishment of new cities around the world. This population growth will produce a significant strain on the food supply. Agricultural land is fast moving towards commercialization and urbanization. Thus, food and edible oil shortages will be the major hurdles for the food frying industries.

The food supply is mostly based on small farmers – in the developing world, they supply 80% of all foods consumed. Economic changes may lead to an increase in the poverty rate, especially among small farmers. Thus, microeconomic indicators will be needed to correct economic inequality.

Geopolitical dynamics are changing rapidly, with wars developing among many countries. A large part of the land in Africa and the Middle East is currently war-torn. These wars have resulted in the highest rate of population displacement seen in recent history. Changes in the political goals of different countries, with rising nationalism and fundamentalism, only increase the risk of global food shortage.

14.5 Key Concepts

- The current challenges for food frying are health issues such as obesity, diabetes, cardiovascular diseases, and cancer.
- Future scenarios will include work on frying media, oil blending, food availability and characterization, after-frying evaluation, mitigation strategies, quality control, and the curbing of health effects.
- Innovations and emerging technologies will greatly benefit food frying.
- There are some important factors that need to be addressed by the fast food industry in order to curb the negative health effects of fried food consumption.
- The tremendous population growth, increasing urbanization and commercialization of agricultural land, and geopolitical issues may significantly affect the food supply for frying.

References

Buttriss, J.L. (2013). Food reformulation: the challenges to the food industry. *Proceedings of the Nutrition Society* 72: 61–69.

Cahill, L.E., Pan, A., Chiuve, S.E. et al. (2014). Fried-food consumption and risk of type 2 diabetes and coronary artery disease: a prospective study in 2 cohorts of US women and men. *American Journal of Clinical Nutrition* 100: 667–675.

Chang, Y.W., Sung, W.C., and Chen, J.Y. (2016). Effect of different molecular weight chitosans on the mitigation of acrylamide formation and the functional properties of the resultant Maillard reaction products. *Food Chemistry* 199: 581–589.

Fiolet, T., Srour, B., Sellem, L. et al. (2018). Consumption of ultra-processed foods and cancer risk: results from NutriNet-Santé prospective cohort. *British Medical Journal* 360: k322.

Gadiraju, T.V., Patel, Y., Gaziano, J.M., and Djoussé, L. (2015). Fried food consumption and cardiovascular health: a review of current evidence. *Nutrients* 7: 8424–8430.

Ganesan, K., Sukalingam, K., and Xu, B. (2017). Impact of consumption and cooking manners of vegetable oils on cardiovascular diseases – a critical review. *Critical Reviews in Food Science and Nutrition* 19: 1–18.

Guallar-Castillon, P., Rodriguez-Artalejo, F., Fornes, N.S. et al. (2007). Intake of fried foods is associated with obesity in the cohort of Spanish adults from the European Prospective Investigation into Cancer and Nutrition. *American Journal of Clinical Nutrition* 86: 198–205.

Guo, L.W., Liu, S.Z., Zhang, M. et al. (2018). Multivariate analysis of the association between consumption of fried food and gastric cancer and precancerous lesions. *Zhonghua Yu Fang Yi Xue Za Zhi* 52: 170–174.

Hashemi, S.M.B., Mousavi Khaneghah, A., Koubaa, M. et al. (2017). Novel edible oil sources: microwave heating and chemical properties. *Food Research International* 92: 147–153.

Hu, P., Li, Y., and Ca Mpos, H. (2018). Fried food intake and risk of nonfatal acute myocardial infarction in the Costa Rica Heart Study. *PLoS One* 13: e0192960.

Jiménez-Carvelo, A.M., González-Casado, A., and Cuadros-Rodríguez, L. (2017). A new analytical method for quantification of olive and palm oil in blends with other vegetable edible oils based on the chromatographic fingerprints from the methyl-transesterified fraction. *Talanta* 164: 540–547.

Kojima, R., Okada, E., Ukawa, S. et al. (2017). Dietary patterns and breast cancer risk in a prospective Japanese study. *Breast Cancer* 24: 152–160.

Prentice, A.M. and Jebb, S.A. (2003). Fast foods, energy density and obesity: a possible mechanistic link. *Obesity Reviews* 4: 187–194.

Qi, Q., Chu, A.Y., Kang, J.H. et al. (2014). Fried food consumption, genetic risk, and body mass index: gene-diet interaction analysis in three US cohort studies. *British Medical Journal* 348: g1610.

Qi, Y., Zhang, H., Wu, G. et al. (2018). Mitigation effects of proanthocyanidins with different structures on acrylamide formation in chemical and fried potato crisp models. *Food Chemistry* 250: 98–104.

Rannou, C., Laroque, D., Renault, E. et al. (2016). Mitigation strategies of acrylamide, furans, heterocyclic amines and browning during the Maillard reaction in foods. *Foodservice Research International* 90: 154–176.

Rossell, J. (2001). *Frying: Improving Quality*. Cambridge: Woodhead Publishing.

Sayon-Orea, C., Bes-Rastrollo, M., Basterra-Gortari, F. et al. (2013). Consumption of fried foods and weight gain in a Mediterranean cohort: the SUN project. *Nutrition, Metabolism and Cardiovascular Diseases* 23: 144–150.

Statista. 2018. U.S. population: Consumption of frozen fried chicken from 2011 to 2020. Available from: https://www.statista.com/statistics/280277/us-households-consumption-of-frozen-fried-chicken-trend [Accessed 14 September 2018].

Tayyem, R.F., Bawadi, H.A., Shehadah, I. et al. (2018). Fast foods, sweets and beverage consumption and risk of colorectal cancer: a case-control study in Jordan. *Asian Pacific Journal of Cancer Prevention* 19: 261–269.

United Nations. 2015. World population prospects. United Nations Department of Economic and Social Affairs, Population Division. Available from: https://esa.un.org/unpd/wpp/ [Accessed 14 September 2018].

Wainwright, R.E. and Lampert, D. (2007). Current and future frying issues. In: *Deep Frying: Chemistry, Nutrition and Practical Application* (ed. M.D. Erickson). Urbana, IL: AOCS Press.

Wallin, A., Di Giuseppe, D., Orsini, N. et al. (2017). Fish consumption and frying of fish in relation to type 2 diabetes incidence: a prospective cohort study of Swedish men. *European Journal of Nutrition* 56: 843–852.

WHO. 2018. Obesity and overweight. Available from: http://www.who.int/news-room/fact-sheets/detail/obesity-and-overweight [Accessed 14 September 2018].

Zeb, A. and Haq, I. (2016). The protective role of tomato powder in the toxicity, fatty infiltration and necrosis induced by oxidized tallow in rabbits. *Journal of Food Biochemistry* 40: 428–435.

Zeb, A. and Hussain, S. (2014). Sea buckthorn seed powder provides protection in the oxidative stress produced by thermally oxidized sunflower oil in rabbits. *Journal of Food Biochemistry* 38: 498–508.

Zeb, A. and Rahman, S.U. (2017). Protective effects of dietary glycine and glutamic acid toward the toxic effects of oxidized mustard oil in rabbits. *Food & Function* 8: 429–436.

Zeb, A. and Ullah, S. (2015). Sea buckthorn seed oil protects against the oxidative stress produced by thermally oxidized lipids. *Food Chemistry* 186: 6–12.

Zuo, S., Zhang, T., Jiang, B., and Mu, W. (2015). Reduction of acrylamide level through blanching with treatment by an extremely thermostable L-asparaginase during French fries processing. *Extremophiles* 19: 841–851.

Index

C

t

u

v